QUANTUM NANOCHEMISTRY

(A Five-Volume Set)

Volume III:
Quantum Molecules and Reactivity

QUANTUM NANOCHEMISTRY

(A Five-Volume Set)

Volume III:
Quantum Molecules and Reactivity

Mihai V. Putz

Assoc. Prof. Dr. Dr.-Habil. Acad. Math. Chem.
West University of Timişoara,
Laboratory of Structural and Computational Physical-Chemistry
for Nanosciences and QSAR, Department of Biology-Chemistry,
Faculty of Chemistry, Biology, Geography,
Str. Pestalozzi, No. 16, RO-300115, Timişoara, ROMANIA
Tel: +40-256-592638; Fax: +40-256-592620

&

Principal Investigator of First Rank, PI1/CS1
Institute of Research-Development for Electrochemistry
and Condensed Matter (INCEMC) Timisoara,
Str. Aurel Paunescu Podeanu No. 144,
RO-300569 Timişoara, ROMANIA
Tel: +40-256-222-119; Fax: +40-256-201-382

E-mail: mv_putz@yahoo.com
URL: www.mvputz.iqstorm.ro

APPLE
ACADEMIC
PRESS

Apple Academic Press Inc. | Apple Academic Press Inc.
3333 Mistwell Crescent | 9 Spinnaker Way
Oakville, ON L6L 0A2 Canada | Waretown, NJ 08758 USA

©2016 by Apple Academic Press, Inc.

First issued in paperback 2021

Exclusive worldwide distribution by CRC Press, a member of Taylor & Francis Group
No claim to original U.S. Government works

ISBN 13: 978-1-77463-101-0 (pbk)
ISBN 13: 978-1-77188-135-7 (hbk)

Library and Archives Canada Cataloguing in Publication

Putz, Mihai V., author
Quantum nanochemistry / Mihai V. Putz (Assoc. Prof. Dr. Dr. Habil. Acad. Math. Chem.) West University of Timişoara, Laboratory of Structural and Computational Physical-Chemistry for Nanosciences and QSAR, Department of Biology-Chemistry, Faculty of Chemistry, Biology, Geography, Str. Pestalozzi, No. 16, RO-300115, Timişoara, ROMANIA, Tel: +40-256-592638; Fax: +40-256-592620, & Institute of Research-Development for Electrochemistry and Condensed Matter (INCEMC) Timişoara, Str. Aurel Paunescu Podeanu No. 144, RO-300569 Timişoara, ROMANIA Tel: +40-256-222-119; Fax: +40-256-201-382, E-mail: mv_putz@yahoo.com, URL: www.mvputz.iqstorm.ro.

Includes bibliographical references and index.
Contents: Volume I: Quantum theory and observability -- Volume II: Quantum atoms and periodicity -- Volume III: Quantum molecules and reactivity -- Volume IV: Quantum solids and orderability -- Volume V: Quantum structure–activity relationships (Qu-SAR).
Issued in print and electronic formats.
ISBN 978-1-77188-133-3 (volume 1 : hardcover).--ISBN 978-1-77188-134-0 (volume 2: hardcover).--ISBN 978-1-77188-135-7 (volume 3 : hardcover).-- ISBN 978-1-77188-136-4 (volume 4 : hardcover).--ISBN 978-1-77188-137-1 (volume 5 : hardcover).--ISBN 978-1-4987-2953-6 (volume 1 : pdf).--ISBN 978-1-4987-2954-3 (volume 2 : pdf).--ISBN 978-1-4987-2955-0 (volume 3 : pdf).--ISBN 978-1-4987-2956-7 (volume 4 : pdf).--ISBN 978-1-4987-2957-4 (volume 5 : pdf) 1. Quantum chemistry. 2. Nanochemistry. I. Title.
QD462.P88 2016 541'.28 C2015-908030-4 C2015-908031-2

Library of Congress Cataloging-in-Publication Data

Names: Putz, Mihai V., author.
Title: Quantum nanochemistry / Mihai V. Putz.
Description: Oakville, ON, Canada ; Waretown, NJ, USA : Apple Academic Press, [2015-2016] | "2015 | Includes bibliographical references and indexes.
Identifiers: LCCN 2015047099| ISBN 9781771881388 (set) | ISBN 1771881380 (set) | ISBN 9781498729536 (set ; eBook) | ISBN 1498729533 (set ; eBook) | ISBN 9781771881333 (v. 1 ; hardcover) | ISBN 177188133X (v. 1 ; hardcover) | ISBN 9781498729536 (v. 1 ; eBook) | ISBN 1498729533 (v. 1 ; eBook) | ISBN 9781771881340 (v. 2 ; hardcover) | ISBN 1771881348 (v. 2 ; hardcover) | ISBN 9781498729543 (v. 2 ; eBook) | ISBN 1498729541 (v. 2 ; eBook) | ISBN 9781771881357 (v. 3 ; hardcover) | ISBN 1771881356 (v. 3 ; hardcover) | ISBN 9781498729550 (v. 3 ; eBook) | ISBN 149872955X (v. 3 ; eBook) | ISBN 9781771881364 (v. 4 ; hardcover) | ISBN 1771881364 (v. 4 ; hardcover) | ISBN 9781498729567 (v. 4 ; eBook) | ISBN 1498729568 (v. 4 ; eBook) | ISBN 9781771881371 (v. 5 ; hardcover) | ISBN 1771881372 (v. 5 ; hardcover) | ISBN 9781498729574 (v. 5 ; eBook) | ISBN 1498729576 (v. 5 ; eBook) Subjects: LCSH: Quantum chemistry. | Chemistry, Physical and theoretical. | Nanochemistry. | Quantum theory. | QSAR (Biochemistry)
Classification: LCC QD462 .P89 2016 | DDC 541/.28--dc23
LC record available at http://lccn.loc.gov/2015047099

When you're a quantum chemist, it's like you're sitting at the top of the mountain. The distinction between biochemistry, inorganic, and organic chemistry are less important, as all of chemistry revolves around what the electrons are doing in molecules, and that domain can be best "seen" by applying theory.
(Rodney J. Bartlett, 2007)

To XXI Scholars

CONTENTS

LIST OF ABBREVIATIONS

AIM	atoms-in-molecules
AM1	Austin model
BB	build-in-bondonic
BEC	Bose-Einstein condensation
CFD	compact finite difference
CM	mass center
CN	coordination number
COSR	chemical orthogonal space and reactivity
CR	chemical reactivity
DFE	density functional electronegativity
DFT	density functional theory
EA	electronic affinity
EFs	extracellular fractions
ELF	electronic localization (super) function
ELISA	enzyme-linked immunosorbent assay
FB	fermionic-bosonic/bondonic space
FD	finite difference
FLU	fluctuation index
GB	Ghosh-Biswas
GP	Gross-Pitaevsky equation
HMO	Hückel molecular orbital theory
HOMO	highest occupied molecular orbital
HSAB	hard and soft acids and bases
HT	Hückel theory
IBD	infectious bursal disease
IBD-AB	infectious bursal disease virus antibodies
IG	internal composition
IP	ionization potential
KS	Kohn-Sham
KS	Kohn-Sham equation

LCAO	linear combination of atomic orbitals
LDA	local density approximation
LFSE	ligand field stabilization energies
LUMO	lowest unoccupied molecular orbital
LYP	Lee, Yang, and Parr
MH	maximization of hardness
MHI	maximum hardness index
MHP	maximum hardness principle
MO	molecular orbitals
MOL	molecular orbitals
NICS	nucleus-independent chemical shift index
NIST	National Institute of Standard and Technology
NO	natural orbitals
PK	Phillips-Kleinman
PM3	parameterized model no. 3
PO	pseudo-orbital
PP	pseudo-potential
QSArR	quantitative structure aromaticity relationship
QSPR	quantitative structure-property relationships
RE	resonance energy
SALC	symmetry adapted linear combinations
SC	semiclassical
SCF	self-consistent field
SLR	spectral-like resolution
SP	standard Padé
TF	Thomas-Fermi theory
TIR	topologic index of reactivity
TOPAZ	topological peripheral paths
TST	transition state theory
VB	valence-bond
VBM	valence bond method
VSEPR	valence-shell electron-pair repulsion
WP	Wang-Parr
XCD	exchange-correlation density
ZPE	zero point energy

PREFACE TO FIVE-VOLUME SET

Dear Scholars (Student, Researcher, Colleague),

I am honored to introduce *Quantum Nanochemistry*, a handbook comprised of the following five volumes:

> *Volume I: Quantum Theory and Observability*
> *Volume II: Quantum Atoms and Periodicity*
> *Volume III: Quantum Molecules and Reactivity*
> *Volume IV: Quantum Solids and Orderability*
> *Volume V: Quantum Structure–Activity Relationships (Qu-SAR)*

This treatise, a compilation of my lecture notes for graduates, post-graduates and doctoral students in physical and chemical sciences as well as my own post-doctoral research, will serve the scientific community seeking information in basic quantum chemistry environments: from the fundamental quantum theories to atoms, molecules, solids and cells (chemical–biological/ligand–substrate/ligand–receptor interactions); and will also creatively explain the quantum level concepts such as observability, periodicity, reactivity, orderability, and activity explicitly.

The book adopts a three-way approach to explain the main principles governing the electronic world:

- firstly, *the introductory principles* of quantumchemistry are stated;
- then, they are analyzed as *primary concepts* employed to understand the microscopic nature of objects;
- finally, they are explained through *basic analytical equations* controlling the observed or measured electronic object.

It explains the first principles of quantum chemistry, which includes quantum mechanics, quantum atom and periodicity, quantum molecule and reactivity, through two levels:

- *fundamental* (or *universal*) character of matter in isolated and interacting states; and
- the primary concepts elaborated for a beginner as well as an advanced researcher in quantum chemistry.

Each volume tells the "story of quantum chemical structures" from different viewpoints offering new insight to some current quantum paradoxes.

- The **first volume** covers the concepts of nuclear, atomic, molecular and solids on the basis of quantum principles—from Planck, Bohr, Einstein, Schrödinger, Hartree–Fock, up to Feynman Path Integral approaches;
- The **second volume** details an atom's quantum structure, its diverse analytical predictions through reviews and an in-depth analysis of atomic periodicities, atomic radii, ionization potential, electron affinity, electronegativity and chemical hardness. Additionally, it also discusses the assessment of electrophilicity and chemical action as the prime global reactivity indices while judging chemical reactivity through associated principles;
- The **third volume** highlights chemical reactivity through molecular structure, chemical bonding (introducing bondons as the quantum bosonic particles of the chemical field), localization from Hückel to Density Functional expositions, especially how chemical principles of electronegativity and chemical hardness decide the global chemical reactivity and interaction;
- The **fourth volume** addresses the electronic order problems in the solid state viewed as a huge molecule in special quantum states; and
- The **fifth volume** reveals the quantum implication to bio-organic and bio-inorganic systems, enzyme kinetics and to pharmacophore binding sites of chemical–biological interaction of molecules through cell membranes in targeting specific bindings modeled by celebrated QSARs (Quantitative Structure–Activity Relationships) renamed here as Qu–SAR (Quantum Structure–Activity Relationships).

Thus, the five-volume set attempts, for the first time ever, to unify the introductory principles, the primary concepts and the basic analytical equations against a background of quantum chemical bonds and interactions (short, medium and long), structures of matter and their properties: periodicity of atoms, reactivity of molecules, orderability of solids, and activity of cells (through an advanced multi-layered quantum structure–activity unifying concepts and algorithms), and observability measured throughout all the introduced and computed quantities (Figure 0.0).

It provides a fresh perspective to the "quantum story" of electronic matter, collecting and collating both research and theoretical exposition the "gold" knowledge of the quantum chemistry principles.

The book serves as an excellent reference to undergraduate, graduate (Masters and PhDs) and post-doctoral students of physical and chemical sciences; for it not only provides basics and essentials of applied quantum theory, but also leads to unexplored areas of quantum science for future research and development. Yet another novelty of this five-volume work is the intelligent unification of the quantum principles of atoms, molecules, solids and cells through the qualitative–quantitative principles underlying the observed quantum phenomena. This is achieved through unitary

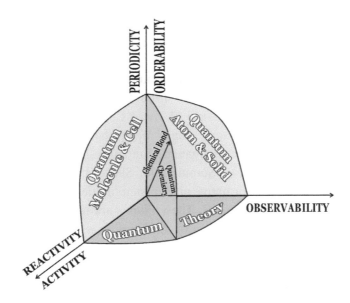

FIGURE 0.0   The featured concepts of the "First Principles of Quantum Chemistry" five-volume handbook as placed in the paradigmatic chemical orthogonal space of atoms and molecules.

analytical exposition of the quantum principles ranging from quanta's nature (either as ondulatory and corpuscular manifestations) to wave function, path integral and electron density tools.

The modern quantum theories are reviewed mindful of their implications to quantum chemistry. Atomic, molecular, solid-state structures along cell/ biological activity are analytically characterized. Major quantum aspects of the atomic, molecular, solid and cellular structure, properties/activity features, conceptual and quantitative correlations are unitarily reviewed at basic and advanced physical-chemistry levels of comprehension.

Unlike other available textbooks that are written as monographs displaying the chapters as themes of interests, this book narrates the "story of quantum chemistry" as *an extended review paper*, where theoretical and instructional concepts are appropriately combined with the relevant schemes of quantization of electronic structures, through path integrals, Bohmian, or chemical reactivity indices. The writing style is direct, concise and appealing; wherever appropriate physical, chemical and even philosophical insights are provided to explain quantum chemistry at large.

The author uses his rich university teaching experience of 15 years in physical chemistry at West University of Timisoara, Romania, along with his research expertise in treating chemical bond and bonding through conceptual and analytical quantum mechanical methods to explain the concepts. He has been a regular contributor to many physical-chemical international journals (*Phys Rev, J Phys Chem, Theor Acc Chem, Int J Quantum Chem, J Comp Chem, J Theor Comp Chem, Int J Mol Sci, Molecules, Struct Bond, Struct Chem, J Math Chem, MATCH*, etc.).

In a nutshell, the book amalgamates an analysis of the earlier works of great professors such as Sommerfeld, Slater, Landau and Feynman in a methodological, informative and epistemological way with practical and computational applications. The volumes are layered such that each can be used either individually or in combination with the other volumes. For instance, each volume reviews quantum chemistry from its level: as quantum formalisms in Volume I, as atomic structure and properties in Volume II, as detailed molecular bonding in Volume III, as crystal/solid state (electronic) in Volume IV, and as pharmacophore activity targeting specific bindings in Volume V.

To the best of my knowledge, such a collection does not exist currently in curricula and may not appear soon as many authors prefer to publish well-specialized monographs in their particular field of expertise. This multiple volumes' work, thus, assists academic and research community as a complete basic reference of conceptual and illustrative value.

I wish to acknowledge, with sincerity, the quantum flaws that myself and many researchers and professors make due to stressed delivery of papers using computational programs and software to report and interpret results based on inter-correlation. I feel, therefore, the need of a new comprehensive quantum chemistry reference approach and the present five-volume set fills the gap:

- *Undergraduate students* may use this work as an *introductory and training textbook* in the quantum structure of matter, for basic course(s) in physics and chemistry at college and university;
- *Graduate (Master and Doctoral) students* may use this work as the *recipe book* for analytical research on quantum assessments of electronic properties of matter in the view of chemical reactivity characterization and prediction;
- *University professors and tutors* may use this work as a *reference textbook* to plan their lectures and seminars in quantum chemistry at undergraduate or graduate level;
- *Research (Academic and Institutes) media* may use this work as a *reference monograph* for their results as it contains many tables and original results, published for the first time, on the atomic-molecular quantum energies, atomic radii and reactivity indices (e.g., electronegativity, chemical hardness, ionization and electron affinity results). It also has a collection of original, special and generally recommended literature, integrated results about quantum structure and properties.
- *Industry media* may use this work as a *working tool book* while assessing envisaged theoretical chemical structures or reactions (atoms-in-molecule, atoms-in-nanosystems), including molecular modeling for pharmaceutical purposes, following the presented examples, or simulating the physical–chemical properties before live production;

- *General media* may use this work as an *information book* to get acquainted with the main and actual quantum paradigms of matter's electronic structures and in understanding and predicting the chemical combinations (involving electrons, atoms and molecules) of Nature, because of its educative presentation.

I hope the academia shares the same enthusiasm for my work as the author while writing it and the professionalism and exquisite cooperation of the Apple Academic Press in publishing it.

Yours Sincerely,

Mihai V. Putz,
Assoc. Prof. Dr. Dr.-Habil. Acad. Math. Chem.
West University of Timişoara
& R&D National Institute for Electrochemistry and Condensed Matter Timişoara
(Romania)

ABOUT THE AUTHOR

Mihai V. PUTZ is a laureate in physics (1997), with an MS degree in spectroscopy (1999), and PhD degree in chemistry (2002), with many post-doctorate stages: in chemistry (2002-2003) and in physics (2004, 2010, 2011) at the University of Calabria, Italy, and Free University of Berlin, Germany, respectively. He is currently Associate Professor of theoretical and computational physical chemistry at West University of Timisoara, Romania. He has made valuable contributions in computational, quantum, and physical chemistry through seminal works that appeared in many international journals. He is Editor-in-Chief of the *International Journal of Chemical Modeling* (at NOVA Science Inc.) and the *New Frontiers in Chemistry* (at West University of Timisoara). He is member of many professional societies and has received several national and international awards from the Romanian National Authority of Scientific Research (2008), the German Academic Exchange Service DAAD (2000, 2004, 2011), and the Center of International Cooperation of Free University Berlin (2010). He is the leader of the Laboratory of Computational and Structural Physical Chemistry for Nanosciences and QSAR at Biology-Chemistry Department of West University of Timisoara, Romania, where he conducts research in the fundamental and applicative fields of quantum physical-chemistry and QSAR. In 2010 Mihai V. Putz was declared through a national competition the Best Researcher of Romania, while in 2013 he was recognized among the first Dr.-Habil. in Chemistry in Romania. In 2013 he was appointed Scientific Director of the newly founded Laboratory of Structural and Computational Physical Chemistry for Nanosciences and QSAR in his alma mater of West University of Timisoara, while from 2014, he was recognized by the Romanian Ministry of Research as Principal Investigator of first rank/degree (PI1/CS1) at National Institute for Electrochemistry and Condensed Matter (INCEMC) Timisoara. He is also a full member of International Academy of Mathematical Chemistry.

FOREWORD TO *VOLUME III: QUANTUM MOLECULE AND REACTIVITY*

The current single-authored volume *Quantum Molecule and Reactivity* is the third book in a set of five with the general theme of "quantum nanochemistry". As befits the interests and research productivity of the volumes' author, Prof. Mihai Putz of the West University of Timişoara, Romania, the contents range from biochemical reactivity to organic chemistry to solid state (condensed matter) physics with a detailed underlying structure of quantum mechanics and mathematics. Each chapter is accompanied by extensive referencing to the research literature; given that there are no problem sets nor an answer key for the reader to calibrate his or her knowledge of the contents, these books are for expert and professional scientists even more than they are for beginning students of theoretical science early in their education.

Reviewing the contents of Volume III brings to mind two well-respected names in the pantheon of intellectual activity, ability and agility immediately. The first is the poet, T. S. Eliot, who wrote near the end of the last part of his multi-page, multi-part poem, "Four Quartets", in the poem entitled "The Little Gidding", the following lines:

We shall not cease from exploration
And the end of all our exploring
Will be to arrive where we started
And know the place for the first time.

Prof. Putz takes the reader back and forth between well-established facts and new insights, between materials from ones earliest education, to that which transcends all earlier education.

The other name that Prof. Putz' work brings to mind is the theoretical physicist J. C. Slater who wrote numerous volumes on the quantum theory of atoms, molecules and solids. Slater's name is now eponymously and, almost inseparably, used as a descriptor of orbitals and determinants, as well as other concepts in the theoretical physical sciences, such as: Slater screening constants, Slater integrals and the Slater–Pauling theory of directed valence. Putz' writing, similar to that of Eliot and Slater, is not for casual perusal, nor do any of these authors "merely" offer text as a refresher course on understanding the nature of the world.

But rewards do await the careful, competent, concerned reader – Putz' books offer much for the neophyte and senior scientist alike, neither of whom should assume that s/he will be able to make numerically more reliable predictions of phenomena than he or she could before reading, whether it be of a melting point, a reaction yield, or an LD_{50}. Understanding is more important than factual knowledge and that is an aspiration shared by the author of these volumes, and that of this foreword.

The author of this foreword remembers the admonition of his thesis advisor, L. C. Allen, "Don't invent a word unless you have to". (Let family lineage now be acknowledged, for whatever it is worth: Allen was a doctoral student of Slater's.) Putz invents a new word "bondon" – actually introduced a few years ago by Putz in the primary research literature. It has yet to be observed by the experimentalist. As such, the reader is left with the question of whether bondons have physical reality such as ions (cations, anions and zwitterions), the photon and the noble gases, and their generic "on"-ending names, or are bondons to be understood as quasi particles that cannot exist in isolation (e.g., excitons, phonons or plasmons)? Reality is less the important issue than objects that allow for conceptual unification and understanding. After all, bonds, resonance structures and molecular orbitals do not really exist either, despite their nearly ubiquitous presence in the textbook and research literature, as well as works such as the current Volume. The current Volume is the sole written book that explains the concept, the uses, the power of bondons and unifies this concept with so, so much else of the molecular sciences.

This foreword closes with a brief mention of a century old concept and then derived neologism "psi" ψ. This symbol is unequivocally more than just a mathematical function of several variables, and saying we should have also written Ψ for time-dependent problems only adds to its worth, its power, its near ubiquity in the discipline of theoretical science. Let us close by noting that ψ and the upper case Ψ letter has two pronunciations, at least in scientific English, namely "sigh" and "see". Now, should we sigh or should we see? Prof. Putz, your five-volume provide much text to read, sigh, but in so reading, there is the hope that we can see more distantly and more clearly than we could before.

Joel F. Liebman

Department of Chemistry and Biochemistry
University of Maryland, Baltimore County (UMBC)
Baltimore, MD 21250, USA
November 2015

PREFACE TO *VOLUME III: QUANTUM MOLECULES AND REACTIVITY*

THE SCIENTIFIC PREMISES

Universe is harmony!

Harmony means proportion. Proportion determines geometry. Geometry allows the construction, the progress and the evolution.

By a "divine" reasoning, Plato (c. 427–347 BC) in *Timeus* deduces the geometric proportions saying that "it is necessary that nature be visible and tangible ... but if the universe had been without depth, a single medium would have been enough to relate all nature as it contains ... so the universe must be sound and no sound can harmonize just with one but two media environment ... but it is impossible for two things to reconcile the absence of the third ... and the most harmonious analogy it is when in three media the middle is proportional to the last as like as the first one to the middle ... therefore fire to the air is like air for water and water to the land."

Therefore, the constant relation between fire/water = air/water = water/land actually involves the constant between successive terms of the series, fire, air, water, and land; thus defining the geometric proportions of the universe (Figure P.1).

Pythagoras (582?–500? BC) observed the stellar distances as proportional to the musical scales, defining the so-called music of the spheres as a "geometry into harmony sound, there is also music in the spheres spatialization".

For Pythagoreans, the solar system comprises spheres around a central fire, and each sphere produces a sound with a specific tonality in their revolution, whose combination generates music of the spheres.

The astronomer Johannes Kepler (1571–1630), 20 centuries later, observed in his paper, Harmonics Mundi – The Harmony of the World

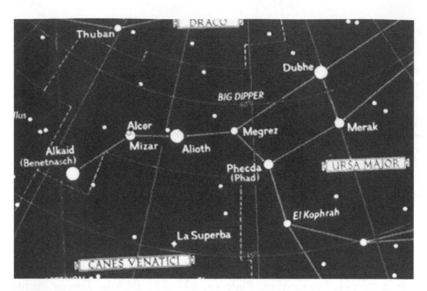

FIGURE P.1   Ursa Major; after National Geografic (Putz, M. V., Mingos, D. M. P., Eds. (2012). Applications of Density Functional Theory to Chemical Reactivity, Structure and Bonding Series Vol. 149, Springer Verlag, Heidelberg-Berlin).

(1619), "I guarantee that the planets movement are in agreement with the harmonic proportions."

Marcus Vitruvius Pollio (70?–25 AD), an engineer in the army of king Augustus, in his De-About Architecture published as The Ten Books on Architecture applied the idea of harmonic proportion to human beings and buildings construction. He said, "without symmetry and proportions there can be no principle in the projection of any temple, just like relationship between members of the human body. Symmetry is a specific agreement between the components of a work and the relationship between different parts of a general scheme in line with some of its parts selected as a standard."

Thus, it anticipates that a unit cell (or primitive cell) by multiplication following certain proportions and rules can generate a whole assembly. Symmetry means proportion; and the proportion is measured by numbers.

Therefore, symmetry is expressing itself through numbers; thus making the power of symbolization, quantification of structures and their properties through numbers meaningful. For instance, the secret of pyramids, among other things, relies on the fact that they are built in harmony with the so-called gold proportion (or ratio).[1]

Euclid defined the gold proportion as the ratio by which a straight line is divided into two parts; and the combination in which the ratio between the whole line and its largest segment is the same with the ratio between the largest (Φ, by convention) and the lowest segment (1, by convention). It is denoted by the Greek letter phi, Φ, after the Athenian sculptor and architect Phideas (490–430, BC), who used the golden ratio in constructing the Parthenon.

What is the value of Φ?

Applying Euclid's definition, we get that the gold ratio $\Phi/1 = (1 + \Phi)\Phi$ is equivalent to the quadratic equation $\Phi^2 - \Phi - 1 = 0$, whose positive solution is $\Phi = 1/2 + \sqrt{5}/2 \approx 1.618$.

For all Egyptian pyramids (Figure 1.2 – left), it was observed that the rectangular triangle formed by half of the base length (the leg lower), the slope height from base to top of the pyramid (the high leg) and edge length from the top of the pyramid at relative corner to the base (hypotenuse) is in the ratio $1 : \sqrt{\Phi} : \Phi$.

These reports were later studied extensively by Kepler in construction of triangles, and the triangles with the sides satisfying these conditions were called the Kepler triangles. In economics, these triangles are also referred as "The Triangles of Price", constructed based on the P/E ratio (price-to-earning ratio) report.[2]

The sounds harmony can also be described in terms of proportion and numbers. It is further substantiated by the famous discussion from Timeus in which Plato introduces the harmonic average of two numbers as the inverse of the arithmetic average of the inverses of basic numbers.

One starts with the monad, 1 – the Creator number, which is then doubled and generates number 2, and moves further by adding its unit once again generating number 3. The monad-point generated by successive additions forming the even or odd line (2 or 3) constitutes the geometric series with even and odd powers: thus, numbers 4 and 9 appear as representatives of the plan, and 8 and 27 to measure the space.

---

[1] Lawlor, R. (1982). Sacred Geometry, Thames & Hudson, New York.
[2] Putz, M. V., Mingos, D. M. P., Eds. (2012). Applications of Density Functional Theory to Chemical Reactivity, Structure and Bonding Series Vol. 149, Springer Verlag, Heidelberg-Berlin.

These seven numbers 1, 2, 3, 4, 8, 9, and 27 are arranged to form the Greek symbol lambda (Figure P.2 – right).

The direct report of the monad with the dyadic 1/2 defines octave musical 1:2 as their arithmetic mean (1+2)/2 generates the inverse fifth 2:3, and their harmonic average 1/((1+1/2)/2) generates the inverse quarter 3:4, which are found at the base of the musical harmony numbers – tetrachlorodibenzo sacred 1, 2, 3, 4 pythagorean.

Furthermore, Plato introduced tone as a basic harmonic relationship between the fourth and fifth (3/4): (2/3) = 9/8. Hence, combining the dyadic's series (2) and triad's (3) generates the full scale of tones on an octave.

Subseries combinations are built for each of the two series, dyadic and triad, in which powers of the number in the opposite series remain constant. Thus, the strings obtained for the 2 subseries is: $(2^1\,3, 2^3, 3^2)$, $(2^2\,3, 2^4, 2^1\,3^2)$, $(2^3\,3, 2^5, 2^2\,3^2)$, $2^4\,3$, and for the 3 subseries is: $3^1\,2, 3^2, (3^1\,2^2, 3^2\,2, 3^3)$, $(3^2 2^2, 3^3\,2, 3^4)$, $3^3\,2^2, 3^4\,2$, respectively, which when re-ordered restore the scale tones on an octave (Figure P.2 – right).

Plato built this scale using only arithmetic calculations and not experimenting with strings stretched to note clear sounds, as proceeded the Pythagoreans.

Again, the numbers' strength lies in serving the harmony. Numbers as a measure of symmetry, and reciprocally, symbolized symmetry quantified by number are the two sides of human knowledge.

The numbers are then ranked, succeed from one to another and combine expressing different reality manifestation of nature to which they are

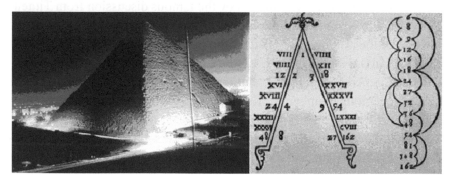

FIGURE P.2 Left: Great Pyramid (from National Geographic, April 1988); right: Theorica Musicae by Gaffurio F., Milano, 1492 (reedited from 1480); after Lawlor, R. (1982). Sacred Geometry, Thames & Hudson, New York).

associated. Similarly, symmetries can rank and combine, modeling and representing structures with different properties and meanings.

Symmetry maybe the primordial nature law and will be referred in the following as mobile and model to the semi-quantitative treat for the hierarchy of quantum nanosystems: atoms, molecules, crystalline network.

With this we arrive at the fundamental problem of the theoretical chemistry, which is to elucidate the answers to questions like: "What is an atom in a molecule?" Or "How many electrons must be transferred between the atoms A and B to form the molecule AB?"

It is clear that an explanation based on superposition of electrons and nuclei of the atoms involved in the formation of a molecule is full of naivety and practically eludes the essence of the private transfers of particles and properties that define a new assembly of electronic fluid and nuclei, found in an external and dynamic interaction.

On the other hand, if the quality and the atomic structure are not preserved in some way, the molecular discrimination for various atomic combinations would practically disappear because at the molecular level we cannot speak about a molecule composed of specific atoms.

Therefore, we have found a theory to render an image of compromise between an atom's loss of specificity and the contribution of the electron in forming the global molecular assembly.

A first compromise image would be the atom in molecule seen as a species surrounded by a number (even fractional) of electrons. This can result from the integral average of the states (species) allowed by the grand canonical molecular assembly. This leads to explanation of the molecular bonds, in fact, the chemical bonds.

Physics has identified the main types of forces that can occur between various macroscopic or microscopic objects in the universe. Therefore, it is natural to define more rigorous and realistic drawing, not of forces (which are exhausted by physics), of bonds by the virtue of forces, which act at the interatomic level in the molecular assemblies.

Ideally, everything should start from the atomic species characterized by the availability or the inertia to participate with the electrons in a molecular combination and the conditions that encourage such participation for a particular species and for a typical bond.

At chemical level, "the measure of all things" is explained by electro-negativity, which acts both for atoms and molecules (see Volumes I and II of this five-volume set). A complete and complex definition is still not provided, but it is certain that it refers to a virtual or potential electronic flow, to or from the outside, when the atomic species is placed in an atomic medium triggering in this direction.[3]

While considering the observability models in chemistry, the density functional theory has at its center the electronic density ρ, which helps in determining the atomic or molecular energy as a density functional as well as its properties. One of these properties is that for a state density coher-ently formulated in relation to an external V potential (Hohenberg–Kohn Theorem), the following succession of identities that define the chemical potential as the virtual flow of the driven electrons and therefore correlated to the electronegativity itself, χ actually validate the reality of chemical reactivity

$$\mu = \frac{\partial E}{\partial N}\Big|_V = \frac{\delta E}{\delta \rho} = V + \frac{\delta F}{\delta \rho} \approx -\frac{1}{2}\left(IP + EA\right) = -\chi \qquad (P.1)$$

where N denotes the electrons number in the system, IP the ionization potential, EA the electron affinity and $F[\rho]$ is the functional density Hohenberg–Kohn (HK), written as the sum of the electronic kinetic energy and the electronic repulsion energy and has a universal character.[4]

The problem is to determine the chemical potential, the electronegativity of a system of N electrons in a given external V potential, with various approximations of the functional HK. This is possible in principle for the individual atoms species, and the part of Euler equation really answers to the Mulliken relation of defining electronegativity, while the bond with chemical potential from the last equality above is due to Parr and Pearson.[5]

---

[3] Schwarz, W. H. E., Mensching, L., Valtazanos, P., Von Niessen, W. (1986). A chemically useful definition of electron difference densities. *Int. J. Quant. Chem.* 29, 909–914; Mortier, W. J., Ghosh, S. K., Shankar, S. (1986). Electronegativity-equalization method for the calculation of atomic charges in molecules. *J. Am. Chem. Soc.* 108(15), 4315–4320; Ferreira, R., de Amorim, A. O. (1981). Electroneg-ativity and the bonding character of Molecular Orbitals. *Theoret. Chim. Acta (Berl.)* 58(2), 131–136; Bader, R. F. W., Becker, P. (1988). Transferability of atomic properties and the theorem of Hohenberg and Kohn. *Chem. Phys. Lett.* 148(5), 452–458.

[4] Hohenberg, P., Kohn, W. (1964). Inhomogeneous electronic gas. *Phys. Rev.* 136, 864–871.

[5] Parr, R. G. (1984). Remarks on the concept of an atom in a molecule and on charge transfer between atoms on molecule formation. *Int. J. Quant. Chem.* 26(5), 687–692.

Coming to the molecular problem, we take two atom species with specific characteristics,[6]

$$A\left(N_A^0, V_A^0 = -\frac{Z_A}{x_A}, \mu_A^0, \rho_A^0 \right) \text{ and } B\left(N_B^0, V_B^0 = -\frac{Z_B}{x_B}, \mu_B^0, \rho_B^0 \right) \quad \text{(P.2)}$$

which are initially considered isolated, each under the potential of their own nucleus. When they come "together" in the molecule AB, they are not retrieved any more as atoms A and B, but as the changed atoms $A^*(N_A^*$ $= N_A^0 + \Delta N$, μ_A^*, $\rho_A^*)$ and $B^*(N_B^* = N_B^0 - \Delta N$, μ_B^*, $\rho_B^*)$ with electrons transfer ΔN between B and A. The new molecule is considered to be in the fundamental state within the meaning of the canonical ensemble average, considered as the limit of zero temperature from the Mermin image of the density functional extension.[7] Thus, in terms of densities and chemical potentials, the atoms in the molecule will satisfy the relation:[8]

$$\rho_{AB} = \rho_A^* + \rho_B^* \quad \text{(P.3)}$$

where the individual atom's density in the molecule must respect the molecular symmetry required by its geometry and relationship:[9]

$$\mu_{AB} = \mu_A^* = \mu_B^* \quad \text{(P.4)}$$

These equations, however simple they may seem, allow several solutions depending on various areas of charge partition where the two atomic species can overlap. To select the most realistic of these solutions, we must add one more criterion, minimizing the energy of density functionals

---

[6] Bamzai, A. S., Deb, B. M. (1981). The role of single-particle density in chemistry. *Rev. Mod. Phys.* 53(1), 95–126; Nalewajski, R. F., Korchowiec, J. (1989). Basic concepts and illustrative applications of the sensitivity analysis of molecular charge distribution. *J. Mol. Catal.* 54, 324–342; Müller-Herold, U. (1984). Algebraic theory of the chemical potential and the condition of reactive equilibrium. *Lett. in Math. Phys.* 8(2), 127–133; Senatore, G., March, N. H. (1994). Recent progress in the field of electron correlation. *Rev. Mod. Phys.* 66(2), 445–479; Spruch, L. (1991). Pedagogic notes on Thomas-Fermi theory (and on some improvements): atoms, stars, and the stability of bulk matter. *Rev. Mod. Phys.* 63(1), 151–209.

[7] Mermin, N. D. (1965). Thermal properties of the inhomogeneous electron gas. *Phys. Rev.* 137, A1441–A1443.

[8] Parr, R. G. (1985). Density functional theory in chemistry. In: *Density Functional Methods in Physics*, Dreizler, R. M., Providencia, J. D. (Eds.), Plenum Press, New York.

[9] Mortier, W. J., Ghosh, S. K., Shankar, S. (1986). Electronegativity-equalization method for the calculation of atomic charges in molecules. *J. Am. Chem. Soc.* 108(15), 4315–4320.

of the atomic species in the molecule toward the corresponding energies from the isolated states:

$$\left\{E_A\left[\rho_A^*\right]-E_A\left[\rho_A^0\right]\right\}+\left\{E_B\left[\rho_B^*\right]-E_B\left[\rho_B^0\right]\right\}=\min \qquad (P.5)$$

which closes the description of the atoms in molecule.

The fundamental problem is to solve the system of the last three equations, where the known quantities are $\rho_{AB}, V_A^0, V_B^0, \mu_{AB}$, whereas the densities ρ_A^*, ρ_B^* remain unknown. It is obvious that after the initialization and the self-consistent solution of the system formed by the last three equations, the resultant densities will be reported to the modified nuclear potential, V_A^*, V_B^*, naturally different from those for the isolated atoms, V_A^0, V_B^0, by the reciprocal of charges exchange around the initial nuclei, which also changes their action in the molecule formed.

An observation of the changed electrons number by the atomic species A and B in the molecular formation process is required:

$$\Delta N = N_B^0 - N_B^* = N_A^* - N_A^0 = \int d\tau\left(\rho_B^0 - \rho_B^*\right)=\int d\tau\left(\rho_A^* - \rho_A^0\right) \qquad (P.6)$$

which in general it is not an integer number.

Parr and Pearson have shown that this electron exchange is related to the chemical potential variation of the atomic species taken individually, referring to a certain "strength" to transfer:[10]

$$\eta = \frac{1}{2}\frac{d\mu}{dN} \approx \frac{1}{2}\left(IP - EA\right) \qquad (P.7)$$

which together with electronegativity constitute the fundamental measures characterizing the reactivity of an atomic or molecular species.

Another fundamental problem following the molecular formation determination from the constituent atoms is the individuality combined with the universality of the constituent species being combined in the molecular assembly, i.e., the transferability of the atomic species properties from one molecule to another. For example, how are they retrieved or how much properties of the atomic species are retrieved in the series of hydrocarbon? One needs to create a methodology to find a criterion on

---

[10] Parr, R. G., Pearson, R. G. (1983). Absolute hardness: companion parameter to absolute electronegativity. *J. Am. Chem. Soc.* 105, 7512–7516.

which this characteristic of the specific atomic species transferability from one molecule to another to can be pursued.[11]

The most convenient criterion is offered by the energy or its density functionals. But which energy? – the one of the atom in the molecule or of the molecule as a whole? Hence there appears the need for nuances in the Hohenberg–Kohn Theorem on the analytical domains and subdomains.[12]

It follows that the atom properties transferability from one molecule to another is natural and the atom bordered in molecule by an area $\partial\Omega$ satisfies the stationary condition of the density flow on the surface bordering the field Ω:[13]

$$\nabla_\Omega \rho(x) \cdot \vec{n}(x) = 0 \quad \forall x \in \partial\Omega \tag{P.8}$$

which defines the notion of "atom in molecule".

Therefore, one can define the medium measures characterizing the atom on this field $M(\Omega)$ at the molecular level as the simple superposition of these averages on various fields:[14]

$$\langle M \rangle = \sum_\Omega M(\Omega) \tag{P.9}$$

However, partitioning on fields remains essential, because a function of atomic state will correlate with the dynamics of G observable in this domain through the Heisenberg stationary equation:

$$\delta L[G\phi,\Omega] = \frac{1}{2}\varepsilon\left\{\frac{i}{\hbar}\langle[H,G]\rangle_\Omega + c.c.\right\} \tag{P.10}$$

where L denotes the Lagrangian, H the atomic Hamiltonian in molecule, ε an infinitesimal number, and $-(i\varepsilon/\hbar)G$ the generator of the unitary infinitesimal transformation, which acting on the atomic state ϕ generates variation in L. Considering the operator G as the position product and an

[11] Del Re, G. (1981). Ground-state charge transfer and electronegativity equalization. *J. Chem. Soc., Faraday Trans. 2* 77, 2067–2076.

[12] McLachlan, A. D., Ball, M. A. (1964). Time-dependent Hartree—Fock theory for Molecules. *Rev. Mod. Phys.* 36, 844–855; Cioslowski, J. (1990). Density functional reformulation of molecular orbital theories. *Adv. in Quant. Chem.* 21, 303–316; Tachibana, A. (1988). Shape wave in density functional theory. *Int. J. Quant. Chem.* 34(4), 309–323; Jones, R. O., Gunnarsson, O. (1989). The density functional formalism, its applications and prospects. *Rev. Mod. Phys.* 61, 689–746

[13] Bader, R. F. W. (1990). *Atoms in Molecules – A Quantum Theory*, Oxford University Press, Oxford.

[14] Mortier, W. J., Ghosh, S. K., Shankar, S. (1986). Electronegativity-equalization method for the calculation of atomic charges in molecules. *J. Am. Chem. Soc.* 108(15), 4315–4320.

electron moment, $G = xp$, results for a stationary state is satisfied by the virial theorem,

$$2T(\Omega) = -V(\Omega) \tag{P.11}$$

where $T(\Omega)$, $V(\Omega)$ are the average of the kinetic and potential terms for the atomic electrons in the field Ω.

Results for the atom energy on the field Ω,

$$E(\Omega) = T(\Omega) + V(\Omega) \tag{P.12}$$

Applying the virial theorem gives the atomic energy on the quantum subsystem, which is also being considered equal to:

$$E(\Omega) = -T(\Omega) = \frac{1}{2}V(\Omega) \tag{P.13}$$

This leads to a very ambitious idea: if you can extend the Hohenberg–Kohn theorem to an atom located in two different molecular systems but with equal electronic densities on two arbitrary quantum subdomains, $\rho_{1\Omega}(x) = \rho_{2\Omega}(x)$, one can conclude that the electron densities of the atom on the quantum space of the two molecules are also equal, $\rho_1(x) = \rho_2(x)$, and the density functionals $E(\Omega)$ can be properly extended to the kinetic energy average for an atom, from an arbitrary molecular subspace to another.

This extension would involve the need to demonstrate that the electron density $\rho(x)$ of the atom in a molecule is a measure in the molecular quantum space, or in other words $\rho(x)$ for the atom in molecule is itself a unique functional with an electronic density $\rho_\Omega(x)$ of the same atom on some space Ω, but bordered.

However, in practical terms, this analyticity means condensation of the atomic properties in the molecule in the associated single-density matrix $\gamma(x,x')$ and its development in Taylor series around the diagonal point, $x = x'$; development performed only to second order. In the correspondent plan of the density functionals of the atomic energies, first order will correspond to electronegativity and second order will correspond to chemical strength.[15]

---

[15] Parr, R. G. (1984). Remarks on the concept of an atom in a molecule and on charge transfer between atoms on molecule formation. *Int. J. Quant. Chem.* 26(5), 687–692.

Thus, we have shown some of the major lines to be approached to formulate an exhaustive and rigorous theory of the atomic phenomenon in the molecular formation and of the properties correlated to this formation. This highlights the very important "card" approach the atom in a molecule has "to play", in terms of the density functionals theory.[16] The present volume is loaded with more details and useful illustrations besides highlighting the "red flow" chemical reactivity and the allied principles, especially those related with electronegativity and chemical hardness forms and principles.

VOLUME LAYOUT

The book is the *third* volume in the five-volume series on *Quantum Nanochemistry*:

> *Volume I: Quantum Theory and Observability*
> *Volume II: Quantum Atoms and Periodicity*
> **Volume III: Quantum Molecules and Reactivity**
> *Volume IV: Quantum Solids and Orderability*
> *Volume V: Quantum Structure-Activity Relationships (Qu-SAR)*

The chapters in **Volume III** are:

Chapter 1 (Modern Quantum Nature of the Chemical Bond: Valence, Orbitals and Bondons): The classical quantum methods in characterizing chemical bonding, i.e., the valence orbital theory (of Pauling) and the molecular orbital theory (of Heitler and London) are detailed (and illustrated at least for H_2 system) for instructional and benchmarking research purpose on fundamental nature of the chemical bonding. Then the "missing link" in characterizing chemical bonding field by its associated quantum particle, the bondon B, is revealed by employing the combined Bohmian quantum formalism within the Schrödinger quantum picture of electron motions and characterized by its mass (m_B), velocity (v_B), charge (e_B), and life-time (t_B); the mass–velocity–charge–time quaternion properties of bondons particles were used in discussing various paradigmatic types of chemical bonds toward assessing their covalent, multiple bonding, metallic and ionic features. The bondonic picture was completed by discussing

[16] Putz, M. V., Mingos, D. M. P., Eds. (2012). *Applications of Density Functional Theory to Chemical Reactivity*, Structure and Bonding Series Vol. 149, Springer Verlag, Heidelberg-Berlin.

the relativistic charge and life-time (the actual *zitterbewegung*) problem, i.e., showing that the bondon equals the benchmark electronic charge through moving with almost light velocity. It carries negligible, although non-zero, mass in special bonding conditions and toward observable femtosecond life-time as the bonding length increases in the nanosystems and bonding energy decreases according with the bonding length–energy relationship of Heisenberg type $E_{bond}[kcal/mol] \times X_{bond}[A] = 182,019$, thus providing the predictive framework in which the *B* particle may be observed when either associate E_{bond} or length X_{bond} is known. However, being the bondon – *a boson*, as the particle carrying the quantum information of the chemical wave/orbital/field, further possibility in assessing the Density Functional Theory (DFT) for of the Kohn–Sham equation within Bose–Einstein Condensation (BEC) and its further reduction to the nonlinear Schrodinger equation under the Gross–Pitaevsky equation is responded in positive through employing specific DFT–BEC exchange functional combined within the Hartree–Fock–Bogoliubov mean-field theory. This non-linear generalization of the Schrödinger equation is then employed in first providing the DFT–BEC connection at the level of the Thomas–Fermi approximation. Such connections are further used toward generalizing the classical Heitler–London formalism and bonding–antibonding equations of homopolar chemical bonding with the aid of the mass quantification of the quantum particle of the chemical bonding field – *the bondons*. Actually, two branches of bosonic–bondonic condensate were identified as physical and chemical bonding BEC, both with bonding and antibonding features. These findings innovatively complete the already "classical" quantum treatment of chemical bonding through the valence orbital theory (dealing with resonances of molecular structures) and then to the orbital molecular theory (providing the hybridization and the bonding/anti-bonding paradox), here also discussed for their main features.

Chapter 2 (Molecular Structure by Quantum Chemistry): The chemical bond is modeled by employing the geometrical skeleton of the binding atoms in molecule, i.e., by the generated symmetry of molecule at its turn imposing the eigen-energetic terms associated and directly correlated with the symmetry basic elements and operations, further quantitatively realized/quantifies by allied matrices – elements of the symmetry group to which they belong; the quantum framework is then completed by means of the superposition principles applied to joint symmetry of a given molecule with the overall eigen-energies, eventually related with the atomic orbitals, and of their repulsive and attractive influences in the crystal field and ligand field theories, respectively, toward considering the electronic pair at the lance shell as the main driving quantum entity in bonding, being this merely associated with a supra-quantum (geometrical) realization of chemical bond, thus complementing at the semi-classical level the previous developed theory of chemical bonding by bondons (by pairing electrons as bosons, with sub-quantum attractive behavior – see Chapter 1).

Chapter 3 (Quantum Chemical Reactivity of Atoms-in-Molecules): Electronegativity and hardness stand within the minimum dimension set of global indices that characterize bonding and reactivity as the electronic density and the effective applied potential function closely relate with the inner structure of atoms and molecules. This chapter advocates that a proper combination between these two sets of global and local indices can generate a whole plethora of density functionals with a role in quantifying many electronic structures and their transformation at the conceptual rather computational quantum level of comprehension. This is proved though applying the obtained electronegativity and hardness atomic scaled to selected problematical chemical reaction to provide the prediction of reactivity and stabilization of bonds in accordance with the main principles of chemistry: equalization and inequality of electronegativity and hardness, known as the electronegativity equalization, inequality of chemical potential, hard and soft acids and base, and maximum hardness principles, respectively. In this context, a novel reactivity index for quantifying the maximum of hardness realization was proposed with reliability proved throughout providing the hierarchy for a series of hard and soft Lewis bases. In all these, once again, the chemical action influence appears to be playing the role of averaged quantum fluctuations that stabilize the molecules at the end of bonding process. There is also for the first time indicated the appropriate complete bonding scenario based exclusively on the correlated quantum quantities and principles of the electronegativity and hardness. This way, the complete set of global electronegativity–hardness indicators of reactivity of atoms and molecules for various physical–chemical conditions is formulated in an elegant analytical manner within the conceptual density functional theory. Therefore, there is still hope that the present scenario will be accompanied by some advanced ultrafast frozen movie of atomic encountering in bonding.

Chapter 4 (Modeling Molecular Aromaticity): The characterization of aromaticity for organic structures is undertaken in order to compare their chemical reactivity by employing common and recent indicators of aromaticity against the chemical hardness computed within the modern density functional theory and the classical Hückel one. One finds that the values of the energetic indices calculated with the two methods correlate with other data presented in the literature and also with the experimental behavior of the studied compounds. On the other hand, the chemical hardness scale determined with Hückel method is in accordance either with the potential contour maps for the sites susceptible for electrophilic attack as well with the computed global values or chemical hardness realized with DFT method. New aromaticity definition is advanced as the compactness formulation through the ratio between atoms-in-molecule and orbital molecular facets of the same chemical reactivity property around the pre- and post-bonding stabilization limit, respectively. Geometrical reactivity index of polarizability was assumed as providing the benchmark aromaticity scale since its observable character; polarizability based aromaticity scale enables introducing the working five referential aromatic rules (Aroma 1-to-5 Rules) with the help of which the aromaticity scales based on energetic reactivity indices of electronegativity and chemical hardness were computed and analyzed within the major semi-empirical and ab initio quantum chemical methods. The best correlations found in modeling the aromaticity criteria by the chemical hardness-based schemes of computations advocate considering further the associate hard-and-soft acids-and-bases and maximum hardness principles as main tools for assessing chemical reactivity and molecular stability. Similar studies are undertaken within the absolute aromaticity framework viewed as the difference contributions between atoms-in-molecules and molecular-orbitals' contribution to chemical reactivity by specific electronegativity or chemical hardness indices. Finally, quantitative Structure Aromaticity Relationship (QSArR) studies on various aromatic, non-aromatic and anti-aromatic molecules are presented aiming for checking the predictor quality of finite-difference electronegativity and chemical hardness descriptors in aromaticity computations. The results show that the "aromaticity of peripheral topological path" may be well described by superior finite difference schemes of electronegativity and chemical hardness indices in certain calibrating conditions.

Accordingly, the special features of the volume are that it:

- Presents in an unitary quantum theory the various aspects of chemical bonding, from classical description of valence bond and molecular orbital theories to the modern bondonic–bosonic picture of chemical field and interaction, to crystal and ligand field related with symmetry properties and eigen-groups of molecules, to Valence Shell Electron Pair Repulsion-VSEPR model;

- Continues the quantum characterization of the atomic structure and periodicity by using the same chemical descriptors as electronegativity and chemical hardness, united in the same parabolic expression of the total/valence energy respecting the total/valence/exchanged number of electrons in a certain chemical bonding;

- Introduces new concepts of chemical bonding and reactivity as the quantum particle of chemical interaction – the bondons and the density functional of chemical action, with their associated principle and quantitative/qualitative realizations, respectively;

- Characterizes molecular structure also by their propensity to engage chemical reactivity, with specific frontier measure appropriately modeled by the Fukui function (the density to number of electrons derivatives) as well as the chemical hardness iteratively evolution among adducts in a complex chemical interaction or through various chemical interactions' channels (so explaining by quantum chemical reactivity principles the Le Châtelier–Braun principle of reactive equilibrium and bonding, or the electronic delocalization by novel sharing index, just to name few preeminent applications);

- Reviews the main principles of chemical reactivity as based on electronegativity and chemical hardness;

- Formulates two complementary pictures of chemical aromaticity, by relative compactness and absolute difference between the atoms-in-molecules and molecular orbital stages of pre- and post-bonding in a molecule, so contributing to deeper understanding of chemical bonding itself as a dynamic competition of atoms in molecule with molecular orbitals as a whole;

- Formalizes chemical bonding and reactivity in a unitary scenario of chemical indices stages including equalization and fluctuation manifestation of electronic systems, as a natural consequence of their inner quantum nature;
- Predicts the chemical bonding and bondons by the (Goldstone type) symmetry breaking mechanism of the second and fourth order of chemical field (the maximum bond order in Nature) driven by electronegativity and chemical hardness indices as contributing to the potential influence in the Schrödinger Lagrangian of a given electronic system.

Kind thanks are addressed to individuals, universities, institutions, and publishers that inspired and supported the topics included in the present volume; a few of them are:

- *Supporting individuals*: Prof. Hagen Kleinert (Free University of Berlin); Priv. Doz. Dr. Axel Pelster (Free University of Berlin); Prof. Nino Russo (University of Calabria); Dr. Ottorino Ori (Actinium Chemical Research, Parma); Prof. Eduardo A. Castro (University La Plata, Buenos Aires); Dr. Eduard Matito (University of Girona); Prof. Pratim K. Chattaraj (Department of Chemistry and Center for Theoretical Studies, Indian Institute of Technology, Kharagpur); Prof. Jan C.A. Boeyens (Center for Advancement of Scholarship, University of Pretoria, South Africa); Prof. Laszlo Szentpaly (Institute of Theoretical Chemistry, University of Stuttgart); Dr. Laszlo Tarko (Institute of Organic Chemistry "Costin Nenitzescu" of Romanian Academy, Bucharest)
- *Supporting universities*: West University of Timişoara (Faculty of Chemistry, Biology, Geography/Biology-Chemistry Department/ Laboratory of Computational and Structural Physical Chemistry for Nanosciences and QSAR); Free University of Berlin (Physics Department/Institute for Theoretical Physics/Research Center for Einstein's Physics, Centre for International Cooperation); University of Calabria (Faculty of Mathematics and Natural Sciences/Chemistry Department);
- *Supporting institutions and grants*: DAAD (German Service for Academic Exchanges) by Grants: 322 A/17690/2004, 322 A/05356/2011; CNCSIS (Romanian National Council for Scientific

Research in Higher Education) by Grant: AT54/2006-2007; CNCS-UEFISCDI (Romanian National Council for Scientific Research) by Grant: TE16/2010-2013;

- *Supporting publishers*: Wiley (New York); Elsevier (Amsterdam); ACS – American Chemical Society (Washington); Nova Science Inc. (New York); World Scientific (Singapore); Multidisciplinary Digital Publishing Institute – MDPI (Basel, Switzerland); University of Kragujevac (Serbia); World Scientific (Singapore); Chemistry Central (London-UK); IGI Global (formerly Idea Group Inc.), Hershey (Pasadena, USA).

The author also expresses his gratitude for his family and his lovely little daughters *Katy & Ela*, for providing him with lighter moments. He hopes to transmit the work-and-play atmosphere to the reader and students too for acquiring their plus-value in scientific knowledge and method of thinking/approaching quantum *molecules* and of their *chemical reactivity* description.

At last but not the least, author especially thanks the publisher, Apple Academic Press (AAP), and in particular to Ashish (Ash) Kumar, the AAP President and Publisher, and Sandra (Sandy) Jones Sickels, Vice President, Editorial and Marketing, for professionally welcoming and supervising the high level production of the *Quantum Nanochemistry* five-volume set in general and of this volume in particular.

Certainly, since aware of the vast field of *quantum (physical-) chemical theory of bonding, molecular reactivity and aromaticity*, as well by its importance in the years to come in science and technology, any constructive observations, corrections and suggestions are welcome for providing corrected, enlarged and updated version of the present volume in its further editions; such peer contribution is kindly thanked in advance for.

Keep close and think high!

Yours Sincerely,

Mihai V. Putz,
Assoc. Prof. Dr. Dr.-Habil. Acad. Math. Chem.
West University of Timişoara
& R&D National Institute for Electrochemistry and Condensed Matter Timişoara
(Romania)

CHAPTER 1

MODERN QUANTUM NATURE OF THE CHEMICAL BOND: VALENCE, ORBITALS AND BONDONS

CONTENTS

ABSTRACT

The classical quantum methods in characterizing chemical bonding, i.e., the valence orbital theory (of Pauling) and the molecular orbital theory (of Heitler and London) are firstly exposed (and illustrated at least for H_2 system) for didactical and benchmarking research purpose on fundamental nature of the chemical bonding. Then the "missing link" in characterizing chemical bonding field by its associated quantum particle, the bondon $\mathcal{B}$, is revealed by employing the combined Bohmian quantum formalism within the Schrödinger quantum picture of electron motions and characterized by its mass ($m_{\mathcal{B}}$), velocity ($v_{\mathcal{B}}$), charge ($e_{\mathcal{B}}$), and life-time ($t_{\mathcal{B}}$); the mass-velocity-charge-time quaternion properties of bondons' particles were used in discussing various paradigmatic types of chemical bond towards assessing their covalent, multiple bonding, metallic and ionic features. The bondonic picture was completed by discussing the relativistic charge and life-time (the actual *zitterbewegung*) problem, *i.e.*, showing that the bondon equals the benchmark electronic charge through moving with almost light velocity. It carries negligible, although non-zero, mass in special bonding conditions and towards observable femtosecond life-time as the bonding length increases in the nanosystems and bonding energy decreases according with the bonding length-energy relationship of Heisenberg type $E_{bond}[kcal/mol] \times X_{bond}[\overset{0}{A}] = 182019$, providing this way the predictive framework in which the $\mathcal{B}$ particle may be observed when either associate E_{bond} or length X_{bond} is known. However, being the bondon *a boson*, as the particle carrying the quantum information of the chemical wave/orbital/field, further, possibility in assessing the density functional theory (DFT) for of the Kohn-Sham equation within Bose-Einstein condensation (BEC) and of its further reduction to the nonlinear Schrodinger equation under the Gross-Pitaevsky equation is responded in positive through employing specific DFT-BEC exchange functional combined within the Hartree-Fock-Bogoliubov mean-field theory. This non-linear generalization of the Schrödinger equation is then employed in first providing the DFT-BEC connection at the level of the Thomas-Fermi approximation; such connections are further used towards generalizing the classical Heitler-London formalism and bonding-antibonding equations of homopolar chemical bonding with the aid of the mass quantification of the quantum particle of the chemical bonding field – *the bondons*. Actually, two branches of bosonic-bondonic condensate were identified as physical and chemical bonding BEC, both with bonding and antibonding features. These findings innovatively complete the already "classical" quantum treatment of chemical bonding through the valence orbital theory (dealing with resonances of molecular structures) and then to the orbital molecular theory (providing the hybridization and the bonding/anti-bonding paradox), here also exposed for their main features.

1.1   INTRODUCTION

Despite the fact that in classical chemistry more types of bonds are distinguished, i.e., the polar semipolar, electrovalent, covalent, and coordinative, nowadays these distinctions are meaningless and should be avoided, their presence inducing many confusions. Actually, the nature of the chemical bond, solved by its quantum manifestation, is based on the fact that the theoretical reasoning is the same, i.e., driven by the Schrödinger's equation, and applied to any molecule, with no exceptions. What is particularly under focus is determining the electronic distribution in the molecular space, the wave functions and the associated electronic density, along with the energetic levels' configuration. Yet, from the qualitatively point of view, there are some differences between the calculation (solving the Schrödinger's equation for the molecular systems) and reality (observing the experimental data), this being *not* due to the principles of Quantum Mechanics itself, but due to the inability of exactly finding the analytical solution(s) of the multielectronic Schrödinger's equation in the multinuclear compounds (beyond the spherical symmetry).

Historically, one of the first attempts to systematically use the electron structure as the basis of the chemical bond is due to the discoverer of the electron itself, J.J. Thomson, who published in 1921 an interesting model for describing one of the most puzzling molecules of chemistry, the benzene, by the aid of C–C portioned bonds, each with three electrons (Thomson, 1921) that were further separated into $2(\sigma) + 1(\pi)$ lower and higher energy electrons, respectively, in the light of Hückel σ-π and of subsequent quantum theories (Hückel, 1931; Doering & Detert, 1951).

On the other side, the electronic theory of the valence developed by (Lewis, 1916) and expanded by (Langmuir, 1919) had mainly treated the electronic behavior like a point-particle that nevertheless embodies considerable chemical information, due to the semiclassical behavior of the electrons on the valence shells of atoms and molecules.

Nevertheless, the consistent quantum theory of the chemical bond was advocated and implemented by the works of Pauling (1931a-c) and Heitler & London (1927), which gave rise to the wave-function characterization of bonding through the fashioned molecular wave-functions (orbitals)–mainly coming from the superposition principle applied on the atomic wave-functions involved. The success of this approach, especially reported by spectroscopic studies, encouraged further generalization toward treating more and more complex chemical systems by the self-consistent wave-function algorithms developed in basic quantum chemistry (Slater, 1928, 1929; Hartree, 1957; Lowdin, 1955a,b, 1960; Roothann, 1951), in subsequent PPP theory (Pariser & Parr, 1953a,b; Pople, 1953), until the turn towards the DFT developed by Hohenberg & Kohn (1964), Kohn & Sham (1965) and by Pople group in the second half of the twentieth century (Pople et al., 1978; Head-Gordon et al., 1989), which marked the subtle feed-back to the earlier electronic point-like view by means of the electronic density functionals and localization functions, see Volume II of the present five-volume work. The compromised picture of the chemical bond may be widely comprised by the emerging Bader's atoms-in-molecule theory (Bader, 1990, 1998, 1994), the fuzzy theory of Mezey's group (Mezey, 1993; Maggiora & Mezey, 1999; Szekeres et al., 2005), along with the chemical reactivity principles (Parr & Yang, 1989) as originating in the electronegativity (Sanderson, 1988) and chemical hardness concepts (Pearson, 1973, 1990), and of their functionality in modern quantum chemistry (Parr et al., 1978; Mortier et al., 1985; Sen & Jørgenson, 1987; Chattaraj et al., 1991; Chattaraj & Schleyer, 1994; Chattaraj & Maiti, 2003).

Within this modern quantum chemistry picture, its seems that the Dirac dream (Dirac, 1929; Putz, 2011a) in characterizing the chemical bond (in particular) and the chemistry (in general) by means of the chemical field related with the Schrödinger wave-function (Schrödinger, 1926) or the Dirac spinor (Dirac, 1928) was somehow avoided by collapsing the undulatory quantum concepts into the (observable) electronic density. To this end, the present chapter, upon exposing the consecrated paradigms of chemical bonding, present the algorithm and the features of the first predicted quantum chemical particle, the bondon – acquiring for the chemical boning information through its propagation among the attractive centers/basins/nuclei; even more, due to its inherent bosonic nature, the bondon can be further involved in the celebrated BEC paradigm too, so that the chemical bonding as a BEC of the bondons stays as the ultimate insight in the chemical boning nature; this is possible due to the conjunction with the DFT adapted for BEC, here advanced by considering BEC specific order parameter and density concepts as variables in the celebrated Kohn-Sham (KS) equations due to their conceptual robustness with respect to both BEC and KS systems alike.

1.2   VALENCE BOND THEORY

1.2.1   ADIABATIC APPROXIMATION

Let's consider a molecule composed of N nuclei and p electrons, within a reference system, connected to the molecule's mass center (CM) for which we will collectively note:

- By (R): the ensemble of all nuclei coordinates and
- By (r): the ensemble of all electrons coordinates.

For the atomic case, for example of a nucleus of mass (M) and an electron of mass (m), once the energy equipartition (as in the classical mechanics) is accepted, then the (quantum) energies of the two particles become equal:

$$\frac{1}{2}Mv_M^2 = \frac{1}{2}mv_m^2 \Rightarrow \frac{v_m}{v_M} = \sqrt{\frac{M}{m}} \gg 1 \qquad (1.1)$$

which corresponds to the fact that the electron speed is much higher than the nucleus speed around the CM point associated to the system. The same happens in localization terms: the associated de Broglie wavelengths are firstly calculated:

$$\lambda_M = \frac{h}{Mv_M} \text{ and } \lambda_m = \frac{h}{mv_m} \qquad (1.2)$$

and then the formed ratio of the proper "volumes" (of undulatory exis-tences) to get

$$\left(\frac{\lambda_M}{\lambda_m}\right)^3 = \left(\frac{mv_m}{Mv_M}\right)^3 = \left(\frac{m}{M}\right)^3 \cdot \left(\frac{M}{m}\right)^{\frac{3}{2}} = \left(\frac{m}{M}\right)^{\frac{3}{2}} << 1 \qquad (1.3)$$

which suggests that the electron is much more delocalized than the nucleus!

This idea is transposed to the molecular systems under the perturbation form:

- by 0th order: through considering the *fixed system of nuclei*; and
- by the first order: through considering the *slow moving of the nuclei system*.

This way the stationary state equation (i.e., the stationary Schrödinger's equation) for the molecule is writing in the general form:

$$\left\{-\frac{\hbar^2}{2m}\sum_{i=1}^{P}\Delta_i - \frac{\hbar^2}{2}\sum_{\alpha=1}^{N}\frac{1}{M_\alpha}\Delta_\alpha + V(r,R)\right\}\Psi(r,R) = E\Psi(r,R) \qquad (1.4)$$

with $V(r;R)$ containing all the interactions' types at the molecular level:

$$V(r,R) = \sum_{\beta > \alpha}^{N}\frac{Z_\alpha \cdot Z_\beta \cdot e^2}{R_{\alpha\beta}} - \sum_{i,\alpha}^{P,N}\frac{Z_\alpha \cdot e^2}{R_{i\alpha}} + \sum_{i > j}^{n}\frac{e^2}{r_{ij}} \qquad (1.5)$$

In the zeroth order approximation (of the fixed-nuclei) the kinetic energy of the nuclei disappears, so remaining the so-called *electronic* equation

$$\left[-\frac{\hbar^2}{2m}\sum_{i=1}^{P}\Delta_i + V(r,R)\right]\varphi_n(r,R) = \varepsilon_n(R)\varphi_n(r,R) \qquad (1.6)$$

where

- $\varphi_n(r,R)$: denote the electronic state functions for the electronic subsystem;
- $\varepsilon_n(R)$: are the electronic energetic values for a spatial configuration for a given nuclei subsystem (R).

Since for each of the parameters' set (R) the nuclear wave function $\phi(R)$ is *independent* of the electronic wave function $\varphi(r,R)$, one can apply the probabilistic interpretation of the wave function which recommends the *quantum events'simultaneity* as a product of associated probabilities with the resulted general wave-function expression $\Psi(r,R)$, as a series over all electronic energetic spectra of wave-function products $\phi(R)\varphi(r,R)$

$$\Psi(r,R)=\sum_{n}\phi_{n}(R)\,\varphi_{n}(r,R) \tag{1.7}$$

Replacing this expression in the Schrodinger general equation (1.4), one can multiply it to the left side with the electronic complex wave-function $\varphi_{m}^{*}(r,R)$ and then to integrate over the electronic coordinates' space/volume dV_{e}:

$$\int \varphi_{m}^{*}\left(-\frac{\hbar^{2}}{2m}\sum_{i=1}^{P}\Delta_{i}\right)\sum_{n}\phi_{n}(R)\varphi_{n}dV_{e} + \int \varphi_{m}^{*}\left(-\frac{\hbar^{2}}{2}\sum_{\alpha=1}^{N}\frac{1}{M_{\alpha}}\Delta_{\alpha}\right)\sum_{n}\phi_{n}(R)\varphi_{n}dV_{e}$$
$$+\int \varphi_{m}^{*}V(r,R)\sum_{n}\phi_{n}(R)\varphi_{n}dV_{e} = \int \varphi_{m}^{*}E\sum_{n}\phi_{n}(R)\varphi_{n}dV_{e}$$

$$\tag{1.8}$$

$$\sum_{i=n}\left(-\frac{\hbar^{2}}{2m_{0}}\right)\phi_{n}(R)\int \varphi_{m}^{*}\Delta_{i}\varphi_{n}dV_{e} + \sum_{\alpha,n}\left(-\frac{\hbar^{2}}{2M_{\alpha}}\right)\int \varphi_{m}^{*}\Delta_{\alpha}\left[\phi_{n}(R)\varphi_{n}\right]dV_{e}$$
$$+\sum_{n}\phi_{n}(R)\int \varphi_{m}^{*}V(r,R)\varphi_{n}dV_{e} = \sum_{n}E\phi_{n}(R)\underbrace{\int \varphi_{m}^{*}\varphi_{n}dV_{e}}_{\delta_{m,n}=\langle\varphi_{m}|\varphi_{n}\rangle}$$

$$\tag{1.9}$$

Now, there appears that one has to solve the operatorial action:

$$\Delta_{\alpha}\left(\phi_{n}\varphi_{n}\right)=\nabla_{\alpha}\left[\nabla_{\alpha}\left(\phi_{n}\varphi_{n}\right)\right]=\nabla_{\alpha}\left[\varphi_{n}\nabla_{\alpha}\phi_{n}+\phi_{n}\nabla_{\alpha}\varphi_{n}\right]$$
$$=\left(\nabla_{\alpha}\varphi_{n}\nabla_{\alpha}\phi_{n}+\varphi_{n}\nabla_{\alpha}\nabla_{\alpha}\phi_{n}\right)+\left(\nabla_{\alpha}\phi_{n}\nabla_{\alpha}\varphi_{n}+\phi_{n}\nabla_{\alpha}\nabla_{\alpha}\varphi_{n}\right)$$
$$=\phi_{n}\Delta_{\alpha}\varphi_{n}+\varphi_{n}\Delta_{\alpha}\phi_{n}+2grad_{\alpha}\varphi_{n}grad_{\alpha}\phi_{n} \tag{1.10}$$

With Eq. (1.10), the Eq. (1.9) is appropriately rearranged as:

$$\sum_{n}\phi_{n}(R)\int \varphi_{m}^{*}\underbrace{\left[-\frac{\hbar^{2}}{2m}\sum_{i}^{n}\Delta_{i}+V(r,R)\right]\varphi_{n}dV_{e}}_{\varepsilon_{n}(R)\varphi_{n}} + \sum_{\alpha,n}-\frac{\hbar^{2}}{2M_{\alpha}}\Delta_{\alpha}\phi_{n}\underbrace{\int \varphi_{m}^{*}\varphi_{n}dV_{e}}_{\delta_{m,n}=\langle\varphi_{m}|\varphi_{n}\rangle}$$

$$= E\phi_m + \sum_{n,\alpha} \frac{\hbar^2}{2M_\alpha} \left[\phi_n \int \varphi_m^* \Delta_\alpha \varphi_n dV_e + 2\int \varphi_m^* \nabla_\alpha \varphi_n \left(\nabla_\alpha \phi_n \right) dV_e \right]$$

(1.11)

which is finally rewritten under the concise form

$$\left[-\frac{\hbar^2}{2} \sum_\alpha \frac{1}{M_\alpha} \Delta_\alpha + \varepsilon_m (R) \right] \phi_m = E\phi_m + \sum_n C_{mn}\phi_n$$

(1.12)

considering the zero-order electronic equation (1.6), and where the non-adiabatic coefficient (the operator) was also introduced

$$C_{mn} = \sum_\alpha \frac{\hbar^2}{2M_\alpha} \left[\int \varphi_m^* \Delta_\alpha \varphi_n dV_e + 2\left(\int \varphi_m^* \nabla_\alpha \varphi_n dV_e \right) \nabla_\alpha \right]$$

(1.13)

The Schrodinger equation with the included non-adiabatic factor C_{mn} is precise. However, if the C_{mn} is small, we actually work within the perturbation approximation of first order. Moreover, if $C_{mn} = 0$ is fully considered, then one only uses the adiabatic approximation (the zero order perturbation of the nuclei system movement respecting the electronic motion one); in this case, the Schrödinger equation becomes (for the whole molecular system):

$$\left[-\frac{\hbar^2}{2} \sum_\alpha \frac{1}{M_\alpha} \Delta_\alpha + \varepsilon_m (R) \right] \phi_{m\upsilon}^0 (R) = E_{m\upsilon}^0 \phi_{m\upsilon}^0 (R)$$

(1.14)

In Eq. (1.14) the role of potential energy is played by the energy of electronic states $\varepsilon_m (R)$ which is obtained by solving the electronic Schrodinger equation, i.e., when assuming the fact that the nuclei are fixed.

Finally, the molecule wave-function can take, in the adiabatic approximation, the following form

$$\Psi_{m\upsilon}^{(r,R)} = \phi_{m\upsilon}^0 (R)\varphi_m (r,R)$$

(1.15)

as the two nuclei-electrons sub-systems would be independent. Actually:

- Each of the electronic state (here quantified by the m – quantum number) will be associated with various motion states further distinguished by the υ set of quantum numbers, so carrying the energy $E_{m\upsilon}$.
- The adiabatic approximation is also justified when the exact solution

of the system with $C_{mn} \neq 0$ differs very little from the zero-order approximation solution.

1.2.2 THE VIRIAL THEOREM ON THE MOLECULAR SYSTEMS

One starts by retaining only the electronic terms from a molecular Hamiltonian system

$$\widehat{H}_e = -\frac{1}{2}\frac{\hbar^2}{m}\sum_{i=1}^{p}\Delta_i + V(r,R) \tag{1.16}$$

with the working potential

$$V(r,R) = -\sum_{i,\alpha}^{p,N}\frac{Z_\alpha \cdot e^2}{R_{i\alpha}} + \sum_{i\rangle j}^{n}\frac{e^2}{r_{ij}} \tag{1.17}$$

with 3D geometrical dependency of the nuclear-electron and electron-electron interaction distances

$$\begin{cases} R_{i\alpha} = \sqrt{\left(R_{\alpha x} - x_i\right)^2 + \left(R_{\alpha y} - y_i\right)^2 + \left(R_{\alpha z} - z_i\right)^2} \\ r_{ij} = \sqrt{\left(x_i - x_j\right)^2 + \left(y_i - y_j\right)^2 + \left(z_i - z_j\right)^2} \end{cases} \tag{1.18}$$

The electronic movement around the nuclei can be introduced through a variational parameter, say η as a scale parameter, and then by applying the quantum postulate of the variational principle (behavior) of the depending energy:

$$\frac{\partial \varepsilon_\eta (R)}{\partial \eta} = \frac{\partial}{\partial \eta}\int \Psi_\eta^* \widehat{H}_e \Psi_\eta dV_e = 0 \tag{1.19}$$

Here Ψ_η will be introduced by "swelling" the available space for the electrons and nuclei movement

$$\begin{cases} dV_e' = d^3r' = \eta^{3p}d^3r \\ dV_{Nucl}' = d^3R' = \eta^{3p}d^3R \end{cases} \Rightarrow \begin{cases} x_i' = \eta x_i; y_i' = \eta y_i; z_i' = \eta z_i \\ R_{\alpha x}' = \eta R_{\alpha x}; R_{\alpha y}' = \eta R_{\alpha y}; R_{\alpha z}' = \eta R_{\alpha z} \end{cases}$$

$$\tag{1.20}$$

from where, one can call the requirement that the electronic normalization condition be equally respected so leaving with the working parametrical wave-function

$$1 = \int \Psi_\eta^*(r',R') \cdot \Psi_\eta(r',R') dV_e' = \frac{1}{\eta^{3p}} \int \Psi_\eta^*(r',R') \cdot \Psi_\eta(r',R') dV_e$$

$$= \int \Psi^*(r,R) \cdot \Psi(r,R) dV_e$$

$$\Rightarrow \Psi_\eta(r',R') = \eta^{\frac{3p}{2}} \Psi(r,R) \tag{1.21}$$

with p being the electrons number.

This way, the electronic Hamiltonian by the new coordinates set will have the form:

$$\widehat{H}_e = -\frac{\eta^2}{2} \sum_{i=1}^{p} \Delta_i' + \eta V(r',R') \equiv \eta^2 \widehat{T} + \eta \widehat{V} \tag{1.22}$$

from where one yields with the parametric result of energy

$$\varepsilon_\eta(R) = \int \Psi_\eta^*(r,R) \cdot \widehat{H}_e \cdot \psi_\eta(r,R) dV_e$$

$$= \frac{1}{\eta^{3p}} \int \Psi_\eta^*(r',R') \cdot \left[\eta^2 \widehat{T} + \eta \widehat{V} \right] \cdot \Psi_\eta(r',R') dV_e'$$

$$= \eta^2 \left(\int \Psi^*(r',R') \widehat{T} \cdot \Psi(r',R') dV_e' \right) + \eta \left(\int \Psi^*(r',R') \widehat{V} \cdot \Psi(r',R') dV_e' \right)$$

$$= \eta^2 \langle \widehat{T} \rangle + \eta \langle \widehat{V} \rangle$$

$$\tag{1.23}$$

Now, the application of the minimum condition (1.19) in relation with η will produce the new equation

$$\frac{\partial \varepsilon(R)}{\partial \eta} = 2\eta \langle \widehat{T} \rangle + \langle \widehat{V} \rangle + \eta^2 \frac{\partial \langle \widehat{T} \rangle}{\partial \eta} + \eta \frac{\partial \langle \widehat{V} \rangle}{\partial \eta} = 0 \tag{1.24}$$

The relationship (1.24) can be further rewritten by considering the translation-vibration-rotation motion of the nuclei system, with 3N-6 cardinal as the associate degree of freedom, so making the derivatives:

$$\begin{cases} \dfrac{\partial \langle \hat{T} \rangle}{\partial \eta} = \sum_{\gamma=1}^{3N-6} \dfrac{\partial \langle \hat{T} \rangle}{\partial R_\gamma{}'} \cdot \dfrac{\partial R_\gamma{}'}{\partial \eta} \\[3mm] \dfrac{\partial \langle \hat{V} \rangle}{\partial \eta} = \sum_{\gamma=1}^{3N-6} \dfrac{\partial \langle \hat{V} \rangle}{\partial R_\gamma{}'} \cdot \dfrac{\partial R_\gamma{}'}{\partial \eta} \end{cases} \tag{1.25}$$

Now, since for the (3N-6) coordinates of the nuclei' motion we have the parametrical scaling

$$R_\gamma{}' = \eta R_\gamma \Rightarrow \frac{\partial R_\gamma{}'}{\partial \eta} = R_\gamma \tag{1.26}$$

the above minimum condition (1.24) further becomes:

$$2\eta \langle \hat{T} \rangle + \langle \hat{V} \rangle + \eta^2 \sum_{\gamma=1}^{3N-6} R_\gamma \frac{\partial \langle \hat{T} \rangle}{\partial R_\gamma{}'} + \eta \sum_{\gamma=1}^{3N-6} R_\gamma \frac{\partial \langle \hat{V} \rangle}{\partial R_\gamma{}'} = 0 \tag{1.27}$$

Finally, this result will be particularized for the exact solution

$$\Psi_{\eta=1}(r,R) = 1\Psi(r,R) \tag{1.28}$$

That is, for $\eta = 1$, so recovering the case in which also the involved coordinates transform identically as $(r') = (r), (R') = (R)$. This way, at the energy level $\varepsilon_\eta(R)$ for $\eta = 1$ we regain the natural condition

$$\varepsilon(R) = \langle \hat{T} \rangle + \langle \hat{V} \rangle \tag{1.29}$$

while from the application of the minimum energy/variation principle (1.27) one will obtain for $\eta = 1$ (implicitly when $R_\gamma{}' = R_\gamma$) the simplified expression

$$2\langle \hat{T} \rangle + \langle \hat{V} \rangle + \sum_{\gamma=1}^{3N-6} R_\gamma \left\{ \frac{\partial \langle \hat{T} \rangle}{\partial R_\gamma} + \frac{\partial \langle \hat{V} \rangle}{\partial R_\gamma} \right\} = 0 \tag{1.30}$$

so providing the *virial equation* for molecules

$$2\left\langle\hat{T}\right\rangle+\left\langle\hat{V}\right\rangle+\sum_{\gamma=1}^{3M-6}R_{\gamma}\frac{\partial\varepsilon}{\partial R_{\gamma}}=0 \qquad (1.31)$$

When the components $\left\langle\hat{T}\right\rangle,\left\langle\hat{V}\right\rangle$ in Eq. (1.31) are combined with the energy expression $\varepsilon(R)=\varepsilon(R_1,...,R_{3N-6})$ in Eq. (1.29) one obtains them respectively as

$$\begin{cases} \left\langle\hat{T}\right\rangle=-\varepsilon-\sum_{\gamma=1}^{3N-6}R_{\gamma}\frac{\partial\varepsilon}{\partial R_{\gamma}} \\ \left\langle\hat{V}\right\rangle=2\varepsilon+\sum_{\gamma=1}^{3M-6}R_{\gamma}\frac{\partial\varepsilon}{\partial R_{\gamma}} \end{cases} \qquad (1.32)$$

Note that for the atomic case, when $\partial\varepsilon/\partial R_{\gamma}=0$, they are further transcribed as:

$$\begin{cases} \left\langle\hat{T}\right\rangle=-\varepsilon \\ \left\langle\hat{V}\right\rangle=2\varepsilon \end{cases} \Leftrightarrow \left\langle\hat{V}\right\rangle=-2\left\langle\hat{T}\right\rangle \qquad (1.33)$$

which generalize the similar expressions from the atomic level, when the nuclei motions after of 3N-6 freedom degrees, also called *internal coordinates*, are likely included.

Essentially, the *virial* (with "filled" meaning here) express the energetic balance of the electronic system (by the nuclear parameter R_{γ}) when the equilibrium is reached (of the observed, measured values, i.e., averaged $\langle\bullet\rangle$ upon the quantum motion and fluctuation whatever).

1.2.3 THE CHEMICAL BOND THROUGH THE VALENCE BOND METHOD (VBM)

Developed by the studies of Pauling, Heitler and London, the valence-bond (VB) method assumes that the atomic orbitals which participate to the chemical bond are almost not at all affected, by keeping in most of the time their electronic configuration, so the bond force being determined by the spin valence electron compensation or of their actually exchange (Pauling, 1960; Murrel et al., 1985; Shaik & Hiberty, 2008). Take as prototype the H-H (H_2) hydrogen molecules for which the so-called limiting/resonances structures are represented in Figure 1.1.

For the working case of Figure 1.1 one has the molecular wave-function as a superposition of the resonances' wave functions

$$\psi(A,B) = c_1\varphi_1 + c_2\varphi_2 + c_3\varphi_3 + c_4\varphi_4 \tag{1.34a}$$

where we recognize:

$$c_1\varphi_1 + c_2\varphi_2 \text{ as covalent structures, along} \tag{1.34b}$$

$$c_3\varphi_3 + c_4\varphi_4 \text{ as ionic, electro-valent structures} \tag{1.34c}$$

All of these wave-functions are forming the so-called *mesomeric* (intermediate) state among all possible wave-functions $\varphi_{i=1,2,3,4}$, yet none of them with a specific significance, nor representing the molecule itself.

However, in the adiabatic approximation only the electronic Hamiltonian is considered and the electronic Schrödinger equation looks like:

$$\widehat{H}_e\Psi(A,B) = \varepsilon_{AB}\Psi(A,B) \tag{1.35a}$$

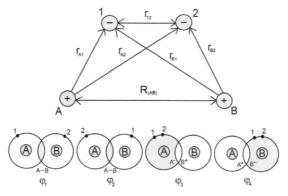

FIGURE 1.1 Schematic description of chemical bond by valence theory (up) and the associate resonance models of bonding (bottom).

$$\widehat{H}_e = -\frac{\hbar^2}{2m}(\Delta_1 + \Delta_2) - \frac{e^2}{4\pi\varepsilon_0}\left[\frac{1}{r_{A1}} + \frac{1}{r_{B1}} + \frac{1}{r_{A2}} + \frac{1}{r_{B2}} - \frac{1}{r_{12}}\right] \quad (1.35b)$$

Here, the ionic structures (the 3rd and 4th instances of Figure 1.1) will be considered as a perturbation case for the covalent one (for $A = B$, they are equal and small, while for $A \neq B$ they are not equal, since one of the atoms being more electronegative and the molecule featuring a dipole moment).

Then we can write that

$$\begin{cases} \Psi_{(AB)} = c_1\varphi_1 + c_2\varphi_2 + f \\ \varepsilon_{AB} = 2E_0 + e \end{cases} \quad (1.36)$$

where e is the perturbation to the system (A and B) energy; the two present co-valent resonances fulfill the quantum working conditions for the first instance of the Figure 1.1:

$$\begin{cases} r_{A2} \to \infty, r_{B1} \to \infty, r_{12} \to \infty, R \to \infty \\ \widehat{H}_0 = \left[-\frac{\hbar^2}{2m}\Delta_1 - \frac{e_0^2}{r_{A1}}\right] + \left[-\frac{\hbar^2}{2m}\Delta_2 - \frac{e_0^2}{r_{B2}}\right] = \widehat{H}_A(1) + \widehat{H}_B(2) \\ \widehat{H}_A(1)\psi_A(r_{A1}) = E_0\psi_A(r_{A1}) \\ \widehat{H}_B(2)\psi_B(r_{B2}) = E_0\psi_B(r_{B2}) \\ \varphi_1 = \psi_A(r_{A1})\psi_B(r_{B2}) \\ \widehat{H}_0\varphi_1 = \left[\widehat{H}_A(1) + \widehat{H}_B(2)\right]\varphi_1 = 2E_0\varphi_1 \end{cases} \quad (1.37)$$

with

$$e_0^2 = \frac{e^2}{4\pi\varepsilon_0} \quad (1.38)$$

and respectively for the second covalent instance of the Figure 1.1 the conditions:

$$\begin{cases} r_{A1} \to \infty, r_{B2} \to \infty, r_{12} \to \infty, R \to \infty \\ \widehat{H}_0 = \left[-\dfrac{\hbar^2}{2m}\Delta_2 - \dfrac{e_0^2}{r_{A2}} \right] + \left[-\dfrac{\hbar^2}{2m}\Delta_1 - \dfrac{e_0^2}{r_{B1}} \right] = \widehat{H}_A(2) + \widehat{H}_B(1) \\ \widehat{H}_A(2)\psi_A(r_{A2}) = E_0\psi_A(r_{A2}) \\ \widehat{H}_B(1)\psi_B(r_{B1}) = E_0\psi_B(r_{B1}) \\ \varphi_2 = \psi_A(r_{A2})\psi_B(r_{B1}) \\ \widehat{H}_0\varphi_2 = \left[\widehat{H}_A(2) + \widehat{H}_B(1) \right]\varphi_2 = 2E_0\varphi_2 \end{cases}$$
(1.39)

Now, the function $\Psi(AB)$ and the energy ε_{AB} satisfy the Schrödinger electronic equation, so resulting into the equation

$$\widehat{H}_e\left[c_1\varphi_1 + c_2\varphi_2 + f \right] = \left(2E_0 + e \right)\left[c_1\varphi_1 + c_2\varphi_2 + f \right]$$
$$\Leftrightarrow c_1\widehat{H}_e\varphi_1 + c_2\widehat{H}_e\varphi_2 + \widehat{H}_ef = 2E_0\left[c_1\varphi_1 + c_2\varphi_2 \right]$$
$$+ e\left[c_1\varphi_1 + c_2\varphi_2 \right] + \left(2E_0 + e \right)f$$
(1.40)

where $\widehat{H}_e$ can be on turns

- For φ_1:

$$\widehat{H}_e(1,2) = \widehat{H}_A(1) + \widehat{H}_B(2) + \left[-\dfrac{e_0^2}{r_{A2}} - \dfrac{e_0^2}{r_{B1}} + \dfrac{e_0^2}{r_{12}} \right]$$
$$= \widehat{H}_A(1) + \widehat{H}_B(2) + \widehat{W}(1,2)$$
(1.41a)

- For φ_2:

$$\widehat{H}_e(2,1) = \widehat{H}_A(2) + \widehat{H}_B(1) + \left[-\dfrac{e_0^2}{r_{A1}} - \dfrac{e_0^2}{r_{B2}} + \dfrac{e_0^2}{r_{12}} \right]$$
$$= \widehat{H}_A(2) + \widehat{H}_B(1) + \widehat{W}(2,1)$$
(1.41b)

and where the products of the type $\left\lfloor \widehat{W}(1,2)f \right\rfloor$, $\left\lfloor \widehat{W}(2,1)f \right\rfloor$, $\left[ef \right]$ can be neglected because of the smallness of $\widehat{W}$ and of e, respectively.

Next, by the replacement of $H_e(1,2)$ with $H_e f$ one has:

$$c_1 \underbrace{\left[\widehat{H}_A(1) + \widehat{H}_B(2)\right] \varphi_1}_{2E_0} + c_1 \left[\widehat{W}(1,2)\right] \varphi_1$$

$$+ c_2 \underbrace{\left[\widehat{H}_A(1) + \widehat{H}_B(2)\right] \varphi_2}_{\widehat{H}_A(2) + \widehat{H}_B(1)} + c_2 \underbrace{\left[\widehat{W}(1,2)\right] \varphi_2}_{\widehat{W}(2,1)}$$

$$+ \left[\widehat{H}_A(1) + \widehat{H}_B(2) + \underbrace{\widehat{W}(1,2)}_{=(2,1)}\right] f = 2E_0 c_1 \varphi_1 + 2E_0 c_2 \varphi_2 + e c_1 \varphi_1 + e c_2 \varphi_2 + 2E_0 f$$

$$\Leftrightarrow \left[\widehat{H}_A(1) + \widehat{H}_B(2)\right] f - 2E_0 f = \left[e - \widehat{W}(1,2)\right] c_1 \varphi_1 + \left[e - \widehat{W}(2,1)\right] c_2 \varphi_2$$

$$(1.42)$$

which is a non-homogeneous equation in f and e. However, the associated homogeneous equation would be

$$\left[\widehat{H}_A(1) + \widehat{H}_B(2)\right] f - 2E_0 f = 0 \qquad (1.43)$$

leaving with the solution $f = \varphi_1$.

Then, according to a notorious math theorem (in the theory of differential equations), the non-homogeneous equation has a solution only in case that the non-homogeneous term is orthogonal to the homogeneous solution, which implies the integral orthogonality relationship:

$$\int \left\{ \left[e - \widehat{W}(1,2)\right] c_1 \varphi_1 + \left[e - \widehat{W}(2,1)\right] c_2 \varphi_2 \right\} \varphi_1 dV_1 dV_2 = 0 \qquad (1.44)$$

Similarly, if we replaced the $\widehat{H}_e(2,1)$ by $\widehat{H}_e f$ we would obtain a non-homogeneous equation of type

$$\left[\widehat{H}_A(2) + \widehat{H}_B(1)\right] f - 2E_0 f = \left[e - \widehat{W}(1,2)\right] c_1 \varphi_1 + \left[e - \widehat{W}(2,1)\right] c_2 \varphi_2$$

$$(1.45)$$

From where the associated homogeneous equation

$$\left[\widehat{H}_A(2) + \widehat{H}_B(1)\right] f - 2E_0 f = 0 \qquad (1.46)$$

with the solution $f = \varphi_2$ and the non-homogeneous equation allows solutions only if the non-homogeneous section is orthogonal to the homogeneous equation solution; i.e.,

$$\int \left\{ \left[e - \widehat{W}(1,2) \right] c_1\varphi_1 + \left[e - \widehat{W}(2,1) \right] c_2\varphi_2 \right\} \varphi_2 dV_1 dV_2 = 0 \qquad (1.47)$$

The existence conditions of the solutions (e, c_1, c_2) can be jointly placed in a system formed through considering also simplifying yet with physical meaning notations:

- *The Coulomb integral*

$$J \ (or \ C \ or \ \alpha) = \int \varphi_1 \widehat{W}(1,2)\varphi_1 dV_1 dV_2 = \int \varphi_2 \widehat{W}(2,1)\varphi_2 dV_1 dV_2 \quad (1.48)$$

representing the energy of an electron on the limiting orbital φ_1 respectively on φ_2, i.e., the ionization energy of the respective orbitals, being identically here by the nature of homo-atomic molecule under discussion, i.e., the particular case of the hydrogen molecules, with the two identical orbitals involved, one for each atom from molecule.

- *The exchange integral*

$$K \ (or \ A \ or \ \beta) = \int \left(\widehat{W}(1,2)\varphi_2 \right) \varphi_1 dV_1 dV_2 = \int \left(\widehat{W}(2,1)\varphi_1 \right) \varphi_2 dV_1 dV_2$$

$$(1.49)$$

representing the energy gain of an electron becoming common between the two atoms in chemical bonding; in the homoatomic bond, it represents ½ from the bonding energy.

- *The overlap Integral*

$$S = \int \varphi_1\varphi_2 dV_1 dV_2 = \int \varphi_2\varphi_1 dV_1 dV_2 \qquad (1.50)$$

represents the measure of the space region where both orbitals have high contribution in bonding (overlapping).

With these notations the working system of equations is formed:

$$\begin{cases} (e-J)c_1 + (eS-K)c_2 = 0 \\ (eS-K)c_1 + (e-J)c_2 = 0 \end{cases} \qquad (1.51)$$

where one has taken into consideration the normalization for each limiting (resonance) molecular orbital:

$$\int \varphi_1 \varphi_2 dV_1 dV_2 = \int \varphi_2 \varphi_1 dV_1 dV_2 = 1 \qquad (1.52)$$

- The previous system is homogeneous and has solutions only if the associate determinant cancels:

$$\begin{vmatrix} e-J & eS-K \\ eS-K & e-J \end{vmatrix} = 0 \qquad (1.53a)$$

$$\Leftrightarrow (e-J)^2 - (eS-K)^2 = 0$$
$$\Leftrightarrow (e-J-eS+K)(e-J+eS-K) = 0$$
$$\Rightarrow e_1 = \frac{J-K}{1-S} \qquad (1.53b)$$

$$\& \ e_2 = \frac{J+K}{1+S} \qquad (1.53c)$$

Turning with solutions (1.53b,c) back for coefficients' determination in whatever of the above system's equation (1.51) one successively gets

$$\left(\frac{J \mp K}{1 \mp S} - J \right) c_1 = \left(K - \frac{J \mp K}{1 \mp S} \cdot S \right) c_2 \qquad (1.54a)$$

$$\Leftrightarrow \frac{J \mp K - J(1 \mp S)}{1 \mp S} c_1 = \frac{K(1 \mp S) - J \mp K \cdot S}{1 \mp S} c_2$$
$$\Leftrightarrow (J \mp K - J \pm JS) c_1 = (K \mp KS - JS \pm KS) c_2$$
$$\Leftrightarrow (\mp K \pm JS) c_1 = (K - JS) c_2$$
$$\Rightarrow \begin{cases} c_1 : c_1 = -c_2 \\ c_2 : c_1 = c_2 \end{cases} \qquad (1.54b)$$

from where the two total forms of the wave-function can be constructed as a symmetric and an anti-symmetric ones, respectively:

$$\begin{cases} \Psi^{(a)}(A,B) = c_1^{(a)}[\varphi_1 - \varphi_2] \\ E_{AB}^{(a)} = 2E_0 + \dfrac{J-K}{1-S} \end{cases}$$

(1.55a)

and

$$\begin{cases} \Psi^{(s)}(A,B) = c_1^{(s)}[\varphi_1 + \varphi_2] \\ E_{AB}^{(s)} = 2E_0 + \dfrac{J+K}{1+S} \end{cases}$$

(1.55b)

The coefficients $c_1^{(a)}$ and $c_1^{(s)}$ results from implementation of normalization condition of the total wave-functions in Eq. (1.55a,b) as following:

$$1 = \int \left(\Psi_{(A,B)}^{(a/s)}\right)^2 dV_1 dV_2 = c_1^{(a/s)2}\int\left[\varphi_1^2 \mp 2\varphi_1\varphi_2 + \varphi_2^2\right]dV_1 dV_2$$
$$= c_1^{(a/s)2}\left[1 \mp 2S + 1\right] = 2(1 \mp S)c_1^{(a/s)2}$$

(1.56a)

$$\Rightarrow \begin{cases} C_1^{(a)} = \dfrac{1}{\sqrt{2(1-S)}} \\ C_1^{(s)} = \dfrac{1}{\sqrt{2(1+S)}} \end{cases}$$

(1.56b)

wherefrom, the two possible states (mesomers) for the electron-nuclei system in the hydrogen molecule finally result as

$$\begin{cases} \Psi_{H_2}^{(a)} = \dfrac{1}{\sqrt{2(1-S)}}[\varphi_1 - \varphi_2] \\ u^{(a)}(R) = \dfrac{e^2}{R} + 2E_0 + \dfrac{J-K}{1-S} \end{cases}$$

(1.57a)

and

$$\begin{cases} \Psi_{H_2}^{(s)} = \dfrac{1}{\sqrt{2(1+S)}}[\varphi_1 + \varphi_2] \\[2mm] u^{(s)}(R) = \dfrac{e^2}{R} + 2E_0 + \dfrac{J+K}{1+S} \end{cases} \qquad (1.57b)$$

with the phenomenological representation in the Figure 1.2.

But how the formed wave-functions are describing the chemical bond itself?

The binding character is determined by calculating the electron probability density $\rho = (\Psi)^2$ in the plane of the points which are equally distant from the two nuclei, i.e., for which the condition $\varphi_1 = \varphi_2$ is fulfilled. Accordingly, one can immediately distinguishing the two forms of chemical bonding character

$$\begin{cases} \rho_{g(erade)}\left(\Psi_{H_2}^{(s)}\right)^2 = \dfrac{4\varphi_1^2}{2(1+S)} \rightarrow BINDING\ character \\[3mm] \rho_{u(ngerade)}\left(\Psi_{H_2}^{(a)}\right)^2 = 0 \rightarrow ANTI-BINDING\ character \end{cases} \qquad (1.58)$$

along recognizing, from comparison of the results of (1.57a,b), that $u^{(s)}(R) < u^{(a)}(R)$, so consecrating the system's stability in the symmetric state ensured over the anti-symmetric one.

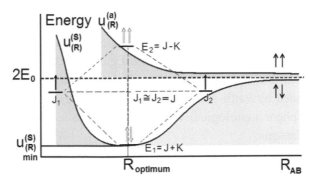

FIGURE 1.2   Typical potential dependence of chemical bonding with the bonding length for anti-bonding (upper curve) and for bonding (bottom curve) chemical realizations, respectively; here the values of J_1 and J_2 are not necessary in scale with respect with the separate atomic sum energies $2E_0$.

In addition, at the infinite distance, the atoms posses only the total energy

$$u^{(a)}(R) = u^{(s)}(R) = 2E_0$$

(1.59)

meaning the sum of their atomic energies in fundamental state along the required limits

$$\lim_{R_{AB} \to \infty} K = \lim_{R_{AB} \to \infty} J = \lim_{R_{AB} \to \infty} S = 0 = \lim_{R_{AB} \to \infty} \left(\frac{e^2}{R} \right)$$

(1.60)

In terms of spin behavior, through applying the Pauli principle, the total spin function for the two molecular mesomeric states is obtained in two basic combinations:

- The triplet state with total spin $S = 1$:

$$\Phi_I = \Psi^{(a)}_{H_2} \Psi^{(symmetry)}_{spin} = \Psi^{(a)}_{H_2} \cdot \begin{cases} \alpha(1)\alpha(2) \\ \beta(1)\beta(2) \\ \dfrac{1}{\sqrt{2}}\left[\alpha(1)\beta(2) + \beta(1)\alpha(2) \right] \end{cases}$$

(1.61a)

- The singlet state with total spin $S = 0$:

$$\Phi_{II} = \Psi^{(s)}_{H_2} \Psi^{(anti-symmetry)}_{spin} = \Psi^{(s)}_{H_2} \cdot \frac{1}{\sqrt{2}}\left[\alpha(1)\beta(2) - \beta(1)\alpha(2) \right]$$

(1.61b)

The case with anti-parallel spins, the singlet state, as being inferior in energy is more stable, so assuring the ligand function, which eventually characterizes the chemical bond on the molecular hydrogen; this result was extended by Pauli and Slater to all molecules; still, the VB method, although phenomenological acceptable is also criticizable because of practical reasons:

- The *limiting states* (also unhappily called as resonance) *are not real states of the molecule*, so the wave-functions φ_i cannot be associated to the concrete atomic orbitals (or concrete contribution of these) in order to causally explaining the chemical bond by atomic contributions;

- The number of involved functions φ_i immediately becomes very large and therefore has, also without involving the excited orbital (states), about 4 millions for the simple ion as SO_4^-, for example;
- Using truncating method (as in Heitler and London method – to be amended in the Section 1.4) and then applying the perturbation approximations (viz. adiabatic coupling), leads to uncertain procedures.

For all these reasons, soon after the mesomerism, Hund and Mulliken had been developed a more consistent theory with the physical significance of molecular orbitals, which confers the atoms as the rooting role of molecular orbitals' formation, from where the molecular orbitals (MO) method was as such nominated.

1.3   MOLECULAR ORBITAL THEORY OF BONDING

The basic idea is simple: next to each nucleus, the most representative terms are those who correspond to the smallest nucleus-electron distances, so the total Hamiltonian is a little different respecting that of the isolated atom. Therefore the wave-functions which describe the electrons around each atom of bonding, called *atomic orbitals*, represent approximate *local solutions* of the Schrödinger's equation in molecule; accordingly, one will search for the molecular function, called *molecular orbital*, as a linear combination of the atomic orbitals, or hybridized, of various atoms getting into the molecule (Hückel, 1934; Coulson, 1938, 1952; Hall, 1950; Griffith & Orgel, 1957; Jensen, 1999; Licker, 2004).

Resuming, the chemical bond which may be formed between two atoms A and B is considered as the resulted MO/the wave-function by "composition" of the two atomic orbitals φ_A and φ_B which by overlapping/superposition constitute the bond

$$\Psi_{AB} = c_1\varphi_A + c_2\varphi_B \qquad (1.62)$$

without other approximations. The effective MO which describe the chemical bond resulted from finding the c_1 and c_2 coefficients, by minimizing the total energy respecting these coefficients (in accordance with the basic postulate of variations in quantum systems – see the Volume I of the present five-volume book):

$$\frac{\partial E}{\partial c_1} = 0, \ \frac{\partial E}{\partial c_2} = 0 \tag{1.63}$$

The quantum energy of chemical bonding will be thus successively written

$$E = \frac{\int \Psi^* \widehat{H} \Psi d\tau}{\int \Psi^* \Psi d\tau}$$

$$= \frac{\int [c_1\varphi_A + c_2\varphi_B] \widehat{H} [c_1\varphi_A + c_2\varphi_B] d\tau}{\int [c_1\varphi_A + c_2\varphi_B][c_1\varphi_A + c_2\varphi_B] d\tau} = \frac{c_1^2 J_1 + 2c_1 c_2 K + c_2^2 J_2}{c_1^2 + c_2^2 + 2c_1 c_2 S} \tag{1.64}$$

given that at the atomic level the normalization of the wave-functions is implicit

$$\int \varphi_A^2 d\tau = \int \varphi_B^2 d\tau = 1 \tag{1.65a}$$

Also here one uses the consecrated notations

$$J_{1(2)} = \int \varphi_{A(B)} \ \widehat{H} \varphi_{B(A)} d\tau \tag{1.65b}$$

$$K = \int \varphi_A \widehat{H} \varphi_B d\tau = \int \varphi_B \widehat{H} \varphi_A d\tau \tag{1.65c}$$

$$S = \int \varphi_A \varphi_B d\tau \tag{1.65d}$$

representing, similarly to the VB method, the respective energetic terms contributing in chemical bonding:

- $J_{1(2)}$: the Coulomb integrals representing the characteristic energy in the atomic nucleus field of $A(B)$;
- K: the exchange integral, specific to the quantum theory approach/ contribution and being due to the *delocalization* properties which result from the quantum undulatory nature of electrons;

- S: The overlapping integral of the atomic orbitals φ_A and φ_B

With these one may unfold the variational principle for the chemical bonding energy respecting the atomic contributions, respectively as:

$$\frac{\partial E}{\partial c_1} = 0:$$

$$\frac{\left(2c_1J_1 + 2c_2K\right) + \left(c_1^2 + c_2^2 + 2c_1c_2S\right) - \left(c_1^2J_1 + 2c_1c_2K + c_2^2J_2\right)\left(2c_1 + 2c_2S\right)}{\left(c_1^2 + c_2^2 + 2c_1c_2S\right)^2} = 0$$

(1.66a)

$$\frac{\partial E}{\partial c_2} = 0$$

$$\Leftrightarrow \frac{\left\{\begin{array}{c}\left(2c_1K + 2c_2J\right) + \left(c_1^2 + c_2^2 + 2c_1c_2S\right)\\ -\left(c_1^2J_1 + 2c_1c_2K + c_2^2J_2\right)\left(2c_2 + 2c_1S\right)\end{array}\right\}}{\left(c_1^2 + c_2^2 + 2c_1c_2S\right)^2} = 0 \qquad (1.66b)$$

So the next system may be formed and successively transformed:

$$\Leftrightarrow \begin{cases} 2\left(c_1J_1 + c_2K\right) - E\left(2c_1 + 2c_2S\right) = 0 \\ 2\left(c_1K + c_2J_1\right) - E\left(2c_2 + 2c_1S\right) = 0 \end{cases} \qquad (1.67a)$$

$$\Leftrightarrow \begin{cases} \left(J_1 - E\right)c_1 + \left(K - SE\right)c_2 = 0 \\ \left(K - SE\right)c_1 + \left(J_2 - E\right)c_2 = 0 \end{cases}$$

$$\Leftrightarrow \begin{vmatrix} \left(J_1 - E\right) & K - SE \\ K - SE & J_2 - E \end{vmatrix} = 0$$

$$\Leftrightarrow \left(J_1 - E\right)\left(J_2 - E\right) - \left(K - SE\right)^2 = 0 \qquad (1.67b)$$

From where worth noting that for $J_1 = J_2 = J$ one regains the previous VB method's solutions, see Eqs. (1.53b,c), namely:

$$\begin{cases} E_1 = \dfrac{J-K}{1-S} \\[2mm] E_2 = \dfrac{J+K}{1+S} \end{cases} \qquad (1.68)$$

Otherwise the new equation is formed, with allied solutions:

$$E^2 - \frac{2KS + J_1 + J_2}{1-S^2} E + \left(J_1 J_2 - K^2 \right) = 0$$

$$\Rightarrow E_{1,2} = \frac{2KS + J_1 + J_2}{2\left(1-S^2\right)} \pm \sqrt{\left[\frac{2KS + J_1 + J_2}{2\left(1-S^2\right)} \right]^2 + K^2 - J_1 J_2} \quad (1.69a)$$

The solutions in Eq. (1.69a) have more approximation levels:

- If one neglects the terms containing S (being in general quite small, usually $S < 0.3$) there is obtained that

$$E_{1,2}(S \rightarrow 0) \cong \frac{J_1 + J_2}{2} \pm \sqrt{\left(\frac{J_1 - J_2}{2} \right)^2 + K^2} \qquad (1.69b)$$

- If $|J_1 - J_2| >> |\beta|$ one can use the approximation series expansion formula $\sqrt{a^2 + x^2} \cong a \left(1 + \dfrac{1}{2} \cdot \dfrac{x^2}{a^2} \right)$ from where the energetic solutions springs out as:

$$\begin{cases} E_1 \cong \dfrac{J_1 + J_2}{2} + \dfrac{J_1 - J_2}{2} \cdot \left[1 + \dfrac{2K^2}{\left(J_1 - J_2\right)^2} \right] = J_1 + \dfrac{K^2}{J_1 - J_2} \\[4mm] E_2 \cong \dfrac{J_1 + J_2}{2} - \dfrac{J_1 - J_2}{2} \cdot \left[1 + \dfrac{2K^2}{\left(J_1 - J_2\right)^2} \right] = J_2 - \dfrac{K^2}{J_1 - J_2} \end{cases} \qquad (1.70)$$

Under the assumption that $J_1 < J_2$ (both are negative values, as K does too) there is observed that the molecule energies' $|E_1 - J_1|$ and $|E_2 - J_2|$ are as higher as:

- the integral K (and implicitly the overlapping integral S) are higher;
- the energies of atomic orbital J_1 and J_2 are closer, being maximum when $J_1 = J_2$, a case opening the third level of approximation as follows.
- If we consider $J_1 = J_2 = J$ one remains with approximate solutions

$$\begin{cases} E_1 \cong J + K \\ E_2 \cong J - K \end{cases} \tag{1.71}$$

being these again in accordance with the previous VB method's findings for full and null overlapping ($S \to 0,1$) respectively.

Next, the c_1 and c_2 coefficients satisfy a relation from the "mother" system (1.67a), for example

$$\frac{c_1^k}{c_2^k} = -\frac{K - S \cdot E_k}{J_1 - E_k}, k = 1, 2 \tag{1.72}$$

which for $S \to 0$ and $J_1 = J$ becomes

$$\frac{c_1^k}{c_2^k} = -\frac{K}{J - E_k} \tag{1.73}$$

Now we can unfold the two cases fro wave-functions' solutions

$$\begin{cases} E_1 \cong J + K \Rightarrow c_1 = +c_2 \Rightarrow \Psi_{AB}^{(s)} = c_1^{(s)}(\varphi_A + \varphi_B) \\ E_2 \cong J - K \Rightarrow c_1 = -c_2 \Rightarrow \Psi_{AB}^{(a)} = c_1^{(a)}(\varphi_A - \varphi_B) \end{cases} \tag{1.74a}$$

while from the normalization condition at the molecular orbital level this time

$$\int \left| \Psi_{AB}^{s/a} \right|^2 d\tau = 1$$

one immediately found the forms

$$c_1^{s/a} = \frac{1}{\sqrt{2(1 \pm S)}} \tag{1.74b}$$

These results brings all phenomenological conclusions from the valence bond method, now having the φ_A and φ_B with the physical significances of the atomic orbitals, for the atoms A and B, respectively. This way, the bonds that a biatomic molecular system may establish are consequently realized through different MO listed as:

a. *The bonding MO* with the energy $E_1 \cong J + K$, characterizing approximately half of the possible molecular orbitals, with energy lower than the average energy of the (separate) atomic orbitals (see also Figure 1.2);

b. *Antibonding MO* with energy $E_2 \cong J - K$ higher than the average energy of the (separate) atomic orbitals (see also Figure 1.2);

c. *Nonbonding MO* for orbitals with the same energy $E \cong J$ as of those of the atomic orbitals wherefrom they appeared.

Considering the spin on MOs, for the bonding orbital $E_1 \rightarrow \Psi_{AB}^{(s)}$ one deals with the asymmetric spin function $S = 0 \left(\uparrow \downarrow \right)$ and the molecule energetic will be driven by the singlet state

$$(S = 0):\begin{cases} u^{S=0}(H_2) = 2(J + K) + \dfrac{e_0^2}{R_{H-H}} \\ E(H) = J \end{cases} \tag{1.75a}$$

so resulting that

$$D_{H_2}^{S=0} = 2E(H) - u^{S=0}(H_2) = -2K - \dfrac{e_0^2}{R_{H-H}} \tag{1.75b}$$

i.e. the dissociation energy can became positive through the existence of $-K$, meaning that $u(H_2)$ showcases a minimum corresponding to the binding MO.

Otherwise, the antibonding MO is activated once the second electron from the bonding MO is promoted so the spins become parallel (one on the bonding MO and the other on the antibonding MO) so the total triplet state is created, with the energetics

$$S = 1:\begin{cases} u^{S=1}(H_2) = (J + K) + (J - K) + \dfrac{e_0^2}{R_{H-H}} \\ E(H) = J \end{cases} \tag{1.76a}$$

so leaving with the dissociation energy as

$$D_{H_2}^{S=1} = 2E(H) - u^{S=1}(H_2) = -\frac{e_0^2}{R_{H-H}} \qquad (1.76b)$$

Accordingly, the molecular system could not exist stably in this state, i.e., since the dissociation energy remains unchanged the molecular system could not reach a minimum in this state so automatically dissociating itself.

In the same way, one can analyze the molecules He_2, He_2^+, HeH, HeH^+:

$$u(He_2) = 2(J+K) + 2(J-K) + \frac{4e_0^2}{R_{He-He}} = 4J + \frac{4e_0^2}{R_{He-He}} \qquad (1.77a)$$

$$E(He) = 2J \qquad (1.77b)$$

$$D_{He_2} = 2E(He) - u(He_2) = -\frac{4e_0^2}{R_{He-He}} < 0 \qquad (1.77c)$$

and independent of K there results that the molecule He_2 is not chemically possible;

$$u(He_2^+) = 2(J+K) + 2(J-K) + \frac{4e_0^2}{R_{He-He}} = 3J + K + \frac{4e_0^2}{R_{He-He}} \qquad (1.78a)$$

$$E(He) = 2J \qquad (1.78b)$$

$$D_{He_2^+} = 2E(He) - u(He_2^+) = -K - \frac{4e_0^2}{R_{He-He}} < 0 \qquad (1.78c)$$

wherefrom there exist the probability of having $D_{He_2^+} > 0$, i.e., $u(He_2^+)$ reaching a minimum, so the molecule He_2^+ should exist;

$$u(HeH) = 2\left[J_{He} + \frac{K^2}{J_{He} - J_H} \right] + \left[J_H - \frac{K^2}{J_{He} - J_H} \right] + \frac{2e_0^2}{R_{He-H}}$$

$$= 2J_{He} + J_H + \frac{K^2}{J_{He} - J_H} + \frac{2e_0^2}{R_{He-H}} \qquad (1.79a)$$

and with

$$E(H) = J_H; E(He) = 2J_{He} \tag{1.79b}$$

produces the result

$$D_{HeH} = E(H) + E(He) - u(HeH) = -\underbrace{\frac{K^2}{J_{He} - J_H}}_{>0} - \underbrace{\frac{2e_0^2}{R_{He-H}}}_{<0} \begin{cases} >0 \\ <0 \end{cases} \tag{1.79c}$$

Analogously for the He-H molecular ion one has

$$u(HeH^+) = 2\left[J_{He} + \frac{K^2}{J_{He} - J_H} \right] + \frac{2e_0^2}{R_{He-H}} \tag{1.80a}$$

and with

$$E(H) = J_H; \quad E(He) = 2J_{He} \tag{1.80b}$$

leaves with

$$D_{HeH^+} = E(H) + E(He) - u(HeH^+) = \underbrace{J_H}_{<0} - \underbrace{\frac{2K^2}{J_{He} - J_H}}_{>0} - \underbrace{\frac{2e_0^2}{R_{He-H}}}_{<0} \begin{cases} >0 \\ <0 \end{cases} \tag{1.80c}$$

Even if apparently both the molecules HeH, HeH^+ would be possible, in the first case we apply to the potential the approximation

$$|J_{He} - J_H| >> K^2 \tag{1.81}$$

wherefrom there result

$$D_{HeH^+} \approx -\frac{2e_0^2}{R_{He-H}} < 0 \tag{1.82}$$

which actually impedes the molecule formation, while for the second case, for the HeH^+ molecule, because of the factor "2" at K^2 in Eq. (1.80c), i.e., $|J_{He} - J_H| \rangle\rangle 2K^2$, the term D_{HeH^+} can assume, for a given R_{He-H}, also positive value, and therefore $u(HeH^+)$ would correspond to a minimum potential with the possibility to sustain the molecular existence.

When the atomic orbitals are involved generating molecules with more than four electrons, an extended molecular diagram is helpful, whose construction is based on two concepts:

1. Orbitals hybridization;
2. Hund's rule of maximum multiplicity of the spin for shielded electrons.

Regarding the orbitals hybridization, one operates with $(\pm)$ on involved atomic orbitals and the symmetry of the MOs is accordingly evaluated (see Figure 1.3).

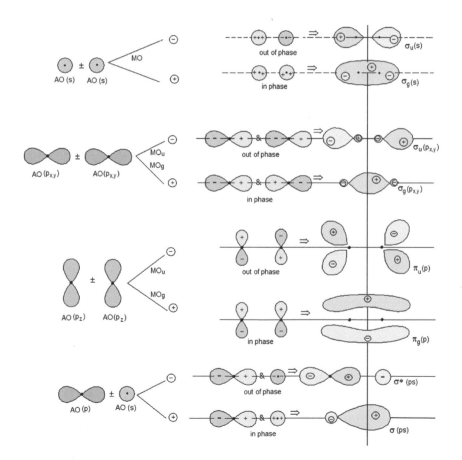

FIGURE 1.3   Molecular orbital formations from atomic orbitals in various s-s, p-p and s-p combinations with distinction between obtained bonding and antibondings (upper) molecular orbitals.

Regarding the Hund's rule, one should taken into consideration that, while for the atoms (with a single nucleus) the *triplet* state is favored, for the diatomic molecules (with two nuclei), the *singlet* state is the most favored to the bond, as shown in the Figure 1.4.

However, for the cases when are more than four electrons involved in a molecule, the first four electrons shield (two from each nucleus) the superior electrons, which actually "see" only a molecular core, behaving therefrom as for having an "atomic core"; therefore, after occupying the first four "binding positions", the following will made up according to the Hund's rule, in order to ensure the maximum multiplicity, from where the OM electronic configuration can be settle as in next exemplified:

- $B_2\left(10e^-\right)$ has the molecular orbital configuration:

$$\sigma_g\left(1s\right)^2 \sigma_u^*\left(1s\right)^2 \sigma_g\left(2s\right)^2 \sigma_u^*\left(2s\right)^2 \pi_u\left(2p\right)^2 \qquad (1.83)$$

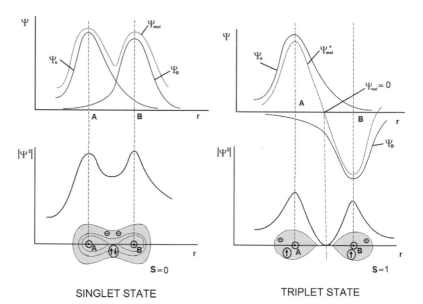

FIGURE 1.4 Quantum mechanical explanation of the bonding and anti-bonding spin behavior as based on the overlapping of atomic wave-functions in bonding (upper representations) following by their probability density realizations towards spectral observation (bottom representations).

With the orbital $\pi_u \left(2p\right)^2$ being occupied parallel spins there is justified the paramagnetic character (permanent magnetic moment), as experimentally observed;

- $Li_2 \left(6e^-\right)$ has the molecular orbital configuration:

$$\sigma_g \left(1s\right)^2 \sigma_u^* \left(1s\right)^2 \sigma_g \left(2s\right)^2 \tag{1.84}$$

having the last two electrons with two spins ($\uparrow\downarrow$) in bonding MO;

- $N_2 \left(14e^-\right)$ has the molecular orbital configuration:

$$\sigma_g \left(1s\right)^2 \sigma_u^* \left(1s\right)^2 \sigma_g \left(2s\right)^2 \sigma_u^* \left(2s\right)^2 \pi_u \left(2p\right)^4 \sigma_g \left(2p\right)^2 \tag{1.85}$$

- $N_2^+ \left(13e^-\right)$ has the molecular orbital configuration:

$$\sigma_g \left(1s\right)^2 \sigma_u^* \left(1s\right)^2 \sigma_g \left(2s\right)^2 \sigma_u^* \left(2s\right)^2 \pi_u \left(2p\right)^4 \sigma_g \left(2p\right)^1 \tag{1.86}$$

- $N_2^- \left(15e^-\right)$ has the molecular orbital configuration

$$\sigma_g \left(1s\right)^2 \sigma_u^* \left(1s\right)^2 \sigma_g \left(2s\right)^2 \sigma_u^* \left(2s\right)^2 \pi_u \left(2p\right)^4 \sigma_g \left(2p\right)^2 \pi_g^* \left(2p\right)^1 \tag{1.87}$$

The N_2 and its ionic cases allows phenomenological operating the *bond order*

$$\vartheta_B = \frac{no.\ bonding\ e^- - no.antibonding\ e^-}{2} \tag{1.88a}$$

so having the actual specializations' hierarchy

$$\vartheta_B \left(N_2\right) = \left(1\sigma, 2\pi\right) > \vartheta_B \left(N_2^+\right) = \left(\frac{1}{2}\sigma, 2\pi\right) > \vartheta_B \left(N_2^-\right) = \left(1\sigma, \frac{3}{2}\pi\right) \tag{1.88b}$$

with the inverse orderings of observed bonding lengths

$$R_{N-N}\left(N_2\right) = 0.10976[nm] < R_{N-N}\left(N_2^+\right)$$

$$= 0.1116384[nm] < R_{N-N}\left(N_2^-\right) = 0.11934[nm] \quad (1.89)$$

Thus consecrating the fact that as the bond is stronger, with ϑ_B higher, as it corresponds to "narrower" bonding space, i.e., with smaller bonding length R_{AB}.

On the other hand, while writing the NO molecule and ionic molecular system electronic configurations

$$\begin{cases} NO\left(13e^-\right): \sigma_g\left(1s\right)^2 \sigma_u^*\left(1s\right)^2 \sigma_g\left(2s\right)^2 \sigma_u^*\left(2s\right)^2 \pi_u\left(2p\right)^4 \sigma_g\left(2p\right)^2 \pi_g^*\left(2p\right)^1 \\ NO^+\left(14e^-\right): \sigma_g\left(1s\right)^2 \sigma_u^*\left(1s\right)^2 \sigma_g\left(2s\right)^2 \sigma_u^*\left(2s\right)^2 \pi_n\left(2p\right)^4 \sigma_g\left(2p\right)^2 \end{cases}$$

$$(1.90)$$

we can infer that the *bonding energy* is directly proportional with the bond order

$$\vartheta_B\left(NO\right) = \left(1\sigma, \frac{3}{2}\pi\right) < \vartheta_B\left(NO^+\right) = \left(1\sigma, 2\pi\right)$$

$$= \left(1\sigma, 2\pi\right) \Rightarrow E_{bond:NO} < E_{bond:NO^+} \quad (1.91)$$

so concluding that as the bond is stronger, i.e., with higher ϑ_B, as it corresponds to a higher bonding energy.

The final present application regards the OH system: by assuming that the energy of the $1s$ orbital from H-atom has the same energy with $2p$ orbital from O-atom, one can represent the diagram of Figure 1.5 for the OH molecule allowing the neutral and ionic electronic configurations OH^+ and OH^- to be worked on it, respectively as:

$$OH : \left(1s\right)^2 \left(2s\right)^2 \sigma\left(sp\right)^2 \left(2p\right)^{3[\uparrow\downarrow...\uparrow]} : 1e^- \left(2p\right) un-paired \quad (1.92a)$$

$$OH^+ : \left(1s\right)^2 \left(2s\right)^2 \sigma\left(sp\right)^2 \left(2p\right)^{2[\uparrow...\uparrow]} : 2e^- \left(2p\right) un-paired \quad (1.92b)$$

$$OH^- : \left(1s\right)^2 \left(2s\right)^2 \sigma\left(sp\right)^2 \left(2p\right)^{4[\uparrow\downarrow...\uparrow\downarrow]} all\ paired \quad (1.92c)$$

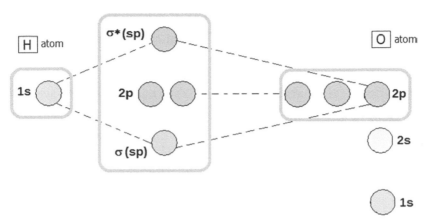

FIGURE 1.5 Illustration of hetero-atomic combination in chemical bonding (H-O bonding here) at the level of s-s- and s-p atomic orbitals' overlapping in resulted bonding and anti-bonding molecular orbitals in bottom and upper parts of the central horizontal axis, respectively.

Note that all species are connected by a σ-bond, wherefrom the bond strength is rendered by the presence (or the absence) of the unpaired e^- towards indicating a slightly rising in the bonding length:

$$R_{O-H}\left(OH^-\right) = 0.09628\,[nm] < R_{O-H}\left(OH\right)$$
$$= 0.09706\,[nm] < R_{O-H}\left(OH^+\right) = 0.1029\,[nm] \quad (1.93)$$

Overall, we may safely conclude so far that the chemical bond is a quantum effect of electronic exchange (see the K-term presence), at its turn raised due to the undulatory character of the electrons and the associated spin functions (in agreement with Pauli Principle).

Note that the bonding length and bonding energy will be also in the central place in the next section presenting the ultimate theory of chemical bonding by the aid of chemical field and the associated bondonic quantum particles.

1.4   SUB-QUANTUM CHEMISTRY: CHEMICAL BONDING FIELD AND BONDONS

Now lets take the paradoxical point: the dispersion of the wave function is historically replaced by the delocalization of density and the chemical bonding information is still beyond a decisive quantum clarification. Moreover, the quantum theory itself is challenged as to its reliability by the Einstein-Podolski-Rosen(-Bohr) entanglement formulation of quantum phenomena (Einstein et al., 1935; Bohr, 1935), qualitatively explained by the Bohm reformulation (Bohm, 1952) of the de Broglie wave packet through the combined de Broglie-Bohm wave-function (de Broglie & Vigier, 1953; Bohm & Vigier, 1954)

$$\Psi_0(t,x) = R(t,x)\exp\left(i\frac{S(t,x)}{\hbar}\right) \tag{1.94}$$

with the R-amplitude and S-phase action factors given, respectively, as

$$R(t,x) = \sqrt{\Psi_0(t,x)^2} = \rho^{1/2}(x) \tag{1.95}$$

$$S(t,x) = px - Et \tag{1.96}$$

in terms of electronic density ρ, momentum p, total energy E, and time-space (t,x) coordinates, without spin, see Volume II/Section 4.1.3 of the present five-volume work.

In this respect, the more deep approaches of quantum description of the chemical bonding advocates in making the required steps toward assessing the quantum particle of the chemical bond as based on the derived chemical field released at its turn by the fundamental electronic equations of motion either within Bohmian non-relativistic (Schrödinger) and to explore the first consequences. If successful, the present endeavor will contribute to celebrate the dream in unveiling the true particle-wave nature of the chemical bond.

1.4.1 INTRODUCING CHEMICAL FIELD

The search for the bondons follows the algorithm (Putz, 2010a-b, 2012a-b):

  i. Considering the de Broglie-Bohm electronic wave-function/ spinor Ψ_0 formulation of the associated quantum Schrödinger/ Dirac equation of motion.
  ii. Checking for recovering the charge current conservation law

$$\frac{\partial \rho}{\partial t} + \nabla \vec{j} = 0 \tag{1.97}$$

that assures for the circulation nature of the electronic fields under study.

iii. Recognizing the quantum potential V_{qua} and its equation, if it eventually appears.

iv. Reloading the electronic wave-function/spinor under the augmented U(1) or SU(2) group form

$$\Psi_G(t,x) = \Psi_0(t,x) \exp\left(\frac{i}{\hbar} \frac{e}{c} \aleph(t,x) \right) \tag{1.98}$$

with the standard abbreviation $e = e_0^2 / 4\pi\varepsilon_0$ in terms of the chemical field $\aleph$ considered as the inverse of the fine-structure order:

$$\aleph_0 = \frac{\hbar c}{e} \sim 137.03599976 \left[\frac{Joule \times meter}{Coulomb} \right] \tag{1.99}$$

since upper bounded, in principle, by the atomic number of the ultimate chemical stable element (Z = 137). Although apparently small enough to be neglected in the quantum range, the quantity (1.99) plays a crucial role for chemical bonding where the energies involved are around the order of 10^{-19} Joules (electron-volts)! Nevertheless, for establishing the physical significance of such chemical bonding quanta, one can proceed with the chain equivalences

$$\aleph_\# \sim \frac{energy \times distance}{charge} \sim \frac{\left(charge \times \dfrac{potential}{difference} \right) \times distance}{charge}$$

$$\sim \left(\frac{potential}{difference} \right) \times distance \tag{1.100}$$

revealing that the chemical bonding field caries *bondons* with unit quanta $\hbar c / e$ along the distance of bonding within the potential gap of stability or by tunneling the potential barrier of encountered bonding attractors.

v. Rewriting the quantum wave-function/spinor equation with the group object Ψ_G, while separating the terms containing the real and imaginary $\aleph$ chemical field contributions.

vi. Identifying the chemical field charge current and term within the actual group transformation context.

vii. Establishing the global/local gauge transformations that resemble the de Broglie-Bohm wave-function/spinor ansatz Ψ_0 of steps (i)–(iii).

viii. Imposing invariant conditions for Ψ_G wave function on pattern quantum equation respecting the Ψ_0 wave-function/spinor action of steps (i)–(iii).

ix. Establishing the chemical field $\aleph$ specific equations.

x. Solving the system of chemical field $\aleph$ equations.

xi. Assessing the stationary chemical field

$$\frac{\partial \aleph}{\partial t} \equiv \partial_t \aleph = 0 \qquad (1.101)$$

that is the case in chemical bonds at equilibrium (ground state condition) to simplify the quest for the solution of chemical field $\aleph$.

xii. The manifested bondonic chemical field $\aleph_{bondon}$ is eventually identified along the bonding distance (or space).

xiii. Checking the eventual charge flux condition of Bader within the vanishing chemical bonding field (Bader, 1990)

$$\aleph_B = 0 \Leftrightarrow \nabla\rho = 0 \qquad (1.102)$$

xiv. Employing the Heisenberg time-energy relaxation-saturation relationship through the kinetic energy of electrons in bonding

$$v = \sqrt{\frac{2T}{m}} \sim \sqrt{\frac{2}{m}\frac{\hbar}{t}} \qquad (1.103)$$

xv. Equate the bondonic chemical bond field with the chemical field quanta (1.99) to get the bondons' mass

$$\aleph_B\left(m_B\right) = \aleph_0 \qquad (1.104)$$

This algorithm will be next unfolded both for non-relativistic as well as for relativistic electronic motion to quest upon the bondonic existence, eventually emphasizing their difference in bondons' manifestations.

1.4.2   PREDICTING THE BONDONIC'S EXISTENCE

For the non-relativistic quantum motion, we will treat the above steps (i)–(iii) at once. As such, when considering the de Broglie-Bohm electronic wave function into the Schrödinger Equation

$$i\hbar\partial_t\Psi_0 = -\frac{\hbar^2}{2m}\nabla^2\Psi_0 + V\Psi_0 \tag{1.105}$$

it separates into the real and imaginary components as (Bohm, 1952; Boeyens, 2005, 2010; Putz, 2010a)

$$\partial_t R^2 + \nabla\left(\frac{R^2}{m}\nabla S\right) = 0 \tag{1.106a}$$

$$\partial_t S - \frac{\hbar^2}{2m}\frac{1}{R}\nabla^2 R + \frac{1}{2m}(\nabla S)^2 + V = 0 \tag{1.106b}$$

While recognizing into the first equation (1.106a), the charge current conservation law with Eq. (1.95) along the identification

$$\vec{j}_S = \frac{R^2}{m}\nabla S \tag{1.107}$$

the second equation helps in detecting the quantum (or Bohm) potential

$$V_{qua} = -\frac{\hbar^2}{2m}\frac{\nabla^2 R}{R} \tag{1.108}$$

contributing to the total energy

$$E = T + V + V_{qua} \tag{1.109}$$

once the momentum-energy correspondences

$$\frac{1}{2m}(\nabla S)^2 = \frac{p^2}{2m} = T \tag{1.110a}$$

$$\partial_t S = -E \tag{1.110b}$$

are engaged.

Next, when employing the associate U(1) gauge wave function of Eq. (1.98) type, its partial derivative terms look like

$$\nabla \Psi_G = \left[\nabla R + \frac{i}{\hbar} R \left(\nabla S + \frac{e}{c} \nabla \aleph \right) \right] \exp \left[\frac{i}{\hbar} \left(S + \frac{e}{c} \aleph \right) \right] \tag{1.111a}$$

$$\nabla^2 \Psi_G = \left\{ \begin{array}{l} \nabla^2 R + 2\dfrac{i}{\hbar} \nabla R \left(\nabla S + \dfrac{e}{c} \nabla \aleph \right) \\[2mm] + \dfrac{i}{\hbar} R \left(\nabla^2 S + \dfrac{e}{c} \nabla^2 \aleph \right) \\[2mm] - \dfrac{R}{\hbar^2} \left[(\nabla S)^2 + \left(\dfrac{e}{c} \nabla \aleph \right)^2 \right] \\[2mm] -2 \dfrac{e}{\hbar^2 c} R \nabla S \nabla \aleph \end{array} \right\} \exp \left[\frac{i}{\hbar} \left(S + \frac{e}{c} \aleph \right) \right] \tag{1.111b}$$

$$\partial_t \Psi_G = \left[\partial_t R + \frac{i}{\hbar} R \left(\partial_t S + \frac{e}{c} \partial_t \aleph \right) \right] \exp \left[\frac{i}{\hbar} \left(S + \frac{e}{c} \aleph \right) \right] \tag{1.111c}$$

Now the Schrödinger Eq. (1.105) for Ψ_G in the form of Eq. (1.98) is decomposed into imaginary and real parts

$$-\partial_t R = \frac{1}{m} \left(\nabla R \cdot \nabla S + \frac{R}{2} \nabla^2 S \right) + \frac{e}{mc} \left(\nabla R \cdot \nabla \aleph + \frac{R}{2} \nabla^2 \aleph \right) \tag{1.112a}$$

$$-R\partial_t S - R \frac{e}{c} \partial_t \aleph = -\frac{\hbar^2}{2m} \nabla^2 R + \frac{R}{2m} \left[(\nabla S)^2 + \left(\frac{e}{c} \nabla \aleph \right)^2 \right]$$

$$+ \frac{e}{mc} R \nabla S \cdot \nabla \aleph + VR \tag{1.112b}$$

that can be rearranged

$$-\partial_t R^2 = \frac{1}{m}\nabla\left(R^2\nabla S\right) + \frac{e}{mc}\nabla\left(R^2\nabla\aleph\right) \qquad (1.113a)$$

$$-\left(\partial_t S + \frac{e}{c}\partial_t\aleph\right) = -\frac{\hbar^2}{2m}\frac{1}{R}\nabla^2 R + \frac{1}{2m}\left[\left(\nabla S\right)^2 + \left(\frac{e}{c}\nabla\aleph\right)^2\right]$$

$$+\frac{e}{mc}\nabla S\cdot\nabla\aleph + V \qquad (1.113b)$$

to reveal some interesting features of chemical bonding.

Firstly, through comparing the Eq. (1.113a) with the charge conserved current equation form (1.97) from the general chemical field algorithm–the step (ii), the conserving charge current takes now the expanded expression:

$$\vec{j}_{U(1)} = \frac{R^2}{m}\left(\nabla S + \frac{e}{c}\nabla\aleph\right) = \vec{j}_S + \vec{j}_\aleph \qquad (1.114)$$

suggesting that the additional current is responsible for the chemical field to be activated, namely

$$\vec{j}_\aleph = \frac{e}{mc}R^2\nabla\aleph \qquad (1.115)$$

which vanishes when the *global gauge* condition is considered

$$\nabla\aleph = 0 \qquad (1.116)$$

Therefore, in order that the chemical bonding is created, the *local gauge* transformation should be used that exists under the condition

$$\nabla\aleph \neq 0 \qquad (1.117)$$

In this framework, the chemical field current $\vec{j}_\aleph$ carries specific bonding particles that can be appropriately called *bondons*, closely related with electrons, in fact with those electrons involved in bonding, either as single, lone pair or delocalized, and having an oriented direction of movement, with an action depending on the chemical field itself $\aleph$.

Nevertheless, another important idea abstracted from the above results is that in the search for the chemical field $\aleph$ no global gauge condition is required. It is also worth noting that the presence of the chemical field does not change the Bohm quantum potential that is recovered untouched in Eq. (1.113b), thus preserving the entanglement character of interaction.

With these observations, it follows that in order for the de Broglie-Bohm-Schrödinger formalism to be invariant under the U(1) transformation (1.98), a couple of gauge conditions have to be fulfilled by the chemical field in Eqs. (1.113a) and (1.113b), namely

$$\frac{e}{mc}\frac{\partial}{\partial x}\left(R^2\nabla\aleph\right)=0 \tag{1.118a}$$

$$\frac{e}{c}\partial_t\aleph+\frac{1}{2m}\left(\frac{e}{c}\nabla\aleph\right)^2+\frac{e}{mc}\nabla S\cdot\nabla\aleph=0 \tag{1.118b}$$

Next, the chemical field $\aleph$ is to be expressed through combining its spatial-temporal information contained in Eq. (1.118). From the first condition (1.118a) one finds that

$$\nabla\aleph=-\frac{R}{2}\frac{\nabla^2\aleph}{\nabla R\cdot\vec{i}}\vec{i} \tag{1.119}$$

where the vectorial feature of the chemical field gradient was emphasized on the direction of its associated charge current fixed by the versor $\vec{i}$ (*i.e.*, by the unitary vector associate with the propagation direction, $\vec{i}^2=1$). We will apply such writing whenever necessary for avoiding scalar to vector ratios and preserving the physical sense of the whole construction as well. Replacing the gradient of the chemical field (1.119) into its temporal equation (1.118b) one gets the unified chemical field motion description

$$\frac{e}{8mc}\frac{R^2}{(\nabla R)^2}\left(\nabla^2\aleph\right)^2-\frac{R}{2m}\frac{\nabla S\cdot\nabla S}{\nabla R\cdot\nabla S}\left(\nabla^2\aleph\right)+\partial_t\aleph=0 \tag{1.120}$$

that can be further rewritten as

$$\frac{e}{2mc}\frac{\rho^2}{(\nabla\rho)^2}\left(\nabla^2\aleph\right)^2 - \frac{\rho\vec{v}\cdot\vec{\imath}}{\nabla\rho\cdot\vec{\imath}}\left(\nabla^2\aleph\right) + \partial_t\aleph = 0 \qquad (1.121)$$

since calling the relations abstracted from Eqs. (1.95) and (1.96)

$$R = \rho^{1/2}; \nabla S = \vec{p} \Rightarrow \begin{cases} \nabla R = \frac{1}{2}\frac{\nabla\rho}{\rho^{1/2}};\ \left(\nabla R\right)^2 = \frac{1}{4}\frac{\left(\nabla\rho\right)^2}{\rho} \\ \frac{\nabla S\cdot\nabla S}{\nabla R\cdot\nabla S} = \frac{2\rho^{1/2}\vec{p}\cdot\vec{\imath}}{\nabla\rho\cdot\vec{\imath}} \end{cases} \qquad (1.122)$$

The (quadratic undulatory) chemical field equation (1.121) can be firstly solved for the Laplacian general solutions

$$\left(\nabla^2\aleph\right)_{1,2} = \frac{\dfrac{\rho\vec{v}\cdot\vec{\imath}}{\nabla\rho\cdot\vec{\imath}} \pm \sqrt{\dfrac{\rho^2 v^2}{\left(\nabla\rho\right)^2} - \dfrac{2e}{mc}\dfrac{\rho^2}{\left(\nabla\rho\right)^2}\partial_t\aleph}}{\dfrac{e}{mc}\dfrac{\rho^2}{\left(\nabla\rho\right)^2}} \qquad (1.123)$$

that give special propagation equations for the chemical field since linking the spatial Laplacian with temporal evolution of the chemical field $\left(\partial_t\aleph\right)^{1/2}$; however, they may be considerably simplified when assuming the stationary chemical field condition (1.101), the step (xi) in the bondons' algorithm, providing the working equation for the stationary bondonic field

$$\nabla^2\aleph = 2\frac{mc}{e}\frac{\vec{v}\cdot\nabla\rho}{\rho} \qquad (1.124)$$

Equation (1.124) may be further integrated between two bonding attractors, say X_A, X_B, toprimarily give

$$\nabla\aleph = 2\frac{mc}{e}v\int_{X_A}^{X_B}\frac{\nabla\rho\cdot\vec{\imath}}{\rho}dx = \frac{mc}{e}v\left[\int_{X_A}^{X_B}\frac{\nabla\rho\cdot\vec{\imath}}{\rho}dx - \int_{X_B}^{X_A}\frac{\nabla\rho\cdot\vec{\imath}}{\rho}dx\right] \qquad (1.125)$$

from where the generic bondonic chemical field is manifested with the form

$$\aleph_{\mathcal{B}} = \frac{mc}{e} v X_{bond} \left(\int_{X_A}^{X_B} \frac{\nabla \rho \cdot \vec{\imath}}{\rho} dx \right) \qquad (1.126)$$

The expression (1.126) has two important consequences. Firstly, it recovers the Bader zero flux condition for defining the *basins* of bonding (Bader, 1990) that in the present case is represented by the zero chemical boning fields, namely

$$\aleph_{\mathcal{B}} = 0 \Leftrightarrow \nabla \rho \cdot \vec{\imath} = 0 \qquad (1.127)$$

Secondly, it furnishes the bondonic (chemical field) analytical expression

$$\aleph_{\mathcal{B}} = \frac{mc}{e} v X_{bond} \qquad (1.128)$$

within the natural framework in which

$$X_B - X_A = X_{bond} \qquad (1.129a)$$

$$\nabla \rho \cdot \vec{\imath} \to \frac{\rho}{X_{bond}} \qquad (1.129b)$$

i.e., when one has

$$\int_{X_A}^{X_B} \frac{\nabla \rho \cdot \vec{\imath}}{\rho} dx = 1 \qquad (1.129c)$$

The step (xiv) of the bondonic algorithm may be now immediately implemented through inserting the Eq. (1.103) into Eq. (1.128) yielding the simple chemical field form

$$\aleph_{\mathcal{B}} = \frac{c\hbar}{e} \sqrt{\frac{2m}{ht}} X_{bond} \qquad (1.130)$$

Finally, through applying the expression (1.104) of the bondonic algorithm–the step (xv) upon the result (1.130) with quanta (1.99) the *mass of bondons* carried by the chemical field on a given distance is obtained (Putz, 2010a, 2012a-b)

$$m_{_\mathcal{B}} = \frac{\hbar t}{2} \frac{1}{X_{bond}^2}$$

(1.131)

Note that the bondons' mass (1.131) directly depends on the time the chemical information "travels" from one bonding attractor to the other involved in bonding, while fast decreasing as the bonding distance increases. This phenomenological behavior has to be in the sequel cross-checked by considering the generalized relativistic version of electronic motion by means of the Dirac equation, Further quantitative consideration will be discussed afterwards.

Further on, by quantum relativistic grounds, when rewriting the chemical field of bonding (1.130) within the de Broglie and Planck consecrated corpuscular-undulatory quantifications

$$\aleph_{_\mathcal{B}}(t, X_{bond}) = \frac{\hbar c}{e} \exp\left[\frac{i}{\hbar}(pX_{bond} - Et)\right]$$

(1.132)

it may be further combined with the unitary quanta form (1.99) in the Eq. (1.104) of the step (xv) in the bondonic algorithm to produce the phase condition

$$1 = \exp\left[\frac{i}{\hbar}(pX_{bond} - Et)\right]$$

(1.133)

that implies the quantification

$$pX_{bond} - Et = 2\pi n\hbar, \ n \in \mathbf{N}$$

(1.134)

By the subsequent employment of the Heisenberg time-energy saturated indeterminacy at the level of kinetic energy abstracted from the total energy (to focus on the motion of the bondonic plane waves)

$$E = \frac{\hbar}{t}$$

(1.135a)

$$p = mv = \sqrt{2mT} \rightarrow \sqrt{\frac{2m\hbar}{t}}$$

(1.135b)

the bondon equation (1.134) becomes

$$X_{bond}\sqrt{\frac{2m\hbar}{t}}=(2\pi n+1)\hbar \tag{1.136}$$

that when solved for the bondonic mass yields the expression (Putz, 2010a)

$$m_{\mathcal{B}}=\frac{\hbar t}{2}\frac{1}{X_{bond}^{2}}(2\pi n+1)^{2}, n=0,1,2... \tag{1.137}$$

which appears to correct the previous non-relativistic expression (1.131) with the full quantification.

1.4.3 CHEMICAL BONDING BY BONDONS

Let us analyze the consequences of the bondon's existence, starting from its mass (1.131) formulation on the ground state of the chemical bond.

At one extreme (Putz, 2010a-b), when considering *atomic* parameters in bonding, *i.e.*, when assuming the bonding distance of the Bohr radius size $a_0 = 0.52917 \cdot 10^{-10}[m]_{SI}$ the corresponding binding time would be given as $t \to t_0 = a_0 / v_0 = 2.41889 \cdot 10^{-17}[s]_{SI}$ while the involved bondonic mass will be half of the electronic one $m_0 / 2$, to assure fast bonding information. Of course, this is not a realistic binding situation; for that, let us check the hypothetical case in which the electronic m_0 mass is combined, within the bondonic formulation (1.131), into the bond distance $X_{bond} = \sqrt{\hbar t / 2m_0}$ resulting in it completing the binding phenomenon in the femtosecond time $t_{bonding} \sim 10^{-12}[s]_{SI}$ for the custom nanometric distance of bonding $X_{bond} \sim 10^{-9}[m]_{SI}$. Still, when both the femtosecond and nanometer time-space scale of bonding is assumed in Eq. (1.131), the bondonic mass is provided in the range of electronic mass $m_{\mathcal{B}} \sim 10^{-31}[kg]_{SI}$ although not necessarily with the exact value for electron mass nor having the same value for each bonding case considered. Further insight into the time existence of the bondons will be reloaded for molecular systems below after discussing related specific properties as the bondonic velocity and charge.

For enlightenment on the last perspective, let us rewrite the bondonic mass (1.137) within the spatial-energetic frame of bonding, *i.e.,* through replacing the time with the associated Heisenberg energy, $t_{bonding} \to \hbar / E_{bond}$, thus delivering another working expression for the bondonic mass

$$m_{\mathcal{B}} = \frac{\hbar^2 (2\pi n + 1)^2}{2 E_{bond} X_{bond}^2}, \quad n = 0, 1, 2\dots \tag{1.138}$$

that is more practical than the traditional characterization of bonding types in terms of length and energy of bonding; it may further assume the numerical ground state ratio form

$$\varsigma_m = \frac{m_{\mathcal{B}}}{m_0} = \frac{87.8603}{\left(E_{bond}[kcal / mol]\right)\left(X_{bond}[\overset{0}{A}]\right)^2} \tag{1.139}$$

when the available bonding energy and length are considered (as is the custom for chemical information) in kcal/mol and Angstrom, respectively. Note that having the bondon's mass in terms of bond energy implies the inclusion of the electronic pairing effect in the bondonic existence, without the constraint that the bonding pair may accumulate in the internuclear region (Berlin, 1951).

Moreover, since the bondonic mass general formulation (1.137) resulted within the relativistic treatment of electron, it is considering also the companion velocity of the bondonic mass that is reached in propagating the bonding information between the bonding attractors. As such, when the Einstein type relationship (Einstein, 1905a)

$$\frac{mv^2}{2} = h\upsilon \tag{1.140}$$

is employed for the relativistic bondonic velocity-mass relationship (Einstein, 1905a,b)

$$m = \frac{m_{\mathcal{B}}}{\sqrt{1 - \frac{v_{\mathcal{B}}^2}{c^2}}} \tag{1.141}$$

and for the frequency of the associate bond wave

$$\upsilon = \frac{v_{\mathcal{B}}}{X_{bond}} \tag{1.142}$$

it provides the quantified searched bondon to light velocity ratio

$$\frac{v_{\cancel{B}}}{c} = \frac{1}{\sqrt{1 + \frac{1}{64\pi^2}\frac{\hbar^2 c^2 (2\pi n + 1)^4}{E_{bond}^2 X_{bond}^2}}}, \quad n = 0,1,2... \quad (1.143)$$

or numerically in the bonding ground state as

$$\varsigma_v = \frac{v_{\cancel{B}}}{c} = \frac{100}{\sqrt{1 + \frac{3.27817 \times 10^6}{(E_{bond}[kcal/mol])^2 (X_{bond}[\overset{0}{A}])^2}}}[\%] \quad (1.144)$$

Next, dealing with a new matter particle, one will be interested also on its charge, respecting the benchmarking charge of an electron. To this end, one re-employs the step (xv) of bondonic algorithm, Eq. (1.104), in the form emphasizing the bondonic charge appearance, namely

$$\aleph_{\cancel{B}}(e_{\cancel{B}}) = \aleph_0 \quad (1.145)$$

Next, when considering for the left-hand side of (1.145), the form provided by Eq. (1.128), and for the right-hand side of (1.145), the fundamental hyperfine value of Eq. (1.99), one gets the working Equation

$$c\frac{m_{\cancel{B}} v_{\cancel{B}}}{e_{\cancel{B}}} X_{bond} = 137.036 \left[\frac{Joule \times meter}{Coulomb}\right] \quad (1.146)$$

from where the bondonic charge appears immediately, once the associate expressions for mass and velocity are considered from Eqs. (1.138) and (1.143), respectively, yielding the quantified form

$$e_{\cancel{B}} = \frac{4\pi \hbar c}{137.036} \frac{1}{\sqrt{1 + \frac{64\pi^2 E_{bond}^2 X_{bond}^2}{\hbar^2 c^2 (2\pi n + 1)^4}}}, \quad n = 0,1,2... \quad (1.147)$$

However, even for the ground state, and more so for the excited states, one may see that when forming the practical ratio respecting the unitary electric charge from (1.147), it actually approaches a referential value, namely

$$\varsigma_e = \frac{e_B}{e} = \frac{4\pi}{\sqrt{1 + \frac{\left(E_{bond}[kcal/mol]\right)^2 \left(X_{bond}[\overset{0}{A}]\right)^2}{3.27817 \times 10^6 \left(2\pi n + 1\right)^4}}} \cong 4\pi \qquad (1.148)$$

for, in principle, any common energy and length of chemical bonding. On the other side, for the bondons to have different masses and velocities (kinetic energy) as associated with specific bonding energy but an invariant (universal) charge seems a bit paradoxical. Moreover, it appears that with Eq. (1.148) the predicted charge of a bonding, even in small molecules such as H_2, considerably surpasses the available charge in the system, although this may be eventually explained by the continuous matter-antimatter balance in the Dirac Sea to which the present approach belongs. However, to circumvent such problems, one may further use the result (1.148) and map it into the Poisson type charge field Equation

$$e_B \cong 4\pi \times e \leftrightarrow \nabla^2 V \cong 4\pi \times \rho \qquad (1.149)$$

from where the bondonic charge may be reshaped by appropriate dimensional scaling in terms of the bounding parameters (E_{bond} and X_{bond}) successively as

$$e_B \sim \frac{1}{4\pi} \left[\nabla_x^2 V\right]_{X=X_{bond}} \rightarrow \frac{1}{4} \frac{E_{bond} X_{bond}}{\aleph_0} \qquad (1.150)$$

Now, Eq. (1.150) may be employed towards the working ratio between the bondonic and electronic charges in the ground state of bonding

$$\varsigma_e = \frac{e_B}{e} \sim \frac{1}{32\pi} \frac{\left(E_{bond}[kcal/mol]\right)\left(X_{bond}[\overset{0}{A}]\right)}{\sqrt{3.27817 \times 10^3}} \qquad (1.151)$$

With Eq. (1.151) the situation is reversed compared with the previous paradoxical situation, in the sense that now, for most chemical bonds (of Table 1.1, for instance), the resulted bondonic charge is small enough to be not yet observed or considered as belonging to the bonding wave spreading among the binding electrons.

TABLE 1.1 Ratios for the Bondon-To-Electronic Mass and Charge and for the Bondon-To-Light Velocity, Along the Associated Bondonic Life-Time for Typical Chemical Bonds in Terms of Their Basic Characteristics Such As the Bond Length and Energy (Oelke, 1969; Findlay, 1955) Through Employing the Basic Formulas (1.139), (1.144), (1.151) and (1.152) for the Ground States, Respectively (Putz, 2010a, 2012a-b)

Bond Type	X_{bond} (Å)	E_{bond} (kcal/mol)	$\varsigma_m = \dfrac{m_{\mathcal{B}}}{m_0}$	$\varsigma_v = \dfrac{v_{\mathcal{B}}}{c}[\%]$	$\varsigma_e = \dfrac{e_{\mathcal{B}}}{e}[\times 10^3]$	$t_{\mathcal{B}}[\times 10^{15}]$ (seconds)
H–H	0.60	104.2	2.34219	3.451	0.3435	9.236
C–C	1.54	81.2	0.45624	6.890	0.687	11.894
C–C[(a)]	1.54	170.9	0.21678	14.385	1.446	5.743
C=C	1.34	147	0.33286	10.816	1.082	6.616
C≡C	1.20	194	0.31451	12.753	1.279	5.037
N≡N	1.10	225	0.32272	13.544	1.36	4.352
O=O	1.10	118.4	0.61327	7.175	0.716	8.160
F–F	1.28	37.6	1.42621	2.657	0.264	25.582
Cl–Cl	1.98	58	0.3864	6.330	0.631	16.639
I–I	2.66	36.1	0.3440	5.296	0.528	26.701
C–H	1.09	99.2	0.7455	5.961	0.594	9.724
N–H	1.02	93.4	0.9042	5.254	0.523	10.32
O–H	0.96	110.6	0.8620	5.854	0.583	8.721
C–O	1.42	82	0.5314	6.418	0.64	11.771
C=O[(b)]	1.21	166	0.3615	11.026	1.104	5.862
C=O[(c)]	1.15	191.6	0.3467	12.081	1.211	5.091
C–Cl	1.76	78	0.3636	7.560	0.754	12.394
C–Br	1.91	68	0.3542	7.155	0.714	14.208
C–I	2.10	51	0.3906	5.905	0.588	18.9131

(a) in diamond;
(b) in CH_2O;
(c) in O=C=O.

Instead, aiming to explore the specific information of bonding reflected by the bondonic mass and velocity, the associated ratios of Eqs. (1.139) and (1.144) for some typical chemical bonds (Oelke, 1969; Findlay, 1955) are computed in Table 1.1. They may be eventually accompanied by the predicted life-time of corresponding bondons, obtained from the bondonic mass and velocity working expressions (1.139) and (1.144), respectively, throughout the basic time-energy Heisenberg relationship—here restrained at the level of kinetic energy only for the bondonic particle; this way one yields the successive analytical forms

$$t_{\not B} = \frac{\hbar}{T_{\not B}} = \frac{2\hbar}{m_{\not B} v_{\not B}^2} = \frac{2\hbar}{\left(m_0 \varsigma_m\right)\left(c\varsigma_v \cdot 10^{-2}\right)^2} = \frac{\hbar}{m_0 c^2} \frac{2 \cdot 10^4}{\varsigma_m \varsigma_v^2}$$

$$= \frac{0.0257618}{\varsigma_m \varsigma_v^2} \times 10^{-15} [s]_{SI} \qquad (1.152)$$

and the specific values for various bonding types that are displayed in Table 1.1. Note that defining the bondonic life-time by Eq. (1.152) is the most adequate, since it involves the basic bondonic (particle!) information, mass and velocity; instead, when directly evaluating the bondonic life-time by only the bonding energy one deals with the working formula

$$t_{bond} = \frac{\hbar}{E_{bond}} = \frac{1.51787}{E_{bond}[kcal/mol]} \times 10^{-14} [s]_{SI} \qquad (1.153)$$

that usually produces at least one order lower values than those reported in Table 1.1 upon employing the more complex Eq. (1.152). This is nevertheless reasonable, because in the last case no particle information was considered, so that the Eq. (1.153) gives the time of the associate *wave* representation of bonding; this departs by the case when the time is computed by Eq. (1.152) where the information of bonding is contained within the *particle* (bondonic) mass and velocity, thus predicting longer life-times, and consequently a more susceptible timescale in allowing the bondonic observation. Therefore, as far as the chemical bonding is modeled by associate bondonic particle, the specific time of Eq. (1.152) rather than that of Eq. (1.153) should be considered.

While analyzing the values in Table 1.1, it is generally observed that as the bondonic mass is large as its velocity and the electric charge lower in their ratios, respecting the light velocity and electronic benchmark charge, respectively, however with some irregularities that allows further discrimination in the sub-bonding types. Yet, the life-time tendency records further irregularities, due to its complex and reversed bondonic mass-velocity dependency of Eq. (1.152), and will be given a special role in bondonic observation—see the Table 1.2 discussion below. Nevertheless, in all cases, the bondonic velocity is a considerable (non-negligible) percent of the photonic velocity, confirming therefore its combined quantum-relativistic nature. This explains why the bondonic reality appears even in the *non-relativistic* case of the Schrödinger equation when augmented with Bohmian entangled motion through the hidden quantum interaction.

Going now to particular cases of chemical bonding in Table 1.1, the hydrogen molecule maintains its special behavior through providing the bondonic mass as slightly more than double of the only two electrons contained in the whole system. This is not a paradox, but a confirmation of the fact the bondonic reality is not just the sum or partition of the available valence atomic electrons in molecular bonds, but a distinct (although related) existence that fully involves the undulatory nature of the electronic and nuclear motions in producing the chemical field. Remember the chemical field was associated either in Schrödinger as well in Dirac pictures with the internal rotations of the (Bohmian) wave function or spinors, being thus merely a phase property—thus inherently of undulatory nature. It is therefore natural that the risen bondons in bonding preserve the wave nature of the chemical field traveling the bond length distance with a significant percent of light.

Moreover, the bondonic mass value may determine the kind of chemical bond created, in this line the H_2 being the most covalent binding considered in Table 1.1 since it is most closely situated to the electronic

TABLE 1.2 Predicted Basic Values for Bonding Energy and Length, Along the associated Bondonic Life-Time and Velocity Fraction From the Light Velocity for a System Featuring Unity Ratios of Bondonic Mass and Charge, Respecting the Electron Values, Through Employing the Basic Formulas (1.152), (1.144), (1.139), and (1.151), Respectively (Putz, 2010a, 2012a-b)

X_{bond} (Å)	E_{bond} (kcal/mol)	$t_{\text{-}B}[\times 10^{15}]$ (SECONDS)	$\varsigma_v = \dfrac{v_{\text{-}B}}{c}[\%]$	$\varsigma_m = \dfrac{m_{\text{-}B}}{m_0}$	$\varsigma_e = \dfrac{e_{\text{-}B}}{e}$
1	87.86	10.966	4.84691	1	0.4827×10^{-3}
1	182019	53.376	99.9951	4.82699×10^{-4}	1
10	18201.9	533.76	99.9951	4.82699×10^{-5}	1
100	1820.19	5337.56	99.9951	4.82699×10^{-6}	1

pairing at the mass level. The excess in H_2 bond mass with respect to the two electrons in isolated H atoms comes from the nuclear motion energy converted (relativistic) and added to the two-sided electronic masses, while the heavier resulted mass of the bondon is responsible for the stabilization of the formed molecule respecting the separated atoms. The H_2 bondon seems to be also among the less circulated ones (along the bondon of the F_2 molecule) in bonding traveled information due to the low velocity and charge record—offering therefore another criterion of covalency, *i.e.*, associated with better localization of the bonding space.

The same happens with the C–C bonding, which is predicted to be more *covalent* for its simple (single) bondon that moves with the *smallest velocity* ($\varsigma_v <<$) or fraction of the light velocity from all C–C types of bonding; in this case also the bondonic *highest mass* ($\varsigma_m >>$), *smallest charge* ($\varsigma_e <<$), and *highest (observed) life-time* ($t_B >>$) criteria seem to work well. Other bonds with high covalent character, according with the bondonic velocity criterion only, are present in N≡N and the C=O bonding types and less in the O=O and C–O ones. Instead, one may establish the criteria for *multiple* (double and triple) *bonds* as having the series of current bondonic properties as: $\{\varsigma_m <, \varsigma_v >, \varsigma_e >, t_B <\}$.

However, the diamond C–C bondon, although with the smallest recorded mass ($\varsigma_m <<$), is characterized by the highest velocity ($\varsigma_v >$) and charge ($\varsigma_e >$) in the CC series (and also among all cases of Table 1.1). This is an indication that the bond is very much delocalized, thus recognizing the solid state or *metallic* crystallized structure for this kind of bond in which the electronic pairings (the bondons) are distributed over all atomic centers in the unit cell. It is, therefore, a special case of bonding that widely informs us on the existence of conduction bands in a solid; therefore the metallic character generally associated with the bondonic series of properties $\{\varsigma_m <<, \varsigma_v >, \varsigma_e >, t_B <\}$, thus having similar trends with the corresponding properties of multiple bonds, with the only particularity in the lower mass behavior displayed—due to the higher delocalization behavior for the associate bondons.

Very interestingly, the series of C–H, N–H, and O–H bonds behave similarly among them since displaying a shrink and medium range of mass (moderate high), velocity, charge and life-time (moderate high) variations for their bondons, $\{\varsigma_m \sim>, \varsigma_v \sim, \varsigma_e \sim, t_B \sim>\}$; this may explain why these bonds are the most preferred ones in DNA and genomic construction of proteins, being however situated towards the *ionic character* of chemical bond by the lower bondonic velocities computed; they have also the most close bondonic mass to unity; this feature being due to the manifested polarizability and inter-molecular effects that allows the 3D proteomic and specific interactions taking place.

Instead, along the series of halogen molecules F_2, Cl_2, and I_2, only the observed life-time of bondons show high and somehow similar values, while from the point of view of velocity and charge realms only the last two bonding types display compatible properties, both with drastic difference for their bondonic mass respecting the F–F bond—probably due the most negative character of the fluorine atoms. Nevertheless, judging upon the higher life-time with respect to the other types of bonding, the classification may be decided in the favor of covalent behavior. At this point, one notes traces of covalent bonding nature also in the case of the rest of halogen-carbon binding (C–Cl, C–Br, and C–I in Table 1.1) from the bondonic life-time perspective, while displaying also the ionic manifestation through the velocity and charge criteria $\{\varsigma_v \sim, \varsigma_e \sim\}$ and even a bit of metal character by the aid of small bondonic mass ($\varsigma_m <$). All these mixed features may be because of the joint existence of both inner electronic shells that participate by electronic induction in bonding as well as electronegativity difference potential.

Remarkably, the present results are in accordance with the recent signalized new binding class between the electronic pairs, somehow different from the ionic and covalent traditional ones in the sense that it is seen as a kind of resonance, as it appears in the molecular systems like F_2, O_2, N_2 (with impact in environmental chemistry) or in polar compounds like C–F (specific to ecotoxicology) or in the reactions that imply a competition between the exchange in the hydrogen or halogen (e.g., HF). The valence explanation relied on the possibility of higher orders of orbitals' existing when additional shells of atomic orbitals are involved such as <f> orbitals reaching this way the *charge-shift bonding* concept (Hiberty et al., 2006); the present bondonic treatment of chemical bonds overcomes the charge shift paradoxes by the relativistic nature of the bondon particles of bonding that have as inherent nature the time-space or the energy-space spanning towards electronic pairing stabilization between centers of bonding or atomic adducts in molecules.

However, we can also made predictions regarding the values of bonding energy and length required for a bondon to acquire either the unity of electronic charge or its mass (with the consequence in its velocity fraction from the light velocity) on the ground state, by setting Eqs. (1.139) and (1.151) to unity, respectively. These predictions are summarized in Table 1.2.

From Table 1.2, one note is that the situation of the bondon having the same charge as the electron is quite improbable, at least for the common chemical bonds, since in such a case it will feature almost the light velocity (and almost no mass–that is, however, continuously decreasing as the bonding energy decreases and the bonding length increases). This is natural since a longer distance has to be spanned by lower binding energy yet carrying the same unit charge of electron while it is transmitted with the same relativistic velocity! Such behavior may be regarded as the present *zitterbewegung* (trembling in motion) phenomena, here at the bondonic level. However one records the systematic increasing of bondonic lifetime towards being observable in the femtosecond regime for increasing bond length and decreasing the bonding energy–under the condition the chemical bonding itself still exists for certain $\{X_{bond}, E_{bond}\}$ combinations.

On the other side, the situation in which the bondon will weigh as much as one electron is a current one (see the Table 1.1); nevertheless, it is accompanied by quite reasonable chemical bonding length and energy information that it can carried at a low fraction of the light velocity, however with very low charge as well. Nevertheless, the discovered bonding energy-length relationship from Table 1.2, based on Eq. (1.151), namely

$$E_{bond}[kcal \, / \, mol] \times X_{bond}[\overset{0}{A}] = 182019 \qquad (1.154)$$

should be used in setting appropriate experimental conditions in which the bondon particle *B* may be observed as carrying the unit electronic charge yet with almost zero mass. In this way, *the bondon is affirmed as a special particle of Nature, that when behaving like an electron in charge it is behaving like a photon in velocity and like neutrino in mass, while having an observable (at least as femtosecond) lifetime for nanosystems having chemical bonding in the range of hundred of Angstroms and thousands of kcal/mol!* Such a peculiar nature of a bondon as the quantum particle of chemical bonding, the central theme of Chemistry, is not as surprising when noting that Chemistry seems to need both a particle view (such as offered by relativity) and a wave view (such as quantum mechanics offers), although nowadays these two physics theories are not yet fully compatible with each other, or even each fully coherent internally. Maybe the concept of 'bondons' will help to improve the situation for all concerned by its further conceptual applications.

1.4.4   CHEMICAL BONDING AS A BOSONIC QUANTUM CONDENSATE

From the December 22, 1995 when the Science magazine declared the Bose condensate as the "molecule of the year", the Bose-Einstein condensation (BEC, in short), basically viewed as *the macroscopic occupation of the same single-particle state in a many-body systems of bosons*, had received new impetus both at theoretical and experimental levels in searching and comprehending new states of matter (Anderson et al., 1995; Ketterle, 2002). However, with the ever increasing number of experiments revealing quantum phase transitions at atomic scales (Yukalov, 2004), the need for accurate models for this new state of matter became imperative. Yet, although powerful variational and perturbation methods are available (Kleinert et al., 2004), a basic approach, centered on the key object of BEC – the bosonic gas density ρ – it is not yet systematically developed and implemented (Vetter, 1997).

1.4.4.1   ψ-Theory of Bose-Einstein Condensation

The DFT affirmed themselves as a formal theory able to encompasses great deal of universality through its density based functional relationship with the total number of particle in a system (Parr & Yang, 1989; March, 1992; Dreizler & Gross, 1990; Putz, 2012a)

$$N = \int |\psi(\mathbf{r})|^2 \, d\mathbf{r}$$ (1.155)

While generalizing the custom quantum normalization condition Eq. (1.155) allows extending the universality class from fermions to bosons as well, with the quantum field $\psi(\mathbf{r})$ representing the collective behavior; namely it unfolds as simple superposition for the individual component particles in *fermionic* case (Kohn & Sham, 1965)

$$\psi(\mathbf{r}) = \sum_i \varphi_i(\mathbf{r}) \tag{1.156}$$

and through the mean (condensed) field + excitation spectra for bosonic case, as the so called Hartree-Fock-Bogoliubov transformations prescribes for direct and conjugate (creation and annihilation) fields (Bogoliubov, 1947):

$$\hat{\psi}(\mathbf{r}) = \widehat{\Psi}(\mathbf{r}) + \widehat{\tilde{\psi}}(\mathbf{r}) = \widehat{\Psi}(\mathbf{r}) + \sum_i \varphi_i(\mathbf{r})\hat{a}_i \tag{1.157}$$

$$\hat{\psi}^+(\mathbf{r}) = \widehat{\Psi}^+(\mathbf{r}) + \widehat{\tilde{\psi}}^+(\mathbf{r}) = \widehat{\Psi}^+(\mathbf{r}) + \sum_i \varphi_i^*(\mathbf{r})\hat{a}_i^+ \tag{1.158}$$

However, at the level of the quantum equations they accomplish, the fermionic case produced the equivalent for the Schrodinger equation, under the form of the so called Kohn-Sham equation for the mono-electronic orbitals (Jones & Gunnarson, 1989)

$$\left[-\frac{1}{2}\nabla^2 + V_{KS} \right]\varphi(\mathbf{r}) = \mu\varphi(\mathbf{r}) \tag{1.159}$$

Equation (1.159) has as the eigen-spectra the mono-orbital chemical potentials (μ) produced through the specific Kohn-Sham potential

$$V_{KS}(\mathbf{r}) = V(\mathbf{r}) + \int \frac{\rho(\mathbf{r}_2)}{|\mathbf{r} - \mathbf{r}_2|} d\mathbf{r}_2 + V_{XC}(\mathbf{r}) \tag{1.160}$$

introducing the celebrated exchange-correlation potential

$$V_{XC}(\mathbf{r}) = \left(\frac{\delta E_{XC}[\rho]}{\delta\rho(\mathbf{r})} \right)_{V(\mathbf{r})} \tag{1.161}$$

that accounts for the fermionic statistical correlation and exchange by means of the "tricky" exchange-correlation energy functional

$$E_{XC}[\rho] = \left\{ (T[\rho] - T_{KS}[\rho]) + (W[\rho] - J[\rho]) \right\} \tag{1.162}$$

constructed as the sum between the kinetic quantum contribution over the referential uniform kinetic Kohn-Sham (X.artificial) energy contribution

$$T_{KS}[\rho] = \sum_{i}^{N} \int n_i \varphi_i^*(\mathbf{r}) \left[-\frac{1}{2}\nabla^2 \right] \varphi_i(\mathbf{r})d\mathbf{r}, n_i \in [0,1] \qquad (1.163)$$

and the quantum excess of the *inter-particle* potential energy density functional $W[\rho]$ respecting the "spherical" or homogeneous classical repulsion inter-electronic Coulombic energy

$$J[\rho] = \frac{1}{2}\iint \frac{\rho(\mathbf{r_1})\rho(\mathbf{r_2})}{r_{12}}d\mathbf{r_1}d\mathbf{r_2} \qquad (1.164)$$

Unlike the fermionic quantum description, the bosonic case is driven by the condensate wave function (the main field) by means of the adapted version of Eq. (1.155), i.e.,

$$N = \int |\psi(\mathbf{r})|^2\, d\mathbf{r} = \int |\Psi(\mathbf{r})|^2\, d\mathbf{r} = \int \rho_{cond}d\mathbf{r} \qquad (1.165)$$

It nevertheless produces a non-linear Schrodinger equation known as Gross-Pitaevsky (GP) equation (Gross, 1961; Pitaevsky, 1961)

$$\left[-\frac{\hbar^2}{2m_B}\nabla^2 + V(\mathbf{r}) + g|\psi|^2 \right]\psi = \mu\psi \qquad (1.166)$$

with the non-linear term being modulated by the strength of the condensate g, controlling the interaction between two particles of the condensate, with the meaning of "scattering length" between particles in condensate, on the custom size of an atomic hard-sphere equivalent diameter.

The natural question arises whether the two type equations, Kohn-Sham and Gross-Pitaevsky, may be seen as equivalent, and in which conditions? The present article likes to address such important issue in the nowadays pioneering developments of DFT of BEC (Vetter, 1997; Nunes, 1999; Albus et al., 2003; Kim & Zubarev, 2003; Brand, 2004; Argaman & Band, 2011; Putz, 2011b-c, 2012c). As such, the main plan of the paper consists in:

- formulating KS-equation within BEC, due to the universality feasibility of the DFT itself;
- exploring the KS-DFT-BEC reductive power to BEC-GP equation.

The eventually positive enterprise of the present project will highly enhanced the possibility of exploring BEC effects in quantum chemical phenomena, and in chemical bonding in special.

1.4.4.2   DFT adapted to Bose-Einstein Condensation

The specificity of DFT treated for BEC implies functionals of the couple $[\rho(\mathbf{r}), \Psi(\mathbf{r})]$ in terms of the overall *super-fluidic density* $\rho(\mathbf{r})$ and of the *order parameter* $\Psi(\mathbf{r})$, both defined through the N-particles' ensemble statistical average (Vetter, 1997; Putz, 2011b-c, 2012c)

$$\begin{cases} \langle \bullet \rangle_T = \dfrac{1}{\mathcal{Z}} \mathrm{Tr}[\bullet e^{-\beta E[\psi]}] := \mathrm{Tr}[\rho_{\mathcal{Z}} \bullet] \\[2mm] \rho_{\mathcal{Z}} = \dfrac{e^{-\beta E[\psi]}}{\mathcal{Z}} = \dfrac{e^{-\beta E[\psi]}}{\mathrm{Tr}[e^{-\beta E[\psi]}]} \end{cases} \qquad (1.167)$$

respectively as:

$$\rho(\mathbf{r}) \equiv \langle \psi^+(\mathbf{r})\psi(\mathbf{r}) \rangle_T \qquad (1.168)$$

$$\Psi(\mathbf{r}) \equiv \langle \psi(\mathbf{r}) \rangle_T \qquad (1.169)$$

with the *thermodynamic sample* average with the quantum average included

$$E[\psi] = \langle \widehat{H} \rangle_\psi = \int d\mathbf{r} \, \psi^+(\mathbf{r}) \widehat{H} \psi(\mathbf{r}) \qquad (1.170)$$

Next, one should learn DFT deals with two kinds of working systems: one real- called Hohenberg-Kohn and one artificial – called Kohn-Sham; the natural or Hohenberg-Kohn system ("*ext. or HK*") includes the external $V_{ext}(X.\mathbf{r})$ and interacting $W(X.\mathbf{r},\mathbf{r}')$ potentials, the last one being included in the HK-functional F_{HK}

$$F_{HK}[\Psi, \rho] = \left\langle T_{HK} + W + \dfrac{\ln \rho_{\mathcal{Z}}}{\beta} \right\rangle_T \qquad (1.171)$$

The specific auxiliary order parameter related potential D_{HK} is just considered formally since at the end it has to be set to zero due to the fact its coupling currents are vanishing ($J_{HK} \rightarrow 0$) for the macroscopic point of view:

$$D_{ext}[\Psi, \rho] = \int dr \left[\Psi^+(\mathbf{r}) J(\mathbf{r}) + J^*(\mathbf{r}) \Psi(\mathbf{r}) \right] \qquad (1.172)$$

The formed grand potential looks therefore as

$$\Omega_{HK}[\Psi, \rho] = F_{HK}[\Psi, \rho] + \int dr \left[V_{HK}(\mathbf{r}) - \mu \right] \rho(\mathbf{r}) + D_{HK}[\Psi, \rho] \quad (1.173)$$

Note that F_{HK} stays for the Hohenberg-Kohn functional, $V(\mathbf{r})$ is the external applied potential, μ is the chemical potential, W: the inter-particle pair interaction, T_{HK}: the kinetic energy of the real system, and $J(\mathbf{r})$ the external source that assure the observable character of the DFT-BEC picture.

The non-interacting or Kohn-Sham (auxiliary) construction employs the artificial KS potential V_{KS} that has the effective potential role, which includes also the external or HK potential already (Kohn & Sham, 1965), see Eq. (1.160), but having no interaction potential any longer within the KS functional

$$F_{KS}[\Psi, \rho] = \left\langle T_{KS} + \frac{\ln \rho_{\mathcal{z}}^{KS}}{\beta} \right\rangle_T^{KS} \qquad (1.174)$$

with the actual KS-statistical average

$$\begin{cases} \left\langle \bullet \right\rangle_T^{KS} = \dfrac{1}{\mathcal{Z}_{KS}} \mathrm{Tr}[\bullet e^{-\beta E_{KS}[\psi]}] = \mathrm{Tr}[\rho_{\mathcal{z}}^{KS} \bullet] \\[2mm] \rho_{\mathcal{z}}^{KS} = \dfrac{e^{-\beta E_{KS}[\psi]}}{\mathcal{Z}_{KS}} = \dfrac{e^{-\beta E_{KS}[\psi]}}{\mathrm{Tr}[e^{-\beta E_{KS}[\psi]}]} \end{cases} \qquad (1.175)$$

Yet, the KS systems carries *the same density and order parameter* as that characterizing the HK system; accordingly, the associate potential D_{HK} does not vanish to any further extent, now being characterized by the non-vanishing currents J_{KS}; therefore the grand canonical potential reads as

$$\Omega_{KS}[\Psi, \rho] = F_{KS}[\Psi, \rho] + \int dr \left[V_{KS}(\mathbf{r}) - \mu \right] \rho(\mathbf{r}) + D_{KS}[\Psi, \rho] \quad (1.176)$$

Equation (1.176) can be specialized, within the Bogoliubov transformations (1.157) and (1.158), for the zero-temperature limit ($\beta \to \infty$) to the
form:

$$\Omega_{KS}[\Psi, \tilde{\psi}] = \int d\mathbf{r}\, \Psi^+(\mathbf{r})\hat{T}\Psi(\mathbf{r}) + \int d\mathbf{r}\, \tilde{\psi}^+(\mathbf{r})\hat{T}\tilde{\psi}(\mathbf{r})$$
$$+ \int d\mathbf{r}\Psi^+(\mathbf{r})[V_{KS}(\mathbf{r}) - \mu]\Psi(\mathbf{r}) + \int d\mathbf{r}\tilde{\psi}^+(\mathbf{r})[V_{KS}(\mathbf{r}) - \mu]\tilde{\psi}(\mathbf{r})$$
$$+ \int d\mathbf{r}\left[\Psi^+(\mathbf{r})J_{KS}(\mathbf{r}) + J^*_{KS}(\mathbf{r})\Psi(\mathbf{r})\right]$$

$$(1.177)$$

The result of Eq. (1.177) is then subject to the extremizing conditions in
order the equilibrium of the BEC is assessed for the KS system

$$\begin{cases} \dfrac{\delta}{\delta\Psi^+}\Omega_{KS} = 0 \\[2mm] \dfrac{\delta}{\delta\rho}\Omega_{KS} = 0 \end{cases}$$

$$(1.178)$$

While the second equation of the system (1.178) produces the Schrödinger
equation associated with the thermal part ($\tilde{\psi}$) of the (superfluid) system,
the first equation brings the <ψ>/KS equation under the form

$$\left(-\frac{\hbar^2\nabla^2}{2m} + V_{KS}(\mathbf{r}) - \mu\right)\Psi(\mathbf{r}) = -J_{KS}(\mathbf{r}) \qquad (1.179)$$

to be next employed for the real system characterized by the Hohenberg-
Kohn grand-potential (1.173). As such, HK and KS systems can be connected by means of the interaction term within the so called *local density
approximation* (LDA)

$$F_{int}[\Psi, \rho] = \int f_{int}(\Psi, \rho)d\mathbf{r} \qquad (1.180)$$

explicitly as:

$$\Omega_{HK}[\Psi, \rho] = \Omega_{KS}[\Psi, \rho] + F_{int}[\Psi, \rho]$$
$$- \int d\mathbf{r}[V_{KS}(\mathbf{r}) - V_{HK}(\mathbf{r})]\rho(\mathbf{r}) - \int d\mathbf{r}\begin{bmatrix} \Psi^+(\mathbf{r})J_{KS}(\mathbf{r}) \\ + J^*_{KS}(\mathbf{r})\Psi(\mathbf{r}) \end{bmatrix} \quad (1.181)$$

Remarkably, the minimization conditions on HK

$$\begin{cases} \dfrac{\delta}{\delta \Psi^+} \Omega_{HK} = 0 \\ \dfrac{\delta}{\delta \rho} \Omega_{HK} = 0 \end{cases} \tag{1.182}$$

combined with those of KS system, Eq. (1.178), applied on Eq. (1.181) provide the equations

$$\begin{cases} \dfrac{\delta}{\delta \Psi^+} f_{int}\left(\Psi, \rho\right) = J_{KS} \\ \dfrac{\delta}{\delta \rho} f_{int}\left(\Psi, \rho\right) = V_{KS}(\mathbf{r}) - V_{HK}(\mathbf{r}) \end{cases} \tag{1.183}$$

Remembering that the HKL system reflects the external influence on the natural system, *ext.=HK*, through once replacing the Eqs. (1.183) back into the Eq. (1.179) the working form for <ψ>/KS is obtained

$$\left(-\frac{\hbar^2 \nabla^2}{2m} + V_{ext}(\mathbf{r}) - \mu + \frac{\delta}{\delta \rho} f_{int}\left(\Psi, \rho\right) \right) \Psi(\mathbf{r}) = -\frac{\delta}{\delta \Psi^+} f_{int}\left(\Psi, \rho\right) \tag{1.184}$$

Once the interaction functional is eventually specified into Eq. (1.184) it may be related with the Gross-Pitaevskii equation of the BEC, as will be in next revealed.

1.4.4.3   Local Density Approximation of Bose-Einstein Condensation

In order Eq. (1.184) may be further worked out the form of the inter-action function should be specified; it act as an inter-particle interaction energy, being thus related with the inter-particle bilocal potential itself; in fact it was found having the functional form (Vetter, 1997; Putz, 2011b-c, 2012a,c):

$$F_{int}[\Psi,\rho]=\frac{1}{2}\int d\mathbf{r}d\mathbf{r}'W(\mathbf{r},\mathbf{r}')\begin{Bmatrix}|\Psi(\mathbf{r})|^2|\Psi(\mathbf{r}')|^2+2|\Psi(\mathbf{r})|^2\,\tilde{\rho}(\mathbf{r}')+\tilde{\rho}(\mathbf{r})\tilde{\rho}(\mathbf{r}')\\-2\Psi(\mathbf{r})\Psi^+(\mathbf{r}')G(\mathbf{r}\tau;\mathbf{r}'\tau)\\+G(\mathbf{r}\tau;\mathbf{r}'\tau)G(\mathbf{r}'\tau;\mathbf{r}\tau)\end{Bmatrix}$$

$$(1.185)$$

However, as usually in DFT, but also due to the special feature of the BEC being localized in a narrow region of space, the custom local density approximation may be also here implemented by imposing the bilocal-to-local transformation of the interaction upon the order parameter's delta function dependency

$$W(\mathbf{r},\mathbf{r}')\to g\delta(\mathbf{r}-\mathbf{r}') \qquad (1.186)$$

In this case, its coupling with Green function turns to be

$$\int d\mathbf{r}'\delta(\mathbf{r}-\mathbf{r}')G(\mathbf{r}\tau;\mathbf{r}'\tau)=-\int d\mathbf{r}'\delta(\mathbf{r}-\mathbf{r}')\left\langle\tilde{\psi}^+(\mathbf{r})\tilde{\psi}(\mathbf{r}')\right\rangle_T^{KS}=-\left\langle\tilde{\rho}(\mathbf{r})\right\rangle_T^{KS} \quad (1.187)$$

and the interaction functional reshapes as

$$F_{int}[\Psi,\rho]=\frac{g}{2}\int d\mathbf{r}\begin{Bmatrix}\Psi^{+2}(\mathbf{r})\Psi^2(\mathbf{r})+2|\Psi(\mathbf{r})|^2\,\tilde{\rho}(\mathbf{r})+\tilde{\rho}^2(\mathbf{r})\\+2\Psi(\mathbf{r})\Psi^+(\mathbf{r})\left\langle\tilde{\rho}(\mathbf{r})\right\rangle_T^{KS}+\left\langle\tilde{\rho}(\mathbf{r})\right\rangle_T^{KS}\left\langle\tilde{\rho}(\mathbf{r})\right\rangle_T^{KS}\end{Bmatrix} \quad (1.188)$$

Further replacement of the fluctuation density KS-average by the global prescription given by Eq. (1.157), while vanishing the fluctuation itself, transforms Eq. (1.188) successively as

$$F_{int}[\Psi,\rho]=\frac{g}{2}\int d\mathbf{r}\begin{Bmatrix}\rho^2(\mathbf{r})+2\Psi(\mathbf{r})\Psi^+(\mathbf{r})\left\langle\left[\rho(\mathbf{r})-|\Psi(\mathbf{r})|^2\right]\right\rangle_T^{KS}\\+\left\langle\left[\rho(\mathbf{r})-|\Psi(\mathbf{r})|^2\right]\right\rangle_T^{KS}\left\langle\left[\rho(\mathbf{r})-|\Psi(\mathbf{r})|^2\right]\right\rangle_T^{KS}\end{Bmatrix}$$

$$=\frac{g}{2}\int d\mathbf{r}\begin{Bmatrix}\rho(\mathbf{r})^2+2\Psi(\mathbf{r})\Psi^+(\mathbf{r})\left\langle\rho(\mathbf{r})\right\rangle_T^{KS}\\-2\Psi(\mathbf{r})\Psi^+(\mathbf{r})\left\langle|\Psi(\mathbf{r})|^2\right\rangle_T^{KS}\\+\left\langle\rho(\mathbf{r})\right\rangle_T^{KS}\left\langle\rho(\mathbf{r})\right\rangle_T^{KS}-2\left\langle\rho(\mathbf{r})\right\rangle_T^{KS}\left\langle|\Psi(\mathbf{r})|^2\right\rangle_T^{KS}\\+\left\langle|\Psi(\mathbf{r})|^2\right\rangle_T^{KS}\left\langle|\Psi(\mathbf{r})|^2\right\rangle_T^{KS}\end{Bmatrix}$$

$$(1.189)$$

so that the actual expression solely depends on the overall and condensate fields, for which the following mean approximations may be formulated:

- the thermodynamic average over the order parameter may leave it unaffected since it characterizes the condensate order anyway; that is the case of the third term of Eq. (1.189)

$$\left\langle \left| \Psi(\mathbf{r}) \right|^2 \right\rangle_T^{KS} = \left| \Psi(\mathbf{r}) \right|^2 \tag{1.190}$$

- the product of the two thermodynamic average over the superfluid density reduces to that of the order parameter in the mean field approximation

$$\left\langle \rho(\mathbf{r}) \right\rangle_T^{KS} \left\langle \rho(\mathbf{r}) \right\rangle_T^{KS} \cong \left\langle \left| \Psi(\mathbf{r}) \right|^2 \right\rangle_T^{KS} \left\langle \left| \Psi(\mathbf{r}) \right|^2 \right\rangle_T^{KS} = \left[\Psi^+(\mathbf{r}) \Psi(\mathbf{r}) \right]^2 \tag{1.191}$$

- in the same framework of main field approximation, for the last term of Eq. (1.189), the fourth power of the mean field is considered as giving the effect of the squared superfluid density, respectively

$$\left| \Psi(\mathbf{r}) \right|^4 \cong \rho^2(\mathbf{r}) \tag{1.192}$$

This way, the expression (1.189) readily simplifies

$$F_{int}[\Psi, \rho] = \frac{g}{2} \int d\mathbf{r} \left\{ 2\rho^2(\mathbf{r}) - \left[\Psi^+(\mathbf{r}) \Psi(\mathbf{r}) \right]^2 \right\} \tag{1.193}$$

allowing for immediate formulation of the local interaction function

$$f_{int}(\Psi, \rho) = g \left\{ \rho^2(\mathbf{r}) - \frac{1}{2} \left[\Psi^+(\mathbf{r}) \Psi(\mathbf{r}) \right]^2 \right\} \tag{1.194}$$

Now, one can uses expression (1.194) to accordingly specify the system (1.183) under the current LDA form:

$$\begin{cases} \dfrac{\delta}{\delta \rho} f_{int}(\Psi, \rho) = 2g\rho(\mathbf{r}) \\[3mm] \dfrac{\delta}{\delta \Psi^+} f_{int}(\Psi, \rho) = -g \left[\Psi^+(\mathbf{r}) \Psi(\mathbf{r}) \right] \Psi(\mathbf{r}) \end{cases} \tag{1.195}$$

With the specific forms of Eq. (1.195), the Eq. (1.184) takes the actual ψ-KS mean field realization

$$\left(-\frac{\hbar^2\nabla^2}{2m}+V_{ext}(\mathbf{r})-\mu+g\left[2\rho(\mathbf{r})-\Psi^+(\mathbf{r})\Psi(\mathbf{r})\right]\right)\Psi(\mathbf{r})=0 \quad (1.196)$$

Finally, one may once more apply the mean field expansion (1.157) in order to evaluate the g-coupling in Eq. (1.196)

$$2\rho(\mathbf{r})-\Psi^+(\mathbf{r})\Psi(\mathbf{r})=\Psi^+(\mathbf{r})\Psi(\mathbf{r})+2\tilde{\psi}^+(\mathbf{r})\tilde{\psi}(\mathbf{r})$$
$$+2\left[\Psi^+(\mathbf{r})\tilde{\psi}(\mathbf{r})+\Psi(\mathbf{r})\tilde{\psi}^+(\mathbf{r})\right] \quad (1.197)$$

The last term of Eq. (1.197) vanishes through the left multiplication of (1.196) with $\Psi^+(\mathbf{r})$ followed by spatial integration since providing, in fact, the average of the fluctuation field, this way leaving the Eq. (1.196) with the so called modified Gross-Pitaevsky (Hartree-Fock) equation

$$\left(-\frac{\hbar^2\nabla^2}{2m}+V_{ext}(\mathbf{r})-\mu+2g\tilde{\rho}(\mathbf{r})+g\Psi^+(\mathbf{r})\Psi(\mathbf{r})\right)\Psi(\mathbf{r})=0 \quad (1.198)$$

Worth observing that Eq. (1.198) formally differs from the Landau version by the thermal density presence, $\tilde{\rho}(\mathbf{r})$, yet noting that the original Gross-Pitaevsky equation is given in general by the terms of the superfluid wave-function

$$\left(-\frac{\hbar^r\nabla^2}{2m}+V_{ext}(\mathbf{r})+g\left|\psi(\mathbf{r})\right|^2\right)\psi(\mathbf{r})=\mu\psi(\mathbf{r}) \quad (1.199)$$

so viewed as the non-linear (X.or generalized) Schrödinger equation for superfluid systems.

Fortunately, the celebrated DFT of atoms and molecules may provide a suitable framework for BEC modeling by exploiting DFT main features (Dreizler and Gross, 1995).

1.4.4.4   Heitler-London Bondonic Condensation

It is a many (N)-body theory whose *main vehicle is the density itself*

$$\rho(\mathbf{r}) = N \int d\mathbf{r}_2 \int d\mathbf{r}_3 \dots \int d\mathbf{r}_N \Psi^*(\mathbf{r}, \mathbf{r}_2 \dots \mathbf{r}_N) \Psi(\mathbf{r}, \mathbf{r}_2 \dots \mathbf{r}_N) \qquad (1.200)$$

fulfilling the normalization condition, as were Eqs. (1.155) and (1.165), here simple becoming as DFT main constraint

$$N = \int d\mathbf{r} \rho(\mathbf{r}) \qquad (1.201)$$

matching in this way the Gross-Pitaevsky uni-particle field ψ which describes the dilute bosonic gases through the nonlinear Schrödinger equation having chemical potential μ as the eigen-value (Pitaevsky & Stringari, 2003):

$$\left[-\frac{\hbar^2}{2m_B} \nabla^2 + V(\mathbf{r}) + g|\psi|^2 \right] \psi = \mu \psi \qquad (1.202)$$

Equation (1.202) accounts for the bosonic condensation through the self-interaction coupling:

$$g = \frac{4\pi a_B \hbar^2}{m_B} \bigg|_{a_B \to R_{bond}} = 8\pi E_{bond} R_{bond}^3 = \frac{8\pi E_{bond}}{\left| \langle \psi(\mathbf{r}) \rangle \right|^2} \qquad (1.203)$$

by considering the mass of the *bondons - the quantum bosonic particles of the chemical bonding* (Putz, 2010a)- defined as:

$$m_B = \frac{\hbar^2}{2} \frac{1}{E_{bond} R_{bond}^2} \qquad (1.204)$$

representing E_{bond} the bonding equilibrium energy. Generalizing the s-scattering length a_B to the bonding length R_{bond}, the uni-bondonic volume is obtained:

$$R_{bond}^3 = V_{bond} = \frac{1}{\rho_{bodnons} (\mathbf{r})} = \frac{1}{\left| \langle \psi(\mathbf{r}) \rangle \right|^2} \qquad (1.205)$$

$$\phantom{R_{bond}^3 = V_{bond} = \frac{1}{\rho_{bodnons}}} {}_{(bosons)}$$

on the basis of mean-field approximation $\rho \sim <\psi>^2$, attributing to the condensate density as the role of order parameter.

The DFT *supports the variational principle* (Dreizler & Gross, 1990; Putz, 2012a):

$$\delta\left(E[\rho] - \mu \int d\mathbf{r}\rho(\mathbf{r})\right) = 0 \tag{1.206}$$

in minimizing the energy functional

$$E[\rho] = C_A[\rho] + F_{HK}[\rho] \tag{1.207}$$

by means of Lagrangian method in terms of chemical potential parameter. Note that Eq. (1.207) expresses the effects of the chemical action $C_A[\rho] = \int d\mathbf{r}V(\mathbf{r})\rho(\mathbf{r})$, directly related with chemical bonding and reactivity (Putz, 2012b), see Volume II/Section 4.6.3 of the present five-volume work, as well as the end of Section 3 of the present Volume III, added to the universal Hohenberg-Kohn functional $F_{HK}[\rho]$ that, by containing the kinetic effects, due to Coulombic and exchange-correlation contributions (Hohenberg & Kohn, 1964), it is therefore susceptible in being related with the bosonic condensation phenomenology.

Overall, by originally employing DFT to model the Bose-Einstein condensate, the analytical DFT-BEC relationships are here primarily reported along their consequences in modeling the chemical bonding. The last item is also motivated by the intrigued case of forbidden He-He bonding within ordinary molecular orbital theory, albeit largely allowed through supefluidic (i.e., within nonlinear Schrödinger) treatment (London, 1938); the hidden BEC side of the fermionic wave functions superposition in general and of chemical bonding in special, it is therefore worth to explore.

The practical implementation of BEC and of its mean field approximation usually makes use of thermodynamically limit constraint which enables the usage of the so called Thomas-Fermi approximation of DFT (Parr and Yang, 1989) for systems with many-to-infinite number of particles ($N \rightarrow \infty$), since in condensation phenomenon "infinite more is the same" (Kadanoff, 2009).

By considering that the potential and the interaction energies are larger than the kinetic energy, the kinetic term can be neglected in the Gross-Pitaevsky equation (1.202), reducing it to the algebraic form:

$$|\psi|^2 \cong \frac{1}{g}[\mu - V(\mathbf{r})] \tag{1.208}$$

On the other side, the variational DFT principle (1.206) provides the working relationship among the chemical potential, external potential and the Hohenberg-Kohn functional:

$$\mu = V(\mathbf{r}) + \frac{\delta F_{HK}[\rho]}{\delta \rho(\mathbf{r})} \tag{1.209}$$

Combining Eqs. (1.208) and (1.209) the *second DFT-BEC connection* is directly obtained:

$$|\psi|^2 \cong \frac{1}{g} \left[\frac{\delta F_{HK}[\rho]}{\delta \rho(\mathbf{r})} \right]_{V(\mathbf{r})} \tag{1.210}$$

However, the *third working DFT-BEC connection* can be established by employing the inter-bosonic quantum average (Putz, 2011b-c, 2012a)

$$\int d\mathbf{r} |\psi(\mathbf{r})|^4 \cong \frac{1}{g^2} \int d\mathbf{r} [\mu - V(\mathbf{r})]^2 \cong -\frac{F_{HK}\eta}{g^2} \tag{1.211}$$

where the chemical hardness η was recognized through its local integrated version (Baekelandt et al., 1995)

$$\eta = \int d\mathbf{r}\eta(\mathbf{r}) \tag{1.212}$$

$$\eta(\mathbf{r}) = \frac{\delta\mu}{\delta\rho(\mathbf{r})} = \frac{\delta^2 F_{HK}[\rho]}{\delta\rho^2(\mathbf{r})} \tag{1.213}$$

involving the Hohenberg-Kohn functional. In this way, the chemical hardness manifests its straight involvement in bosonic condensates in general, while its particular contribution in chemical bonding will be treated below.

Based on previously emphasized fermionic-bosonic (DFT-BEC) relationships, one can consider the *build-in-bondonic* (BB) superposition of fermionic (real) wave functions coming from the two atoms labeled as *"1"* and *"2"*

$$\underbrace{\psi_{BB}(\mathbf{r})}_{BONDONIC} = \underbrace{c_1\psi_1(\mathbf{r}) + c_2\psi_2(\mathbf{r})}_{FERMIONIC\ SUPERPOSITION}, \quad c_1, c_2 \in \Re \tag{1.214}$$

in providing the bonding energy of the homopolar system "1 and 2" (i.e., a system in which all atoms are of the same type with different wave-functions contributing to the bond). The corresponding eigen-value is written as the expectation value of the Hamiltonian of Eq. (1.202) emphasizing on Schrodinger (Sch) and non-linear (BB) entries:

$$
\begin{aligned}
E_{bond} &= \frac{\int dr \left(c_1 \psi_1(\mathbf{r}) + c_2 \psi_2(\mathbf{r}) \right)^* H_{Sch} \left(c_1 \psi_1(\mathbf{r}) + c_2 \psi_2(\mathbf{r}) \right) + g \int dr \left| \psi_{BB}(\mathbf{r}) \right|^4}{\int dr \left(c_1 \psi_1(\mathbf{r}) + c_2 \psi_2(\mathbf{r}) \right)^* \left(c_1 \psi_1(\mathbf{r}) + c_2 \psi_2(\mathbf{r}) \right)} \\[2ex]
&\cong \frac{\left(c_1^2 + c_2^2 \right) \left(H_{11} + \dfrac{\eta_{Molec}}{g} V_{11} \right) + 2 c_1 c_2 \left(H_{12} + \dfrac{\eta_{Molec}}{g} V_{12} \right)}{c_1^2 + c_2^2 + 2 c_1 c_2 S + \dfrac{\eta_{Molec}}{g}}
\end{aligned}
$$

(1.215)

The notations in Eq. (1.215) represent the integrals of the intra-atomic $H_{11} = \int \psi_{1/2} H_{Sch} \psi_{1/2}$, inter-atomic $H_{12} = \int \psi_{1/2} H_{Sch} \psi_{2/1}$, and overlapping $S = \int [\psi_{1/2}]^2$ contributions, along the potential integrals paralleling the Coulombic and exchange contributions for electrons in bonding, namely $V_{11} = \int \psi_{1/2} V \psi_{1/2}$ and $V_{12} = \int \psi_{1/2} V \psi_{2/1}$, respectively. Also note that in deriving the result of Eq. (1.215) the universal Hohenberg-Kohn functional was implemented with the expression:

$$
F_{HK} = E_{bond} - C_A = E_{bond} - \left(c_1^2 + c_2^2 \right) V_{11} - 2 c_1 c_2 V_{12} \qquad (1.216)
$$

Now, the variational principle respecting the MO-coefficients of Eq. (1.214) upon the trial energy (1.215) yields that the *bonding (+)* and *antibonding (–)* wave-functions display the same forms as in the classical Heitler-London homopolar model (Heitler and London, 1927), while the corresponding energies are corrected by the *bondonic* BEC contribution

$$
E_{bond}^{\pm} = \frac{H_{11} \pm H_{12}}{1 \pm S} + \left| \langle \psi(\mathbf{r}) \rangle \right|^2 \frac{\eta_{Molec}}{8 \pi E_{bond}^{\pm}} \frac{V_{11} \pm V_{12}}{1 \pm S} \qquad (1.217)
$$

In Eq. (1.217) one recognizes the inverse of the bosonic coupling parameter $(1/g)$ considered with its bondonic-order parameter form of Eq. (1.203);

nevertheless, unlike the classical treatment, variational result (1.217) shapes as an additional quadratic equation of type $x = a + b/x$ of which the basic solutions $x_{1,2} = \frac{1}{2}[a \pm \sqrt{(a^2 + 4b)}]$ provides in the limit of small bosonic interaction, i.e., $b \sim \langle \psi \rangle \to 0$, as is the case of bondons in chemical bonding, the actual BEC-DFT two-fold bonding and antibonding energies:

$$E^{\pm}_{bond-BEC-I} = -\left|\left\langle \psi\left(\mathbf{r}\right)\right\rangle\right|^2 \frac{\eta_{Molec}}{8\pi} \frac{V_{11} \pm V_{12}}{H_{11} \pm H_{12}} \qquad (1.218)$$

$$E^{\pm}_{bond-BEC-II} = \frac{H_{11} \pm H_{12}}{1 \pm S} + \left|\left\langle \psi\left(\mathbf{r}\right)\right\rangle\right|^2 \frac{\eta_{Molec}}{8\pi} \frac{V_{11} \pm V_{12}}{H_{11} \pm H_{12}} \qquad (1.219)$$

The results of Eqs. (1.218) and (1.219) considerably enlarge (and enrich) the chemical bonding paradigm by including condensation and ordering phenomena, especially those associated with ^{4}He or alkali gases (Leggett, 2001).

The DFT formulation of BEC is currently a very vital topic due to innumerous recent theoretical advances in describing both fermionic and bosonic matter, based on the main DFT and BEC concepts, namely the non-linear Schrödinger equation, the condensate density, the Hohenberg-Kohn universal functional, the Thomas-Fermi limit, the self-interacting bosonic strength, and chemical hardness, enabling the bosonic (bondonic)-fermionic mixture formulation of the chemical bonding as a direct conceptual consequence. The results show that the bosonic nature of the chemical bonding primarily depends on the degree with which the atoms involved in bonding display the bosonic condensation when considered as single gas. As such, when the small non-zero order parameter ($\langle \psi \rangle \neq 0$, $\langle \psi \rangle \to 0$) is assumed upon the quantum particles of the chemical bonding – the bondons, see Eqs. (1.203)–(1.205), the BEC influence is manifested over Heitler-London bonding and antibonding energetic terms in Eqs. (1.218) and (1.219). Equally relevant, the presence of the molecular chemical hardness η_{Molec} on the nominator of Eqs. (1.218) and (1.219) further supports the condensation phenomenology by its consecrated maximum hardness principle for a stable molecule (Chattaraj et al., 1991, 1995; Ayers and Parr, 2000; Putz, 2012a-b): *the higher the hardness the higher the stability of the chemical system.*

The model is conceptually verified by the prediction of interdiction of the He-He chemical bond, even through the bondonic particle, specifying instead new bonding states for the H_2 molecule suitable to be experimentally detected in condition of BEC (temperatures of nano-Kelvin degree), see next section.

1.4.4.5 Molecular Orbitals' Bondonic Condensation

The atomic systems of units, which that correspond to atomic dimensions, are adopted and eliminate the physical constants from the Schrödinger equation; specifically, the length and energy units become the first Bohr radius $a_0 = 0.529\text{Å}$ and the interaction energy between two units of charge separated by the Bohr radius (Hartree) $\varepsilon_0 = 27.246eV$.

For homopolar chemical binding, the fermionic diatomic potential, in atomic units, is considered with the working form (Putz, 2012d)

$$V(A,B) = -\frac{2}{r_A} - \frac{2}{r_B} + \frac{1}{R} \tag{1.220}$$

such that the actual quantum condensation paradigm of chemical bonding views *the inter-electronic interaction as a condensed bondon superimposed on the two independent/equivalent electron-nuclei fermionic interactions.*

By introducing the specific *Coulombic* integral (Mulliken & Riecke, 1949; Roberts & Jaffe, 1957)

$$I = \int d\tau \psi_A^*(r_A) \frac{1}{r_B} \psi_A(r_A) = \left\langle \psi_A \left| \frac{1}{r_B} \right| \psi_A \right\rangle \tag{1.221}$$

and the *exchange*

$$J = \int d\tau \psi_A^*(r_A) \frac{1}{r_A} \psi_B(r_B) = \left\langle \psi_A \left| \frac{1}{r_A} \right| \psi_B \right\rangle \tag{1.222}$$

one has for the associate fermionic energy integrals the representative form

$$H_{11} = \int d\tau \psi_A^*(r_A) \left[-\frac{1}{2}\nabla^2 - \frac{2}{r_A} - \frac{2}{r_B} + \frac{1}{R} \right] \psi_A(r_A)$$

$$= \left\langle \psi_A \left| -\frac{1}{2}\nabla^2 - \frac{1}{r_A} \right| \psi_A \right\rangle - \left\langle \psi_A \left| \frac{1}{r_A} \right| \psi_A \right\rangle + \left\langle \psi_A \left| \frac{1}{R} \right| \psi_A \right\rangle - \left\langle \psi_A \left| \frac{2}{r_B} \right| \psi_A \right\rangle$$

$$= E_0 - U + \frac{N_0}{R} - 2I$$

$$(1.223)$$

by considering the atomic DFT-type normalization relationships for the valence electrons

$$\begin{cases} \langle \psi_A | \psi_A \rangle = \langle \psi_B | \psi_B \rangle = N_0 \\ N_0 = 1,2 \end{cases}$$

$$(1.224)$$

the *atomic* eigen-energy equation in the ground state

$$\left(-\frac{1}{2}\nabla^2 - \frac{1}{r_A} \right) | \psi_A \rangle = E_0 | \psi_A \rangle$$

$$(1.225)$$

and the self-interaction (a sort of internal) energy

$$U = \left\langle \psi_A \left| \frac{1}{r_A} \right| \psi_A \right\rangle$$

$$(1.226)$$

Analogously, one successively yields for the integral

$$H_{12} = \int d\tau \psi_A^*(r_A) \left[-\frac{1}{2}\nabla^2 - \frac{2}{r_A} - \frac{2}{r_B} + \frac{1}{R} \right] \psi_B(r_B)$$

$$= \underbrace{\left\langle \psi_A \left| -\frac{1}{2}\nabla^2 - \frac{1}{r_B} \right| \psi_B \right\rangle}_{E_0 | \psi_B \rangle} - \left\langle \psi_A \left| \frac{1}{r_B} \right| \psi_B \right\rangle + \frac{1}{R}\underbrace{\langle \psi_A | \psi_B \rangle}_{S} - \left\langle \psi_A \left| \frac{2}{r_A} \right| \psi_B \right\rangle$$

$$= E_0 S + \frac{S}{R} - 3J$$

$$(1.227)$$

while recognizing the influence of the overlap integral

$$\langle \psi_A | \psi_B \rangle = \langle \psi_B | \psi_A \rangle = S \tag{1.228}$$

and the atomic energy (1.225), which is derived from the B-atomic contribution.

Additionally, following the same arguments, the newly developed integrals unfold, respectively:

$$V_{11} = \int d\tau \psi_A^*(r_A) \left[-\frac{2}{r_A} - \frac{2}{r_B} + \frac{1}{R} \right] \psi_A(r_A)$$

$$= -2 \underbrace{\left\langle \psi_A \left| \frac{1}{r_A} \right| \psi_A \right\rangle}_{U} - 2 \underbrace{\left\langle \psi_A \left| \frac{1}{r_B} \right| \psi_A \right\rangle}_{I} + \frac{\langle \psi_A | \psi_A \rangle}{R}$$

$$= -2U - 2I + \frac{N_0}{R} \tag{1.229}$$

$$V_{12} = \int d\tau \psi_A^*(r_A) \left[-\frac{2}{r_A} - \frac{2}{r_B} + \frac{1}{R} \right] \psi_B(r_B)$$

$$= -2 \underbrace{\left\langle \psi_A \left| \frac{1}{r_A} \right| \psi_B \right\rangle}_{J} - 2 \underbrace{\left\langle \psi_A \left| \frac{1}{r_B} \right| \psi_B \right\rangle}_{J} + \frac{S}{R}$$

$$= -4J + \frac{S}{R} \tag{1.230}$$

With these, one employs the results

$$H_{11} + H_{12} = E_0(1 + S) + \frac{N_0 + S}{R} - U - 2I - 3J \tag{1.231}$$

$$H_{11} - H_{12} = E_0(1 - S) + \frac{N_0 - S}{R} - U - 2I + 3J \tag{1.232}$$

$$V_{11} + V_{12} = -2U - 2I - 4J + \frac{N_0 + S}{R} \tag{1.233}$$

$$V_{11} - V_{12} = -2U - 2I + 4J + \frac{N_0 - S}{R} \qquad (1.234)$$

to re-express the physical and chemical bonding condensates' Eqs. (1.118) and (1.119), respectively as:

$$E^{\pm}_{bond-BEC-I} = -\left|\langle \psi(\mathbf{r})\rangle\right|^2 \frac{\eta_{Molec}}{8\pi} \frac{(-2U - 2I \mp 4J)R + N_0 \pm S}{[E_0(1 \pm S) - U - 2I \mp 3J]R + N_0 \pm S}$$

$$(1.235)$$

$$E^{\pm}_{bond-BEC-II} = E_0 + \frac{N_0 \pm S}{(1 \pm S)R} - \frac{U + 2I \pm 3J}{1 \pm S} - E^{\pm}_{bond-BEC-I} \qquad (1.236)$$

Further evaluations of expressions (1.235) and (1.236) are possible by calculating the involved integrals using elliptic coordinates (see Figure 1.6), i.e., by coordinate transformations:

$$\lambda = \frac{r_A + r_B}{R} \in [1, \infty) \qquad (1.237)$$

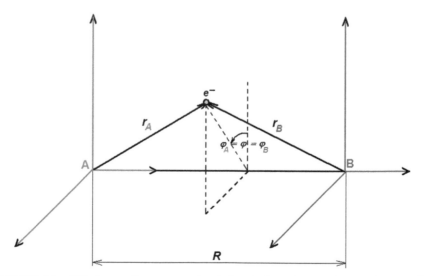

FIGURE 1.6 Elliptic coordinate system for the bondonic equivalent electron "immersed" within a diatomic molecular system (Putz, 2012d).

$$\delta = \frac{r_A - r_B}{R} \in [-1, +1] \tag{1.238}$$

$$\varphi = \varphi_A = \varphi_B \in [0, 2\pi] \tag{1.239}$$

producing the elementary volume

$$d\tau = \frac{R^3}{8}\left(\lambda^2 - \delta^2\right)d\lambda d\delta d\varphi \tag{1.240}$$

and the inverse relationships

$$r_A = \frac{R(\lambda + \delta)}{2} \tag{1.241}$$

$$r_B = \frac{R(\lambda - \delta)}{2} \tag{1.242}$$

These geometric conditions are applied to the special 1s atomic orbitals, suitable for modeling the bosonic interactions as well, with semi-classical variational wave functions (Parr, 1972; Putz, 2003) in atomic units

$$\psi(r) = \sqrt{\frac{\gamma}{\pi}}\exp(-\alpha r) \tag{1.243}$$

with the variational parameters

$$\alpha = \xi, \ \gamma = N_0\xi^3, \ \xi = \frac{21 - 5N_0}{16} \tag{1.244}$$

which features the following important properties:

- Integrates the atomic (or valence) number of electrons, according to a DFT-like normalization condition (1.224);
- Includes the orbital contribution (1.244), which recovers the 1s hydrogen wave function (when $N_0 = 1$).

Under these conditions, one may proceed with the analytical unfolding of the contributing integrals in Eqs. (1.235) and (1.236) as follows:

- The overlap integral (1.227) produces

$$
\begin{aligned}
S = \langle \psi_A | \psi_B \rangle &= \frac{\gamma}{\pi} \int e^{-\alpha(r_A + r_B)} d\tau = \frac{\gamma}{\pi} \int e^{-\alpha R \lambda} d\tau \\
&= \frac{\gamma R^3}{8\pi} \int_1^\infty \int_{-1}^{+1} \int_0^{2\pi} e^{-\alpha R \lambda} \left(\lambda^2 - \delta^2 \right) d\lambda d\delta d\varphi \\
&= \frac{\gamma R^3}{4} \left[\int_{-1}^{+1} d\delta \int_1^\infty \lambda^2 e^{-\alpha R \lambda} d\lambda - \int_{-1}^{+1} \delta^2 d\delta \int_1^\infty e^{-\alpha R \lambda} d\lambda \right] \\
&= \frac{\gamma R^3}{2} \left[\left(\frac{1}{\alpha R} + \frac{2}{\alpha^2 R^2} + \frac{2}{\alpha^3 R^3} \right) e^{-\alpha R} - \frac{1}{3\alpha R} e^{-\alpha R} \right] \\
&= \gamma \left(\frac{1}{\alpha^3} + \frac{R}{\alpha^2} + \frac{R^2}{3\alpha} \right) e^{-\alpha R} \\
&= N_0 \left(1 + \xi R + \frac{1}{3} \xi^2 R^2 \right) e^{-\xi R}
\end{aligned}
\tag{1.245}
$$

- The exchange integral (1.222) analytically yields

$$
\begin{aligned}
J = \left\langle \psi_A \left| \frac{1}{r_A} \right| \psi_B \right\rangle &= \frac{\gamma}{\pi} \int \frac{e^{-\alpha(r_A + r_B)}}{r_A} d\tau \\
&= \frac{\gamma R^3}{8\pi} \int_1^\infty \int_{-1}^{+1} \int_0^{2\pi} e^{-\alpha R \lambda} \left(\lambda^2 - \delta^2 \right) \frac{2}{R(\lambda + \delta)} d\lambda d\delta d\varphi \\
&= \frac{\gamma R^2}{2} \left[\int_{-1}^{+1} d\delta \int_1^\infty \lambda e^{-\alpha R \lambda} d\lambda - \int_{-1}^{+1} \delta d\delta \int_1^\infty e^{-\alpha R \lambda} d\lambda \right] \\
&= \gamma \left(R + \frac{1}{\alpha} \right) e^{-\alpha R} \\
&= N_0 \xi^2 \left(1 + \xi R \right) e^{-\xi R}
\end{aligned}
\tag{1.246}
$$

- The Coulombic integral (1.221) equivalently becomes

$$I = \left\langle \psi_A \left| \frac{1}{r_B} \right| \psi_A \right\rangle = \frac{\gamma}{\pi} \int \frac{e^{-2\alpha r_A}}{r_B} d\tau$$

$$= \frac{\gamma R^3}{8\pi} \int_1^\infty \int_{-1}^{+1} \int_0^{2\pi} e^{-\alpha R(\lambda+\delta)} \frac{2}{R(\lambda-\delta)} \left(\lambda^2 - \delta^2\right) d\lambda d\delta d\varphi$$

$$= \frac{\gamma R^2}{2} \int_1^\infty \int_{-1}^{+1} e^{-\alpha R(\lambda+\delta)} \left(\lambda+\delta\right) d\lambda d\delta$$

$$= \frac{\gamma R^2}{2} \left[\left(\int_1^\infty \lambda e^{-\alpha R \lambda} d\lambda \right) \left(\int_{-1}^{+1} e^{-\alpha R \delta} d\delta \right) + \left(\int_1^\infty e^{-\alpha R \lambda} d\lambda \right) \left(\int_{-1}^{+1} \delta e^{-\alpha R \delta} d\delta \right) \right]$$

$$= \frac{\gamma}{2} \left[-\frac{1}{\alpha} \left(1 + \frac{1}{\alpha} \right) \left(1 + \frac{1}{\alpha R} \right) e^{-2\alpha R} + \frac{1}{\alpha^2 R} \left(1 + \frac{1}{\alpha} \right) + \frac{1}{\alpha} \left(1 - \frac{1}{\alpha} \right) \right]$$

$$= \frac{N_0}{2} \left\{ \xi(\xi-1) + (1+\xi) \left[\frac{1}{R} - \left(\frac{1}{R} + \xi \right) e^{-2\xi R} \right] \right\}$$

$$(1.247)$$

- The self-energy integral (1.226) unfolds as

$$U = \left\langle \psi_A \left| \frac{1}{r_A} \right| \psi_A \right\rangle = \frac{\gamma}{\pi} \int \frac{e^{-2\alpha r_A}}{r_A} d\tau$$

$$= \frac{\gamma R^3}{8\pi} \int_1^\infty \int_{-1}^{+1} \int_0^{2\pi} e^{-\alpha R(\lambda+\delta)} \frac{2}{R(\lambda+\delta)} \left(\lambda^2 - \delta^2\right) d\lambda d\delta d\varphi$$

$$= \frac{\gamma R^2}{2} \int_1^\infty \int_{-1}^{+1} e^{-\alpha R(\lambda+\delta)} \left(\lambda-\delta\right) d\lambda d\delta$$

$$= \frac{\gamma R^2}{2} \left[\left(\int_1^\infty \lambda e^{-\alpha R \lambda} d\lambda \right) \left(\int_{-1}^{+1} e^{-\alpha R \delta} d\delta \right) - \left(\int_1^\infty e^{-\alpha R \lambda} d\lambda \right) \left(\int_{-1}^{+1} \delta e^{-\alpha R \delta} d\delta \right) \right]$$

$$= \frac{\gamma}{2\alpha} \left[\left(\frac{1}{\alpha} - 1 \right) \left(1 + \frac{1}{\alpha R} \right) e^{-2\alpha R} + \frac{1}{\alpha R} \left(1 - \frac{1}{\alpha} \right) + \left(1 + \frac{1}{\alpha} \right) \right]$$

$$= \frac{N_0}{2} \left\{ \xi(\xi+1) + (\xi-1) \left[\frac{1}{R} - \left(\frac{1}{R} + \xi \right) e^{-2\xi R} \right] \right\}$$

$$(1.248)$$

Before moving on to numerical evaluation, the bonding energetic integrals in the short- and long-long range interaction ($R \to \infty$) can still be evaluated analytically. Thus, by inspection of Eqs. (1.245)–(1.248), one finds the limits

$$S \to \begin{cases} 0...R \to \infty \\ N_0...R \to 0 \end{cases} \quad (1.249)$$

$$J \to \begin{cases} 0...R \to \infty \\ \dfrac{N_0}{a_0} \xi^2 ...R \to 0 \end{cases} \quad (1.250)$$

$$I \to \begin{cases} \dfrac{N_0}{a_0} \dfrac{\xi(\xi-1)}{2}...R \to \infty \\ -\dfrac{N_0}{a_0} \xi ...R \to 0 \end{cases} \quad (1.251)$$

$$U \to \begin{cases} \dfrac{N_0}{a_0} \dfrac{\xi(\xi+1)}{2}...R \to \infty \\ \dfrac{N_0}{a_0} \xi ...R \to 0 \end{cases} \quad (1.252)$$

Under these conditions, by substituting Eqs. (1.249)–(1.252) back into the basic *physical and chemical bonding condensate formulations* (1.235) and (1.236), one notes the equalization of the bonding and antibonding BEC levels under the long-range conditions, respectively, as follows:

$$\begin{aligned} E_{bond-BEC-1}^{\pm(R\to\infty)} &= -\left|\langle \psi(\mathbf{r}) \rangle\right|^2 \frac{\eta_{Molec}}{8\pi} \frac{-2U - 2I \mp 4J}{E_0(1\pm S) - U - 2I \mp 3J} \\ &= \left|\langle \psi(\mathbf{r}) \rangle\right|^2 \frac{\eta_{Molec}}{2\pi} \frac{\xi^2 N_0}{2a_0 E_0 + \xi N_0 - 3\xi^2 N_0} \end{aligned} \quad (1.253)$$

$$E_{bond-BEC-II}^{\pm(R\to\infty)} = E_0 - \frac{U + 2I \pm 3J}{1 \pm S} - E_{bond-BEC-I}^{\pm(R\to\infty)}$$

$$= E_0 + \frac{N_0\xi(1-3\xi)}{2a_0} - E_{bond-BEC-I}^{\pm(\infty)} \qquad (1.254)$$

However, for the short-range interaction, one obtains a *physical condensate* with equal bonding-antibonding energies

$$E_{bond-BEC-I}^{\pm(R\to 0)} = -\left|\langle\psi(\mathbf{r})\rangle\right|^2 \frac{\eta_{Molec}}{8\pi} \qquad (1.255)$$

while the *chemical condensate appears forbidden* due to the infinite limits for the bonding and antibonding cases

$$E_{bond-BEC-II}^{\pm(R\to 0)} \to \infty \qquad (1.256)$$

Now, to evaluate the BEC-bonding potential curves for bonding and antibonding energies as provided by the present model, one must recall the repulsive-attractive potential in a diatomic molecule of the form

$$V(r) = \frac{A}{r^n} - \frac{B}{r^m} \ , \ n > m \qquad (1.257)$$

For the equilibrium binding distance R and the energy E_{bond}, one has the system

$$\begin{cases} \left(\dfrac{\partial V(r)}{\partial r}\right)_{r=R} = 0 \\ V(r = R) = E_{bond} \end{cases} \qquad (1.258)$$

with the solution

$$A = \frac{m}{m-n} E_{bond} R^n \ , \ B = \frac{n}{m-n} E_{bond} R^m \qquad (1.259)$$

leading to the so-called Mie equation (Borg & Diens, 1992)

$$V(r) = \frac{E_{bond}}{m-n}\left[m\left(\frac{R}{r}\right)^n - n\left(\frac{R}{r}\right)^m\right] \qquad (1.260)$$

or specialized to the celebrated Lennard-Jones 12–6 variant (Lennard-Jones, 1924, 1931; Barron & Domb, 1955; Smit, 1992; Atkins & de Paula, 2006; Frenkel & Smit, 2002; Zhen & Davies, 1983)

$$V^{12-6}(r) = -E_{bond}\left[6\left(\frac{R}{r}\right)^{12} - 12\left(\frac{R}{r}\right)^{6}\right] \tag{1.261}$$

suitable for modeling H_2 and the inert gas interactions (He-He as well).

For practical implementation, one needs to set a few parameters, namely

- The bosonic order parameter may be, as a first approximation, set to unity, $|\langle\psi(\mathbf{r})\rangle| = 1$, within the Tisza's phenomenological two-fluid model of BEC (Tisza, 1938);
- The molecular chemical hardness may be set equal to that of atomic species entering homopolar binding, according to molecule softness (s) based formulation rooted in the Bratsch atoms-in-molecule approach (Bratsch, 1985)

$$\eta_{AIM} = \left.\frac{\sum_A s_A \eta_A}{\sum_A s_A}\right|_{all\ A\ equal} = \eta_A \tag{1.262}$$

- The equilibrium length R and the atomic ground state energy E_0 are obtained from experimental or accurate quantum mechanical computations;
- The bonding energy E_{bond} is determined by physical and/or chemical BEC formulations for diatomic systems with one or two electrons in the valence shell; hydrogen and helium are the main prototypes.

Molecular hydrogen BEC bonding (thick line) and antibonding (thin line) potentials compared with ordinary or "normal" chemical bonding potentials (dashed line) as plotted from Eq. (1.261) using data presented in Table 1.3, respectively.

Specifically, for atomic systems such as H and He as well as H-H and He-He bindings, the working and calculated values are summarized in Table 1.3. The results presented in Table 1.3 reveal

TABLE 1.3 Atomic and Molecular Data (Number of Electrons in Atomic Valence Shell, Atomic Chemical Hardness η, Atomic Ground-State Energy E_0, Inter-Atomic Homopolar Binding Length with Respect to the First Bohr Radius a_0, Along the Specific Physical and Chemical Bonding BEC Energies) for Hydrogen and Helium Toward Assessing Their BEC-Bonding Lennard-Jones-Type Potential (1.261), All in Atomic Units (Putz, 2012d)

Atoms	N_0	η[a]	E_0	Molecules	R	E_{bond}	E^+_{BEC-I}	E^-_{BEC-I}	E^+_{BEC-II}	E^-_{BEC-II}
H	1	0.237	-0.5	H-H	1.3[c]	-1.324[c]	-0.011	-0.017	-2.054	-1.513
He	2	0.458	-2.9[b]	He-He	5.67	-2.85×10^{-5}[d]	-8.15×10^{-3}	-1.3×10^{-2}	-3.695	-4.072

a) Computed using finite-difference formula $0.5(E_{LUMO}-E_{HOMO})$ for highest occupied and lowest unoccupied molecular orbital (HOMO and LUMO, respectively) energies, spectroscopically determined in electron volts and then converted in atomic units;

b) From Bransden & Joachain (2003); Drake & Yan (1994); Yan & Drake (1995); Drake (1999); Baker et al. (1990); Scott et al. (2007); and Drake (2006);

c) From Condon (1927);

d) Converted into Hartree from the graphical values shown in Figure 1.7, with the conversion factor $x=k_B/E_h = 3.1668114(29)\times10^{-6}$.

interesting features about the physical and chemical bonding BEC at the level of hydrogen and helium systems, such as

- The inversion of the bonding (+) and antibonding (−) states, in the sense that the latter appears more "deeply" energetic than the former, is a sign of the non-bonding of the respective states; this is the case for the *physical bonding BEC(I)* of H_2 as well as for both the physical and chemical bonding BECs (I and II) of He_2, thus confirming that in this sophisticated way the He-He interaction is not possible in the sense of inter-atomic bonding, either in the physical or chemical sense;
- However, the present BEC approach provides correction to chemical bonding of molecular hydrogen (BEC-II), showcasing the correct bonding vs. anti-bonding hierarchy; nevertheless, it reveals the interesting feature of having both of these states below the "normal" H_2 bonding. This is a sign that the BEC of chemical bonding provides an even more stable system and should be further studied at both theoretical and experimental levels to ultimately reconcile the associated 12–6 potentials abstracted from Eq. (1.261) with the data presented in Table 1.3, see Figure 1.7.

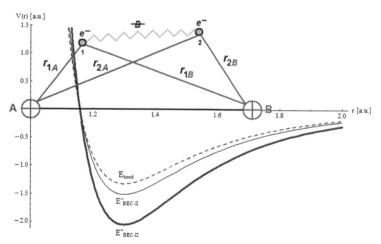

FIGURE 1.7   Illustration of the bi-atomic chemical bond through the bondonic-bosonic contribution at inter-electronic interaction; the model is overlapped with the states of the molecular hydrogen thus treated, with the highlighting of the BEC levels of bonding (thick line) and antibonding (thin line), both placed under the level of "normal" bonding of the molecule in gaseous state, adapted from (Putz, 2012d).

Still, even for the hydrogen molecular system, the difference between the physical and chemical bonding behavior with respect to the bondonic-bosonic condensate emphasizes the specificity of natural law in the chemical realm, which is not necessarily viewed as a reducible physical counterpart. This affirms once more that chemical structure and bonding are special manifestations of nature at the level of atoms and molecules or as the surrealist poet Lucian Blaga writes "an unveiled mystery more mysteries revealed".

1.5   CONCLUSION

Chemical bond, perhaps the greatest challenge in theoretical chemistry, had attempt over the years many inspiring thesis, although none definitive; few of the most preeminent regard the orbitalic based explanation of electronic pairing, in valence shells of atoms and molecules, rooting in the hybridization concept (Pauling, 1931b) then extended to the valence-shell electron-pair repulsion (VSEPR) (Gillespie, 1970). Alternatively, when electronic density in considered the atoms-in-molecule paradigms were formulated through the geometrical partition of forces by Berlin (1951), or in terms of core, bonding, and lone-pair lodges by Daudel (1980), or by the zero local flux in the gradient field of the density $\nabla \rho$ by Bader (1990), until the most recent employment of the chemical action functional in bonding (Putz, 2012a,b).

Yet, all these approaches do not depart significantly from the undulatory nature of electronic motion in bonding, either by direct wave-function consideration or through its probability information in electronic density manifestation (for that is still considered as a condensed – observable version – of the undulatory manifestation of electron).

In other terms, while passing from the Lewis point-like ansatz to the undulatory modeling of electrons in bonding, the reverse passage was still missing in an analytical formulation. Only recently the first attempt was formulated, based on symmetry broken approach of the Schrödinger Lagrangean with the electronegativity-chemical hardness parabolic energy dependency, showing that a systematical quest for the creation of particles from the chemical bonding fields is possible, see the Chapter 3 of the present volume.

treatments of many-atomic molecular systems by means of generalizing DFT-BEC-Hückel theory of chemical bonding is envisaged.

Still, even for the hydrogen molecular system, the difference between the physical and chemical bonding behavior with respect to the bondonic-bosonic condensate emphasizes the specificity of natural law in the chemical realm, which is not necessarily viewed as a reducible physical counterpart.

KEYWORDS

- **adiabatic approximation**
- **Bondonic condensation**
- **bondons**
- **Bose-Einstein condensation (BEC)**
- **chemical field**
- **DFT adapted BEC**
- **Heitler-London**
- **local density approximation**
- **Valence Bond Theory**
- **virial theorem**

REFERENCES

AUTHOR'S MAIN REFERENCES

Putz, M. V. (2012a). *Quantum Theory: Density, Condensation, and Bonding*, Apple Academics, Toronto, Canada.

Putz, M. V. (2012b). *Chemical Orthogonal Spaces*, in Mathematical Chemistry Monographs, Vol. 14, University of Kragujevac.

Putz, M. V. (2012c). From Kohn-Sham to Gross-Pitaevsky equation within Bose-Einstein condensation y-theory. *Int. J. Chem. Model.* 4(1), 1–11.

Putz, M. V. (2012d). Density functional theory of Bose-Einstein condensation: road to chemical bonding quantum condensate. *Structure and Bonding* 149: 1–50 (DOI: 10.1007/978–3–642–32753–7_1)

Putz, M. V. (2011a). Fulfilling the Dirac's promises on quantum chemical bond, In: *Quantum Frontiers of Atoms and Molecules*, Putz, M. V. (Ed.), NOVA Science Inc., New York, Chapter 1, pp.1–20.

Putz, M. V. (2011b). Conceptual density functional theory: from inhomogeneous electronic gas to Bose-Einstein condensates. In: *Chemical Information and Computational Challenges in 21ˢᵗ Century,* Putz, M. V. (Ed.), NOVA Science Publishers, Inc., New York, Chapter 1, pp.1–60.

Putz, M. V. (2011c). Hidden side of chemical bond: the bosonic condensate. In *Advances in Chemistry Research. Volume 10*, Taylor, J. C. (Ed.), NOVA Science Publishers, Inc., New York, Chapter 8, pp. 261–298.

Putz, M. V. (2010a). The bondons: the quantum particles of the chemical bond. *Int. J. Mol. Sci.* 11(11), 4227–4256.

Putz, M. V. (2010b). Beyond quantum nonlocality: chemical bonding field. *Int. J. Environ. Sci.* 1: 25–31.

Putz, M. V. (2003). *Contributions within Density Functional Theory with Applications to Chemical Reactivity Theory and Electronegativity.* Dissertation.com, Parkland.

SPECIFIC REFERENCES

Albus, A. P., Illuminati, F., Wilkens, M. (2003). Ground-state properties of trapped Bose-Fermi mixtures: Role of exchange correlation. *Phys. Rev. A* 67, 063606.

Andersen, J. O. (2004). Theory of the weakly interacting Bose gas. *Rev. Mod. Phys.* 76, 599–639..

Anderson, M. H., Ensher, J. R., Matthews, M. R., Wieman, C. E. Cornell, E. A. (1995). Observation of Bose-Einstein condensation in a dilute atomic vapor. *Science* 269, 198–201.

Argaman, N., Band, Y. B. (2011). Finite-temperature density-functional theory of Bose-Einstein condensates. *Phys. Rev. A* 83, 023612.

Atkins, P., de Paula, J. (2006). *Atkins' Physical Chemistry*, 8ᵗʰ Edn. Oxford University Press, Oxford.

Ayers, P. W., Parr, R. G. (2000). Variational principles for describing chemical reactions: the Fukui function and chemical hardness revisited. *J. Am. Chem. Soc.* 122, 2010–2018.

Bader, R. F. W. (1990). *Atoms in Molecules - A Quantum Theory*, Oxford: Oxford University Press.

Bader, R. F. W. (1994). Principle of stationary action and the definition of a proper open system. *Phys. Rev. B* 49, 13348–13356.

Bader, R. F. W., Gillespie, R. J., MacDougall, P. J. (1988). A physical basis for the VSEPR model of molecular geometry. *J. Am. Chem. Soc.* 110, 7329–7336

Baekelandt, B. G., Cedillo, A., Parr, R. G. (1995). Reactivity indices and fluctuation formulas in density functional theory: Isomorphic ensembles and a new measure of local hardness. *J. Chem. Phys.* 103, 8548–8556.

Bagnato, V., Pritchard, D. E., Kleppner, D. (1987). Bose-Einstein condensation in an external potential. *Phys. Rev. A.* 35, 4354–4358.

Kadanoff, L. P. (2009). More is the same mean field theory and phase transitions. *J. Stat. Phys.* 137, 777–797.

Ketterle, W. (2002). Nobel Lecture: When atoms behave as waves: Bose-Einstein condensation and the atom laser. *Rev. Mod. Phys.* 74, 1131–1151.

Kim, Y. E., Zubarev, A. L. (2003). Density-functional theory of bosons in a trap. *Phys. Rev. A* 67, 015602.

Kleinert H (2003). Five-loop critical temperature shift in weakly interacting homogeneous bose-einstein condensate. *Mod. Phys. Lett. B* 17, 1011.

Kleinert, H., Schmidt, S., Pelster, A. (2004). Reentrant phenomenon in the quantum phase transitions of a gas of bosons trapped in an optical lattice. *Phys Rev. Lett.* 93, 160402.

Kohn, W., Sham, L. J. (1965). Self-consistent equations including exchange and correlation effects. *Phys. Rev.* 140, 1133–1138.

Langmuir, I. (1919). The arrangement of electrons in atoms and molecules. *J. Am. Chem. Soc.* 41, 868–934.

Leggett, A. J. (2001). Bose-Einstein condensation in the alkali gases: some fundamental concepts. *Rev. Mod. Phys.* 73, 307–356.

Lennard-Jones, J. E. (1924). On the Determination of Molecular Fields. II. From the Equation of State of a Gas. *Proc. R. Soc. Lond. A* 106 (738), 463–477.

Lennard-Jones, J. E. (1931). Cohesion. *Proc Phys Soc.* 43(5), 461–482.

Lewis, G. N. The atom and the molecule. *J. Am. Chem. Soc.* (1916). 38, 762–785.

Licker, M. J. (2004). *McGraw-Hill Concise Encyclopedia of Chemistry*, McGraw-Hill, New York.

Lima, A. R. P., Pelster, A. (2011). Quantum fluctuations in dipolar Bose gases. *Phys. Rev. A* 84: 041604(R.)

London, F. (1938). On the Bose-Einstein condensation. *Phys. Rev.* 54, 947–954.

Löwdin, P. O. (1955a). Quantum theory of many-particle systems. I. Physical interpretations by means of density matrices, natural spin-orbitals, and convergence problems in the method of configurational interaction. *Phys. Rev.* 97, 1474–1489.

Löwdin, P. O. (1955b). Quantum theory of many-particle systems. II. Study of the ordinary Hartree-Fock approximation. *Phys. Rev.* 97, 1474–1489.

Löwdin, P. O. (1955c). Quantum theory of many-particle systems. III. Extension of the Hartree-Fock scheme to include degenerate systems and correlation effects. *Phys. Rev.* 97, 1509–1520.

Maggiora, G. M., Mezey, P. G. (1999). A fuzzy-set approach to functional-group comparisons based on an asymmetric similarity measure. *Int. J. Quantum Chem.* 74, 503–514.

March, N. H. *Electron Density Theory of Atoms and Molecules*. Academic Press, NY, 1992.

Mezey, P. G. (1993). *Shape in Chemistry: An Introduction to Molecular Shape and Topology*, VCH Publishers: New York.

Mortier, W. J., Genechten, K. V., Gasteiger, J. (1985). Electronegativity equalization: application and parametrization. *J. Am. Chem. Soc.* 107, 829–835.

Mulliken, R. S., Riecke, C. A., Orloff, D., Orloff, H. (1949). Formulas and Numerical Tables for Overlap Integrals. *J. Chem. Phys.* 17(12), 1248.

Murrel, J. N., Kettle, S. F. A., Tedder, J. M. (1985). *The Chemical Bond* (2nd ed.). John Wiley & Sons.

Nunes, G. S. (1999). Density functional theory of the inhomogeneous Bose–Einstein condensate. *J. Phys. B: At. Mol. Opt. Phys.* 32, 4293–4299.

Oelke, W. C. (1969). *Laboratory Physical Chemistry*, Van Nostrand Reinhold Company, New York.

Pariser, R., Parr, R. (1953a). A semi-empirical theory of the electronic spectra and electronic structure of complex unsaturated molecules. I. *J. Chem. Phys.* 21, 466–471.

Pariser, R., Parr, R. (1953b). A semi-empirical theory of the electronic spectra and electronic structure of complex unsaturated molecules. II. *J. Chem. Phys.* 21, 767–776.

Parr, R. G. (1972). *The Quantum Theory of Molecular Electronic Structure*, W. A. Benjamin Inc., Reading-Massachusetts.

Parr, R. G., Donnelly, R. A., Levy, M., Palke, W. E. (1978). Electronegativity: the density functional viewpoint. *J. Chem. Phys.* 68, 3801–3808.

Parr, R. G., Yang, W. (1989). *Density Functional Theory of Atoms and Molecules*. Oxford University Press, New York.

Pauling, L. (1931b). The nature of the chemical bond. I. Application of results obtained from the quantum mechanics and from a theory of paramagnetic susceptibility to the structure of molecules. *J. Am. Chem. Soc.* 53, 1367–1400.

Pauling, L. (1931c). The nature of the chemical bond II. The one-electron bond and the three-electron bond, *J. Am. Chem. Soc.* 53, 3225–3237.

Pauling, L. (1960). *The Nature of the Chemical Bond*, Cornell University Press, Ithaca (NY).

Pauling, L. Quantum mechanics and the chemical bond. *Phys. Rev.* (1931a). 37, 1185–1186.

Pitaevsky, L., Stringari, S. (2003). *Bose-Einstein Condensation*, Clarendon Press, Oxford.

Pitaevsky, L. P. (1961). Vortex lines in an imperfect Bose gas. *Zh. Eksp. Teor. Fiz.* 40, 646–651 [(1961). *Sov. Phys. JETP* 13, 451–454].

Pople, J. A. (1953). Electron interaction in unsaturated hydrocarbons. *Trans. Faraday Soc.* 49, 1375–1385.

Pople, J. A., Binkley, J. S., Seeger, R. (1976). Theoretical models incorporating electron correlation. *Int. J. Quantum Chem.* 10, 1–19.

Roberts, J. L., Jaffe, H. (1957). Some Overlap Integrals Involving d Orbitals. II. *J. Chem. Phys.* 27, 883–886.

Roothaan, C. C. J. (1951). New developments in molecular orbital theory. *Rev. Mod. Phys.* 23, 69–89.

Sanderson, R. T. (1988). Principles of electronegativity Part, I. General nature. *J. Chem. Educ.* 65, 112–119.

Schrödinger, E. (1926). An undulatory theory of the mechanics of atoms and molecules. *Phys. Rev.* 28, 1049–1070.

Scott, T. C., Lüchow, A., Bressanini, D., Morgan, J. D. III (2007). Nodal surfaces of helium atom Eigen functions. *Phys. Rev. A* 75: 060101(R).

Sen, K. D., Jørgensen, C. K. (Eds.) (1987). Electronegativity, *Structure and Bonding*, Springer Verlag Berlin, Vol. 66.

Shaik, S. S., Hiberty, P. C. (2008). *A Chemist's Guide to Valence Bond Theory*. Wiley-Interscience, New Jersey.

Slater, J. C. (1928). The self consistent field and the structure of atoms. *Phys. Rev.* 32, 339–348.

Slater, J. C. (1929). The theory of complex spectra. *Phys. Rev.* 34, 1293–1322.

Smit, B. (1992). Phase diagrams of Lennard–Jones fluids. J. Chem. Phys. 96, 8639–8640.

Stevenson, P. M. (1981). Optimized perturbation theory. *Phys. Rev. D.* 23, 2916–2944.

Stringari, S. (1996). Collective excitations of a trapped bose-condensed gas. *Phys. Rev. Lett.* 77, 2360–2363.

Szekeres, Z., Exner, T., Mezey, P. G. (2005). Fuzzy fragment selection strategies, basis set dependence and HF–DFT comparisons in the applications of the ADMA method of macromolecular quantum chemistry. *Int. J. Quantum Chem.* 104, 847–860.

Thomson, J. J. On the structure of the molecule and chemical combination. *Philos. Mag.* (1921). 41, 510–538.

Tisza, L. (1938). Transport Phenomena in Helium II. *Nature (Lond)* 141, 913.

Vetter, A. (1997). *Density Functional Theory for BEC* (in German), Thesis, Institute of Theoretical Physics, Bayerische Julius-Maximilians University Würzburg.

Whitney, C. K. (2007). Relativistic dynamics in basic chemistry. *Found. Phys.* 37, 788–812.

Yan, Z.-C., Drake, G. W. F. (1995). High Precision Calculation of Fine Structure Splittings in Helium and He-Like Ions. *Phys. Rev. Lett.* 74(24), 4791–4794.

Yi, S., You, L. (2001). Trapped condensates of atoms with dipole interactions. *Phys. Rev. A* 63, 053607.

Yukalov, V. I. (2004). Principal problems in Bose-Einstein condensation of dilute gases. *Laser Phys. Lett.* 1, 435–461.

Zhen, S., Davies, G. J. (1983). Calculation of the Lennard-Jones n–m potential energy parameters for metals. *Phys. Status Solidi A* 78, 595–605.

CHAPTER 2

MOLECULAR STRUCTURE BY QUANTUM CHEMISTRY

CONTENTS

ABSTRACT

The chemical bond is here modeled by employing the geometrical skeleton of the binding atoms in molecule, i.e., by the generated symmetry of molecule at its turn imposing the eigen-energetic terms associated and directly correlated with the symmetry basic elements and operations, further quantitatively realized/quantifies by allied matrices – elements of the symmetry group to which they belong; the quantum framework is then completed by means of the superposition principles applied to joint the symmetry of a given molecule with the overall eigen-energies, eventually related with the atomic orbitals, and of their repulsive and attractive influences in the crystal field and ligand field theories, respectively, toward considering the electronic pair at the lance shell as the main driving quantum entity in bonding, being this merely associate with a supra-quantum (geometrical) realization of chemical bond, thus complementing at the semi-classical level the previous developed theory of chemical bonding by bondons (by pairing electrons as bosons, with sub-quantum attractive behavior – see Chapter 1).

2.1   INTRODUCTION

The impact of the quantum view upon the nature of the chemical bond was considerable, as it offered both qualitatively and quantitatively a scheme of structure analysis together with the chemical-physical transformations, being accurately confirmed by both the computational and the experimental expertise in all branches of chemistry, and whenever the electronic structure is about. As a consequence, the chapters of the structural physical chemistry should be classified within an *intensive, localization, and reactivity levels* of chemical bonding assessment (Allendoerfer, 1990; Boeyens, 1995; Breneman, 1988; Novak, 1999; Sen & al. 1993; Summerfield et al., 1999; Page et al., 1999; Ringe & Petsko, 1999; Bendazzoli, 1992; Buckingham &

Rowlands, 1991; Dias, 1989; Foresman, 1997; Gallup, 1988; Gaines & Page, 1980; Klein & Trinajstić, 1990; Mebane et al., 1999; Morrison et al., 1993; Nordholm, 1988; Sannigrahi & Kar, 1988; Shusterman & Shusterman, 1997; Simons & Smith, 2000; Putz & Chiriac, 2008).

At the intensive level it was established that, for an adequate treatment in the quantum space of the polyatomic combinations, the electronic density $\rho(r)$ rather than the already historical wave function $\psi(r_1,...r_N)$ stays as the main variable for a system with N electrons. This is because, contrary to the wave function, the electronic density is an experimental detectable quantity, defined in the real three dimensional space, and not within a $3N$ Hilbert abstract one, being also directly related to the total number of electrons in the concerned system through the functional relation $N=\int\rho$.

However, since the quantum existence of the atoms in molecules represents the key to chemical bonding description depending on how much of the individuality of an atom is preserved and how much of it is transferred to the bond, the localization level appears as a compulsory next stage in chemical bond characterization. In this context the idea of molecular partitioning in terms of the domains of stability of molecular electronic density was advanced.

The results consist in the emergence of the so called *atomic basins* that include all the atomic nuclei but also their interspaces until the surface delimited by fulfillment of the zero flux condition of electronic density $(\nabla\rho(r)\cdot\vec{n}=0)$ according with Bader and co-workers' atoms-in-molecules approaches.

Nevertheless, the electronic localization complements, at the local level, the quantum information comprised in the reactivity indices, being ultimately described through the so called *localization functions*. These should express the balance between the local stability and the delocalization tendency of the involved electrons in the chemical bond in the view of the forthcoming transformations.

So, the localization functions indicate the ratio of the non-uniformly localized electronic distribution to the uniform delocalization of the electronic gas, accordingly with the Heisenberg quantum principle of delocalization and that of the Pauli indiscernibility. It was however proved, through a series of hydracids molecules that the atoms-in-molecule electronic localization function in its exponential form and with error interpretation, as recently recommended by the authors' recipe, see the Volume II of the present five-volume book, stands as a viable quantum tool for identifying bonds within the bonding space.

On the other way, at the global level, the reactivity indices' studies are essential for indicating the propensity of a multielectronic system to participate into a chemical reaction. At the molecular level, these indices are defined so as to quantitatively measure the chemical reactivity, while at the biomolecular level they are associated with the biological activity, being the best candidates to be correlated in the context of quantitative structure activity (property) relationships, QSA(P)Rs, see Figure 2.1 and the Volume V of the present five-volume book set.

Thus, since the reactivity indices are placed at the informational interface between the electronic systems' stability and their tendency to transform and combine they are mathematically introduced as the integral functions of the electronic density function, releasing the so called *electronic density functionals* as the efficient tool for the global prediction of the electronic properties of the investigated nanosystems.

In this respect the electronegativity and chemical hardness seems to provide the minimal set of descriptors to be considered for characterizing

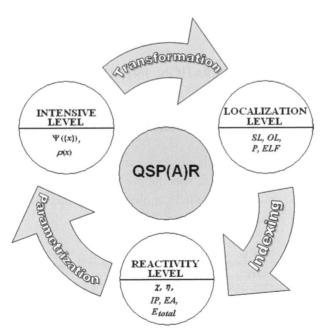

FIGURE 2.1   Synopsis of the intensive, local, and global levels of quantum chemistry and their inter-relation respecting the chemical bond characterization towards QSP(A)R (quantitative structure-property(activity) relationship] modeling the chemical and chemical-biological complex interaction; after (Putz & Chiriac, 2008).

the bond involvements in chemical reactions, see Chapters 3 and 4 of the present Volume. Such an analysis was also performed on specific hard-and-soft-acids-and-bases reactions characteristic for the hydracids considered.

However, all above phenomenologically identified levels of quantum nature of the chemical bond can be analytically or computationally inter-related as well through particular QSA(P)R models, see Figure 2.1. This way the inner circle of quantum chemistry is opened to include application to biological interaction in a unitary manner.

Overall, searching for the unity of the manifestation forms of the chemical bonding at various levels of mater organization has become a very active interdisciplinary field in the last years, being one of the main goals in the frame of the nanosciences. The quantum paradigm of bonding unification through the formulation of a minimal set of concepts and quantities having as much universal multielectronic relevance as possible represents a real challenge for the conceptualization and prescription of the viable applicative directions of the nanosystems, from atoms to bio-molecules in the years to come.

2.2   INTER-ATOMIC VIBRATION, INTERACTION, AND BONDING LOCALIZATION

2.2.1   HARMONIC OSCILLATOR IN BIATOMIC (H2) MOLECULE

A direct application of the Schödinger equation consists in determining the harmonic oscillator spectrum, specific for the biatomic molecular vibrations (with the typical case of the H_2 molecule). For this case, the quantum basic equation is expressed as (Putz, 2006):

$$\frac{d^2\psi}{dx^2} + \frac{2m}{\hbar^2}\left(E - \frac{1}{2}m\omega^2 x^2\right)\psi = 0 \qquad (2.1)$$

However, although we will not directly solve this equation, we will determine the solutions by the test functions method. Thus, based on the stationary wave functions properties, to be continuous, derivable and tend to zero when the variable tends to infinite, one will "try" the wave function with the right form, which corresponds to the first vibration mode:

$$\psi_0 = e^{-\alpha x^2}, \ \alpha = \frac{m\omega}{2\hbar} \tag{2.2}$$

Then, by effectuating the first derivative

$$\frac{d\psi_0}{dx} = -2x\alpha e^{-\alpha x^2} \tag{2.3}$$

and the second spatial derivative, respectively

$$\frac{d^2\psi_0}{dx^2} = -2\alpha e^{-\alpha x^2} + 4\alpha^2 x^2 e^{-\alpha x^2}$$

$$= -\frac{2m\omega}{2\hbar} e^{-\alpha x^2} + 4\frac{m^2\omega^2}{4\hbar^2} x^2 e^{-\alpha x^2}$$

$$= -\frac{2m}{\hbar^2}\left(\frac{1}{2}\hbar\omega - \frac{1}{2}m\omega^2 x^2\right)\psi_0 \tag{2.4}$$

the identity results:

$$\frac{d^2\psi_0}{dx^2} + \frac{2m}{\hbar^2}\left(\frac{1}{2}\hbar\omega - \frac{1}{2}m\omega^2 x^2\right)\psi_0 = 0 \tag{2.5}$$

which implies, by comparison with the basic equation (2.1), the value of fundamental energy,

$$E_0 = \frac{1}{2}\hbar\omega \tag{2.6}$$

The quality of the absolute energetic minimum resides in the fact that the wave function associated with Eq. (2.2) has no nodes, it is not annulated anywhere in the space, thus corresponding to the fundamental energy of the quantum harmonic oscillator.

Further considering the wave function on the first excited state

$$\psi_1 = xe^{-\alpha x^2} \tag{2.7}$$

there is observed that *at x = 0* this present a node, a zero value. Repeating the above operations, also in this case is got an identity of type (2.1)

$$\frac{d^2\psi_1}{dx^2} + \frac{2m}{\hbar^2}\left(\frac{3}{2}\hbar\omega - \frac{1}{2}m\omega^2 x^2\right)\psi_1 = 0 \tag{2.8}$$

wherefrom follows the energy of the first excited state, under the form

$$E_1 = \frac{3}{2}\hbar\omega \tag{2.9}$$

By induction, there is relatively easy to show that the general form of the *eigen*-solutions of the quantum harmonic oscillator, satisfies the spectral shape

$$E_n = \left(n_v + \frac{1}{2}\right)\hbar\omega = \left(n_v + \frac{1}{2}\right)h\nu \,, n_v \in \mathbb{N} \tag{2.10}$$

quantified by the vibrational quantum number, n_v, with natural values.

2.2.2 VAN DER WAALS BIATOMIC (HE-HE) INTERACTION

A very useful application of the formula above of the energy of the quantum oscillator resides in calculation of the energy of the weak intramolecular van der Walls interaction, specific of the biatomic molecules of noble gases (with the typical case of the He_2 molecule). Assuming a unidimensional model, or two atoms of Helium, each of them with the two valence electrons (say nos. 1 & 2), oscillating in the nuclei positive potential field. Then, the potential energy exerted by a system over the other (the nuclei are considered as fixed and separated at the R distance) will be (Putz, 2006):

$$V = \frac{e^2}{R} + \frac{e^2}{R+x_1+x_2} - \frac{e^2}{R+x_1} - \frac{e^2}{R+x_2} \tag{2.11}$$

The four interaction terms (2.11) can be compactly rewritten,

$$V = \frac{e^2}{R}\left[1 + \frac{1}{1+\frac{x_1+x_2}{R}} - \frac{1}{1+\frac{x_1}{R}} - \frac{1}{1+\frac{x_2}{R}}\right] \tag{2.12}$$

so for the electronic distance close to the nuclear centers to which the electrons belong, $x_{1,2} \ll R$, can be developed in series the truncated second orders potential (2.11):

$$V \cong \frac{e^2}{R}\left[1+\left(1-\frac{x_1+x_2}{R}+\left(\frac{x_1+x_2}{R}\right)^2\right)-\left(1-\frac{x_1}{R}+\frac{x_1^2}{R^2}\right)-\left(1-\frac{x_2}{R}+\frac{x_2^2}{R^2}\right)\right]$$

$$=\frac{e^2}{R}\left[\frac{2x_1x_2}{R^2}\right]=\frac{2e^2x_1x_2}{R^3}$$

(2.13)

In general, in addition to the inter-atomic potential (2.13), will also appear the intra-atomic forces, exerted by the nuclei on the electrons from the belonging atoms. To evaluate the intra-atomic associated potential, will be used the shape of the wave functions (2.2) for the energy of the electrons in the fundamental vibrational quantum state, here rewritten as:

$$\psi_0 = \exp\left(-\alpha'x^2/2\right), \ \alpha'=2\alpha=\frac{4\pi^2 m v_0}{h}$$

(2.14)

Then, in the electronic states, characterized by wave function (2.14) the vibrational deviation (amplitude) is calculated through the average values formula of the observable (by quantum averaging) postulate of the quantum mechanics:

$$\overline{\delta x^2} = \overline{x}^2 = \frac{\int\limits_{-\infty}^{+\infty} x^2\psi_0^2 dx}{\int\limits_{-\infty}^{+\infty} \psi_0^2 dx} = \frac{\int\limits_{-\infty}^{+\infty} x^2 e^{-\alpha'x^2} dx}{\int\limits_{-\infty}^{+\infty} e^{-\alpha'x^2} dx} = \frac{1}{2\alpha'}$$

(2.15)

However, as based on the deviation (2.15) the intra-atomic potential of the quantum vibration has the form $e^2 x_1^2 / 2\alpha'$ and $e^2 x_2^2 / 2\alpha'$, for the electrons 1 and 2, respectively.

Finally, the total potential of the electrons-nuclei system from a He atom, in the electron-nucleus field of the other He atom, has the form:

$$V = \frac{1}{2m}\left(P_1^2 + P_2^2\right) + \frac{e^2}{2\alpha'}\left(x_1^2 + x_2^2\right) + \frac{2e^2 x_1 x_2}{R^3}$$

(2.16)

where P_1 and P_2 are the impulses of the electrons from a He atom, the term

$$\frac{e^2}{2\alpha'}\left(x_1^2 + x_2^2\right)$$ (2.17)

corresponds to the dipole energy, while the term

$$\frac{2e^2 x_1 x_2}{R^3}$$ (2.18)

recording the dipole–dipole interaction, between the two systems of He. In addition, while based on the symmetry of the He-He system, when a variable change of the type is considered

$$\begin{cases} x_b = \dfrac{1}{\sqrt{2}}(x_1 + x_2) & P_b = \dfrac{1}{\sqrt{2}}(P_1 + P_2) \\ x_a = \dfrac{1}{\sqrt{2}}(x_1 - x_2) & P_a = \dfrac{1}{\sqrt{2}}(P_1 - P_2) \end{cases}$$ (2.19)

the energy variables (2.16) in the new homogeneous coordinates look like:

$$\begin{cases} x_1 = \dfrac{1}{\sqrt{2}}(x_a + x_b) & P_1 = \dfrac{1}{\sqrt{2}}(P_a + P_b) \\ x_2 = \dfrac{1}{\sqrt{2}}(x_b - x_a) & P_2 = \dfrac{1}{\sqrt{2}}(P_b - P_a) \end{cases}$$ (2.20)

Then, the interaction energy (2.16) will be successively written:

$$\begin{aligned} V &= \frac{1}{2m}\left[\frac{P_a^2 + P_b^2 + 2P_a P_b}{2} + \frac{P_a^2 + P_b^2 - 2P_a P_b}{2}\right] \\ &+ \frac{e^2}{2\alpha'}\left[\frac{x_a^2 + x_b^2 + 2x_a x_b}{2} + \frac{x_a^2 + x_b^2 - 2x_a x_b}{2}\right] + \frac{2e^2}{R^3}\frac{x_b^2 - x_a^2}{2} \\ &= \frac{1}{2m}\left(P_a^2 + P_b^2\right) + \frac{e^2}{2\alpha'}\left(x_a^2 + x_b^2\right) + \frac{e^2}{R^3}\left(x_b^2 - x_a^2\right) \\ &= \left[\frac{1}{2m}P_b^2 + \left(\frac{e^2}{2\alpha'} + \frac{e^2}{R^3}\right)x_b^2\right] + \left[\frac{1}{2m}P_a^2 + \left(\frac{e^2}{2\alpha'} - \frac{e^2}{R^3}\right)x_a^2\right] \end{aligned}$$ (2.21)

In the form (2.21) one recognizes the total energy as the sum of two vibrational energies corresponding of two non-coupled harmonic oscillators, from where it can be extracted the frequencies of associated vibration:

$$v_a = \frac{1}{2\pi}\sqrt{\frac{e^2}{m}\left(\frac{1}{\alpha'}-\frac{2}{R^3}\right)}, \quad v_b = \frac{1}{2\pi}\sqrt{\frac{e^2}{m}\left(\frac{1}{\alpha'}+\frac{2}{R^3}\right)} \tag{2.22}$$

Based on the frequency (2.22) it can be immediately written the energetic spectrum associated to the vibrational motion, as (2.10)

$$E_{n_{v(a,b)}} = hv_{(a,b)}\left(n_{v(a,b)}+\frac{1}{2}\right) \tag{2.23}$$

This way results the fundamental electronic vibrational energy He-He:

$$E_0 = \frac{1}{2}h(v_b+v_a) = \frac{h}{4\pi}\left[\sqrt{\frac{e^2}{m}\left(\frac{1}{\alpha'}+\frac{2}{R^3}\right)}+\sqrt{\frac{e^2}{m}\left(\frac{1}{\alpha'}-\frac{2}{R^3}\right)}\right]$$

$$= \frac{h}{4\pi}\sqrt{\frac{e^2}{m\alpha'}}\left[\sqrt{1+\frac{2\alpha'}{R^3}}+\sqrt{1-\frac{2\alpha'}{R^3}}\right]$$

$$\cong \frac{h}{2\pi}\sqrt{\frac{e^2}{m\alpha'}}\left[1-\frac{1}{2}\frac{\alpha'^2}{R^6}\right] \tag{2.24}$$

where it had been used the developments in the truncated series:

$$(1+x)^a \cong 1+ax+\frac{a(a-1)}{2}x^2+... \tag{2.25}$$

$$(1-x)^a \cong 1-ax+\frac{a(a-1)}{2}x^2+... \tag{2.26}$$

$$(1+x)^{-a} \cong 1-ax+\frac{a(a+1)}{2}x^2+... \tag{2.27}$$

$$(1-x)^{-a} \cong 1+ax+\frac{a(a+1)}{2}x^2+.... \tag{2.28}$$

From Eq. (2.24) it can be noted the presence of the additional term, appeared thanks to the electronic oscillations of the inverse sixth power order of the internuclear distance, the negative terms, typically attractive for the van der Waals interaction.

2.2.3 ORBITAL VS. DENSITY ELECTRONIC LOCALIZATION IN BONDING

Despite the fact that Hartree-Fock or Kohn-Sham self-consistent field (*SCF*) equations provide in principle the complete set of electronic orbitals that describe the multi-electronic-poly-center bonds, their main drawback is that of providing the delocalized description over an entire molecular space. Such an analysis has to be accomplished with special techniques through which the localized orbitals and localized chemical bond are to be recovered (Coulson&Longuet-Higgins,Julg,1967;Daudeletal.,1983;Feynman,1939; Schwinger, 1951; Rüdenberg, 1951; Edmiston & Rüdenberg, 1963; Clintonetal.,1969;Clinton&Massa,1972;Clintonetal.,1973;Frishberg&Massa, 1981; Massa et al., 1985; Henderson & Zimmerman, 1976; Harriman, 1978; Harriman, 1979; Harriman, 1983, 1984; Casida & Harriman, 1986; Harriman, 1986; Malmqvist, 1986; Koga & Umeyama, 1986; Velders & Feil, 1993; Kryachko et al., 1987; Savin et al., 1983; Gangi & Bader, 1971; Srebrenik & Bader, 1975; Srebrenik et al., 1978; Bader et al., 1979; Bader & Essén, 1984; Bader et al., 1987; Bader & Becker, 1988; Bader, 1990; Bader & Heard, 1999; Bader et al., 2000; Bader, 2001; Cassam-Chenaï & Jayatilaka, 2002; Bader, 2002, 2003; Becke & Edgecombe, 1990; Schmidera & Becke, 2002; Berski et al., 2003; Kohout et al., 2004; Nesbet, 2002; Putz & Chiriac, 2008). Only this way can quantum mechanics provide a viable rationale, i.e., quantum chemistry, in chemical bond characterization. Nevertheless, such a rationale can be achieved in two ways: one of them involves the orbital transformation producing the localized set of orbitals and indices, the present approach; the other one, based on electronic density, includes the electronic density, to a certain degree, into an *electronic localization (super) function* (ELF) so as to generate a local, analytical indication of the electronic pair of the chemical bond, see the Volume II of the present five-volume work. In what follows, we are going to outline the first of these major approaches.

Let's assume that, for instance, after the Hartree-Fock *SCF* computation is undertaken the set of *canonical molecular one-electronic orbitals* are determined so that the Slater determinantal wave function for $2N$

electrons with N doubly occupied orthonormal real orbitals $\chi_1, \chi_2, ..., \chi_N$ is laid down (Ponec, 1982):

$$X = \frac{1}{\sqrt{2N!}} \left| (\chi_1 \alpha)^{(1)} (\chi_1 \beta)^{(2)} ... (\chi_N \alpha)^{(2N-1)} (\chi_N \beta)^{(2N)} \right| \qquad (2.29)$$

It has to be transformed into the corresponding localization wave function

$$\Lambda = \frac{1}{\sqrt{2N!}} \left| (\lambda_1 \alpha)^{(1)} (\lambda_1 \beta)^{(2)} ... (\lambda_N \alpha)^{(2N-1)} (\lambda_N \beta)^{(2N)} \right| \qquad (2.30)$$

on the basis of the *strictly localized orbitals* $\lambda_1, \lambda_2, ..., \lambda_N$.

Fortunately, such orbital transformation is allowed by the flexibility carried by the determinantal wave function respecting a unitary transformation that takes the canonical into localized orbitals through the matrix $(\mathbf{T})$

$$\lambda_i = \chi_j T_{ij} \qquad (2.31)$$

In order to complete the localization picture some "physical" criterion has to be assumed in order that matrix $(\mathbf{T})$ to be determined. Such constraints may refer to the maximization of the distance between the electrons of the *spatially non-localized* orbitals, i.e., the so called Boys condition (Chakraborty, 2007):

$$SL = \sum_i \langle \chi_i | r | \chi_i \rangle^2 \rightarrow \max \qquad (2.32)$$

in terms of χ_i or, reversely, asking that the energy interaction that equally appears in Coulomb and exchange terms be minimized in localized orbitals, i.e., the Edmiston-Reudenberg, or *orbital localization*, condition (Roothan, 1960)

$$OL = \sum_i \langle \chi_i^2 | r^{-1} | \chi_i^2 \rangle \rightarrow \min \qquad (2.33)$$

for orbitals χ_i^2, where we have least localization.

When employing the last condition one firstly gets:

$$0 = \delta OL = 4 \sum_i \langle \chi_i^2 | r^{-1} | \chi_i \delta \chi_i \rangle \qquad (2.34)$$

Next, remembering that a change in delocalized towards localized orbitals follows the orthogonal transformation rules:

$$\chi_n + \delta\chi_n = \sum_i \chi_i T_{in} \tag{2.35}$$

$$\sum_n T_{in} T_{jn} = \delta_{ij} \tag{2.36}$$

Then, assuming that

$$T_{ij} = \delta_{ij} + t_{ij} \tag{2.37}$$

the relations (2.35) and (2.36) lead to the first order connections:

$$\delta\chi_n = \sum_i \chi_i t_{in} \tag{2.38}$$

$$t_{in} + t_{ni} = 0 \tag{2.39}$$

respectively.

It is worth noting that the condition (2.39) accounts for the anti-symmetries of the spatial orbitals in fulfilling Pauli principle. With these, the least localization principle (2.34) can be successively written as:

$$0 = \delta OL = 4\sum_{ni} \left\langle \chi_i^2 \left| r^{-1} \right| \chi_i \chi_n \right\rangle t_{ni}$$
$$= 4\sum_{n>m} \left\{ \left\langle \chi_n^2 \left| r^{-1} \right| \chi_m \chi_n \right\rangle - \left\langle \chi_m^2 \left| r^{-1} \right| \chi_m \chi_n \right\rangle \right\} t_{mn} \tag{2.40}$$

leaving with the delocalization orbital condition:

$$\left\langle \chi_n^2 \left| r^{-1} \right| \chi_m \chi_n \right\rangle = \left\langle \chi_m^2 \left| r^{-1} \right| \chi_m \chi_n \right\rangle \tag{2.41}$$

that, nevertheless is identical in nature to that associated with the localized orbitals:

$$\left\langle \lambda_n^2 \left| r^{-1} \right| \lambda_m \lambda_n \right\rangle = \left\langle \lambda_m^2 \left| r^{-1} \right| \lambda_m \lambda_n \right\rangle \tag{2.42}$$

due to the fact that the optimum condition $\delta OL = 0$ do not distinguish between the minimal and maximum constraints, respectively.

Yet, such dual behavior of localization measure OL represents the quantum mechanical basis for *hybridization* and *bonding*. For instance, one can check that while the simple set of $\{\chi_A, \chi_B\}$ orbitals provide the interaction measure

$$OL_I = \left\langle \chi_A^2 \left| r^{-1} \right| \chi_A^2 \right\rangle + \left\langle \chi_B^2 \left| r^{-1} \right| \chi_B^2 \right\rangle \qquad (2.43)$$

once the linear combination between them is considered, namely $\{(\chi_A + \chi_B)/\sqrt{2}$ and $(\chi_A - \chi_B)/\sqrt{2}\}$, they provide the interaction augmented measure

$$OL_{II} = \frac{1}{2}OL_I + \left\langle \chi_A^2 \left| r^{-1} \right| \chi_B^2 \right\rangle + 2\left\langle \chi_A \chi_B \left| r^{-1} \right| \chi_A \chi_B \right\rangle \qquad (2.44)$$

Now, the difference between the terms OL_I and OL_{II} can be easily visualized since the pure covalent homo-orbitals case is consider giving:

$$OL_{II}^{cov} = 2OL_I^{cov} = 4\left\langle \chi_A^2 \left| r^{-1} \right| \chi_A^2 \right\rangle \qquad (2.45)$$

The lesson is clear: the hybridized orbitals have provided the maximum localization measure whereas the simple pair of orbitals associates with minimum localization measure-maximum delocalization behavior. This way, the old concept of Pauling regarding hybridization is quantum mechanically restored through localization recipe. As well the bonding and anti-bonding concepts of Coulson find here their full power of interpretation in the light of symmetrical/anti-symmetrical spatial orbitals that contribute to localization of chemical bonding.

The exposed localization aspects were refined over the last 60 years through all available quantum mechanical scheme of computation: from joining with semiempirical schemes (Coulson, 1939; Coulson & Longuet-Higgins, 1947; Berlin, 1951; Roothaan, 1991; Rüdenberg, 1951; Mulliken, 1955; Kolos et al., 1960), unitary transformations of operators and bases (Löwdin, 1954, 1955, 1960; Clinton et al., 1969; Clinton et al., 1973; Frishberg & Massa, 1981; Massa et al., 1985; Henderson & Zimmerman, 1976; Harriman, 1978; Harriman, 1979; Harriman, 1983, 1984; Harriman, 1986; Malmqvist, 1986; Koga & Umeyama, 1986; Kryachko et al., 1987) until the most accurate population analyses (Mulliken, 1955; Clinton & Massa, 1972; Casida & Harriman, 1986; Velders & Feil, 1993; Hirshfeld & Rzotkiewics, 1974).

In this respect it is worth mentioning that the Mulliken population analysis produces an alternative way of looking at chemical bond in terms of charge localization by means of the sum of partial populations that participate in the i-th bond, core, or lone pair,

$$P = \sum_i P_i \qquad (2.46)$$

with

$$P_i = \sum_{(v,\mu) \subset \Gamma_i} C_{iv} C_{i\mu} S_{v\mu} \qquad (2.47)$$

when the bonds are recognized (by chemical intuition or by preliminary simplified quantum analysis, for example, Hückel analysis) from the beginning in order to define the sets Γ_i of bonding atomic orbitals.

Finally, we need to briefly discuss the way the cornerstone chemical concept of *valence* can be equally recovered by means of localization procedure. At this point the developed theory stands as the *pseudo-potential formalism* (Pickett, 1989) since its main purpose is to provide the *valence-only* theory for atoms and molecules. In fact, the pseudo-potential techniques aim to substitute the Pauli Exclusion Principle with specific operators and potentials. The main advantage relays on the reduced number of orthogonal conditions, namely those related with valence (say χ^{val}) and core orbitals (say $\{\chi_i^{core}\}$) through the pseudo-potential wave function or pseudo-orbital (*PO*) ϑ:

$$\vartheta = \chi^{val} + \sum_i \alpha_i \chi_i^{core} \qquad (2.48)$$

with

$$\alpha_i = \left\langle \chi_i^{core} \middle| \vartheta \right\rangle \qquad (2.49)$$

which is widely recognized as being of Gram-Schmidt orthogonalization type (Daudel et al., 1983).

Formally, it appears that the valence orbital is localized respecting the rest of core orbitals, and one can assumes the basic Schrödinger equation exclusively for it, so that as all other core orbitals would not exist, i.e., dividing the intrinsic eigen-problem into two, possible disjoint, regions associated with valence:

$$\widehat{H} \chi^{val} = \varepsilon^{val} \chi^{val} \qquad (2.50)$$

and core

$$\widehat{H}\chi_i^{core} = \varepsilon_i \chi_i^{core} \tag{2.51}$$

orbitals.

However, this localization is achieved through the pseudo-orbitals of above type. This way the valence orbitals are localized once the pseudo-orbital is determined and the core orbitals are properly subtracted from it. That is, the pseudo-orbital eigen-equation has to be solved, namely:

$$\left(\widehat{H} + \widehat{V}_{PP}\right)\vartheta = \varepsilon^{val}\vartheta \tag{2.52}$$

where the so called Phillips-Kleinman (PK) pseudo-potential (PP) (Phillips & Kleinman, 1959, 1960)

$$\widehat{V}_{PP} = \sum_i \alpha_i \frac{\left(\varepsilon^{val} - \varepsilon_i\right)\chi_i^{core}}{\vartheta} \tag{2.53}$$

results from combining of Eqs. (2.49)–(2.52).

More, generally, since the PK pseudo-potential is rewritten in its linear form:

$$\widehat{V}_{PP} = \sum_i \left(\varepsilon^{val} - \varepsilon_i\right)\left|\chi_i^{core}\right\rangle\left\langle\chi_i^{core}\right| \tag{2.54}$$

it follows that any transformation of the PO

$$\tilde{\vartheta} = \vartheta + \sum_i a_i \chi_i^{core} \tag{2.55}$$

with a_i's are arbitrary constants, leads to solutions of PP equation (2.52) with the same eigen-value:

$$
\begin{aligned}
\left(\widehat{H} + \widehat{V}_{PP}\right)\tilde{\vartheta} &= \left(\widehat{H} + \widehat{V}_{PP}\right)\vartheta + \left(\widehat{H} + \widehat{V}_{PP}\right)\sum_j a_j\left|\chi_j^{core}\right\rangle \\
&= \varepsilon^{val}\vartheta + \sum_j a_j\widehat{H}\left|\chi_j^{core}\right\rangle + \sum_{i,j} a_j\left(\varepsilon^{val} - \varepsilon_i\right)\left|\chi_i^{core}\right\rangle\left\langle\chi_i^{core}\middle|\chi_j^{core}\right\rangle \\
&= \varepsilon^{val}\vartheta + \sum_j a_j\varepsilon_j\chi_j^{core} + \sum_j a_j\left(\varepsilon^{val} - \varepsilon_j\right)\chi_j^{core} \\
&= \varepsilon^{val}\left(\vartheta + \sum_j a_j\chi_j^{core}\right) \\
&= \varepsilon^{val}\tilde{\vartheta}
\end{aligned}
$$

$$\tag{2.56}$$

Such a feature of equivalent pseudo-orbitals in establishing the localization of the valence orbital and eigen-value consecrates the reality of the valence reality, on the one hand, and corresponds to those involving localization measures through unitary orthogonal transformations, described before, on the other hand.

With the advent of the density functional theory, i.e., with the growing recognition of the role that electronic density plays in describing quantum states of atoms and molecules, there also appears the possibility of visualizing bonds and electronic localization through procedures applied on electronic densities.

Basically, the theory of atoms in molecules (*AIM*) was born with the Hellmann-Feynman theorem formulation (Feynman, 1939),

$$\frac{dE}{dQ} = \left\langle \Psi \left| \frac{d\widehat{H}}{dQ} \right| \Psi \right\rangle \tag{2.57}$$

prescribing the variation of the total energy E respecting an arbitrary parameter Q of the system, for example, the inter-nuclear distance, from its quantum-mechanically average. Since $Q=R$, the resulted force on particular nucleus, dE/dR, yields, in fact, the electronic localization measure in that molecular region; it can be easily visualized by further connection between the force and density by means of Poisson equation:

$$\vec{F}(r) = -\nabla V(r) = 4\pi \frac{\vec{r}}{r} \int\limits_{+\infty}^{r} \rho(\tau) d\tau \tag{2.58}$$

This way the chemical bond is classically partitioned into binding and anti-binding regions.

However, other approaches have also been formulated, aiming to more accurately exploit the electronic bond by using electronic density directly, so as to include the Pauli Exclusion Principle – the vital ingredient when it comes to electronic pairs. In this respect, the next quoted contribution comes from the Daudel's lodges (Daudel et al., 1983), resulting in the difference density between the actual molecular density ρ_{mol} and the so called *reference density* ρ_{ref}, a hypothetical entity associated with the obtained molecular density when at each nuclear position neutral spherical ground states atoms are placed:

$$\Delta\rho(r) = \rho_{mol}(r) - \rho_{ref}(r) \tag{2.59}$$

Although useful among crystallographers, where it is known as the *standard deformation density concept* (Julg, 1967; Daudel et al., 1983; Anderson, 1968, 1969), the Daudel localization measure of bonding seems to disagree with Pauli Exclusion Principle due to the reference density concept that allows atomic charges to overlap unchanged.

A step forward is made with considering the topological issues associated with electron density. In this context, the bond finds both an in-depth and geometrical interpretations once the so called *critical points* of bonds are employed to describe the wild variety of chemical compounds, especially those categorized as electron deficient or posing hypervalences (Gangi & Bader, 1971; Srebrenik & Bader, 1975; Srebrenik et al., 1978; Bader et al., 1979; Bader & Essén, 1984; Bader et al., 1987; Bader & Becker, 1988; Bader, 1990; Bader & Heard, 1999; Bader et al., 2000; Bader, 2001). This way, Bader developed a theory according which AIM are seen now as open systems forming basins of attractors and repellors, bounded by a surface $\Sigma(r_\Sigma)$ of local zero flux in the gradient vector field defined by the so called *zero-flux partitioning condition of electron densities* (Bader, 2001):

$$\nabla\rho(r)\cdot\vec{n}=0 \ , \forall r \in \Sigma(r_\Sigma) \qquad (2.60)$$

A close consequence of this condition is assuming the Laplacian of the electron density $\nabla^2\rho(r)$ as the associate localization measure of bonds, obtained by functional integration of the last condition:

$$\delta\int_{\Omega(\Sigma)} d\tau\nabla^2\rho(r)=0 \qquad (2.61)$$

locally, on the domain $\Omega(\Sigma)$.

However, despite the physical background of Bader's approach, its local zero flux has been found with some limitation in defining bonding (Schwinger, 1951; Cassam-Chenaï & Jayatilaka, 2002; Bader, 2002, 2003; Mohallem, 2002; Kryachko, 2002; Delle Sie, 2002). For instance, it was established that a bond path between two nearby helium atoms in forming He_2 exists quite analogue to that appeared in forming H_2, although the He-He bond has been spectroscopically detected only at very low temperature. Fortunately, another route for defining a localization measure with the help of electronic density was explicated in the context of Thom's theory of catastrophe combined with quantum theory. It leads to the so called *ELFs*, see Volume II/Section 5.5 of the present five-volumes work (Becke & Edgecombe, 1990; Schmidera & Becke, 2002; Berski et al., 2003; Kohout et al., 2004; Nesbet, 2002; Putz & Chiriac, 2008).

2.3   MOLECULAR STRUCTURE BY SYMMETRY ANALYSIS

2.3.1   MOLECULAR SYMMETRY GROUPS

2.3.1.1   Symmetry Elements and Operations

The relation *number-symmetry* is identified with the human knowledge itself, over the centuries, preceding, overcoming, but always enriching any science or *multitude* of natural laws. Thus, the number one, the monad or the Platonic point, is symbolized by a *circle and its center*, by the perfect symmetry, the infinite line, self overlapped, without beginning or end, God or the Creator, Figure 2.2.

Therefore, it is not arbitrary that the symmetry toward the central point of a structure is a very important symmetry in nature, the inversion symmetry. By extrapolating the inversion concept, the inversion symmetry,

FIGURE 2.2   Left: Shiva Nataraja, Lord of the Dance, Upper India, XI century; after (Campbell & Moyers, 1988); right: the roof of the Baptist church in Florence.

abbreviated "*i*" is not confined to the symmetry points on a circle or sphere (see also Figure 2.3) toward its center, but can be applied whenever the analyzed structure allows it. In general, the distinct components of a structure to analyze from the symmetry perspective, will be called *structural points* and may be atoms, molecules or other structural groups. The molecular level provides numerous examples, where the inversion center may or may not be an atom, and the symmetrical structure relative to the center *i*, may also vary from case to case, Figure 2.3. There is not an insignificant thing the way in which the existence of symmetry center decides the trans-type (presence of *i*) or the cis-type (absence of *i*) for a given structure, Figure 2.3 middle.

Precisely in this generalization and representation power, and in the end of reduction and classification, of the molecular and natural structure, depending on the symmetry possessed, resides the power of this concept and make it functional for specific analysis.

In Figure 2.4-left is shown the translational symmetry, while in the right side is exemplified the rotational symmetry, this time not toward a central point, but toward the *axis of symmetry*. Also, from the analysis of the Figure 2.4-right, is remarked how the symmetry to the rotation around the axis of rotation coexists with the reflection symmetry, toward a plan that contains the relative axis.

In Figure 2.5 are theorized these symmetries as the form of symmetry on making it's own rotation operations (taken by *p*-times) of *n* order:

$$C_n^p = p\frac{2\pi}{n} \tag{2.62}$$

with p and n natural numbers, different from 0, and respectively of the reflection symmetry of the plane m ("mirror") or σ("spiegel"= mirror, in German), called the mirror symmetry.

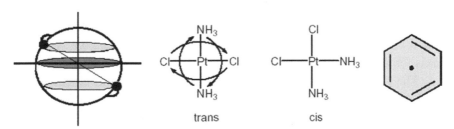

FIGURE 2.3   Left: the inversion center representation (i); middle: the inversion center is an atom; right: the inversion center is a point; after (Heyes, 1999).

FIGURE 2.4 Left: The bison's painted in the Altamira and Spain caves, 15,000–10,000 BC; right: The Kiss, by Constantin Brâncuși (1908).

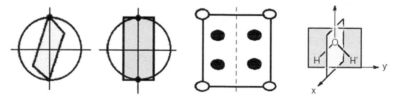

FIGURE 2.5 Left: the representation of global rotation (C_1) and the one with 180° (C_2); middle: the representation of the reflection toward the median plane (σ or m); right: the reflection planes and the axis of rotation z – C_2 for the molecule H_2O; after (Heyes, 1999).

Thus, for a complete rotation of 2π, in order to restore the initial position of a structure, is generated the rotation C_1, while for the "Kiss" of Brancusi or for the water molecules ($H - O - H$) from the Figure 2.4 and Figure 2.5-right, the initial position is regained through a rotation around the axis $0z$ with π radians (180°), which is corresponding to a rotational symmetry of C_2 order. Noteworthy is the fact that the rotational symmetry of C_2 produces the same effect with the reflection toward the plan m which contains the axis C_2.

Before continuing the symmetry discourse, it is worth to be mentioned the usual convention for choosing the reference system and the symmetry axis for a given system (molecule).

The positive direction of rotation is clockwise or anti-clockwise, so that, in the Cartesian system (x, y) the rotation of axis $+x$ to $+y$ generates a negative rotation C_n^{-p} (see the example in Figure 2.6-top).

Then, choosing the main rotation axis, is based on the maximum symmetry principle: as the symmetry is more elevated, the order of rotation

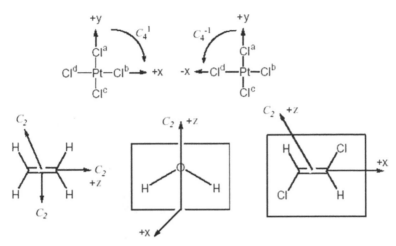

FIGURE 2.6 The exemplification of the conventions for the rotation convention (above) and for the symmetry axis system (below); after (Heyes, 1999).

n is larger, and the angle with which the structure must rotate around the axis of order n (for regain identity) is smaller.

The principal rotation axis will correspond to the axis with the higher order. If there are more rotation axes, any of them can be the main axis. For example, the circle has the maximum (infinite) symmetry on the rotation around the axis that pass by its center and is perpendicular on the circle plane, but has the symmetry of second order (involves the rotation with π in order to regain a state identic with the initial one) toward any axis which, passing through the center of the circle, is included in the circle surface, therefore, the first axis (with the infinite order) is the main axis of rotation of the circle.

Further, for a structure (molecule) is very important a significant selection of the reference axes in relation with the ones of symmetry, Figure 2.6 (bottom).

If the molecule has only an axis of symmetry C_n that will be the axis $0z$.

If there are more C_n axes, with different n, then the main axis will be the $0z$ axiz. If there are more possible main axes, then the axis that connects most atoms (or structural points) will be the axis $0z$.

If the molecule is planar, the $0z$ axis will be the main included in that plane and the $0x$ axis will be perpendicular on that plane.

Vice versa, if the molecule is planar and the $0z$ axis is selected as perpendicular on the molecule plane, it will be chosen the $0x$ axis, which connects the most atoms (or structural points) of the plane, see the Figure 2.6-bottom.

The Figures 2.7 and 2.8, illustrates the case of the symmetry on the rotation around the axis C_3 with $2\pi/3 = 120°$, at symbolic level, theoretic and molecular.

There is noteworthy how the presence of the third order rotation axis (C_3) involves also the multiplication of the plane, for the mirror symmetry.

FIGURE 2.7 Left: three fish, after (Prospero, 1944); right; The Greek Triskelion, victory and progress; after (Lehner, 1950).

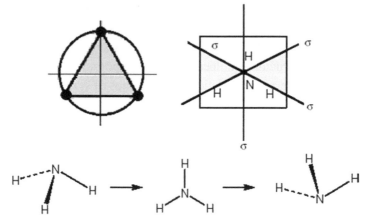

FIGURE 2.8 Left: the representation of the rotation by 120^0 (C_3); middle: the reflections plans of the ammonia molecule (NH_3); right: mirror inversion of the ammonia molecule; after (Heyes, 1999).

Further, by increasing the rotational symmetry order, in Figure 2.9 the allegorical representations of the number 4 are shown as an order of the rotation axes $C_4 = 2\pi/4 = 90°$, the last Pythagorean sacred numbers.

Figure 2.9 theorizes the rotational symmetry C_4 and exemplified it in Figure 2.10, at the molecular level, being noted some essential observations.

The first one asserts that the presence of the C_4 symmetry, automatically leads the second order rotational symmetry, C_2.

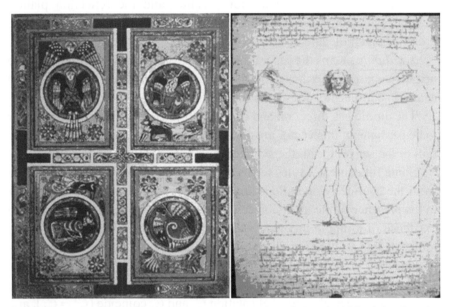

FIGURE 2.9   Left: The four Evangelists, after (Mackworth-Praed, 1993); right: Vitruvian Man, the presence of human in circle and square, according with Leonardo da Vinci (cca.1490).

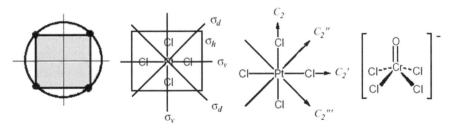

FIGURE 2.10   Left: the C_4 symmetry representation; middle: reflection planes of the molecule $[PtCl_4]^{2-}$; right: the axes of rotation C_4 form the sheet plane of the $[PtCl_4]^{2-}$ molecule and the dipolar molecule $[CrOCl_4]^-$; after (Heyes, 1999).

The second observation, already existing in case C_3, regards the appearance of new plans of mirror symmetry.

In the present case, is noticed even a differentiation between the mirror planes, precisely in the horizontal plane (σ_h) included by the sheet, and the perpendicular planes, as vertical planes (σ_v) – containing the structural points, and as diagonal or dihedral plane (σ_d) – which "diagonally" pass through the structural points.

As an observation, the inversion center and the reflection planes, together with the proper rotation axes, also give informations about the presence in a molecule of the *dipole moments.*

The dipole means the separation of the positive charges center from the negative charges one in a molecule, in other words, the electrons gravity center in a molecule, does not coincide with the nuclei of gravity center.

The dipole has direction and orientation, therefore is a vector. In order to be permanent, this vector must be coincidental (do not modify the direction or the sense) for all the symmetry operations that a molecule allows.

Some fundamental rules for establishing the dipole moment existence from the symmetry operations are: the nonexistence of a dipole moment in the presence of the inversion center (which would annul the vector); the existence of a dipole moment if all the axes C_n are overlapped, or if the molecule has a reflection plane but no C_n axes, or if all the mirror planes include the C_n overlapped axes.

For example, the molecule $[CrOCl_4]^-$, Figure 2.10-right, has the dipole moment, because all its axes of rotation C_2 and C_4 coincide $\left(C_4^2 = C_2\right)$, and moreover are included in all the mirror planes.

The third observation derives from the first one: the symmetry of order 4 (but not only) shows how through the rotation around the axis C_4 at each quarter from interval 2π, is founded the initial structure, their components being in the same spatial arrangement, as in the identical previous position.

Practically says how the rotational symmetry had been produced the rotation operations, meaning the permitted successive rotation which preserve the initial symmetry, see Figure 2.11.

Here is worth making also the difference between the equivalence points (which can be interchanged through a symmetry operation existing in structure) and the identical points (witch are identical overlapped through a symmetry operation).

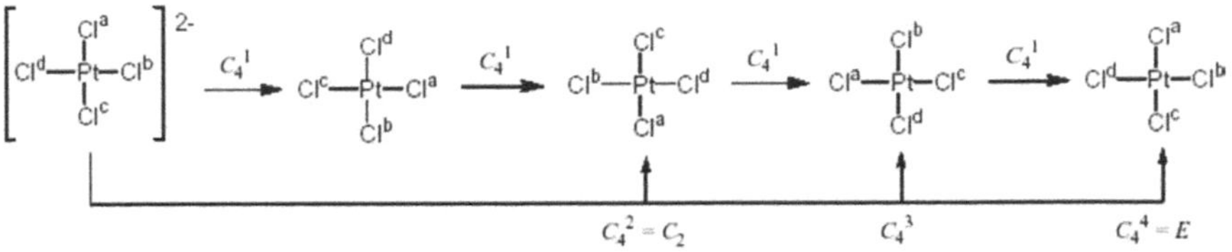

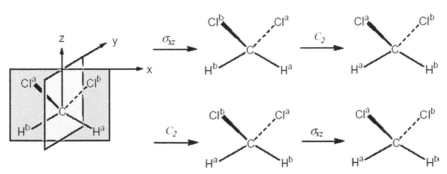

FIGURE 2.11   The successive rotations C_4 of the molecule $[PtCl_4]^{2-}$; after (Heyes, 1999).

Obviously, the identical points are associated with the identical operation, noted with E ("einheit" = unity, in German) corresponding, for the rotation case, to the $p = n$ case in the definition equation (2.62), see also the Figure 2.11.

The fourth observation regards the symmetry combinations, when those do not refer to the simple sequence of the symmetry operations derived from one another, as in the previous case presented for the successive rotations.

For instance, through the combinations (are noted with "x") between the rotations and the reflections are obtained new symmetry operations, the *roto-reflections*, which generate spatial changes, so that the final results overlapped to the initial structure.

Figure 2.12 exposes a situation in which the combination of symmetry operations at the rotation is a commutative one $(C_2 \times \sigma_{xz} = \sigma_{xz} \times C_2)$, no matter the order in which the symmetry operations are applied.

A contrary example, of non-commutative combinations of the symmetry operations is presented in Figure 2.13. Generally, the further

FIGURE 2.12   The plane and the $z - C_2$ symmetry axis of the CH_2Cl_2 molecule and the commutative combinations between them; after (Heyes, 1999).

pairs of symmetry combination always commute: two rotations around the same axis of rotation; the reflexions toward the mirror planes mutually perpendicular; the inversion and any reflection or rotation; two rotations C_2 around the rotation axes reciprocally perpendicular; the rotation around a rotation axis and the reflection across a mirror plane perpendicular to the rotation axis.

Passing to the number 5 and to the 5th order symmetry both means an ontological leap, as long as the pentagram represents the microcosmos (the universe from human beings) and also a symbolic-geometric one, as long as the space cannot be uniformly overlapped with pentagons, regular geometric structures with 5-equivalent sides Figure 2.14.

The last idea is vital for defining the so-called crystallographic symmetry (see the further section). Figure 2.15 exemplifies the molecular case of the $2\pi / 5$ radians rotation around the axis. This time, is noted a very interesting fact, i.e., the rotations C_5 are not the only one which product symmetry at the rotation of 5th order.

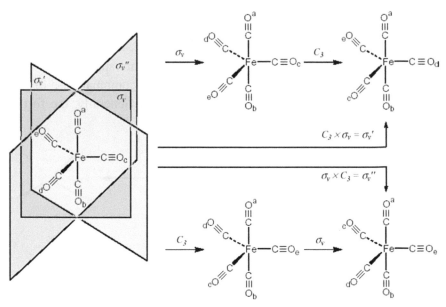

FIGURE 2.13   The symmetry planes and the C_3 symmetry axis of the bipyramidal $Fe(CO)_5$ and the non-commutative operations between them; after (Heyes, 1999).

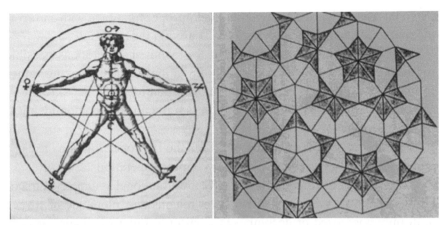

FIGURE 2.14 Left: the symbolic representation of the microcosmos by including the human body in a pentagram, from Agripa, after (Lehner, 1950); right: pentagrams marked by polygons, after (Kappraff, 1990).

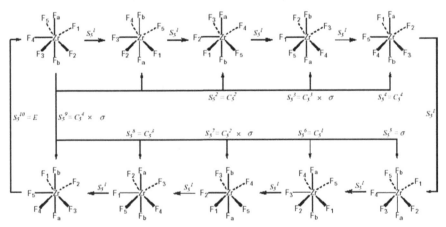

FIGURE 2.15 The successive rotations S_5 of the bipiramidal molecules $[Z_r F_7]^{3-}$; after (Heyes, 1999).

If each C_5 rotation is ulterior combined with a mirror symmetry on the perpendicular space on the axis C_5, then there are obtained new symmetry stages, but from the combinations of new elements.

And exactly this kind of combination as rotoreflection, with the reflection toward the perpendicular plane on the axis, around which was previously made the rotation, operation different from the rotoreflection shown in Figure 2.12, will generate the so-called improper rotation: the rotation followed by the perpendicular mirror, or shortened $S_n = C_n \times \sigma$.

In this case, Figure 2.15, is about the improper rotations, with a mirroring that follows each rotation with $2\pi / 5$ around the axis C_5.

It is worth to mention the fact that the symmetry operations possible for the S_5 axis are $10(5 \times 2)$, 5 for each of rotations C_5^p with p taking values from 1 to 5, until when is regained the operation identical at the rotation: $C_5^5 = E$ and 2, for each reflection σ until is refined the operation identical on the mirroring: $\sigma^2 = E$.

This observation is generally valid for the improper rotations axes of odd orders S_{2n+1}: generating $2(2n + 1)$ symmetry operations.

The situation is different for the 6th order rotational symmetry. First of all, the rotational symmetry of 6th order regains the "plenitude of space" uniformly composed or decomposed from/in hexagons or their variants, Figure 2.16, argument that will be reopened in the further sections, much more systematically.

Regarding the effective rotations behavior, especially the improper ones, Figure 2.17 theorizes and exemplifies the improper rotation of 6th order around the C_6 rotation axis with angles of $2\pi/6 = 60°$ followed by a mirroring in the perpendicular plane on the rotation axis.

Contrary to the S_5 case, the 6th order symmetry does not generate the duplication of the symmetry operations, also see the Figure 2.18, observation that can be also extended for the all improper rotations of even order, S_{2n}.

This fact explanation resides in the improper rotation definition, which for the even orders ($2n, n = 3$ *for the current case*) generates the $S_{2n}^1, S_{2n}^2, S_{2n}^3, \ldots, S_{2n}^{2n} = C_{2n}^{2n} \times \sigma^{2n} = E$ actually regaining the initial state after $2n$ operations. This situation does not happens in the case S_{2n+1} where after $2n + 1$ improper operations it is obtained: $S_{2n+1}^{2n+1} = C_{2n+1}^{2n+1} \times \sigma^{2n+1} = E \times \sigma = \sigma$,

FIGURE 2.16 Left: snow flakes, after (Bentley, 1962); right: the Moise Church dome, in San Marco district of Venice, after (Demus, 1988).

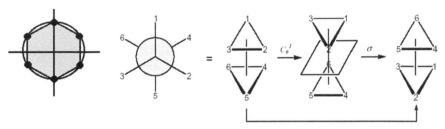

FIGURE 2.17   Left: the representation of the of 60^0 (C_6) rotation; right: the axis $(1-5)$ of 6th order rotation, combined with the reflection on the perpendicular plane, generates the improper rotation (S_6) for the C_2H_2 molecule; after (Heyes, 1999).

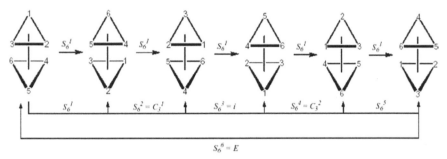

FIGURE 2.18   The S_6 successive rotations of ethane molecule; after (Heyes, 1999).

instead of the identical operation E; this fact implies the necessity of dupli-cation of $2n + 1$ operations, becoming $2(2n + 1)$ operations, in order to reobtain the identity $\sigma^2 = E$.

Moreover, from the same reasons explicated here, in the even improper rotations case is always recommended the presence of (proper) rotations axes order

$$2n/2 = n : C_n \qquad (2.63)$$

In other words, the even improper rotation operations can be always rewritten through the equivalent operartions. For the improper axis of 6th order rotation, can be written the equivalence:

$$S_6^2 = C_6^2 \times \sigma^2 = C_3 \times E = C_3 \qquad (2.64)$$

$$S_6^3 = C_6^3 \times \sigma^3 = C_2 \times \sigma = i \qquad (2.65)$$

$$S_6^4 = C_6^4 \times \sigma^4 = C_3^2 \times E = C_3^2 \tag{2.66}$$

which implies that the complete set of the symmetry operations, generated by the improper axis S_6, to be written as a multitude of operations:

$$S_6 = \left\{ S_6, C_3, i, C_3^2, S_6^5, E \right\} \tag{2.67}$$

wherefrom it is remarked the rotations subset generated by the presence of axis

$$C_3 = \left\{ SC_3, C_3^2, E \right\} \tag{2.68}$$

A last observation here, the absence of the improper axes of rotation symmetry is the real test for assigning chirality to a molecule. As long as the chirality characterizes the structures whose mirror image do not overlap, often is erroneously considered the absence of the mirror planes σ or of the inversion center i as sufficient signals for attributing the chirality.

Figure 2.19 exemplifies a molecular case in which are missing the reflections planes and the inversion center, but still the molecule is not chiral, because of the presence of the improper S_4 rotations axis.

The symmetry superior to the 6th order, are treated similar. Noteworthy is the influence of the platonic solids for symmetry characterization, also see the Figure 2.20.

Remarkably, the platonic solids are the only possible regular polyhedral (the front faces are regular polygons, tops and edges are equivalent – interchangeable by symmetry operations).

This situation can be easily demonstrated if we take into consideration that for the construction of a polyhedron is necessary that at least 3 of the polygon faces to have a mutual point, in a compact pyramidal construction (non-planar).

Therefore, the faces with 3, 4, and even 5 equilateral triangles with a common top are possible, while when about 6 equilateral triangles top-to-top the 360° round is made, yet in a planar manner, i.e., non-spatial.

As squares, it can be used only the combination of 3 squares with a common top, considering that 4 squares would already fulfill the construction spatiality. Also, 3 pentagons (108° at each top) around a common top can be coupled.

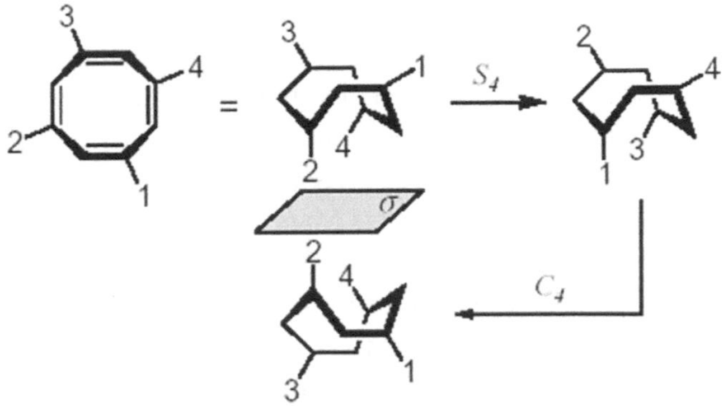

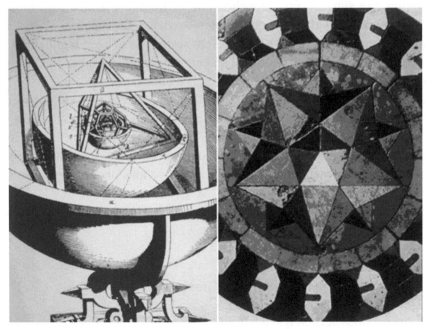

FIGURE 2.19   The non-chirality test for the molecule 1, 3, 5, 7-tetramethyl tetrametil cyclooctatetraene; after (Heyes, 1999).

FIGURE 2.20   Left: Kepler's model of the Universe, after (Lawlor, 1982); right: mosaic from the San Marco Cathedral, Venice, 1425–1430, after (Emmer, 1993).

Any other combinations generate planar structures or beyond a complete rotation of 360°.

Therefore, results the tetrahedron, the cube, the octadedron, the dodecahedron and the icosahedron. Worth to remind that this platonic solids were imagined by Kepler, exactly by their ampleness symmetry operations (as like as the model of which the planetar system itself had been formed), by mutual registrations of these structure, Figure 2.20-left. Thus, the thetrahedron separates the planets Mars and Jupiter, the cube the planets Jupiter and Saturn, the octadedron divides Mercury from Venus, the dodecahedron dissociated the Earth from Mars and the icosahedron was interposed between Venus and Earth.

Although spectacular, the idea had been proved to be erroneous once with the discovery of the other three planets from the solar system, but still had been generated the viable notion of symmetry structures packing, this time under the form or mutual registrations. Not accidentally, Kepler had subsequently launched one of the most difficult geometric conjectures, but only very recently demonstrated, the theorem of crystalline packing (Hales, 2012). The present section have been highlighted to the significance of both historical and mystical of the number and also its representation by the symmetry which is associated.

At this point, we have to do the difference between the symmetry elements (plane, symmetry or inversion center, proper or improper axes of rotation) and the associated symmetry operations (reflection in a plane, the inversion toward the center, the successive rotations around the axis of rotation, the rotations followed by the reflections on the perpendicular plane on the rotation axis).

Table 2.1 synthesizes the notations and the significances of the elements and of the symmetry operations.

2.3.1.2   Building the Symmetry Groups

In the previous section it had been exposed how the symmetry elements of a structure generate associated operations.

The relation number → symmetry is obviously at the levels of the rotations axes, but less clear in relation with the other operations and even lesser immediate when it comes to be global characterized, through a unique number, all the symmetry operations allowed by a structure.

TABLE 2.1   The Symmetry Elements and Operations Associated Types (Putz, 2006)

Element	Operation	
Inversion center	I	The inversion transformation (identical with the operation S_2). In Cartesian systems generates the coordinates inversion: $(x,y,z) \rightarrow (-x,-y,-z)$.
Proper axis of rotation	C_n	The rotation (positive in clockwise sense) with the angle of $2\pi / n$ radians, with n natural number. The axis of the biggest n is called the main rotation axis.
	C_n^p	The rotation (positive in clockwise sense) with the angle $2\pi / n$ radians, with n natural and p integer.
Improper axis of rotation	S_n	The rotation (positive in clockwise sense) with the angle $2\pi / n$ radians, followed by reflection in the perpendicular plane on the rotation axis.
	σ_h	The horizontal reflection – in the plane which pass through the origin and is perpendicular on the rotation main axis.
Mirror plane	σ_v	The vertical reflection – in the plane which pass through the origin and contains the main rotation axis.
	σ_d	The diagonal (dihedral) reflection – an especial case of vertical reflection σ_v but toward the plane that in addition also bissextiles the angles between the axes C_2 perpendicular on the main rotation axis.
-	E	The identical transformation.

In other words, starting from the pythagorean numbers $(1, 2, 3, 4)$ these have been identified with the correlated order of the main rotations axes and toward which have been also identified and introduced the other possible symmetry operations.

How can be now recupered the inverse relation?

A rigorous representation of the inverse connection of symmetry operations ™ numbers, will be further exposed and regards the notion of the *group*.

The group, G, is composed by a collection (multitude) of objects $a, b, c,\ldots$ which are interrelated through a law (it will be noted by "×") of mutual transformation called transformation law (multiplication or composition "×") of the group and which satisfies four conditions (Cotton, 1990; Putz, 2006).

IG). Whichever would be the objects a and b from the G the law × transforms them in the product

$$a \times b = c \qquad\qquad (2.69)$$

also an object from G. Obviously, this rule is also applied for the $a = b$ case.

This property attests the quality of internal composition law of the operation "×".

Another observation is the fact that the law × does not have to be necessary commutative, meaning that the inverse product $b \times a = d$ does not have to be essentially equal with $c = a \times b$, but both objects (c and d) must be objects of G. If $c = d$ it will be considered that the law × is also commutative (or abelian) in relation with the objects of G.

EG). In the multitude G there is always an object e (called identical or neutral) which, by multiplication with anyone else object (for example o) to the group (and with itself), lets invariant the relative object of the group, independently by the order in which are considered the products (the compositions)

$$e \times o = o \times e = o \qquad\qquad (2.70)$$

AG). Whichever would be the objects a, b and c from the G, these own the property of associativity in relation with of composition law

$$a \times (b \times c) = (a \times b) \times c \qquad\qquad (2.71)$$

This property have to be satisfied independently of the commutativity one, exposed above.

RG). However, any object o of G must have an unique correspondent, called the inverse (or the reciprocal) of o, noted with o^{-1} it also from G, so the product with this can be able to generate the above identical element:

$$o \times o^{-1} = o^{-1} \times o = e \qquad\qquad (2.72)$$

This property, called the existence of the Universe must be verified, again, independently of the commutativity of the objects from G in relation with the law ×.

In brief, a set G of objects is called group in relation with the composition law × if the relative law of internal composition (IG) generates the identical or neutral object (EG), is associated (AG) and allows the existence of the inverse or reciprocal (RG) for any objects from G (i.e., for all of them).

Moreover, if the law $\times$ is also commutative for the objects of G, it will be considered that G is a commutative or abelian group.

ThRG). Also very useful is a consequence of those definitions, called the theorem of the inverse (or reciprocal) product, which establishes that the products reverse of as many objects from G is equal with the product in inverse order, of the reverse of relative objects

$$\left(a \times b \times c \ldots v \times w\right)^{-1} = w^{-1} \times v^{-1} \ldots \times c^{-1} \times b^{-1} \times a^{-1} \qquad (2.73)$$

The demonstration is easy and is based on the mathematical induction. We expose here only the first step, the case of the two elements product from the group G. Given a and b objects from the G with the product $a \times b = c$, in relation with the law of group "$\times$".

Therefore, being discussed about the group G, there are the inverses a^{-1} and b^{-1}, so that

$$a \times a^{-1} = a^{-1} \times a = e \qquad (2.74)$$

and respectively

$$b \times b^{-1} = b^{-1} \times b = e \qquad (2.75)$$

with e the neutral element of the group. The group should not necessary be an abelian group.

By multiplying c on its right with the product $b^{-1} \times a^{-1}$ is obtained

$$a \times b \times \left(b^{-1} \times a^{-1}\right) = c \times \left(b^{-1} \times a^{-1}\right) \qquad (2.76)$$

from where, by applying on the left of the equal the associativity condition of the group (AG), the previous identity can be rewritten as:

$$a \times e \times a^{-1} = c \times \left(b^{-1} \times a^{-1}\right) \qquad (2.77)$$

wherefrom, further, applying it on the left of the equal, the action (twice) of the neutral action of the group (EG), is finally obtained:

$$e = c \times \left(b^{-1} \times a^{-1}\right) \qquad (2.78)$$

The last step consist in the multiply the last identity, on the left, with c^{-1} and again taking into consideration the effects of the action of the neutral object, on the objects of the group (EG), results the equality:

$$c^{-1} = b^{-1} \times a^{-1} \qquad (2.79)$$

Substituting now the c with its initial form, results the identity

$$\left(a \times b\right)^{-1} = b^{-1} \times a^{-1} \qquad (2.80)$$

Analogically, are demonstrated similar identities, for as many similar products which are formed between objects of group G.

The objects, together with the law of composition of the group, form the so-called group multiplication table, see below the example of the group G_3 with three objects: e, a, b.

G_3	e	a	b
e	?	?	?
a	?	?	?
b	?	?	?

The multiplication table must overlap all the possible combinations between the group objects in relation with its law of internal composition.

ThTG). The completion of the table of the group, must take into consideration another fundamental theorem, the theorem of rearrangement of the table of the group: each line and column of the table of a group contain objects of the group, each taken by one, so that cannot form 2 columns or 2 lines with object identically arranged.

In other words, each line or column is a different rearrangement of the objects of the group.

The demonstration is very simple and takes into consideration the internal composition law of the group, which cannot produced "neither more nor less" objects than the group has.

The number of the total objects of a group is called the order of the group.

Thus, for G_3 can be immediately completed the second line and column, based on the neutral object action from the group (EG), obtaining the intermediate table that follow.

G_3	e	a	b
e	e	a	b
a	a	?	?
b	b	?	?

Further, we have to take into account the existence of the reverse objects in group (*RG*) and also the fact that two different objects from the group can be composed in a way so that their product have an unique result, different from any other combinations, of any of objects of the product with the rest of the objects of the group.

Under these conditions, the above last table can be completed in a single mode, below reproduced.

G_3	e	a	b
e	e	a	b
a	a	b	e
b	b	e	a

This uniqueness of each table of group renders to the group the power of representation of any physico-chemical reality, by which the type objects and law of composition can be identified.

For example, for the molecular structures, in the previous section, it had been analyzed the presence of the symmetry elements and operations. Remarkable, the symmetry operations can be constituted as objects of a group, therefore generating the associated group.

Moreover, once identified the group associated to the symmetry, results the quantified group order (the total number of symmetry operations), by associating a number. Is reestablished the characterization of symmetry in number, in a global manner, which through the associated group, take into consideration all the symmetry operations – the objects of the group.

Now it can be understood why it had been avoided the use of the elements groups title and had been preferred their nomination as the group objects.

In order to avoid the confusion, are identified the symmetry operations with the objects of the group, because the symmetry makes the distinction between elements and operations, see the Table 2.1, the conceptual difference being notable: the symmetry elements generate the symmetry operations.

Therefore, it has been avoided a formulation as "the symmetry operations are elements of the associated group", so that still kept the difference between elements and operations at the level of symmetry.

Thus, by convention, come out the notion of the objects of the group which are identified with the symmetry operations.

As example of work, it will be considered the ammonia molecule, Figure 2.8, where the whole set of symmetry operations contain the objects of so-called group

$$C_{3v} : E, C_3, C_3^2, \sigma_v, \sigma_v', \sigma_v'' \tag{2.81}$$

with the composition table, further presented.

C_{3v}	E	C_3	C_3^2	σ_v	σ_v'	σ_v''
E	E	C_3	C_3^2	σ_v	σ_v'	σ_v''
C_3	C_3	C_3^2	E	σ_v''	σ_v	σ_v'
C_3^2	C_3^2	E	C_3	σ_v'	σ_v''	σ_v
σ_v	σ_v	σ_v'	σ_v''	E	C_3	C_3^2
σ_v'	σ_v'	σ_v''	σ_v	C_3^2	E	C_3
σ_v''	σ_v''	σ_v	σ_v'	C_3	C_3^2	E

The table of C_{3v}, above exposed, corresponds to all the symmetry operations and their compositions, also satisfying the theorem of the table rearrangement, ThTG, so that in each line and column are refound all the operations, but in a different arrangement.

Therefore, the order of the group C_{3v} is $g = 6$. But, many times, is difficult to manage the whole group of symmetry operations, being useful to appeal to its subunities.

Two group subunities are essential, the subgroups and the classes of the group, and will be separately analyzed, in the following.

The subgroup is a "small group" included in the total group, in other words a submultitude of objects (in this case symmetry operations) which respects all the four group conditions, toward the same law of composition as like as the large group in which is included.

For example, for the group C_{3v} are identified two subgroups: the one formed only from the unit operation E and the subgroups of pure rotations of C_{3v},

$$C_3 = \left\{ C_3, C_3^2, C_3^3 = E \right\} \tag{2.82}$$

with the tables of composition marked as subtables (tables included) in the table of the large group C_{3v}.

Moreover, thanks to the symmetry operations cyclicality from the pure subgroups of the rotations C_3, that will be called cyclic group. Any cyclic group is abelian, natural property derived from the of the group objects nature: only rotations, by their nature commutative.

An important property of the subgroups, PS, says that the order s of the subgroup is a natural multiple of the order g of the large group, $g / s = n(atural)$, but is not generally valid the reverse, i.e., the large group G has as many subgroups as his order g has divisors.

For example, is clearly that $g(C_{3v}) = 2s(C_3)$ because $(C_{3v}) = 6$ and $s(C_3) = 3$, but C_{3v} does not have four subgroups, even if its order 6 has four natural divisors: 1, 2, 3 and 6.

The class is another important subunit of a group and expresses the complete multitude of conjugate operations (in this case symmetry operations.

Two (symmetry) operations a and b are called conjugate in the same class, if there is another (symmetry) operation q in group (not necessarily from the same class with a and b), which transforms a in b, through an operation called similarity operation, expressed as:

$$b = q^{-1} \times a \times q \tag{2.83}$$

In the following, will be exposed some properties of the classes.

PC1). Any object a from the group is its own conjugate. Very true, as long as for any object a from the group exists the neutral object e, so that

$$a = e^{-1} \times b \times e \tag{2.84}$$

according to the group definitions (EG, RG, AG).

PC2). If a is conjugated with b, only if exists q from group, so that

$$a = q^{-1} \times b \times q \tag{2.85}$$

then exists also an another object from group x so that

$$b = x^{-1} \times a \times x \tag{2.86}$$

This fact is demonstrated searching x from the group.

If there are combined the two previous forms of similarity, could be obtained, for example

$$b = x^{-1}\left(q^{-1} \times b \times q\right) \times x \tag{2.87}$$

expression which, regrouped in accord with the associated rule in group (AG), becomes

$$b = \left(x^{-1} \times q^{-1}\right) \times b \times \left(q \times x\right) \tag{2.88}$$

equality satisfied if and only if q and x are reciprocal reverse in group, and as long as any q from group has a reverse in group (RG), it had been found the searched $x = q^{-1}$.

PC3). The similarity operation (the conjugate) is transitive, i.e., if a is conjugated with b exists

$$x : a = x^{-1} \times b \times x \tag{2.89}$$

and separately with c also exists

$$y : a = y^{-1} \times c \times y \tag{2.90}$$

then b and c are conjugated as well, i.e., there is

$$z : b = z^{-1} \times c \times z \tag{2.91}$$

Once again, the demonstration means to find the object z from group. From the similarity of a with b and c, can be written the equality:

$$(a =) x^{-1} \times b \times x = y^{-1} \times c \times y \qquad (2.92)$$

wherefrom – applying the rules of definition of the group, the inverse law RG and the associativity AG – is successively multiplied, on the left and on the right, the previous equality with x and with x^{-1}, obtaining – after regrouping – the transformation of similarity

$$b = \left(x \times y^{-1} \right) \times c \times \left(y \times x^{-1} \right) \qquad (2.93)$$

The RG law says that always exists the inverse objects from a group and together with the rule IG which recommends any composition from the group as being internal, results the researched object:

$$z = y \times x^{-1} \qquad (2.94)$$

which ensures the conjugation (similarity) between b and c.

Remains only the exemplification on a concrete case, and will be analyzed the group C_{3v} with the multiplication table, previously exposed.

Therefore, it will be considered each group operation and will be operated similarity transformations, with all the operations existing in group. Finally, will be formed classes, from those multitude of symmetry operations, that turn one to another.

For example, it will be considered the rotation operation C_3, for which will be formed all the similarity transformations, with all the group operations, and will be evaluated the compositions, based on the multiplication table of C_{3v}.

Thus, is obtained:

$$E^{-1} \times \left(C_3 \times E \right) = E^{-1} \times C_3 = E \times C_3 = C_3$$
$$C_3^{-1} \times \left(C_3 \times C_3 \right) = C_3^{-1} \times C_3^2 = C_3^2 \times C_3^2 = C_3$$
$$\left(C_3^2 \right)^{-1} \times \left(C_3 \times C_3^2 \right) = \left(C_3^2 \right)^{-1} \times E = C_3 \times E = C_3$$
$$\sigma_v^{-1} \times \left(C_3 \times \sigma_v \right) = \sigma_v^{-1} \times \sigma_v'' = \sigma_v \times \sigma_v'' = C_3^2$$
$$\sigma_v'^{-1} \times \left(C_3 \times \sigma_v' \right) = \sigma_v'^{-1} \times \sigma_v = \sigma_v' \times \sigma_v = C_3^2$$
$$\sigma_v''^{-1} \times \left(C_3 \times \sigma_v'' \right) = \sigma_v''^{-1} \times \sigma_v' = \sigma_v'' \times \sigma_v' = C_3^2 \qquad (2.95)$$

Therefore, results that C_3 is transformed only form C_3 in C_3^2, through any similarity action with the group operations.

Similar relations result when are analyzed the similarity of the operation C_3^2, and is also find that C_3^2 is transformed only in C_3 and C_3^2. Thus, C_3 and C_3^2 are conjugated and belong to the same class in the group C_{3v}.

Also based on the multiplicity table in C_{3v}, are proved the similarities of the mirror symmetry operations, for each mirroring $\sigma_v, \sigma_v', \sigma_v''$ in part, and is finally revealed how any of this operations are transformed, by any similarity action, also in a mirror operation, so being conjugated and belonging to the same class.

Consequently, the establishment of the classes of a group narrow the total number of operations of the group to the conjugated one, similar or equivalent. For example, the group C_{3v} allows three classes: E, $2C_3$ and $3\sigma_v$, which mark the separation by equivalent operations classes: the identity (always present, like the distinct classes as well), proper rotations and mirroring's, see Table 2.2.

Similar analysis, i.e., the identification of the symmetry elements, of symmetry operations, the construction of the multiplication table, the groupment in classes of conjugated operations (equivalent), can be performed for any of five platonic structures.

Note that for the tethraedric case, from the pure rotations group

$$T = \left\{ E, 4C_3, 4C_3^2, 3C_2 \right\} \tag{2.96}$$

can be generated another tethraedric group, by composing the objects (operations) of T with the one of the operation σ_h (which contain pairs of axes C_2, for which are let invariant the rotations around them):

$T_h = 3\sigma_h \times T$

$$= \left\{ 3\sigma_h, 4[S_6], 3C_2 \right\} = \begin{cases} 3\sigma_h, 4S_6^1, 4C_3 \left(= S_6^2 \right), \\ i \left(= C_2 = S_6^3 \right), 4C_3^2 \left(= S_6^4 \right), S_6^5, E \left(= S_6^6 \right), 3C_2 \end{cases}$$

$$\tag{2.97}$$

More generally, in the desire to systematize the symmetry information, in terms of representations of group, is worth noting that all the symmetry

TABLE 2.2 Types, Notations and Characterization of Point Groups (Putz, 2006)

Type	Notation	Characteristics
Groups with a single symmetry element	C_1	only E
	C_s	E and σ
	C_i	E and i
	C_n	E and C_n
	S_n	$n = 2k$, k natural
Groups with more than one symmetry element	D_n	axis C_n and n axes $C_2 \perp C_n$
	C_{nh}	axis C_n and a plane $\sigma \perp C_n$
	C_{nv}	axis C_n and two or more σ which contain C_n
	D_{nd}	axis C_n, n axes $C_2 \perp$ pe C_n, n diedral planes parallel with C_n and which bissextile the angles between the n axes $\perp C_n$
	D_{nh}	axis C_n, n axes $C_2 \perp$ on C_n, and a plane $\sigma \perp C_n$
Special groups	$C_{\infty v}$	linear structures without i
	$D_{\infty h}$	linear structures with i
	T_d	Tethraedric groups, include T_h and T
	O_h	Orthoaedic groups, include O
	I_h	Icosaedric groups, include I

operations which are introduced, exemplified and analyzed in the previous section, are performed in relation to the symmetry elements (axes, planes, center), all intersect in a point. Therefore, incorporating these groups of symmetry operations, will designate punctual groups of symmetry.

These point groups, can have 3 basic types: containing symmetry operations around a single symmetry element, around more symmetry elements and the so-called "special" groups (which include also the symmetry groups of the platonic structures).

The enumeration and description of groups contained in each of these types of punctual groups, are given in Table 2.2 Schöneflies's notation. This notation has both an advantage and a disadvantage; the disadvantage is that the whole group is marked with symbols, that often coincide with one operation group (object); but this is also an advantage, because symbolizes the general feature of the group and the symmetry elements which contains.

In Table 2.3, on the contrary, are even more detailed the types of groups from Table 2.2 specifying the symmetry operations (already grouped into

TABLE 2.3 The Symmetry Operations Are Organized in Point Groups, the Notations and the Isomorphisms Associated (Putz, 2006)

Symmetry operations	Order	Schönflies notation	International notation	Symmetry notation	Isomorphisms
$E, 4C_3, 4C_3^2, 3C_2$	12	T	23	23	
$E, 8C_3, 3C_2, 3\sigma_v, i, 8S_6$	24	T_h	$m3$	$\dfrac{2}{m}\bar{3}$	
$E, 6C_4, 8C_3, 3C_2, 6C_2{'}$	24	O	432	432	T_d
$E, 8C_3, 3C_2, 6S_4, 6\sigma_d$	24	T_d	$\bar{4}3m$	$\bar{4}3m$	O
$E, 8C_3, 6C_2, 6C_4, 3C_2{'}, i, 6S_4, 8S_6, 3\sigma_h, 6\sigma_d$	48	O_h	$m3m$	$\dfrac{4}{m}\bar{3}\dfrac{2}{m}$	
E, C_4, C_2, C_4^3	4	C_4	4	4	S_4
E, S_4, C_2, S_4^3	4	S_4	$\bar{4}$	$\bar{4}$	C_4
$E, C_4, C_2, C_4^3, i, S_4^3, \sigma_h, S_4$	8	C_{4h}	$4/m$	$\dfrac{4}{m}$	
$E, 2C_4, C_2, 2C_2{'}, 2C_2{''}$	8	D_4	422	422	C_{4v}, D_{2d}
$E, 2C_4, C_2, 2\sigma_v, 2\sigma_d$	8	C_{4v}	$4mm$	$4mm$	D_4, D_{2d}
$E, 2S_4, C_2, 2C_2{'}, 2\sigma_d$	8	$D_{2d}(V_d)$	$\bar{4}2m$	$\bar{4}2m$	D_4, C_{4v}
$E, 2C_4, C_2, 2C_2{'}, 2C_2{''}, i, 2S_4, \sigma_h, 2\sigma_v, 2\sigma_d$	16	D_{4h}	$4/mmm$	$\dfrac{4}{m}\dfrac{4}{m}\dfrac{4}{m}$	
$E, C_2, C_2{'}, C_2{''}$	4	$D_2(V)$	222	222	C_{2v}, C_{2h}
$E, C_2, \sigma_v, \sigma_v{'}$	4	C_{2v}	$mm2$	$mm2$	D_2, C_{2h}
$E, C_2, C_2{'}, C_2{''}, i, \sigma, \sigma{'}, \sigma{''}$	8	$D_{2h}(V_h)$	mmm	$\dfrac{2}{m}\dfrac{2}{m}\dfrac{2}{m}$	
E, C_2	2	C_2	2	2	C_s, C_i
E, σ_h	2	$C_s(C_{1h})$	m	m	C_2, C_i
E, C_2, i, σ_h	4	C_{2h}	$2/m$	$\dfrac{2}{m}$	D_2, C_{2v}
E	1	C_1	1	1	
E, i	2	$C_i(S_2)$	$\bar{1}$	$\bar{1}$	C_s, C_2

TABLE 2.3   Continued

Symmetry operations	Order	Schönflies notation	International notation	Symmetry notation	Isomorphisms
E, C_3, C_3^2	3	C_3	3	3	
$E, C_3, C_3^2, i, S_6^5, S_6$	6	$S_6 (C_{3i})$	$\bar{3}$	$\bar{3}$	C_6, C_{3h}
$E, 2C_3, 3C_2$	6	D_3	32	32	C_{3v}
$E, 2C_3, 3\sigma_v$	6	C_{3v}	3m	3m	D_3
$E, 2C_3, 3C_2, i, 2S_6, 3\sigma_d$	12	D_{3d}	$\bar{3}m$	$\bar{3}\dfrac{2}{m}$	C_{6v}, D_6, D_{3h}
$E, C_6, C_3, C_2, C_3^2, C_6^5$	6	C_6	6	6	$S_6 \cdot C_{3h}$
$E, C_3, C_3^2, \sigma_h, S_3, S_3^2$	6	$C_{3h} (S_3)$	$\bar{6}$	$\bar{6}$	$S_6 \cdot C_6$
$E, C_6, C_3, C_2, C_3^2, C_6^5, i, S_3^2, S_6^5, \sigma_h, S_6, S_3$	12	C_{6h}	6/m	$\dfrac{6}{m}$	
$E, 2C_6, 2C_3, C_2, 3C_2', 3C_2''$	12	D_6	622	622	C_{6v}, D_{3d}, D_{3h}
$E, 2C_6, 2C_3, C_2, 3\sigma_v, 3\sigma_d$	12	C_{6v}	6mm	6mm	D_6, D_{3d}, D_{3h}
$E, 2C_3, 3C_2, \sigma_h, 2S_3, 3\sigma_v$	12	D_{3h}	$\bar{6}m2$	$\bar{6}m2$	D_6, D_{3d}, C_{6v}
$E, 2C_6, 2C_5, C_2, 3C_2', 3C_2'', i, 2S_3, 2S_6, \sigma_h, 3\sigma_d, 3\sigma_v$	24	D_{6h}	6/mmm	$\dfrac{6\ 2\ 2}{m\ m\ m}$	
$E, 2C_\infty$	∞	C_∞	∞		-
$E, 2C_\infty, i, 2S_\infty^\phi$	∞	$C_{\infty h}$	∞/m		-
$E, 2C_\infty^\phi, \infty\sigma_v$	∞	$C_{\infty v}$	∞m		
$E, 2C_\infty^\phi, \infty\sigma_v, i, 2S_\infty^\phi, \infty C_2$	∞	$D_{\infty h}$	∞/mm		-
$E, C_5, C_5^2, C_5^3, C_5^4$	5	C_5	5		
$E, S_8, C_4, S_8^3, C_2, S_8^5, C_4^3, S_8^7$	8	S_8	-	-	
$E, 2C_5, 2C_5^2, 5C_2$	10	D_5	-	-	C_{5v}
$E, 2C_5, 2C_5^2, 5\sigma_v$	10	C_{5v}	-	-	D_5
$E, C_5, C_5^2, C_5^3, C_5^4, \sigma_h, S_5, S_5^7, S_5^3, S_5^9$	10	C_{5h}	-	-	

TABLE 2.3 Continued

Symmetry operations	Order	Schönflies notation	International notation	Symmetry notation	Isomorphisms
$E, 2S_8, 2C_4, 2S_8^3,$ $C_2, 4C_2', 4\sigma_d$	16	D_{4d}	-	-	
$E, 2C_5, 2C_5^2, 5C_2,$ $i, 2S_{10}, 2S_{10}^3, 5\sigma_d$	20	D_{5d}	-	-	D_{5h}
$E, 2C_5, 2C_5^2, 5C_2,$ $\sigma_h, 2S_5, 2S_5^2, 5\sigma_d$	20	D_{5h}	-	-	D_{5d}
$E\ 2S_{12}\ 2C_6\ 2S_4$ $2C_3\ 2S_{12}^5\ C_2\ 6C_2'$ $6\sigma_d$	24	D_{6d}	-	-	
$E, 12C_5, 12C_5^2,$ $20C_3, 15C_2$	60	I	-	-	
$E, 12C_5, 12C_5^2,$ $20C_3, 15C_2, i,$ $12S_{10}, 12S_{10}^3,$ $20S_6, 15\sigma$	120	I_h	-	-	

classes), that each punctual group in part allows, the order of the group, the notations Shönflies together with the International one (Hermann-Mauguin) and the symmetry one, and the isomorphisms of the relative group as well.

The isomorphism expresses the equivalence (mathematics) between two or more groups. The isomorphic groups have different symmetry operations, but have the same order. Therefore, the isomorphism helps rearranging the groups, according to the groups orders as in Table 2.4 (Novak, 1995): to equal orders, correspond isomorphic groups. Collections of the isomorphic groups can be grouped, in turn, in the so-called abstract groups, as in Table 2.4.

The last group-symmetry item resides in the presentation of algorithm for identifying the associated group of a molecular structure, from the analysis of symmetry elements (Figure 2.21)

By analyzing the Figure 2.21, it is clear how in the assumption of the symmetry group of a molecular structure, must be followed, generally, four-stage selection:

TABLE 2.4 The Point Groups, Classified by the Group Order (by Isomorphism) – Vertically, and By Belonging to the Same Abstract Group – Horizontally; After (Novak, 1995)

Order	S_n	C_n	C_{nh}	C_{nv}	D_n	D_{nd}	D_{nh}	Cubical	Icosahedral
1		C_1							
2	S_2	C_2	C_1						
3		C_3							
4	S_4	C_4	C_{2h}	C_{2v}	D_2				
5		C_5							
6	S_6	C_6	C_{3h}	C_{3v}	D_3				
7		C_7							
8	S_8	C_8	C_{4h}	C_{4v}	D_4	D_{2d}	D_{2h}		
9		C_9							
10	S_{10}	C_{10}	C_{5h}	C_{5v}	D_5				
12	S_{12}		C_{6h}	C_{6v}	D_6	D_{3d}	D_{3h}	T	
14	S_{14}		C_{7h}	C_{7v}	D_7				
16	S_{16}		C_{8h}	C_{8v}	D_8	D_{4d}	D_{4h}		
18	S_{18}		C_{9h}	C_{9v}	D_9				
20	S_{20}		C_{10h}	C_{10v}	D_{10}	D_{5d}	D_{5h}		
24						D_{6d}	D_{6h}	O, T_d	
								T_h	
48								O_h	
60									I
120									I_h

i) establishing the belonging to one of the special groups;

ii) establishing of non-existence of proper rotation axis, but the outstanding presence of the mirror planes, of the inversion center or of no other additional element (in which case it is the group C_1);

iii) establishing the existence of the improper axis of rotation S_{2n}: if yes is identified the group S_{2n}, if no, are identified the group C_n;

iv) the existence of the main axis C_n is examined in relation to the mirror planes (parallel and perpendicular to it).

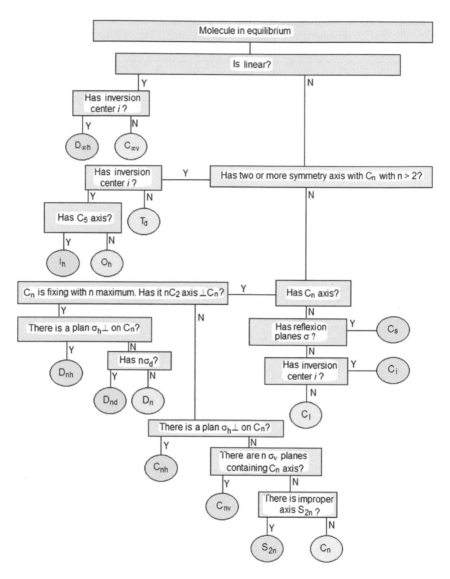

FIGURE 2.21 The algorithm of the molecular classification according to the symmetry group; after (Cotton, 1990; Putz, 2006).

In Figure 2.22, a succession of molecular structures are illustrative presented, where the point group to which they belong had been identified, as based on the algorithm of Figure 2.21 and above explained.

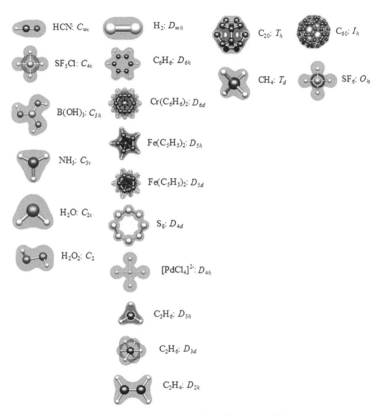

FIGURE 2.22   Examples of molecules, with identification of the point-groups, to which they belong; after (Duch, 2002).

In conclusion, the natural and molecular structures can be systematized by successive identifying the elements, the symmetry operations and the associated groups, by whose order is found the number, not only as a symbol, but also enriched by the quality of measurement and representation.

2.3.2 SYMMETRY ADAPTED LINEAR COMBINATIONS (SALC) WAVE FUNCTIONS

C.A. Coulson, in 1972, said that "the behavior of a molecule cannot be understood until its structure is known", referring to the importance of quantum approaches for studying molecules.

In fact, by considering the molecules, the treated systems become not only multi-electronic, but also multi-nuclei, so that must be studied the electronic state in the common field of nuclei. The case of the periodic field of nuclei will correspond to the metallic state and bonding. These sections, detail the quantum image of the multi-electronic multi-nuclei state.

First of all, the Schrödinger equation for a system of electrons (mass m_0) and nuclei (the core "a" having the mass m_a) will be considered neglecting the spin and the electrons correlation (the quantum correction to the classical Coulomb interaction, electron-electron) as:

$$\widehat{H} \bullet \Psi \equiv \left\{ -\sum_a \frac{\hbar^2}{2m_a} \nabla_a^2 \bullet - \sum_i \frac{\hbar^2}{2m_0} \nabla_i^2 \bullet + \sum_{i<j} \frac{e^2}{r_{ij}} + V(a,r) + U(a) \right\} \Psi = E\Psi$$

(2.98)

where $\nabla^2 \equiv \Delta = \partial^2 / \partial x^2 + \partial^2 / \partial y^2 + \partial^2 / \partial z^2$ corresponds to the Laplacian operator (the squared gradient), $V(a, r)$ corresponds to the electron-nuclei potential, and $U(a)$ to the nuclear-nuclear one, with r being position vectors of the electrons in a Cartesian system.

As in the case of quantum atomic problem, also in the molecular case the rewriting solution of the Eq. (2.98) is used as a partitioning of Eigen function on the subsystems of the electrons $\psi(a, r)$ and of nuclei $\phi(a)$:

$$\Psi = \psi(a,r)\phi(a)$$

(2.99)

At the base of the approximation (2.99) is the fact that the electron mass is considerably lower (by approx. 2000 times!) than the nuclei one, thus being reasonable to assume that "the orbital cycles" of electrons is developed without significant change to the nuclei reciprocal arrangement (named the nuclei configuration) and is called the adiabatic or Born-Oppenheimer approximation (Smith, 1969).

But the reciprocal also applies, i.e., the mutual arrangement of the nuclei or their configuration, directly determine the electronic subsystem state, reason for which the electronic Eigen function has a parametric dependence on the configuration of nuclei: $\psi(a, r)$.

The last idea has a huge importance for the molecular systems description; this because the partitioning (2.99) even by producing a problem reformulation (2.98), is far from exhausted it, requiring a further approximation, the uni-electronic approximation, and even so the solution of the molecular problem can be formulated only self-consistent (iterative) and not in the compact analytical form.

For a quantum analytical characterization, however, the idea of the electronic orbitals dependence (eigenfunctions) on the nuclei configuration opens a possibility of analysis, both interesting and rigorous: appeal to the molecular symmetry!

Which is the relation between molecular symmetry and quantum mechanics? The symmetry operations!

By definition, the symmetry operations (R) "operate" on the molecules nuclei, interchanging them, in other words:

(i) have an operator role $\left(\widehat{R}\right)$ of the global system (electron-nuclei) and

(ii) do not change the Eigen states of the molecule, so that switch with the molecular Hamiltonian $\left(\widehat{H}\right)$: $\widehat{R}\widehat{H} = \widehat{H}\widehat{R}$. By combining the last relation with the Eq. (2.98), successively results (Putz, 2006)

$$\widehat{H}\left(\widehat{R}\bullet\Psi\right)=\left(\widehat{H}\widehat{R}\right)\bullet\Psi=\left(\widehat{R}\widehat{H}\right)\bullet\Psi=\widehat{R}\bullet\left(\widehat{H}\bullet\Psi\right)=\widehat{R}\bullet\left(E\Psi\right)=E\left(\widehat{R}\bullet\Psi\right)$$

(2.100)

hence, from the identity of the extreme terms, results that also $\widehat{R}\bullet\Psi$ is an Eigen function state of the relative molecular system. But a state Eigen function must satisfy the quantum normalization condition, simultaneously with the Eigen function Ψ. This simultaneity (as a mathematical system) can be accomplished if and only if $\widehat{R}\bullet\Psi$ and Ψ are connected through the relation:

$$\widehat{R}\bullet\Psi=\left(\pm1\right)\Psi$$

(2.101)

The result (2.101) has an enormous importance. First of all, appears as an eigenvalues equations for the operator of symmetry operation $\widehat{R}$, which gives the absolute quantum identity for symmetry operations applied to the molecular systems.

Secondly, even if it does not solve the molecular eigenfunctions $\widehat{R}\bullet\Psi$, being as difficult to find them as to solve the initial equation (2.100), it solves the eigenvalues of a symmetry operation (2.101). These are two, +1 and −1, integer numbers, discreet, therefore quantified. These Eigenvalues of the symmetry operators (operations) are special called *characters of the operation R* in the symmetry group of which it belongs (the symmetry operation R) – and is marked with $\Gamma(R)$.

Thirdly, but no less importantly, these Eigenvalues, or characters correspond to the irreducible representations (which cannot be further reduced by an operation of similarity) of the symmetry operations in a group.

For the matrix representation of the symmetry operations of symmetry groups, the most common representation, especially in crystallography, see the Volume IV of the present five-volume work, (Chiriac-Putz-Chiriac, 2005), the character of a symmetry operation corresponds to the sum of the numbers on the main diagonal of its irreducible matrix.

In this sense, by the matrix representation of the symmetry groups (associated to the molecules) is obvious how the eigenvalues (characters) +1 and −1 correspond to the one-dimensional matrix and [−1], respectively.

Thus, is extrapolated the idea that the Eq. (2.101) corresponds to the non-degenerate eigenfunctions (when the main quantum level is not energetic separated on its orbital sublevels) and to the one-dimensional irreducible representations.

These one-dimensional irreducible representations are noted according to the *Mulliken symbols*: with "A" if the character +1 are registered to the symmetry operation around the main axis C_n and with "B" for the situation of antisymmetry around the axis C_n specified by the character −1.

For the two-dimensional irreducible representations (the matrix of symmetry operation diagonalized around the main diagonal in blocks of irreducible one-dimensional matrix) is used the notation "E" (do not be confused with the operation of symmetry identity!) and for the tri-dimensional irreducible representations the symbol "T" is used

To all these symbols sub-indices "1" and "2" are added if the symmetry, respectively the antisymmetry is registered in relation to an axis $C_2 \perp C_n$ or $\sigma_v \perp C_n$ (if C_2 coincides with C_n).

Similarly, the super-indices $\langle ' \rangle$ and $\langle '' \rangle$ are added for the symmetry, respectively antisymmetry toward the plane σ_h (if applicable) and again as sub-indices are added the symbol "g" (gerade = even, in German) and "u"

(un-gerade = odd, in German) for the symmetry/antisymmetry, toward the inversion center "*i*" (if applicable).

Because of the elevate quantum importance of the symmetry operations character for characterizing the molecular structures, it seems a natural need to formulate some general rules, to establish the characters (eigenvalues) of symmetry operations in a symmetry group G of order h (the total number of symmetry operations of the group).

These rules derive from a very general expression that bonds the characters of the symmetry operations from a group, called the great orthogonally theorem

$$\sum_R \left[\Gamma_i(R)_{mn} \right] \left[\Gamma_j(R)_{m'n'} \right]^* = \frac{h}{\sqrt{l_i l_j}} \delta_{ij} \delta_{mm'} \delta_{nn'} \qquad (2.102)$$

and which rigorously says that "in a group, the matrix elements in irreducible representations $\left[\Gamma_i(R)_{mn} \right]$ are components of a vector in the h-dimensional space, that all these vectors are mutually orthogonal and each vector is normalized, so that the square of its <<length>> has the value h/l_i", with l_i the size of the irreducible representation i and $\delta_{ij} = 1$ for $i=j$ and equal to 0 otherwise (Kronecker delta symbol) (Cotton, 1990; Putz, 2006).

The Eq. (2.102) can be better understood by its consequences: the five rules for determining the symmetrical operations characters from a group.

But, to be even clearer, the exposure will be accompanied by an example, the one of group C_{2v}. The Table 2.3 says that this group has four symmetry operations: E, C_2, σ_v, σ_v'. It is searching for the irreducible representations of this group.

First of all, how many irreducible representations are?

R1. Here, the first rule says that the number of irreducible representations equals the number of classes in the group. For the group C_{2v} are four classes, each with a single symmetry operation, the one above.

Therefore, there must be four irreducible representations, currently denoted as unknown $X_1, X_2, X_3,$ and X_4

Each of these four irreducible representations will be associated with the four classes (in this case even operations) of the C_{2v} symmetry group, so that the connection between them to be the characters of symmetry operation in the relative representations. In other words, will be represented the symmetry operations in each of the four irreducible representations through the corresponding characters.

This way it creates the so-called character table, as a quantum correspondent of the group table.

C_{2v}	E	C_2	$\sigma_{v(xz)}$	$\sigma_{v(yz)}$
$X_1(?)$	?	?	?	?
$X_2(?)$	?	?	?	?
$X_3(?)$	?	?	?	?
$X_4(?)$	?	?	?	?

R2. Another rule of the character table construction prescribes that all the symmetry operations from a class have the same character in a representation (irreducible or reducible), a self-satisfied rule of table construction for the C_{2v} group above. Next are following the rules for completing the characters table.

R3. In a group, the dimensions squares sum of the irreducible representations equals the order of group h. As long as the dimension of an irreducible representation is fixed on the character module of the identity operation, in that representation the first rule will be mathematically written.

$$\sum_i \left[\Gamma_i(E) \right]^2 = h \tag{2.103}$$

In case of the group C_{2v} the rule application (2.103) with $h = 4$ allow the completion of the first column, in a unique way, knowing that the dimension of any representation should be minimum 1 (one).

C_{2v}	E	C_2	$\sigma_{v(xz)}$	$\sigma_{v(yz)}$
$X_1(?)$	$1:(1)^2$	?	?	?
$X_2(?)$	$1:(1)^2$	?	?	?
$X_3(?)$	$1:(1)^2$	?	?	?
$X_4(?)$	$1:(1)^2$	?	?	?

R4. The next rule says that "the squares sum of all characters for any irreducible representation equals the order of the group", in other words, which prescribes R3 "vertical" R4 re-formulates on the "horizontal" and is mathematically written in the general case:

$$\sum_R g\left[\Gamma_i\left(R\right)\right]^2 = h \qquad (2.104)$$

including the case when there are g symmetry operations in the symmetry class of operation R. For the group C_{2v} all symmetry classes have $g=1$ and the "horizontal" completion can be already done for the first irreducible representation, which is therefore identified with A_1 representation, according to the Mulliken symbols:

C_{2v}	E	C_2	$\sigma_{v(xz)}$	$\sigma_{v(yz)}{}'$
A_1	1	1	1	1
$X_2(?)$	1	?	?	?
$X_3(?)$	1	?	?	?
$X_4(?)$	1	?	?	?

R5. The rest of the characters table is completed based on the rule that absolute bearing the condition of orthogonally between two different irreducible representations, under the general mathematical form:

$$\sum_R g\left[\Gamma_i\left(R\right)\right]\left[\Gamma_j\left(r\right)\right] = 0 \qquad (2.105)$$

This rule mostly applies "visual", but it can also be rigorously rewritten in a system of algebraical equations. The case of C_{2v} group is a visual one, being obvious that in the remaining empty spaces from the last table on each horizontal line (unknown irreducible representation) should exist two positions (two characters) such as –1, and 2 as +1, and in different positions, from representation to representation, in order to verify the condition (2.105) to the product of characters between any two different irreducible representations.

There results the table:

C_{2v}	E	C_2	$\sigma_{v(xz)}$	$\sigma_{v(yz)}{}'$
A_1	1	1	1	1
A_2	1	1	-1	-1
B_1	1	-1	1	-1
B_2	1	-1	-1	1

where, again, depending on the character position −1, were denominated the irreducible representation, in agreement with Mulliken terminology. Similar analysis can be extended (Cotton, 1990; Lancashire, 2002), to any of the molecular punctual groups from Table 2.3.

Regarding the irreducible representations of a molecule symmetry group are advanced two fundamental questions:

(i) what is actually means? and
(ii) what are useful to? for characterizing the molecular structure.

We have been seen that the characters can be associated with the eigenvalues of the symmetry operations operators, the Eq. (2.101).

Also, from the previous discussion, it has been established that the horizontal character development, after the associated symmetry operations (more generally, classes) from a group, $\sum_{R} g\left[\Gamma_i(R)\right]$ characterizes an irreducible representation.

In the same time, the product of two different sizes satisfies the orthogonal relation type established by the great orthogonally theorem (2.102).

Yet, according to the [P3] postulate of quantum mechanics, such orthogonal dimensions like the irreducible representations of the symmetry groups, may be formed as a complete basic set (i) for eigenfunctions of the analyzed system (in this case molecular) (ii).

So we answered to both questions at once (i) and (ii), see above.

However, in practice, a molecular system is analyzed starting from its basic atoms, by admitting as a starting point the atomic orbitals of the constituent atoms as the basic set of primordial importance.

From the atomic set to the molecular one, through the symmetries of the molecular group, is passed through four main stages, further explained.

S1. The atomic basic set will generate, in relation to the symmetry operations of the respective molecule group, a reducible representation, whose determination will be made by setting of the each character from the representation according to the rule:

$$\Gamma_{total}(R) = N_{Immovable\ Atoms}\Gamma_{xyz}(R) \tag{2.106}$$

which prescribes that for each symmetry operation is obtained the total character, establishing the number of atoms that are not moving in any way by the symmetry operation performed ($N_{Immovable\,Atoms}$) and is multiplied by the character $\Gamma_{xyz}(R) = +1$ or -1 associated to an atom that changes its position symmetric and antisymentric, relatively to the considered symmetry operation.

S2. The reducible representation (2.106) will be reduced to the irreducible representations of the molecular group, using the molecular formula of detached decomposition from the orthogonally theorem (2.102)

$$a_i = \frac{1}{h}\sum_R g\left[\Gamma_i(R)\right]\left[\Gamma_{total}(R)\right] \tag{2.107}$$

The result of the decomposition with the coefficients of (2.107) indicates:

(1) the total number of irreducible representations that contributes to the reducible representation, which further indicates the number of the molecular orbitals (eigenfunctions) allowed by the electronic system with the considered nuclear geometry (symmetry), and

(2) the irreducible representations type that contribute to the reducible representation decomposition (2.106).

S3. The atomic projection basic set will be considered in relation with the irreducible representations of the molecule group which contributes to the reducible representation decomposition (2.106), using the operators of incomplete projectors (unnormalized):

$$\hat{P}^{X_i}\bullet = \frac{l_i}{h}\sum_R\left[\Gamma_i(R)\right]\hat{R}\bullet \tag{2.108}$$

and applies to each member of the atomic basis set, eliminating the null projection cases (trivial cases) and the equivalent one (by symmetry operations of the group).

Finally, for each irreducible representation that contributing (even in multiple) to the initial decomposition of the reducible representation, will be noted the result of the unequivalent and nontrivial null projection, result which will be an Eigen function for the analyze molecular system, and will be called *SALC* (*Symmetry-Adapted Linear Combination* = the Linear Combination Adapted by Symmetry).

There can be verified how the obtained SALC functions are eigenfunctions of the group symmetry, if applying to each SALC any symmetry operation from the group, are obtained identities of (2.101) type!

S4. The molecular eigenfunctions are normalized, the SALC obtained and there are formed the molecular orbitals, thanks to their linear combinations (sum, difference), so that being as many as indicated by the number (1) from (S4).

It can be verified the correctness of the analysis, if for each molecular orbital obtained is formed the associated reducible representation and further is decomposed in the same irreducible representations as for the atoms (atomic orbitals) from the primordial basic set, the steps (S1) and (S2).

For example, the allyl molecule, C_3H_5, is considered with representation in the Cartesian coordinates in Figure 2.23, for which, presenting the group symmetry C_{2v}, can be used the character table, previously determined.

(S1). Is decided the set of basic atomic orbitals as being the orbitals p_z ($\phi 1$, $\phi 2$ and $\phi 3$), one for each atom in the molecule. This atomic orbital are subjected to the symmetries molecule, to the C_{2v} group, in order to establish the symmetry reducible representation that generates this primordial atomic set.

In this case, a phase change of the orbitals p_z lobes (shown in Figure 2.23 by the white/black) toward the initial state after a symmetry operation, equals to the character −1 in (2.106).

By proceeding in the same way for all the C_{2v} group operations, is obtained the reducible representation of C_3H_5:

C_{2v}	E	C_2	$\sigma_{v(xz)}$	$\sigma_{v(yz)}{}'$
X	3	-1	1	-3

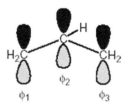

FIGURE 2.23   The atomic orbital set p_z in describing the π bond from the C_3H_5 molecule; after (Putz, 2006).

(S2). Further, will be "reduced" the total reducible representation, previously obtained for the irreducible representations of the C_{2v} group, which produces – by applying (2.107) – the results: $a_{A1} = 0$, $a_{A2} = 1$, $a_{B1} = 2$, $a_{B2} = 0$, generating the decomposition (reduction): $X = A_2 + 2B_1$.

The latest result already indicates the existence of three molecular orbitals that must come from the combinations of SALC normalized eigenfunctions, after the projections (2.108) (non-trivial and non-equivalent) of the atomic orbitals on the irreducible representations (S3). & (S4).

The projections (2.108) are first obtained and then they are normalized to the SALC eigenfunctions:

$$\hat{P}^{A_2} \bullet \phi_1 = (\phi_1 - \phi_3)/2 \to \mathrm{SALC}(A_2) = (\phi_1 - \phi_3)/\sqrt{2} \qquad (2.109a)$$

$$\hat{P}^{B_2} \bullet \phi_1 = (\phi_1 + \phi_3)/2 \to \mathrm{SALC}_1(B_1) = (\phi_1 + \phi_3)/\sqrt{2} \qquad (2.109b)$$

$$\hat{P}^{B_2} \bullet \phi_2 = \phi_2 \to \mathrm{SALC}_2(B_1) = \phi_2 \qquad (2.109c)$$

Finally, the three molecular orbitals of the allyl molecule are formed, considering the linear combinations of the eigenfunctions SALC with the same symmetry (in this case for the B_2 representation), ignoring the normalization factors:

$$\Psi_1 \approx (\phi_1 + \phi_3) + \phi_2 = \phi_1 + \phi_2 + \phi_3 \qquad (2.110a)$$

$$\Psi_2 \approx (\phi_1 + \phi_3) - \phi_2 = \phi_1 - \phi_2 + \phi_3 \qquad (2.110b)$$

$$\Psi_3 \approx \phi_1 - \phi_3 \qquad (2.110c)$$

The molecular orbitals representation (2.110), following the orbital phases (+ or - in front of the p_z atomic orbitals from the primordial bases), is shown in Figure 2.24.

This way, based on the symmetry groups theory and their representations, and in particular by the SALC exposed procedure, it can be formed a quantum picture of the molecular bonds, without approximations, but also without the possibility of energetic "ordering" of the obtained orbitals and without the capacity to assess the real order of combination of atomic orbitals in the molecular orbitals. Energetic and hybridization informations of

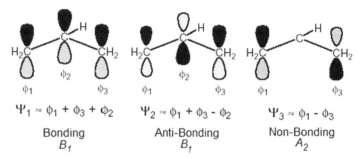

FIGURE 2.24 The molecular orbitals for the allyl molecule; after (Putz, 2006).

the molecular orbitals can be obtained, however, appealing to the supplementary approximations, expressed in the next Section.

2.3.3 LINEAR COMBINATIONS OF ATOMIC ORBITALS (LCAO): HЬCKEL APPROXIMATION

Based on the quantum postulate of basis set the SALC formalism for the construction of the molecular orbitals, from the atomic orbitals, by the molecular symmetry group, can be generalized to the construction of the molecular orbital (Ψ_k) of the linear combination of atomic orbital (ϕ_i), abbreviated LCAO – "Linear Combinations of Atomic Orbitals"

$$\Psi_k = \sum_{i=1}^{n} c_{ik}\phi_i \tag{2.111}$$

appealing, to determine the coefficients c_{ik}, to the postulate of the Schrödinger equation, rewritten under integral form (Putz, 2006):

$$\int \Psi_k \left[\left(\hat{H} - E \right) \bullet \Psi_k \right] d\tau = 0 \tag{2.112}$$

With the expression (2.111) in (2.112) appear typical integral terms, which are renoted and interpreted as:

$$H_{ii} = \int \phi_i \hat{H} \bullet \phi_i d\tau \tag{2.113}$$

as the integral that defines the energy of atomic orbital ϕ_i and which is called the Coulomb integral;

$$H_{ij} = \int \phi_i \, \hat{H} \bullet \phi_j d\tau \tag{2.114}$$

as the integral that defines the energy of the interaction between the atomic orbitals ϕ_i and ϕ_j, and describes the binding energy between two atomic orbitals, being called the resonance integral;

$$S_{ij} = \int \phi_i \phi_j d\tau \tag{2.115}$$

as being the overlap integral between the atomic orbitals ϕ_i and ϕ_j and which becomes identical with the value Kronecker delta $S_{ij} = \delta_{ij}$, if the considered atomic orbitals are orthogonal.

With these notations, the equation of variation (2.112) becomes the system of equations:

$$\begin{cases} c_{11}(H_{11} - ES_{11}) + ... + c_{1n}(H_{1n} - ES_{1n}) = 0 \\ \vdots \\ c_{n1}(H_{n1} - ES_{n1}) + ... + c_{nn}(H_{nn} - ES_{nn}) = 0 \end{cases} \tag{2.116}$$

There is important to note how the number of Eq. (2.116) is equal with the number of the considered atomic orbitals (n in this case) and equal to the number of molecular orbitals that will be generated: the system solutions (2.116) are the orbital coefficients from the Eq. (2.111).

However, from the inability to analytical and numerical general assess the integrals H_{ii}, H_{ij} and S_{ij} is still used the rationing by approximation.

The simplest is the Hückel approximation, which although crude, gives remarkable results in some specific situations (and especially for the π molecular orbitals of planar hydrocarbons) the Hückel approximation considers:

1. for the atoms i which are not directly involved in the bond: $H_{ij}=0$;
2. if all the atoms and the bonds in the molecule are identical, then all the terms H_{ij} are identical and equal for the atoms directly involved in the bonds, and are symbolized through the β-value; the same for the terms H_{ii} and are symbolized through the α-value.

3. there are considered the orthonormal atomic orbitals, i.e., satisfying the identity of overlap integral with Kronecker delta value: $S_{ij}=\delta_{ij}$.

Under these condition, the secular system (2.116) is considerably simplified, whose solution exists if and only if the Hückel secular determinant associated is zero, generating the equation:

$$
\begin{vmatrix}
(\alpha - E) & \cdots & \beta_{1n} \\
\vdots & & \vdots \\
\beta_{n1} & \cdots & (\alpha - E)
\end{vmatrix} = 0 \tag{2.117}
$$

Even with the simplification (2.117) remains the problem of solving the equation by this determinant.

Fortunately, even for the general case there is the possibility to simplify through the factorization of the determinant in diagonal blocks (2.117), which solutions must be determined from equalize to zero each factor from block-diagonal.

Wherefrom this possibility?

From the bond of the atomic orbitals with the SALC representation of molecular orbitals, they also formed thanks to the linear combinations of the atomic orbitals, like LCAO.

How it works?

Is symbolically listed the molecular orbitals and is colligated with the irreducible representations, associated to the decomposition of reducible representation of the primordial set of atomic orbitals, grouped (by one, by two etc.) according to the multiplicities with which the irreducible representations of symmetry group of the molecule appear in the decomposition of reducible representation of the atomic orbitals.

This list is used to properly naming the lines and the columns in the Hückel secular determinant.

Only where are intersected the lines and the columns from the determinant, belonging of the same irreducible representation, will be non-null elements, based on the great orthogonality theorem (2.102) which recommend SALC as being orthogonal eigenfunctions. It also takes into account the adjacency of "×" SALC type which prescribes unitary contributions for the neighboring atomic orbitals and zero for the rest. Thus, the determinant is factorized in diagonal blocks which are resolved, one by one, by the properly equalization to zero.

For better understanding the processes exposed, we will resume (or continue) the example of the allyl molecule from the previous chapter, Figure 2.23. The LCAO representation prescribes how the molecular eigenfunctions of the C_3H_5 molecule have the general form:

$$\Psi_k = c_{1k}\phi_1 + c_{2k}\phi_2 + c_{3k}\phi_3 \ , k = 1,2,3. \tag{2.118}$$

The coefficients $c_{1,2,3}$ from (2.118) will be determined appealing to the Hückel approximation, the factorization of the associated secular determinant according to the SALC phenomenology and, using in this framework, the condition of normalization, the quantum postulate through the equation of type (1.56a), of the Eigen orbitals (2.118), which generates the additional equations (if applicable):

$$\left(c_{1k}\right)^2 + \left(c_{2k}\right)^2 + \left(c_{3k}\right)^2 = 1 \ , k = 1,2,3. \tag{2.119}$$

It customized the Hückel secular determinant (2.117) for the allyl molecule where is recognized $\beta_{13} = \beta_{31} = 0$, the atoms "1" and "3" is not being directly related, and is obtained the equation:

$$\begin{vmatrix} \alpha - E & \beta & 0 \\ \beta & \alpha - E & \beta \\ 0 & \beta & \alpha - E \end{vmatrix} = 0 \tag{2.120}$$

The Eq. (2.120) is not difficult to solve in this case, but it will be illustrated the SALC factorization model in diagonal blocks, which in many cases reduces the volume of calculation.

The molecular eigenfunctions are inserted (2.118) and grouped by decomposition in irreducible representations of the C_{2v} group for the p_z atomic orbitals of the allyl molecule $X=A_2+2B_1$, for which, by considering the composition of "×" SALC from (2.109), are obtained also the adjacent factors, for example:

$$\begin{aligned}
SALC_1(B_1) \times SALC_1(B_1) &= 2^{-1/2}(\phi_1 + \phi_2) \times \phi_2 \\
&= 2^{-1/2}(\phi_1 \times \phi_2 + \phi_3 \times \phi_2) \\
&= 2^{-1/2}(\phi_1 \times \phi_2 + \phi_3 \times \phi_2) \\
&= 2^{-1/2}(1+1) \\
&= \sqrt{2} \\
&= SALC_1(B_1) \times SALC_1(B_1)
\end{aligned} \tag{2.121}$$

Thus, the correspondences are generated:

$$\underbrace{\Psi_1, \Psi_2, \Psi_3}_{B_1 \quad A_2} \rightarrow \begin{array}{c|ccc} & \Psi_1 & \Psi_2 & \Psi_3 \\ \hline \Psi_1 & \alpha - E & \sqrt{2}\beta & 0 \\ \Psi_2 & \sqrt{2}\beta & \alpha - E & 0 \\ \Psi_3 & 0 & 0 & \alpha - E \end{array} \tag{2.122}$$

producing the factorization of the Eq. (2.120) in two simplified equations, properly reduced with the associated irreducible representations

$$B_1 : \begin{vmatrix} \alpha - E & \sqrt{2}\beta \\ \sqrt{2}\beta & \alpha - E \end{vmatrix} = 0 \ \& \ A_2 : |\alpha - E| = 0 \tag{2.123}$$

with the simple solutions of the relative Eigen energies

$$E_1 = \alpha + \sqrt{2}\beta; E_2 = \alpha - \sqrt{2}\beta; E_3 = \alpha \tag{2.124}$$

For each Eigen energy the associated secular system of equations is solved (Putz, 2006):

$$\begin{pmatrix} \alpha - E_k & \beta & 0 \\ \beta & \alpha - E_k & \beta \\ 0 & \beta & \alpha - E_k \end{pmatrix} \begin{pmatrix} c_{1k} \\ c_{2k} \\ c_{3k} \end{pmatrix} = 0, k = 1,2,3 \tag{2.125}$$

in conjunction with the normalization condition (2.119) for generating the LCAO coefficients and, at the end, the molecular orbital (2.118) for the allyl molecule.

The results are obtained as follows:

$$E_1 = \alpha + \sqrt{2}\beta \rightarrow \Psi_1 = \frac{1}{2}\phi_1 + \frac{1}{\sqrt{2}}\phi_2 + \frac{1}{2}\phi_3 : \text{bonding molecular orbital}$$

(2.126a)

$$E_2 = \alpha - \sqrt{2}\beta \rightarrow \Psi_2 = \frac{1}{2}\phi_1 - \frac{1}{\sqrt{2}}\phi_2 + \frac{1}{2}\phi_3 :: \text{nonbonding molecular orbital}$$

(2.126b)

$$E_3 = \alpha \rightarrow \Psi_3 = \frac{1}{\sqrt{2}}\phi_1 - \frac{1}{\sqrt{2}}\phi_3 : \text{antibonding molecular orbital}$$

(2.126c)

Firstly, there is noted the close similarity of the results (2.126) with those obtained by the exclusive SALC method, the Eq. (2.110), unless the coefficients and the fact that in this approach have been deducted also the associated Eigen energies, so that the spectral arrangement of these molecular orbitals can be formed.

Considering the α integral as referential, there is convenient its fixation to zero, $\alpha = 0$, hence the molecular Eigen energies result in β units, as shown in Figure 2.25.

As an observation, the coefficients of atomic orbitals in (2.126) are the real ones, i.e., those that fix the size of the lobes of atomic orbitals in associated molecular orbital, therefore having the role of describing the population of those atomic orbitals in molecule.

If in the relations (2.110) were considered also the coefficients prescribed by the SALC of provenance, the Eq. (2.109), the molecular orbitals $\Psi_{1,2}$ would not correspond with the ones determined in (2.126); therefore, the equi-orbital modeling is preferred, in Figure 2.24.

The phases and the nodes (the points of change of the phase) of atomic orbitals in molecular orbitals were, however, kept.

Therefore, it is noteworthy how the properties of molecular symmetry determine the quantum structure of the molecules, even if, for an elaborate characterization (the Eigen energies, the population of the molecular orbital etc.) this approach should be completed with additional approximations, such as the Hückel ones, here considered.

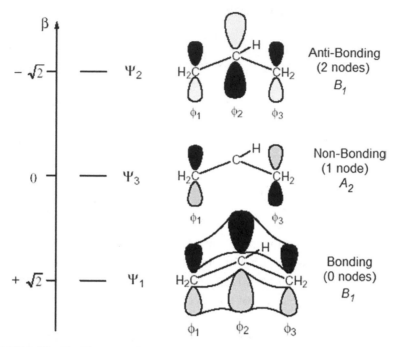

FIGURE 2.25   The Eigen energies and the orbitals for the allyl molecule; after (Putz, 2006).

2.4   ATOMS-IN-MOLECULE COORDINATION BY CRYSTALINE FIELDS' ANALYSIS

2.4.1   THE SEPARATION ORIGIN OF THE FOUR-LOBED ORBITALS

"Further we must accept the universe as a continuous creation, evolution and abolition of forms and that the purpose of science is to probe these changes of the forms and, where possible, to explain them" (Thom, 1975). It is hard to give a more general definition of the natural sciences; the chemistry between them is a real ferrule. The Universe, in terms of chemistry, is "seen" through the chemical bonds between the various components of the matter: atoms, molecules, supramolecules, complexes etc. Therefore, one of the primordial science " cheos-chimos" missions (essence-fluid, in ancient Greek) is to imagine the models which support the transformation of the components of the matter through the bonds, and by the transformation of the bonds.

The typical covalent bond, for example, from the methane molecule between carbon and hydrogen, requires that each bond contributes with an electron to the so-called two-center bond, which is formed. There is also a second type of bond, same with two centers, but where both electrons come only from an atom or molecule, the donor from the bond generically being called ligand (from ligare = to bind in Latin language). This type of covalent bond has the equivalent denominations as the bond of coordination, dative covalent bond or donor-acceptor type bond. If the acceptor is a metal, the compound is called complex.

Remaining in the terminology area, is specified a complex as being homoleptic if contains the same type of ligands, heterolytic in the other cases, and ambidental if the ligands (usually nitrite or thiocyanate groups) coordinate with the metal ion in many ways, case in which the ligand atom is italicized, and being directly involved in the metal contact (e.g., $[Cr(NH_3)_5(NCS)]^{2+}$).

The denticity indicates the number of the donor atoms involved in the coordinative and the ligands can be *monodental* (e.g., O in $[MnO_4]^-$, C in $[Fe(CN)_6]^{3-}$, H_2O in $[Mn(H_2O)_6]^{2+}$, NH_3 in $[Co(NH_3)_6]^{3+}$) or polydental (*bi-dental*: O & O' in $[Fe(ox)_3]^{3-}$ or S & S' in $[Re(S_2C_2Ph_2)_3]$, *tridental*: N, N', N" in $[Cr(typ)_2]^{3+}$, or *polydental* macrocyclic: N, N', N", N''' in $[Ni(cyclam)]^{2+}$).

Even if it is not absolutely necessary that the acceptor atom from the coordination to be a metal, they are the ideal candidates and the most frequent in such types of bonds. Why is happening this way? Alfred Werner, had been postulated at the beginning of the 20th century, also being awarded the Nobel Prize in Chemistry in 1913 for this principle, that a metal ion can manifest two types of valence! The primary valence is associated with the oxidation state, a formal concept for the assignment of the charge distribution in a compound, and indicated by Roman numerals, in parentheses, after the name of the element. It should be highlighted that the oxidation state does not imply and does not indicate the real charges distribution in a compound; for example, WO_3 is written from the oxidation states point of view as (W^{6+}, $3O^{2-}$) being an tungsten oxide (VI) but which does not request that it contains ions of W^{6+}!

A more relevant example is the one of Fe_2O_4: if the compound is ionic exists $4O^{2-}$ which, from the condition of the global neutrality, each iron atom should be in oxidation state of $+8/3$, a not very suggestive picture. Instead, the compound can be written as a mixed-oxide state: $FeO+Fe_2O_3=Fe(II)$ $Fe(III)_2O_4$. But, the secondary valence, introduced by Werner, indexes the number of the attached groups (both anion and neutral groups) to a metal center through the *coordination number of the first coordination sphere*.

In addition, atoms or groups can be further attached to the metal ion, but less bound by it and at a greater distance, forming the secondary or external coordination sphere. For example, $Co(NH_3)_4Cl_3$ has four ammonium groups and two chlorine atoms in the first coordination sphere and a chlorine atom to the second one. A classic example, to describe the coordinate bond, toward the ionic one and the covalent one, was pronounced by Pauling relative to the complex $[Fe(H_2O)_6]^{3+}$, Figure 2.26.

A purely ionic vision of the complex $[Fe(H_2O)_6]^{3+}$ assigns the charge 3+ on the metallic center, Figure 2.26-(a), while the purely covalent approach recommends the charge 3– the iron atom, and the oxygen atoms have each of them a 1+ charge, Figure 2.26-(b). Neither of these extremes is not in agreement with the electronegativity ("the attraction tendency between the electrons of an atom in a molecule) of those elements (Fe: 1.8, O: 3.5, in Pauling scale) which suggests a Fe-O bond, polarized in the direction $Fe^{\delta+} - O^{\delta-}$.

Under these conditions, Pauling had been applied the principle of the local electroneutrality expressed as: the *real charges distribution in a molecule is made so that the charge on a single atom varies between –1 and +1*.

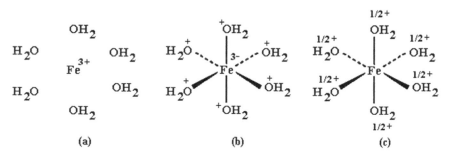

FIGURE 2.26 The model for the ionic bond (a), covalent (b) and coordinate (c) for the complex $[Fe(H2O)6]^{3+}$ (Putz, 2006).

For the current example, the electroneutrality principle suggests that the ideal charge on Fe is zero, which can be achieved only from a combination between the pure ionic state and the pure covalent one; Pauling called it 50% covalent or 50% ionic, so that also results as simple arithmetic average between the charges from the positions (a) and (b) of Figure 2.26, a charge distribution so that the Fe atom accepts three electrons from 6 atoms of oxygen donors, Figure 2.26-(c) in *coordination*.

This was the "macro" vision of the coordination bond. There is again asked the question: why the metal atoms, and especially the transition metals, coordinate most frequently?

At this point it can be qualitatively answered by quantum paradigm. A close look at elements, the Periodic Table, Figure 2.27, confirms that in an overwhelming proportion the metals are quasi-omnipresent from the 3rd period, i.e., from the atoms that have at least ($n = 3$) quantum main levels present at least orbitals with the orbital moment quantified from the orbital quantum number ($l = 3$), i.e., the d-type orbitals! What is special to these orbitals?

There is known that the occupation of the energetic levels in the atoms is made, besides the principle of minimum energy, also using the principle of maximum orbital penetration, Figure 2.28.

It had been established and illustrated for the Na atom in Figure 2.29-right, for example, how the orbitals s (and once with the increasing of the

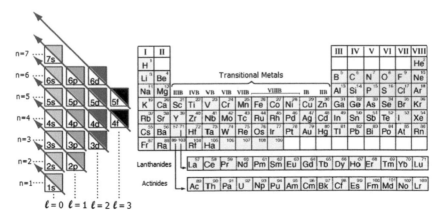

FIGURE 2.27 The occupancy rule for the electronic configuration of the atomic quantification levels (in left) so generating the ordering of elements in the Periodic Table (right); after (HyperPhysics, 2010).

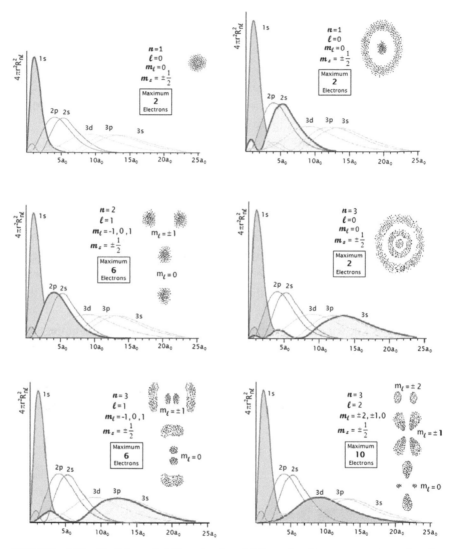

FIGURE 2.28 Radial probability density (bidimensional) representations for the (tridimensional) 1s, 2s, 2p, 3s, 3p, and 3d eigen-states of the Hydrogen atom, from left to right and up to bottom, respectively; after (HyperPhysics, 2010).

period for the transition metals) and the p-orbitals tend to have more negative energy, compared with the orbital $3d$, from larger radial penetration, so that resulting for the transition metals the orbital $3d$ with the energy above the one of $4s$, $4p$.

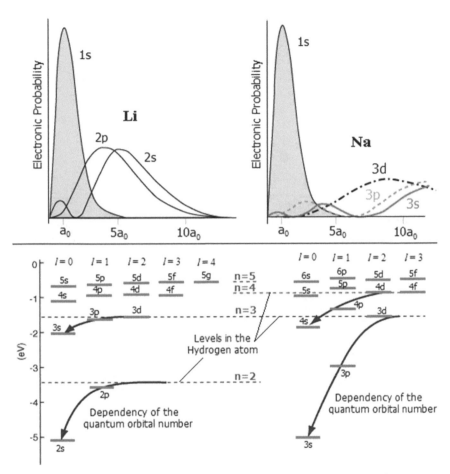

FIGURE 2.29 The orbital penetrations for the Li (left) and Na (right) atomic structures along their energetic levels orderings, respectively; after (HyperPhysics, 2010).

Therefore, it is normal that the orbitals $3d$ will be the first engaged in bond with the ligands, while the orbitals $4s$ and $4p$ will accomplish the orbital overlaps with ligands orbitals, Figure 2.30. This double character of the orbitals d, i.e. of *energetic interaction* but with a *small orbitalic overlap* with the ligands, is the key of the understanding the argument of the coordinative bond with the transition metals.

Moreover, one can develop even a quantum picture for the qualitative effect the interaction with the ligands has on the energetic level of the orbitals d.

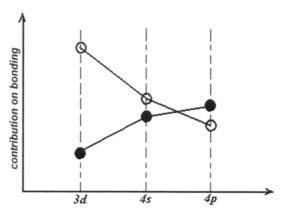

FIGURE 2.30 The relative contribution of 3d, 4s, and 4p orbitals to the contact with ligands, from energetic viewpoint (O) and the one of overlap with the ligands orbitals (•) (Putz, 2006).

For this, we will appeal first to the eigenfunctions related to the d-orbitals, but rewritten so that the eigen-solution to be separated in the pure radial contribution $R_{n,l}(r)$ and in the pure spatial contribution through the so-called spherical harmonics $Y_l^{m_l}(\theta,\varphi) = P_{l,m_l}(\theta)F_{m_l}(\phi)$:

$$\psi_{n,l,m_l}(r,\theta,\varphi) = R_{n,l}(r)Y_l^{m_l}(\theta,\varphi) \equiv (m_l) \qquad (2.127)$$

where the last denotation $\psi_{n,l,m_l}(r,\theta,\varphi) \equiv (m_l)$ helps for the writing simplification and does not cause confusion, as long as the analysis is limited to a fixed orbital, like d orbitals, here quantified through $l = 2$, see Table 2.5.

Under these conditions, the relative spherical harmonics will be given according to the Table 2.5 and to the literature data, see for instance (Barrow, 1963; Dicke & Wittke, 1960; Eisberg & Resnick, 1985; Giancoli, 1995; Greene, 1999; Haken & Wolf, 1996; Herzberg, 1944; Jones & Childers, 1990; Kaufmann, 1991; Keller, 1983; Leighton, 1959; Merzbacher, 1998; Ohanian, 1989; Pauling & Wilson, 1935; Richtmyer et al., 1969; Rohlf, 1994; Serway, 1990; Serway et al., 1997; Thornton & Rex, 1993; Tipler, 1992; Tipler & Llewellyn, 1999; White, 1972; Young, 1968):

$$Y_2^0(\theta,\varphi) = \sqrt{\frac{5}{16\pi}}\left(3\cos^2\theta - 1\right) \qquad (2.128a)$$

TABLE 2.5   The Stationary Schrödinger Equation Solution for the Hydrogen Atom, At a Glance; After (Putz, 2006)

Type of Solution	Form of the Solution	Conditions of Existence	Quantification Consequence
Radial: $R(r)$	$R_{n,l}(r) = r^l L_{n,l}(r) e^{-r/(na_0)}$ $L_{n,l}$: Laguerre functions $a_0 = \dfrac{\hbar^2 4\pi\varepsilon_0}{m_0 e^2}$ the first Bohr radius	If and only if: $n=1, 2, 3,...$ n: principal quantum number	*Quantification of Energetically levels* $E = -\dfrac{m_0 e^4}{8\varepsilon_0^2 h^2}\dfrac{1}{n^2}$ $= -\dfrac{13.6}{n^2}[eV]$ $n=1$: level "K"; $n=2$: level "L"; $n=3$: level "M"...
Orbitalic: $P(\theta)$	$P_{n,l,m}(\theta) = N_{l,m} P_n^m(\cos\theta)$ P_n^m: associated Legendre polynomials; $N_{l,m} = \left[\dfrac{(2l+1)(l-m)!}{2(l+m)!}\right]^{1/2}$: Normalization factor	If an only if: $l = 0, 1, 2, 3, ...$ $n-1$ l: orbitalic quantum number	*Quantification of orbital kinetic momentum*: $L^2 = l(l+1)\hbar^2$ $l=0$:orbital s (sharp); $l=1$:orbital p (principal); $l=2$:orbital d (diffuse); $l=3$:orbital f (fundamental); $l=4$:orbital g ... (alphabetically).
Azimuthal: $F(\phi)$	$F_m(\phi) = A\exp(im_l\phi)$; A: normalization factor	If an only if: $m_l = -l, -l+1, ..., +l$ m_l: magnetic quantum number	*Quantification on $0Z$ projection of the orbital kinetic momentum*: $L_z = m_l \hbar$

$$Y_2^{\pm 1}(\theta,\varphi) = \sqrt{\frac{15}{8\pi}}\sin\theta\cos\theta\exp(\pm i\phi) \qquad (2.128b)$$

$$Y_2^{\pm 2}(\theta,\varphi) = \sqrt{\frac{15}{32\pi}}\sin^2\theta\exp(\pm 2i\phi) \qquad (2.128c)$$

The angular Eigen solutions transformation (2.128) in the real space can be accomplished by appealing to the "stationarization" procedure of the phases through the normalized combination of the Eigen function (2.127) for the d-orbitals. There is obtaining:

$$d_{z^2} = (0) \propto z^2 \tag{2.129a}$$

$$d_{yz} = \frac{1}{\sqrt{2}}[(1)-(-1)] \propto yz \ \& \ d_{zx} = \frac{1}{\sqrt{2}}[(1)+(-1)] \propto xz \tag{2.129b}$$

$$d_{xy} = \frac{1}{\sqrt{2}}[(2)-(-2)] \propto xy \ \& \ d_{x^2-y^2} = \frac{1}{\sqrt{2}}[(2)+(-2)] \propto x^2 - y^2$$

$$\tag{2.129c}$$

where the coordinate transformation from the spherical system and the Euler's relations were considered. Pictorially, the spatial d-orbitals from (2.129) are represented in Figure 2.31.

These are the d orbitals of the central metal surrounded by the ligands. In the ligands *absence*, the valence electrons of a free ion are, essentially, characterize by three energetic constraints:

i. have kinetic energy;
ii. are attracted from the nucleus;
iii. are mutual rejected;
iv. in the ligands presence there appears an energetic constraint, in addition besides the above 3 already mentioned, thanks to the non-spherical field of valence electrons of the ligands ions- *ligands'field*. If the ligands field action toward the electrons of the central metal toward a constraint which is *limited only to the electrons repelling*, it is called the *crystalline field*. Therefore, the crystalline field is a particular case of the ligands field, considering only the constraint of rejection between the valence electrons of the central ion and surrounding (called as the *crystalline field*).

In the ligands crystalline field approximation it can be considered as point of negative charge that surrounding the central ion (with the d-orbitals in Figure 2.31).

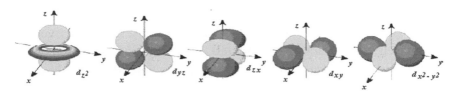

FIGURE 2.31   The spatial orbitals "d"; after (Putz, 2006).

Thanks to the very large frequency of the octahedral coordination (the symmetry group O_h) and tetrahedral (the symmetry group T_d) – also review the Table 2.2 – will be treated especially as the cases of the relative crystalline fields so generated, and also with the interaction effect with the orbitals d of the central ion (the so-called *interstitials*). For the octahedral crystalline field, Figure 2.32-left, there is noted how the orbitals d from the Figure 2.31 can be classified in two interaction groups with the charges of the ligands L.

One of the groups is formed by the orbitals d_{xy}, d_{yz}, d_{zx}, together marked as eigenfunctions t_{2g}, according to the Mulliken notations – and whose interaction (rejection) with the punctual charges of the ligands are equivalent, but smaller as intensity, compared to the one of the orbitals d_{z^2}, $d_{x^2-y^2}$ – together marked as eigenfunctions e_g – those presenting the orbital lobs right to the lines that bind the ligands charges with the central metal.

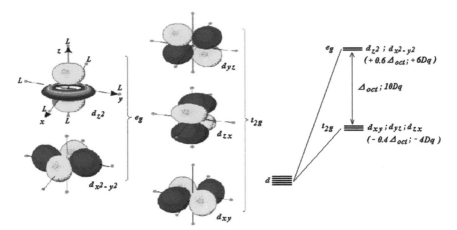

FIGURE 2.32   The separation types (left) and the energetic levels (right) of the orbitals d-interaction with the crystalline field O_h; after (Putz, 2006).

Therefore, the crystalline field effect is to "abolish the degeneration", i.e. the initial one, of the five types of d-orbitals in Figure 2.31; Yet, not completely, but separating the orbitals of the free ion in two groups of three and two equivalent d-orbitals, still degenerate, respectively corresponding to the groups t_{2g} and e_g above.

Also it has to be noted how, beyond the partial rising of orbital degeneration d of the free ion, the perturbation induced by the ligand crystalline field moving energetic global both groups t_{2g} and e_g is in the way indicated in Figure 2.33-right. But, regarding the energy separation of the groups t_{2g} and e_g had been noted, this time in terms of quality, the width Δ_{oct} or $10Dq$ (symbol whose meaning will be revealed in the next section) is distributed between the two groups after the so-called *barycenter rule*: the interval Δ distribution in its $0.2=1/5$ fractions is made so that to equalizes the energy distributed "up" and "down of this interval gravity center (the barycenter)".

For example, if the orbitals d are separated in the crystalline field O_h, of Figure 2.32-right, the superior energetic classification (double degenerate) had been designated as e_g by the value $+0.6\Delta_{oct}$, which heuristically is equivalent to $2\times(0.6) = +1.2$, while to the inferior energy group (triple degenerate) t_{2g} had been assigned the value $0.4\Delta_{oct}$, which heuristically is equivalent to $3\times(-0.4) = -1.2$! The case of the separation of the free ion d-orbitals in case of perturbation induced by the crystalline field of

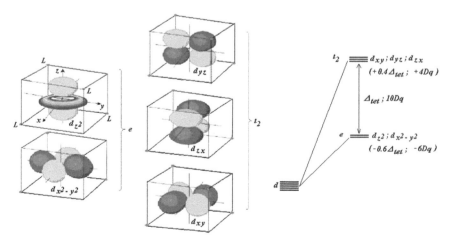

FIGURE 2.33 Separation on the types (left) and energetic levels (right) of the orbitals d-interaction with the crystal field T_d; after (Putz, 2006).

tetrahedral T_d is analyzed in Figure 2.33, as previously for the crystalline field O_h, but generating reverse separation.

This is happening because in this case are the orbitals from the group t_2 (the lower-indices "g" disappears because, although the d-orbitals retain the inversion symmetry, the crystal field T_d does not present it, see Table 2.3) are the closest to the ligands of the electric charges, thus supporting the most significant repulsion and the perturbation and the most consistent energetic displacement, toward the orbitals of the group e (again without the lower-indices "g").

One also notes how the barycenter rule recommends in this case the energy allocation $+0.4\Delta_{tet}$ to the group t_2 and respectively $0.6\Delta_{tet}$ to the group e. However, for the same type of metal found, in turn, in the crystal field O_h and T_d and for the same ligands, it can be written, between the relative separation interval units $(1/\Delta)$, by the simple inspection of the representations from right of Figure 2.32 and Figure 2.33, through the relation:

$$\frac{6}{4}\frac{1}{\Delta_{oct}} = \frac{4}{6}\frac{1}{\Delta_{tet}} \Rightarrow \Delta_{tet} = \frac{4}{9}\Delta_{oct} \qquad (2.130)$$

which indicates (as observed in practice as well) an absorption line $v=\Delta_{tet}/h$ (with h – the Planck constant) more moved to the read (inferior frequencies) than in the coordination case, with the same metal and the same ligands, in the octahedral symmetry.

These were the qualitative considerations regarding the coordinative bond, in the approximation of the ligands' crystal field with the symmetry O_h and T_d toward the central metal ion, especially the transitional one, and its d-orbitals.

A quantitative quantum approach will be presented in the next section.

2.4.2 THE QUANTUM MODEL OF THE CRYSTAL FIELD

The quantum approach of the ligands field interaction with the orbitals (in this case of d-type) of the central metal supposes to solve the Schrödinger equation, see the Volume I of the present five-volume book (Putz, 2016a)

$$\frac{\hbar^2}{2m}\frac{d^2\psi}{dx^2} + \left[E - V(x)\right]\psi = 0 \qquad (2.131)$$

for the Eigen energies and wave eigenfunctions of this system, as a given symmetry interaction (in octahedral and tetrahedral cases).

However, if the ligand crystal field is considered as a weak disturbance toward the central metal orbitals, it can be applied the first order perturbation method, for which the first order corrections to the free metal orbitals Eigen energies are calculated using the eigenfunctions from the undisturbed free-state, as follows.

This can be easily demonstrated, if the Schrödinger equation is written (2.131) in perturbation regime with λ-perturbation factor that disturbs the Hamiltonian, the eigenfunctions and the free eigenfunctions (symbolized by the upper-index 0") by the respective contributions of the first order (the upper-indices "1"), and so on.

There results the Eigen function:

$$\left(\widehat{H}^{(0)} + \lambda \widehat{H}^{(1)} + ... \right) \bullet \left(\Psi^{(0)} + \lambda \Psi^{(1)} + ... \right) = \left(E^{(0)} + \lambda E^{(1)} \right) \left(\Psi^{(0)} + \lambda \Psi^{(1)} + ... \right)$$

$$(2.132)$$

which, if reduced to the first order perturbations (are not considered the terms of λ with powers higher than 1), can be rewritten as the system:

$$\begin{cases} \widehat{H}^{(0)} \bullet \Psi^{(0)} = E^{(0)} \Psi^{(0)} \\ \widehat{H}^{(0)} \bullet \Psi^{(1)} + \widehat{H}^{(1)} \bullet \Psi^{(0)} = E^{(0)} \Psi^{(1)} + E^{(1)} \Psi^{(0)} \end{cases} \quad (2.133)$$

If one multiplies on the left the system second equation (2.133) with the Eigen function $\Psi^{(0)*}$ and then taking the spatially integral one obtains:

$$\int \Psi^{(0)*} \left[\widehat{H}^{(0)} \bullet \Psi^{(1)} \right] dv$$

$$+ \int \Psi^{(0)*} \left[\widehat{H}^{(1)} \bullet \Psi^{(0)} \right] dv = E^{(0)} \int \Psi^{(0)*} \Psi^{(1)} dv + E^{(1)} \int \Psi^{(0)*} \Psi^{(0)} dv \quad (2.134)$$

from where, by applying the normalization condition (3.3) of the Eigen function $\Psi^{(0)*}$ for the last term on the right side,

$$\int \Psi^{(0)*} \Psi^{(0)} dv = 1 \quad (2.135)$$

and considering the hermiticity relation for the Hamiltonian $\hat{H}{}^{(0)}$ for the first term on the left-side, when, combined with the first system equation (2.133),

$$\int \Psi^{(0)*} \left[\widehat{H}^{(0)} \bullet \Psi^{(1)} \right] dv = \int \Psi^{(1)} \left[\widehat{H}^{(0)} \bullet \Psi^{(0)} \right]^* dv$$

$$= E^{(0)} \int \Psi^{(1)} \Psi^{(0)*} dv = E^{(0)} \int \Psi^{(0)*} \Psi^{(1)} dv \quad (2.136)$$

the simple expression comes out:

$$E^{(1)} = \int \Psi^{(0)*} \left[\widehat{H}^{(1)} \bullet \Psi^{(0)} \right] dv \quad (2.137)$$

which is particularly important.

The relation (2.137) clearly shows how the perturbation first order of the first order energy correction is calculated based on the free system eigenfunctions, unperturbed. Meanwhile, here will be generalized the treatment to the orbital d-case Ψ_i $(2l+1)$-degenerated in the atomic orbitals bases ψ_j:

$$\Psi_i = \sum_{j=-l}^{l} c_{ij} \psi_j, \sum_j c_{ij}^* c_{ij} = 1 \quad (2.138)$$

according to the superposition and normalization postulates of the quantum mechanics, respectively.

Under these conditions, the Eq. (2.137) becomes the *secular determinant of the first-order perturbation*

$$\left| H_{ij} - E_{(i)}^{(1)} \delta_{ijk} \right| = 0, \ H_{ij} = \int \psi_i^* \left[\widehat{H}^{(1)} \bullet \psi_j \right] dv; \ i, j = \overline{-l, +l} \quad (2.139)$$

with Eigen energies, as solution. Further, there will be proceeded to the determination of the coefficients and of the total eigen-functions (2.138) by solving, for each case, a perturbed Eigen energy secular system of perturbation:

$$[H_{ij} - E_{(i)}^{(1)} \delta_{ij}][c_{ij}] = 0; \ i, j = \overline{-l, +l} \quad (2.140)$$

Next, the octahedral symmetry O_h will be considered as working model for quantitative-quantum treatment of the d-orbitals separation in the ligands crystal field, Figure 2.34.

Within this picture, the octahedral perturbation potential will be obtained from the Coulomb potentials summation, associated to each charge point (z) associated to the ligand L, found at the distance a from the coordination center, acting on a point $r(x, y, z)$ inside the octahedral formed:

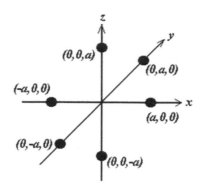

FIGURE 2.34 The symmetry and the charge points ligands producing the crystal potential O_h (Putz, 2006).

$$V_{(x,y,z)}(r) = \sum_{L=1}^{6} \frac{ez}{r_{L-(x,y,z)}} = \sum_{L=1}^{6} ez \left(\sum_{l=0}^{\infty} \sum_{m_l=-l}^{+l} \frac{4\pi}{2l+1} \frac{r^l(x,y,z)}{a^{l+1}} Y_{l(L)}^{m_l} Y_{l(x,y,z)}^{m_l *} \right)$$

(2.141)

where, in the last equality (2.141) the series resumation of spherical harmonics the function $1/r_{L-(x,y,z)}$ has been considered.

Now arises the question: until which order will be developed the series (2.141)? Taking into consideration that the octahedral potential (2.141) will be further combined in the matrix elements of type H_{ij}, as in (2.139), with the d-type orbital eigenfunctions, as in (2.127), i.e., forming integral of harmonic functions products, it can be used the fact that *the products angular integral of three spherical harmonic functions is canceled for l'>2l* (Cotton et al., 1995; Dou, 1990; Douglas et al., 1983; Figgis et al., 1966; Greenwood & Earnshaw, 1984; Huheey, 1983; Kettle, 1998; Konig, 1971; Lever, 1984; Levine, 1970; Meissler & Tarr, 1998; Nicholls, 1971; Purcell & Kotz, 1977; Schlafer & Gliemann, 1969; Shriver & Atkins, 1999; Tanabe & Sugano, 1954; Putz, 2006). As long as the current interest is the orbital d with $l = 2$, for the results in the series (2.141) will contribute only the terms which contain spherical harmonics until the maximum order $l = 4$.

Further will be analytical explained the octahedral potential (2.142), based on the previous observation, employing the expressions of the spherical harmonics functions until the order $l=4$, and also based on the geometry of Figure 2.34. This analysis, together with the results obtained, are summarized in Table 2.6, along – in order to exemplify the calculation – also the null contributions of the spherical harmonics $Y_{2(L)}^{m_l}$. By substituting the results from the Table 2.6 in (2.141), we can write the total perturbation potential from the ligands (L) on a point $r(x,y,z)$ from the octahedral space as such formed:

$$V_{(x,y,z)}(r) = 6\frac{ze}{a} + V_{oct}(r) \tag{2.142}$$

$$V_{oct}(r) = \frac{7}{\sqrt{\pi}}\frac{zer^4}{a^5}\left[Y_4^0 + \sqrt{\frac{5}{14}}\left(Y_4^4 + Y_4^{-4}\right)\right] \tag{2.143}$$

Notice how the first term of the perturbation potential (2.142) corresponds to the spherical harmonics Y_0^0, in Table 2.6, and is presented as the simple sum of the ligands charges toward the point at the distance a from them, meaning that it records the effect of the global displacement of the central ion energetic levels, an effect also noted in the previous section and schematized in Figure 2.32-right.

TABLE 2.6   The Ligands Charges Points Contributions to the Spherical Harmonics; After (Putz, 2006)

Spherical Harmonic	(0,0,a)	(0,0,−a)	(0,a,0)	(0,−a,0)	(a,0,0)	(−a,0,0)	Total $Y_{l(L)}^{m_l}$
	$\forall \phi$		$\theta = \pi/2$		$\theta = \pi/2$		
	$\theta = 0$	$\theta = \pi$	$\phi = \pi/2$	$\phi = -\pi/2$	$\phi = 0$	$\phi = \pi$	
$Y_0^0 = (4\pi)^{-1/2}$	1	1	1	1	1	1	$6(4\pi)^{-1/2}$
$Y_2^0 \propto (3\cos^2\theta - 1)$	2	2	-1	-1	-1	-1	0
$Y_2^{\pm 1} \propto \cos\theta\sin\theta$	0	0	0	0	0	0	0
$Y_2^{\pm 2} \propto \sin^2\theta\, e^{\pm 2i\phi}$	0	0	-1	-1	1	1	0
$Y_4^0 \propto P_4^0(\cos\theta)*$	8	8	3	3	3	3	$28\sqrt{\dfrac{9}{256\pi}}$
$Y_4^{\pm 4} \propto \sin^4\theta\, e^{\mp 4i\phi}$	0	0	1	1	1	1	$4\sqrt{\dfrac{315}{512\pi}}$

* $P_4^0(\cos\theta) = 35\cos^4\theta - 30\cos^2\theta + 3$.

The second term of the perturbation potential (2.142) contains superior spherical harmonics, the 4th order, and will be the responsible term for the separation and the partial raising of the degeneration of the orbitals d of the central ion in the O_h symmetry. Therefore, there will be identified $\widehat{H}^{(1)}$ from (2.139) with V_{oct} from (2.143) to evaluate the first order energetic corrections, induced by the crystal field in the orbitals d of the central metal. The next step is the evaluation of the H_{ij} elements from (2.139).

To this aim one appeals, firstly, to the atomic eigenfunctions as spherical harmonics (2.127), for which, by customizing the d-type orbitals $\psi_{n,2,m_l} = R_{n,2} Y_2^{m_l} := (m_l)$ and by separating the spatial dependence (by r) in the potential (2.143), it can be proceeded to the assumption of the *average distance at 4th power of the d-electrons* toward the central ion (Putz, 2006)

$$\int_0^\infty \left(R_{n,2}^* r^4 R_{n,2} \right) r^2 dr := \overline{r_2^4} \tag{2.144}$$

a quantity that will accompany any term of the H_{ij} elements from Eq. (2.139). Therefore, the remaining elements to evaluate are of the type:

$$\int (m_l[i])^* V_{oct}(m_l[j]) dv = \frac{7}{3} \sqrt{\pi} \frac{ze}{a^5} \overline{r_2^4} \int_0^\pi \int_0^{2\pi} \left[\begin{array}{c} \left[\begin{array}{c} Y_2^{m_l*[i]} Y_4^0 Y_2^{m_l[j]} + \\ \sqrt{\dfrac{5}{14}} \left[\begin{array}{c} Y_2^{m_l*[i]} Y_4^4 Y_2^{m_l[j]} \\ + Y_2^{m_l*[i]} Y_4^{-4} Y_2^{m_l[j]} \end{array} \right] \end{array} \right] \\ \sin\theta \, d\theta \, d\phi \end{array} \right] \tag{2.145}$$

The integrals of Eq. (2.145) are considerably simplified if one notes that, firstly, the spherical harmonics from Eq. (2.127) can be explained when based on the normalized Legendre polynomials

$$P_l^{m_l}(\theta) = \left[\frac{2l+1}{2} \frac{(l-|m_l|)!}{(l-|m_l|)!} \right]^{1/2} \left[\frac{(-1)^l}{2^l l!} (\sin\theta)^{|m_l|} \frac{d^{l+|m_l|}(\sin\theta)^{2l}}{d(\cos\theta)^{l+|m_l|}} \right] \tag{2.146}$$

and by the normalized azimuthally composition (Table 2.5)

$$F_{m_l}(\phi) = (2\pi)^{-1/2} \exp(im_l\phi) \tag{2.147}$$

Then, independently of the orbital quantum number l values, the certainly non-null combinations are just that containing products of the spherical harmonics fully satisfying the condition:

$$\int_0^{2\pi} Y_{l[i]}^{m_l[i]} Y_{l[j]}^{m_l[j]} Y_{l[k]}^{m_l[k]} d\phi \neq 0$$

$$\Leftrightarrow \int_0^{2\pi} F_{m_l[i]} F_{m_l[j]} F_{m_l[k]} d\phi \neq 0$$

$$\Leftrightarrow m_{l[i]} + m_{l[j]} + m_{l[k]} = 0 \qquad (2.148)$$

Applying the prescription (2.148) to the integral terms from Eq. (2.145), there results that for the term containing the spherical harmonic Y_4^0 the next condition is applied:

$$m_l^*[i] + 0 + m_l[j] = 0 \Leftrightarrow -m_l^*[i] = m_l[j] \Leftrightarrow m_l[i] = m_l[j] \qquad (2.149)$$

that is affecting only the contributions to the diagonal elements from the secular determinant:

$$\int (m_l[i])^* V_{oct}(m_l[i]) dv = \frac{7}{3\sqrt{2}} \frac{ze}{a^5} \overline{r_2^4} \left[\int_0^{\pi} P_2^{m_l\,*[i]} P_4^0 P_2^{m_l[i]} \sin\theta\, d\theta \right] \qquad (2.150)$$

Considering now the custom analytical expressions from Eq. (2.146), the relative diagonal elements for the d-orbitals in the crystal field O_h are obtained as:

$$\int (0)^* V_{oct}(0) dv = \frac{ze}{a^5} \overline{r_2^4}$$

$$\int (\pm 1)^* V_{oct}(\pm 1) dv = -\frac{2}{3} \frac{ze}{a^5} \overline{r_2^4}$$

$$\int (\pm 2)^* V_{oct}(\pm 2) dv = \frac{1}{6} \frac{ze}{a^5} \overline{r_2^4} \equiv Dq \qquad (2.151)$$

Similarly, the condition (2.149) is applied for the terms (2.145) containing the spherical harmonics Y_4^4 & Y_4^{-4} and the next restrictions are obtained:

$$\begin{cases} Y_4^4 : m_l^*[i] + 4 + m_l[j] = 0 \\ Y_4^{-4} : m_l^*[i] - 4 + m_l[j] = 0 \end{cases}$$

$$\Leftrightarrow \begin{cases} -m_l[i] + m_l[j] = -4 \\ -m_l[i] + m_l[j] = 4 \end{cases}$$

$$\Rightarrow \begin{cases} -m_l[i] = m_l[j] = -2 \\ -m_l[i] = m_l[j] = 2 \end{cases} \tag{2.152}$$

Based on these conditions, form (2.145) only the properly integrated terms are retained, with the results:

$$\int (\pm 2)^* V_{oct}(\mp 2) dv = \frac{\sqrt{35}}{6} \frac{ze}{a^5} \overline{r_2^4} \left[\int_0^\pi P_2^{2*} P_4^{\pm 4} P_2^2 \sin\theta\, d\theta \right] = \frac{5}{6} \frac{ze}{a^5} \overline{r_2^4} \tag{2.153}$$

Therefore, one reaches a position where the secular determinant associated to the energetic perturbations of the d-orbitals eigenfunctions' separation in the crystal field O_h can be formed under the extended form (2.139) by considering the elements (2.151) and (2.153) rewritten in terms of the final notation of Eq. (2.151), that gives

$$\begin{vmatrix} Dq - E^{(1)}_{(-2)} & 0 & 0 & 0 & 5Dq \\ 0 & -4Dq - E^{(1)}_{(-1)} & 0 & 0 & 0 \\ 0 & 0 & 6Dq - E^{(1)}_{(0)} & 0 & 0 \\ 0 & 0 & 0 & -4Dq - E^{(1)}_{(+1)} & 0 \\ 5Dq & 0 & 0 & 0 & Dq - E^{(1)}_{(+2)} \end{vmatrix} = 0 \tag{2.154}$$

To solve this determinant, one observes that by successive lines and columns interchanging's (when at every change only the determinant sign, and not its value, is modified!) the *associate diagonal form* is obtained, from where the equations along the diagonal blocks are respectively retained with the formed equations:

$$(0) : \left| 6Dq - E^{(1)}_{(0)} \right| = 0 \Rightarrow E^{(1)}_{(0)} = 6Dq$$

$$(1) \wedge (-1) : \left| -4Dq - E^{(1)}_{(\pm 1)} \right| = 0 \Rightarrow E^{(1)}_{(\pm 1)} = -4Dq$$

$$(2) \wedge (-2) : \begin{vmatrix} Dq - E^{(1)}_{(-2\&+2)} & 5Dq \\ 5Dq & Dq - E^{(1)}_{(+2)} \end{vmatrix} = 0 \Rightarrow \begin{cases} \left[E^{(1)}_{(-2\&+2)} \right]_1 = -4Dq \\ \left[E^{(1)}_{(-2\&+2)} \right]_2 = 6Dq \end{cases}$$

$$\tag{2.155}$$

The last stage of the algorithm consists in assigning the eigenfunctions to the energetic corrections (2.155) by solving the attached secular equations of type (2.140).

But, from the solutions form of Eq. (2.155) it is noted that for the energetic corrections $E_{(0)}^{(1)}$ and $E_{(\pm1)}^{(1)}$ there is no doubt in assigning the orbitals, these being directly identified from Eq. (2.129).

But, for the assignment of the ultimate energies of Eq. (2.155) to the orbital combinations $(2) \wedge (-2)$, one should solving the secular systems attached to each energetic value, namely as

$$\begin{cases} \left(Dq - E_{(-2\&+2)}^{(1)}\right)c_{(+2)} + 5Dqc_{(-2)} = 0 \\ 5Dqc_{(+2)} + \left(Dq - E_{(-2\&+2)}^{(1)}\right)c_{(-2)} = 0 \end{cases} \tag{2.156}$$

along the normalization condition (2.138) casting here as

$$c_{(+2)}^{*}c_{(+2)} + c_{(-2)}^{*}c_{(-2)} = 1 \tag{2.157}$$

For example, for the energy case

$$E_{(-2\&+2)}^{(1)} = -4Dq \tag{2.158}$$

when replaced in the system (2.156) there results the relation:

$$c_{(+2)} = -c_{(-2)} \tag{2.159}$$

which together with Eq. (2.157) provide the solution as the coefficients

$$c_{(+2)} = -c_{(-2)} = 2^{-1/2} \tag{2.160}$$

and which finally fixes the orbitalic combination type

$$\Psi = 2^{-1/2}[(2) - (-2)] \tag{2.161}$$

meaning in regaining just the d_{xy}-type orbital from Eq. (2.129) as being the one affected by the energy $-4Dq$.

Similarly, there is proceeded for the energy case $E_{(-2\&+2)}^{(1)} = 6Dq$ for which results the association with the orbital type $d_{x^2-y^2}$ as in Table 2.7,

TABLE 2.7 The Correspondents of the Orbitals d Separation Energies in the Crystal Field O_h (Putz, 2006)

Energy	Orbitals	Group
$-4\,Dq$	$d_{yz} = \dfrac{1}{\sqrt{2}}[(1)-(-1)] \propto yz$	t_{2g}
	$d_{yz} = \dfrac{1}{\sqrt{2}}[(1)-(-1)] \propto yz$	
	$d_{xy} = \dfrac{1}{\sqrt{2}}[(2)-(-2)] \propto xy$	
$6\,Dq$	$d_{z^2} = (0) \propto z^2$	e_g
	$d_{x^2-y^2} = \dfrac{1}{\sqrt{2}}[(2)+(-2)] \propto x^2 - y^2$	

obtaining the same group of orbitals and energy sets, as deduced also by the qualitative considerations in the previous section.

However, one can made an evaluation in terms of $10Dq$ on the energetic orbital separation parameter (2.151), yet quite roughly, by using the value $\overline{r_2^4} \approx 20a_0^4$, with a_0 being the first Bohr's radius (see Table 2.5). The approximation $a \approx 5a_0 \approx 2.5\,Å$, for the distance ligands-central ion, holds as long as the physical constants and the ligands' charges are reduced to unity:

$$10Dq \approx \frac{10}{6}\frac{20}{5^5}[a.u.] \approx 10^{-2}[a.u.] \approx 10^3[cm^{-1}] \approx 100[nm] \qquad (2.162)$$

although the experimentally observed value for the first transition series of the octahedral complex corresponds to about 10^4 [cm^{-1}].

Regarding the quantum treatment of the d-orbitals separation of the central ion (M) in the crystal field of a *tetrahedral coordination symmetry* ML$_4$ with the ligands L, one similarly proceeded as for the octahedral case ML$_6$, by starting from the crystal tetrahedral potential evaluation, this time produced by the points $\left(\pm a/3^{1/2}, \mp a/3^{1/2}, a/3^{1/2}\right)$ and $\left(\pm a/3^{1/2}, \mp a/3^{1/2}, -a/3^{1/2}\right)$. This way, the tetrahedral potential results identical with the octahedral one (2.143) until an analytical factor, namely:

$$V_{tet} = -\frac{4}{9}V_{oct} \qquad (2.163)$$

This relation will inverse the energies sign of the all elements of the secular determinant, while further producing the relationship

$$Dq_{tet} = \frac{4}{9} Dq_{oct} \qquad (2.164)$$

by a term as (2.151), however in accordance with the relation (2.130), as deduced in the previous section for the orbital separations energetic gap in the T_d and O_h symmetries, respectively.

This way, all the qualitative results developed in the previous section are here quantum chemically founded in the current quantitative approach. Thus, the possibility of the phenomenological extension at the d-orbitals levels multiple occupied with the electrons in the crystal field is here validated and will be further analyzed.

2.4.3 MULTI-ELECTRONIC ORBITALS IN THE CRYSTAL FIELD

The treatment exposed till now should be naturally extended to the case of the multi-electron orbitals placed in the crystal field (in the sense of symmetry) of the ligands, atoms or groups of atoms that coordinate the central ion (Cotton et al., 1995; Dou, 1990; Douglas et al., 1983; Figgis et al., 1966; Greenwood & Earnshaw, 1984; Huheey, 1983; Kettle, 1998; Konig, 1971; Lever, 1984; Levine, 1970; Meissler & Tarr, 1998; Nicholls, 1971; Purcell & Kotz, 1977; Schlafer & Gliemann, 1969; Shriver & Atkins, 1999; Tanabe & Sugano, 1954; Putz, 2006).

However, the multi-electronic treatment brings in an additional effect: the field applied to an electron by the other electrons from the *multi-electrons orbitals*.

Therefore, in order to satisfy the Pauli principle of the orbital occupancies, for the multi-electrons orbitals in the crystal field there appears two effects of electronic repelling:

i. the repelling from the valence orbitals electrons of the ligands (ligand field itself) and previously treated and quantified by the (octahedral or tetrahedral) measure Dq;

ii. the inter-electronic repelling, appeared for the central ion orbitalic electrons, quantified by the (Racah) measure of the so-called factor B.

Therefore, the consideration of the multi-electronic treatment of the resulting terms depending on the relation Dq/B between these factors appears as natural next step.

For the case $Dq>>B$ one has the situation of the strong crystal field, but in relation with B, because Dq still remains as a perturbation! On the contrary, the case $Dq<<B$ is treated as the *weak crystal field*.

The practical difference in considering these cases is the order in which the two influences are considered: firstly the strong one and subsequently the weak one.

Finally, both treatments should generate the same terms, the orbitals of multi-electrons - including the degenerations, yet located at different energetic levels, corresponding to the specific terms of the considered influences order.

However the 1:1 correspondence of the terms will be applied to both approaches and the so-called correlation diagram (Tanabe-Sugano) will result indicating their energetic shift paralleling the crystal field.

One will therefore starts from the analysis of the weak crystal field and the working model will be for the d^2 – orbitals type. This choice is natural, and moreover, it is not limitative for the crystal field perturbations because the case d^2 is analogous, however with the sign changed by energetic considerations, with the treatment of the d^{5-2} case in the so-called *hole*-formalism, based on the simple fact that two holes are repelled with the same intensity as two electric charges in their place, while the holes are attracted with the ligands charges with the same intensity, but with opposite sign, to that of which the electrons of the central ion are repelled by the same ligands charges.

Similarly, in the cases d^1 and d^{5-1} are follow the same type of relation, and so on it can be extracted information about the central ion terms in the crystal field, for the same stoichiometry.

For different stoichiometric the effect of crystal field toward the orbitals d^n will be the same with that one on the orbitals d^{10-n}.

One starts with the analysis of the weak crystal field; therefore, the major influence and the first one will be the inter-electronic repulsion.

For the configuration d^n, $1<n<10$, the occupant electrons can be coupled to each other in such manner so that the inter-repulsion B according to the Pauli principle is satisfied.

At this point, one should take into consideration that each electron is characterized by an orbital moment m_l and by a spin orbital moment m_s, so that the interactions that may occur appear as combinations of these moments: spin-spin coupling, orbital-orbital, and spin-orbital, altogether being called as *Russell-Saunders coupling* (LS).

However, the coupling strength is assumed in the order: spin-spin> orbital-orbital> spin-orbital.

This scheme works very well for the transition metals of the first row, and these will be considered in the following. The coupling spin-spin generates the resultant spin S with the associated spin momentum

$$M_S = \sum_i m_{s[i]} \tag{2.165}$$

while expressing the total contribution of the individual spin projections for the separated electrons on the axis 0z.

Similarly, the orbital-orbital coupling produces a resultant orbital moment L for which the terminology in capitals for the atomic correspondences is adopted, see Table 2.8, and whose projection on the 0z axis considers all the orbital moments as summed for the individual electrons:

$$M_L = \sum_i m_{l[i]} \tag{2.166}$$

Finally, the spin-orbital coupling, abbreviated as L-S, gives rise to a *total orbital moment J=L+S*, with total multiplicity (2S+1)(2L+1) and generating the multi-electronic terms symbolized by the form $^{2S+1}L_{(J)}$.

As an example, there is considered the d^2 case of 2 electrons that can occupy the 5 d-orbitals, but actually having available 10 spin-orbitals (including the possibility that an orbital is occupied by an electron with the momentum spin projection quantified as ±1/2).

TABLE 2.8   The Total Orbital Moment Nomenclature for the Multi-Electronic Configurations; After (Putz, 2006)

L	0	1	2	3	4	5
Notation	S	P	D	F	G	H

Thus, the possibilities are given by $C_{10}^2 = 10 \cdot 9/2 = 45$, saying that the d^2 orbitals are of 45 times degenerated, i.e., containing 45 sublevels to the same orbital energy (when the separating action of the crystal field is still not considered).

But, which are the terms of this multi-electronic d^2orbital, which are separated due to the inter-electronic repulsion, under the B-factor influence?

At first instance, the terms types which arise from this configuration of spin-orbital are written, such as $^{(2S+1)}L$, by using the vector coupling rule, for which the terms L which arise from the combination of the orbital moments l_1 (of the electron no. 1) and l_2 (of the electron no. 2) are laying between $(l_1 + l_2)$ and $| l_1 - l_2 |$, and analogously possess a total moment of spin S with values between $(s_1 + s_2)$ and $| s_1 - s_2 |$, successively decreasing by one unit.

In the present case: $l_1 = l_2 = 2$, and since for the electronic spin always $s = 1/2$, the terms possessing L=4 (G), 3 (F), 2 (D), 1 (P), 0 (S) – combined with S = 1 (triplet terms) and S = 0 (singlet terms) are generated, meaning the terms ^{1}G, ^{3}G, ^{1}F, ^{3}F, ^{1}D, ^{3}D, ^{1}P, ^{3}P, ^{1}S, ^{3}S, respectively.

The problem now arises from the fact that performing the sum of multiplicities (2S+1)(2L+1) for all these terms an equal degeneration to 90 instead of 45 (as above) is obtained.

There is clear that in each case an additional term was produced. Then, how can be determined the free ion d^2 terms (still not subject to crystal field action)? In Figure 2.35 there is illustrated such a method. In the left of the Figure 2.35 were symbolically modeled all the states (M_L, M_S) that may occur for the d^2 configuration from all the associated micro-states (m_l, m_s). Then, these resulted states were considered in the right of Figure 2.35 in a rectangular arrangement $M_L \times M_S$, by marking them with an "x". Then, the real terms which appear for this configuration are identified, as based on the principle according which the *inter-electronic B repulsion is stabilized by a maximum multiplicity*, i.e., for the electronic states with *maximum* L & M_L and S & M_S.

In order to identify the maximum successive rectangles $M_L \times M_S$, once identified a rectangle with complete rows and columns, the "x"-es from the successive rows and columns are deleted and one is moving to the next rectangle identification and so on, therefore indicating the maximum values of M_L and M_S for the existing terms, i.e., implicitly identifying also where L and S are associated.

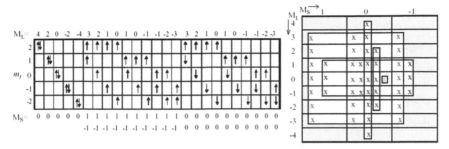

FIGURE 2.35 The microstates d^2 arrangement (left) and the spectral terms determination (right); after (Lancashire, 2002; Putz, 2006).

There results the d^2 terms: ^1G, ^3F, ^1D, ^3P, ^1S.

Last question here: which of these terms represent the fundamental state, i.e., of lowest energy, or the "permanent" state as Bohr called it?

The answer comes from the application of the above criterion and consists of two stages, by the celebrated Hund's rules:

- **(Hund1)** the terms are identified with maximum S;
- **(Hund2)** if the confusion persists (e.g., there are two terms with the same S) the fundamental state corresponds to the term with maximum L.

The resulted order is such as: maximum S and then maximum L under the necessity of stabilization of the large B-inter-repulsion since by maximum S the standing energy of two electrons with the same spin orientation being firstly "consumed".

In this case the term ^3F at parity with ^3P results by applying the Hund1 rule, while through the Hund2 rule ^3F is identified as the fundamental term of the free d^2 ion.

In Table 2.9, the terms generated by the consideration of the B interelectronic repulsion for the orbital d^n configurations are listed with the identification of the fundamental term. Note that the d^n terms are identical with the ones that correspond to the d^{10-n} configurations.

The next step, after considering the B factor of inter-electronic repulsion effect, is to consider the central ions terms in the crystal field, i.e., treating the electronic repulsion from these terms toward the valence electrons of the coordination ligands.

TABLE 2.9   The Free Ions Terms, in the d^n Configuration; After (Lancashire, 2002; Putz, 2006)

Configuration	Fundamental Term	Terms for the Excited States
d^1,d^9	^2D	-
d^2,d^8	^3F	$^3P, {}^1G, {}^1D, {}^1S$
d^3,d^7	^4F	$^4P, {}^2H, {}^2G, {}^2F, 2 \times {}^2D, {}^2P$
d^4,d^6	^5D	$^3H, {}^3G, 2 \times {}^3F, {}^3D, 2 \times {}^3P, {}^1I, 2 \times {}^1G, {}^1F, 2 \times {}^1D, 2 \times {}^1S$
d^5	^6S	$^4G, {}^4F, {}^4D, {}^4P, {}^2I, {}^2H, 2 \times {}^2G, 2 \times {}^2F, 3 \times {}^2D, {}^2P, {}^2S$

At this point the working case of the d^2 configuration terms will be considered and the coordinative symmetry picture of octahedral (O_h) type for the ligands distributed in the corners of a regular octahedron, around the central ion will be chosen.

In this case, the method of separation, or of the removing degeneracy, or of the total multiplicity $(2S+1)(2L+1)$ in the crystal field O_h must be considerate term-by-term of the Table 2.9.

Before we advance in further considerations, we should specified how the terms, as the multi-electronic configurations energies, resulting from the application of the symmetry operations of a symmetry group (here O_h) combined with the inter-electronic repulsion, are denoted with *capital letters* and correspond to the group irreducible representations and to the orbital terms, while the *small letters* represent the corresponding orbitals of the separation field, as in Figure 2.32 and Figure 2.33, including the presence of the B inter-electronic repulsion.

Another essential observation is the fact that in the current hierarchy of action, one has firstly the B action effect and then that one of the crystal field *Dq, the spin multiplicity (2S + 1) of the free ion terms does not change under the action of the weak crystal field.*

Starting from these observations, for each free ion term, one can list the terms generated by the separations caused by the superimposed presence of a crystal field O_h, as in Table 2.10, for which the spin multiplicity of each term in the crystal field has not been specified anymore, being the same with that one of the ion source, for a given d^n configuration.

TABLE 2.10 The Separated Terms by the Crystal Field O_h (on Right) frOm the Russell-Sounders d^n Terms (On Left); After (Lancashire, 2002; Putz, 2006)

S	A_{1g}
P	T_{1g}
D	E_g, T_{2g}
F	A_{2g}, T_{1g}, T_{2g}
G	$A_{1g}, E_g, T_{1g}, T_{2g}$
H	$E_g, 2 \times T_{1g}, T_{2g}$
I	$A_{1g}, A_{2g}, E_g, T_{1g}, 2 \times T_{2g}$

Note that the terms' separation in the crystal field of the free ion from Table 2.10 can be analytical deduced, as in the previous Section – with quantum results identical as in the imaginary reasoning (Gedanken-in literary German) by which the terms' grouping of the equivalent orbitals takes place for a certain type of repulsion for the ligands' charges involved in coordination.

For example, the orbitals s and $p_{x/y/z}$ are relative equivalent on the repulsion strength toward the ligands charges from the O_h symmetry (and the same for T_d) and, while keeping their degeneracy, they will be represented by the irreducible terms corresponding to a_{1g} and t_{1g}, for the symmetry T_d with skipping of the lower-index "g"; the multi-electronic terms corresponding to L=l will showcase similar behavior, being instead marked by capital letters in Table 2.10, so becoming the A_{1g} and T_{1g} terms, respectively.

Similarly, for the D terms in Table 2.10 they are directly written in the O_h symmetry of the crystal field from the uni-electronic terms e_g and t_{2g}.

Analogously there is proceeded also for the other orbitals types. Finally, from the Table 2.10 only the next terms decompositions will be retained:

$$^3F \rightarrow {}^3A_{2g} + {}^3T_{1g} + {}^3T_{2g} \tag{2.167}$$

$$^3P \rightarrow {}^3T_{1g} \tag{2.168}$$

$$^1G \rightarrow {}^1A_{1g} + {}^1E_g + {}^1T_{1g} + {}^1T_{2g} \tag{2.169}$$

$$^1D \rightarrow {}^1E_g + {}^1T_{2g} \tag{2.170}$$

$$^1S \rightarrow {}^1A_{1g} \tag{2.171}$$

Further, the reverse procedure will be applied: firstly the crystal field Dq action will be considered as being the largest (for the strong crystal field) interaction, and then the B inter-electronic repulsion effect will be involved.

For the multi-electronic placed in the strong crystal field case, the central ion terms can be determined in the considered crystal symmetry by applying the *direct product* between the irreducible representations, from the groups, theory.

Before presenting the method, worth defining the direct product here (denoted by "x") between two irreducible representations X_i and X_j of a symmetry group, as being the product of the correspondent characters, on classes of symmetry operations in a group, resulting a new group representation, usually reducible to the group of irreducible representations, through a linear combination of them

$$X_i \times X_j = \sum_n q X_n \tag{2.172}$$

The decomposition (2.172) can be evaluated, as the direct product reduction to the group irreducible representations, from case to case, according to the procedure explained in above Section, yet the systematization rules with the aid of the decomposition rules in Table 2.11 help a lot.

Once introduced the concept of direct product, the multi-electronic terms determination rule can be established, which arise due to the strong action of crystal field (based on its symmetry): *if a term in an electronic configuration is known, then the terms appearing in the configuration having an additional electron will have the irreducible representations terms which occur as a result of the direct product between the initial term and the symmetry term of the added electron.*

Specifically, for the analyzed case of the d^2 configuration in the O_h field, one starts from the d^1 configuration which in the crystal O_h field is separated in terms t_{2g} and e_g. The second electron can be added to any of these terms and resulting in the terms' combinations:

$$t_{2g} \times t_{2g} \rightarrow T_{2g} \times T_{2g} \tag{2.173}$$

$$t_{2g} \times e_g \rightarrow T_{2g} \times E_{2g} \tag{2.174}$$

$$e_g \times e_g \rightarrow E_{2g} \times E_{2g} \tag{2.175}$$

TABLE 2.11   Decomposition Rules for the Direct Product in the Symmetry Groups; After (Lancashire, 2002; Putz, 2006)

A	B	E	T	g & u	'&"	$_1\&_2$
A×A=A	B×A=B	E×A=E	T×A=T	g×g=g	'×'='	1×1=1
A×B=B	B×B=A	E×B=E	T×B=T	g×u=u	'×"="	1×2=2
A×E=E	B×E=E	E×E=*	T×E=T$_1$+T$_2$	u×g=u	"×'="	2×1=2
A×T=T	B×T=T	E×T=T$_1$+T$_2$	T×T=**	u×u=g	"×"='	2×2=1

*for O, T$_d$, C$_{3v}$ & D$_6$: E$_{1(2)}$×E$_{2(1)}$= B$_1$+B$_2$+E$_1$; E$_{1(2)}$×E$_{1(2)}$=A$_1$+A$_2$+E$_2$;
for C$_{4v}$, D$_4$, etc.: E$_{1(2)}$×E$_{1(2)}$= A$_1$+A$_2$+B$_1$+B$_2$;
** T$_{1(2)}$×T$_{1(2)}$= A$_1$+E+T$_1$+T$_2$; T$_{1(2)}$×T$_{2(1)}$= A$_2$+E+T$_1$+T$_2$

At this point the crystal field Dq influence is exhausted from the analysis and the effect of the B inter-electronic repulsion "come into play", which will be stabilized by the decompositions of the direct products obtained. Using the rules from the Table 2.11 the relative irreducible decompositions are immediately founded:

$$T_{2g} \times T_{2g} \to {}^{a}A_{1g} + {}^{b}E_{g} + {}^{c}T_{1g} + {}^{d}T_{2g} \tag{2.176}$$

$$T_{2g} \times E_{g} \to {}^{u}T_{1g} + {}^{v}T_{2g} \tag{2.177}$$

$$E_{g} \times E_{g} \to {}^{p}E_{g} + {}^{q}A_{1g} \times {}^{r}A_{2g} \tag{2.178}$$

for which, the associated multiplicities still remain to be determined. The multiplicities are determined taking into account, for each case, the combination of allowed spin by the Pauli principle.

This way, for the product $T_{2g} \times T_{2g}$ the two electrons have available 6 spin-orbitals, so the combinations are $C_6^2 = 6 \times 5/2 = 15$, equaling the allowed multiplicity; for $T_{2g} \times E_g$ the two electrons can be combined in the two spin orientations and with 6 orbitals available, so occurring 24 occupancy possibilities, corresponding to the total multiplicity of this product; finally, for $E_g \times E_g$ there are two electrons on 4 spin-orbitals thus producing $C_4^2 = 4 \times 3/2 = 6$ occupancy possibilities, as associated multiplicities. In total, there are amounted 15+24+6=45 possibilities, i.e., there was recovered the previous degeneration from the weak crystal field. Still, the problem of the spin multiplicities assignment occurs again, i.e., the identification of the real terms which are separated at the end.

For the case $t_{2g} \times e_g \rightarrow T_{2g} \times E_{2g}$ there is about two electrons distributed between two distinct orbitals, therefore, they may have independent spin orientations, and the total terms generated can be both triplets (the electron from t_{2g} has the same spin orientation as for e_g, resulting $S = 3$) or singlet (the electron from t_{2g} has the spin orientation opposed to that of e_g, resulting $S = 1$).

Therefore, this direct product generates the terms

$$T_{2g} \times E_g \rightarrow {}^3T_{1g} + {}^3T_{2g} + {}^1T_{1g} + {}^1T_{2g} \qquad (2.179)$$

To determine the other terms' multiplicities of the other two direct products, the general method, due to Bethe, of the symmetry descent is applied: it is based on the observation according which such operation of decreasing symmetry in the crystal field (or of ligands) does not affect the spin, and therefore neither its multiplicity.

So being, for each kind of product, the inferior O_h symmetry will be searched, where the double (E) and triple (T) degenerated terms are founded as a sum of non-degenerate terms (irreducible representations) of type A or B. Accordingly, in Table 2.12 the correlations between the irreducible representations of the O_h group and the appropriate amount of the irreducible representations, are shown for the low symmetry groups.

How it works? From the Table 2.12 analysis there is immediately observed how E_g from O_h corresponds to $A_{1g} + B_{1g}$ from the group D_h, and respectively T_{2g} from O_h corresponds to $2A_g + B_g$ from the group C_{2h}.

So, the product of $E_g \times E_g$ type is "lowered" to the direct products between the non-degenerated orbital occupancies a_{1g} and b_{1g} from Figure 2.36(left), while the product $T_{2g} \times T_{2g}$ will be "lowered" to the direct products between the non-degenerated orbitals a_{1g}, a_{2g}, and b_g from the Figure 2.36(right).

Further employment of the direct products for the orbital occupancies types of Figure 2.36, in accordance with the rules of Table 2.11, yields for $E_g(D_{4h})$ the obtained terms:

$$(1i): a_{1g} \times a_{1g} = a_{1g} \rightarrow {}^1A_g \qquad (2.180)$$

$$(2i): b_{1g} \times b_{1g} = a_{1g} \rightarrow {}^1A_g \qquad (2.181)$$

$$(3i)\&(4i): a_{1g} \times b_{1g} = b_{1g} \rightarrow \left({}^1B_{1g} \ \& \ {}^3B_{1g} \right) \qquad (2.182)$$

TABLE 2.12 The Correlation of the Irreducible Representations of the Group O_h With Those of the Lower Symmetry Group; After (Lancashire, 2002; Putz, 2006)

O_h	O	T_d	D_{4h}	D_{2d}	C_{4v}	C_{2v}	D_{3d}	D_3	C_{2h}
A_{1g}	A_1	A_1	A_{1g}	A_1	A_1	A_1	A_{1g}	A_1	A_g
A_{2g}	A_2	A_2	B_{1g}	B_1	B_1	A_2	A_{2g}	A_2	B_g
E_g	E	E	$A_{1g}+B_{1g}$	A_1+B_1	A_1+B_1	A_1+A_2	E_g	E	A_g+B_g
T_{1g}	T_1	T_1	$A_{2g}+E_g$	A_2+E	A_2+E	$A_2+B_1+B_2$	$A_{2g}+E_g$	A_2+E	A_g+2B_g
T_{2g}	T_2	T_2	$B_{2g}+E_g$	B_2+E	B_2+E	$A_1+B_1+B_2$	$A_{1g}+E_g$	A_1+E	$2A_g+B_g$
A_{1u}	A_1	A_1	A_{1u}	B_1	A_2	A_2	A_{1u}	A_1	A_u
A_{2u}	A_2	A_2	B_{1u}	A_1	B_2	A_1	A_{2u}	A_2	B_u
E_u	E	E	$A_{1u}+B_{1u}$	A_1+B_1	A_2+B_2	A_1+A_2	E_u	E	A_u+B_u
T_{1u}	T_1	T_1	$A_{2u}+E_u$	B_2+E	A_1+E	$A_1+B_1+B_2$	$A_{2u}+E_u$	A_2+E	A_u+2B_u
T_{2u}	T_2	T_2	$B_{2u}+E_u$	A_2+E	B_1+E	$A_2+B_1+B_2$	$A_{1u}+E_u$	A_1+E	$2A_u+B_u$

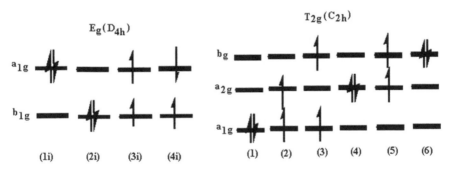

FIGURE 2.36 The orbital occupancies with 2 electrons for E_g "lowered" in the group D_{4h} (left) and respectively for T_{2g} "lowered" in the group C_{2h} (right); after (Lancashire, 2002; Putz, 2006).

In this last case may exists a triplet or a singlet combination, as indicated in the variants (3i) and (4i) of Figure 2.36.

For the terms $T_{2g}(C_{2h})$ analogously are obtained the terms:

$$(1): a_{1g} \times a_{1g} = a_{1g} \to {}^1A_g \tag{2.183}$$

$$(2): a_{1g} \times a_{2g} = a_{2g} \to \left({}^1A_g \ \& \ {}^3A_g\right) \tag{2.184}$$

$$(3): a_{1g} \times b_g = b_g \rightarrow \left({}^1B_g \ \& \ {}^3B_g \right) \tag{2.185}$$

$$(4): a_{2g} \times a_{2g} = a_{2g} \rightarrow {}^1A_g \tag{2.186}$$

$$(5): a_{1g} \times b_g = b_g \rightarrow \left({}^1B_g \ \& \ {}^3B_g \right) \tag{2.187}$$

$$(6): b_g \times b_g = a_g \rightarrow {}^1A_g \tag{2.188}$$

Then, for each envisaged direct product $E_g \times E_g$ and $T_{2g} \times T_{2g}$, the existing terms will be directly obtained by their re-composition from the terms with the same spin multiplicity, obtained in lower symmetry, D_{4h} and C_{2h}, respectively, taken as once, with respecting the de-composition rule of the terms from O_h, according to the Table 2.12.

This correspondence of re-composing terms is schematically shown in Figure 2.37, wherefrom the terms in O_h are directly obtained with the searched spin multiplicity.

By collecting the results and obtained terms in the strong crystal field framework, the next decompositions are released:

$$T_{2g} \times E_g \rightarrow {}^3T_{1g} + {}^3T_{2g} + {}^1T_{1g} + {}^1T_{2g} \tag{2.189}$$

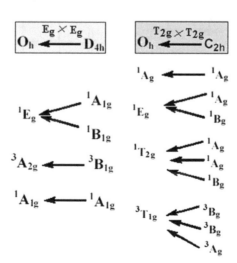

FIGURE 2.37   The one-to-one correspondence between the direct products' terms $E_g \times E_g$ (left) and $T_{2g} \times T_{2g}$ (right) from the lower symmetries of O_h, D_{4h} – left and C_{2h} – right, toward O_h one; after (Lancashire, 2002; Putz, 2006).

$$T_{2g} \times T_{2g} \rightarrow {}^{1}A_{g} + {}^{1}E_{g} + {}^{1}T_{2g} + {}^{3}T_{1g} \tag{2.190}$$

$$E_{g} \times E_{g} \rightarrow {}^{1}E_{g} + {}^{3}A_{2g} + {}^{1}A_{1g} \tag{2.191}$$

These terms of the strong crystal field should be identical with those similarly obtained in the weak crystal field workflow, as long as both effects (Dq and B) were considered at the end, even if in a different order.

As a consequence, these terms will be correlated, so that the lines which connect the similar terms from the two approaches do not be crossing. The so-called correlation diagram of the d^2 configuration terms in the crystal (or coordinative) O_h symmetry results as in Figure 2.38-left.

This diagram can be more rationalized in a graphic of the increasing Eigen energies for the free ion terms when is subjected to a more and more stronger crystal field, from weak to strong, resulting the so-called *Tanabe-Sugano diagram,* Figure 2.38-right.

Therefore, all the current presentation leads to the Tanabe-Sugano diagrams; what are they for? They are used in order to prescribe and analyze

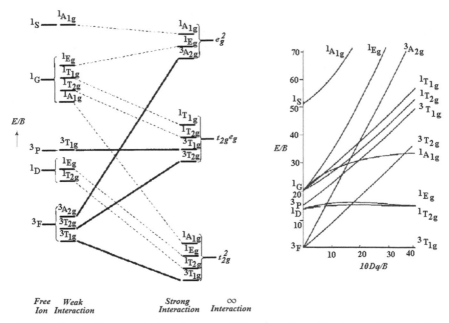

FIGURE 2.38 The correlation diagram of the d^2 configuration terms (left) and the associated Tanabe-Sugano diagram (right); after (Tanabe & Sugano, 1954).

the spectral transitions between the terms of an electronic configuration, in a given coordination symmetry.

To understand these transitions there is useful to introduce the so-called *selection rules of transitions*, which select those satisfying:

1. **the spin rule** $\Delta S = 0$ between the transition terms (i.e., there are allowed transitions only between the terms of the same multiplicity, without changing the spin orientation) and

2. **the orbital rule** (of Laporte) which allows only the transitions that satisfy the jump of the orbital moment $\Delta \lambda = +/-1$ (in other words, there is forbidden the electronic redistribution inside the same subshell or sub-level of $g \to g$ and $u \to u$ type).

Relaxations of these rules can be also possible by the spin-orbital couplings, but the respective transitions have a low intensity and occur through the vibrations of complex (octahedral) molecular asymmetries, or when the π-acceptor and π-donor orbitals of the ligands are mixed (hybridized) with the d-orbitals of the central ion, i.e., the *d-d* pure transitions not appearing anymore.

However, the charge exchanges, from ligand to metal or from metal to ligand, are generally very intense and are identified in the spectrum UV region, see Figure 2.39. Note that the *d-d* transitions from the UV and V regions, generally being forbidden, have a low intensity.

Typical examples of electronic transitions are listed in Table 2.13, depending on the degree of "violation" of the spectral transition rules above and with the extinction ε-coefficient, directly related to the spectral bands' intensity.

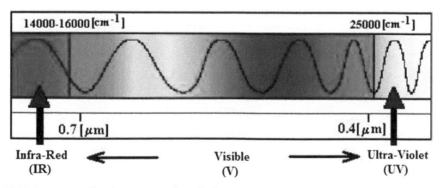

FIGURE 2.39   The electromagnetic radiation spectrum from IR to UV; after (Putz, 2006).

TABLE 2.13   Typical Electronic Transitions in a Ionic Complex, After (Lancashire, 2002; Putz, 2006)

Transition type	Example	ε [m² mol⁻¹]
Spin forbidden, Laporte forbidden	$[Mn(H_2O)_6]^{2+}$	0.1
Spin forbidden, Laporte forbidden	$[Ti(H_2O)_6]^{3+}$	1
Spin forbidden, Laporte partially allowed through the mix d-p (T_d)	$[CoCl_4]^{2-}$	50
Spin allowed, Laporte allowed (charge transfer bands)	$[TiCl_6]^{2-}$ and MnO_4^{-}	1000

Returning to the Tanabe-Sugano diagram from Figure 2.38-right, there is clear that because the selection rules the electronic transitions that violate both the spin rule and the Laporte rules are prohibited.

The term which represents the fundamental state of the d^2 configuration in the crystal O_h field is the term with the lowest energy, so $^3T_{1g}$, and therefore the transitions which occur necessarily involve this term. There appears, for instance, the problem of establishing the parameter value of the crystal field for $10Dq = \Delta_0$ of the complex $[V(H_2O)_6]^{2+}$ from the $d-d$ spectral registration in Figure 2.40-(a)

The procedure is simple:

(a) the two frequencies from the spectrum are numerically identified and v_1=17.200 [cm⁻¹] and v_2=25.600 [cm⁻¹] results;

(b) the Tanabe Sugano diagram for $d^2(O_h)$ is used, see Figure 2.40-(b), and there is accordingly identified how these frequencies correspond to the transitions between the terms: v_1: $^3T_{1g}(F) \rightarrow {}^3T_{2g}(F)$, and between the terms: v_2: $^3T_{1g}(F) \rightarrow {}^3T_{1g}(P)$.

Notice that in this case the spin rule is satisfied, but not the Laporte one, for the spectral transitions which justifies the values for extinction coefficient ε[m²mol⁻¹], on the vertical axis, of being of order unity for the intensities of these transitions.

Further on, the Tanabe-Sugano diagram is used to identify the values and the ratio for the two transitions types the grid as in Figure 2.40-(c) is formed. With the aid of this grid the current report v_2/v_1=256/172=1.49 is discovered as corresponding to a report $10Dq/B \approx 28$.

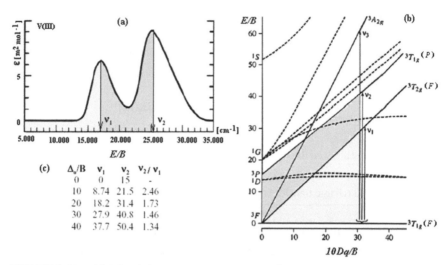

FIGURE 2.40   *(a):* The *d-d* spectrum of the [V(H$_2$O)$_6$]$^{2+}$complex; (b) the Tanabe Sugano diagram $d^2(O_h)$ with the spectral transitions identification from the diagram (a); (c) the corresponding grid *E/B versus 10Dq/B* for the first two spectral frequencies and their ratio; after (Lancashire, 2002; Putz, 2006).

However, to find the size $10Dq$ one also needs to determine the B-factor: it is determined by collocating the value $10Dq/B=28$ on the vertical axis in diagram of Figure 2.40(b), on turns to the frequencies $v_{1,2}$ and to the values *E/B* on the corresponding axis.

Thus, one finds the identification $v_1 \to E_1/B_1 \approx 25.9$ and $v_2 \to E_2/B_2 \approx 38.7$; transforming in atomic units the values for the frequencies identified from the spectrum; instead of the energies E_1 and E_2, one finds respectively the factors $B_1=664.093$ [cm^{-1}] and $B_2=661.499$ [cm^{-1}], i.e., producing an average value about $B \approx 662.771$ [cm^{-1}].

Finally, the crystal field strength $10Dq=\Delta_o=(2.8)(662.771)= 18557.6$ [cm^{-1}] immediately results, while indicating a field of medium intensity. Moreover, once the factor B is determined also the frequency of the third transition v_3: $^3T_{1g}(F) \to ^3A_{2g}(F)$ can be predicted as based on the Tanabe-Sugano diagram of Figure 2.40-(b).

For this, the value of the *E/B* ratio will be further identified at the point where the term-line of $^3A_{2g}$ (F) is crossed by the vertical at $10Dq/B=28$: there is obtained $E_3/B \approx 53 \to v_3=(53)(662.771) \approx 35126.9$ [cm^{-1}].

This type of analysis can be performed for any type of spectrum, when the Tanabe-Sugano diagram is available for the relative electronic configuration type, with a symmetry field imposed by the analyzed coordination. Other effects of the ligands field, more general than the crystal one, are to be presented in the next Section.

2.5   MOLECULAR COORDINATIVE COMPLEXES BY LIGAND FIELDS' ANALYSIS

In describing the crystal field and allied effects, one remark that the *attractive forces* of bonding were not even mentioned, while only the repulsive effects being invariable considered.

To this aim, the ligand field theory comes to complete the presence of the attractive forces in chemical bonding, by combining the crystal field theory developed by Bethe and van Vleck (about 1930) with the contemporary theory of the valence bond of Pauling, which assumes the molecular complex formation as a reaction between the Lewis bases (ligands) and the Lewis acids (the metals or the metallic ions).

This way the coordinative or dative covalent bond is resulting, likely as "molecular orbitals" ingredient of the crystal field theory. More specifically, the ligands treatment is supplemented with the molecular orbital picture, precisely by admitting both the terms and the orbitals rising from the hybridization of the ligand orbitals.

Essentially, for the metal or the central ion its valence state is considered with the valence orbitals (s, p, d) written as prescribed by the separation of this type of orbital in the crystalline field; the ligands, instead, are not anymore considered only as point charges, but they also achieve orbital identity, as s and p, for which there are also considered their hybridization, such as σ and π.

These hybridizations are then reduced to the irreducible representations of the coordination symmetry group and the terms that correspond to the bond orbitals and the anti-bond orbitals of the ligands toward the central ion are therefore formulated.

Next, the correlation diagram of one-to-one for the terms associated to the metals with those of the ligand system is constructed, being this process equivalent with the final hybridization s-p-d and the complex orbitalic determination. These diagrams characterize the complex and the dative bonding at the quantum level.

Here the O_h coordination will be exemplified on the arbitrarily chosen complex $[Co(CO)_6]^{3+}$ as representative example, Figure 2.41 (left-top).

For the Co atom the valence state of the orbitals s, p, d, is considered which, by the virtue of the crystalline field theory from the previous Section, are associated with the terms $s \rightarrow A_{1g}$, $p \rightarrow T_{1u}$, $d \rightarrow E_g + T_{2g}$ in the O_h field. The central metal terms analysis stops here.

Next, one moves to the ligands s & p orbitals analysis within the O_h symmetry: the possible hybridizations are those of σ(s-s & s-p) and π (p-p), Figure 2.41 (left-bottom). With the aid of the Table of characters of the group O_h, these hybrid orbitals will be characterized by the symmetries with the reducible representation $X_{(s\&p)\sigma}$ and $X_{p\pi}$ of Table 2.14, with the projections in the irreducible base of the group O_h. If, at first, only the ligands hybrid orbitals σ are considered, the ligands terms result respectively

$$X_{s\sigma} \rightarrow A_{1g} + E_g + T_{1u} \tag{2.192}$$

$$X_{p\sigma} \rightarrow A_{1g} + E_g + T_{1u} \tag{2.193}$$

By one-to-one correlating these terms of ligands with those of the central ion above the diagram of the complex molecular orbitals is drawn as in Figure 2.41 (right).

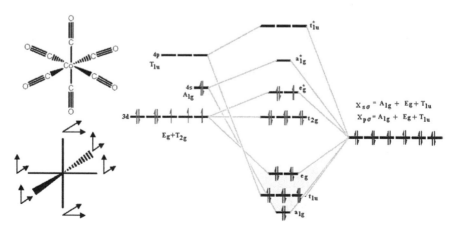

FIGURE 2.41 Left-top: the complex $[Co(CO)_6]^{3+}$; left-bottom: the p-orbitals orientation of the hybridization π (p-p) from the O_h coordination ligands; right: the diagram of the molecular orbitals for the complex or the molecule AB_6; after (Lancashire, 2002; Putz, 2006).

TABLE 2.14   The Table of Characters and the Reducible Representations $X_{(s\&p)\sigma}$ and $X_{p\pi}$ of the Ligands' Hybrid Orbitals for the O_h Symmetry; After (Lancashire, 2002; Putz, 2006)

Oh	E	$8C_3$	$6C_2$	$6C_4$	$3C_2$	i	$6S_4$	$8S_6$	$3\sigma_h$	$6\sigma_d$	
$X_{(s\&p)\sigma}$	+6	0	0	2	2	0	0	0	4	2	$=A_{1g}+E_g+T_{1u}$
$X_{p\pi}$	12	0	0	0	-4	0	0	0	0	0	$=T_{1g}+T_{2g}+T_{1u}+T_{2u}$

Form the correlation diagram of Figure 2.41 (right) clearly results that the last two occupied levels correspond to the orbitals e_g^* and t_{2g}, i.e., to the same type of degenerate orbitals as prescribed for the d-orbitals separation (of the central metal) in the crystalline O_h field (see Figure 2.32), however with the only difference that now the orbital e_g^* instead of e_g is prescribed.

This way the approach of the ligand field as a generalized picture of the crystalline field for the dative bonding is confirmed.

In case that also the influence of the ligand hybrid π-orbital is considered it will generate the decomposition terms

$$X_{p\pi} \rightarrow T_{1g}+T_{1u}+T_{2g}+T_{2u} \tag{2.194}$$

in accordance with the Table 2.14.

However, because the π-bonding is weaker than σ-bonding in general, by considering the ligand hybrid π_L orbital will not radically change the diagram of the ligands correlation of the terms for the σ-bonding with the central ion terms but rather will have an effect on such diagram.

Among the terms of the π_L-bonding only the T_{2g} term appears properly for the correlation with the T_{2g} term of the central ion, within the molecular t_{2g} orbital of the Figure 2.41-right. Yet such correlation can be made in two stages: the acceptor (acid) ligand in the Figure 2.42-top, or the donor (base) ligand in the Figure 2.42-bottom, both producing an increasing respectively a decreasing of the last occupied levels' separations.

These two possibilities involving the π_L orbital energies are jointly treated through the phenomenon of synergic ligand-metal-ligand bonding, when, the ligand fulfills the (cyclic) role of the acceptor (the back-bonding hypothesis) and respectively that of donor toward the central metal, see Figure 2.42-right-middle.

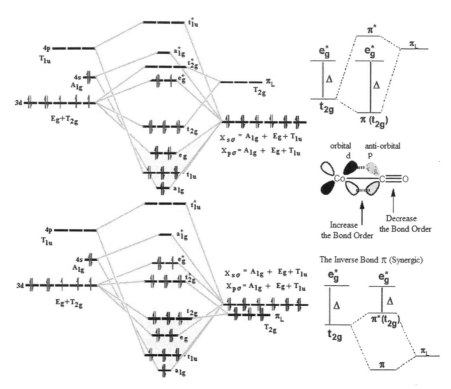

FIGURE 2.42 The energy levels' correlation diagram for AB$_6$ complex with the ligand hybrid π_L orbital of unsaturated (top) and saturated (bottom, along the details of the highest occupied orbitals (right - top and bottom) showing the effect of the synergistically π bonding as detailed at the back-bonding level (right-middle); after (Lancashire, 2002; Putz, 2006).

Such a ligand synergic effect abolishes the dogma of the crystalline field according which the ligands are reduced to the punctual negative charges and can fulfill only the role of donor. Moreover, the existence of the orbital separation as a back-bonding was experimentally confirmed, by recording the absorption spectrum of the complexes while the geometry and the ligands are kept constant and having only the central ion as changeable, or vice versa.

This way the so-called *spectrochemical series* have been constituted. For instance, when keeping constant the geometry and the ligands type, the orbital separation decreases from the strong-field ions to the weak-field ions in the series:

$Pt4+$ > Ir^{3+} > Rh^{3+} > Co^{3+} > Cr^{3+} > Fe^{3+} > Fe^{2+} > Co^{2+} > Ni^{2+} > Mn^{2+}
Strong-field ions   *Weak-field ions*

Similarly, while keeping constant the coordination geometry and the central ion the ligands series from those of strong-field to those of weak-field looks like:

CO > CN > NO_2^- > NH_3 > NHS^- > H_2O > OH^- > F^- > SCN^- > Cl^- > Br^-
Strong-field ions   *Weak-field ions*

Until now only the complexes with high symmetry were considered, as those of O_h or T_d. But, what happens in the coordination with a lower symmetry? The electronic spectra will become much more complicated because the orbitals are much less degenerated (less grouped).

However, even in the complexes with high symmetry distortions can be occurring having as effect the descending of the symmetry, of the coordination and of the energy for the highest occupied molecular orbital (HOMO) or of the fundamental state term.

The behavior/response of descending the symmetry and the energy by distortions is rationalized by the *Jahn-Teller Theorem*, firstly announced in 1930 and then being formalized by Orgel in 1950, simply affirming that *"in a nonlinear molecule in a degenerated electronic state occur distortions which will lower the symmetry, will remove the degeneration and will reduce the energy"*.

However, the Jahn-Teller theorem does not predict the distortions type which will occur, but only the fact that the complexes symmetry's center must remains constant, because it is the reference respecting which the initial symmetry is lowered, Figure 2.43-left. And yet, what is the phenomenological basis for the appearance of these distortions?

The plain answer is: the spin electronic configuration! The distortion's promoting force is a steric effect of the open d-shells (with electrons having the not-pairing spin), see the Figure 2.43-right.

For the species with a high spin configurations the Jahn-Teller effect for the d^4 and d^9 configurations is favored because pose the orbital degeneration in the e_g configuration, and respectively for the d^1, d^2, d^6, and d^7 configurations because a t_{2g} orbital degeneration is present. For the species with

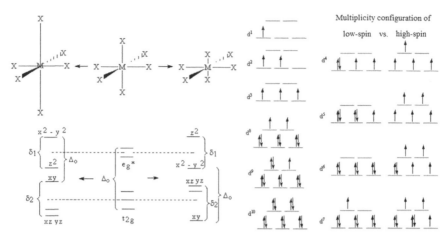

FIGURE 2.43 Jahn-Teller distortions (left) and d^n configurations of high and low spin (right); after (Lancashire, 2002; Putz, 2006).

a low spin configuration the Jahn-Teller effect was consecrated as "static", being observed in many complexes, as those exposed in Table 2.15.

Nevertheless, a cooperative *Jahn-Teller effect* also occurs, with temperature variation, for which, over a certain limit, the molecule distortion centers become independent one each other, having a dynamical manifestation.

Such a documented type of complexes corresponds to the structure $M_2PbCu(NO_2)_6$, which, for M=Cs and below 285K displays a tetragonal symmetry, a phenomena that occurs for M = K for temperatures below 273 K, and for M = Rb at less than 276K, while for M = Tl below reaches the limit of 245K. Above these temperatures the complex has an octahedral symmetry.

Another steric consequence of the electronic configuration of the d-orbitals placed in the ligand field consists in the energetic stabilization towards a certain configuration of the coordination symmetry.

The calculation of these *ligand field stabilization energies* (LFSE) is a simple exercise, for example, for the symmetry O_h in the high spin configuration picture (Figure 2.43-right) in relation with the width Δ_0 for the sets of degenerated t_{2g} and e_g orbitals' separation, see the Figure 2.32, with the results displayed in the column 4 of Table 2.16.

The LFSE/Δ_0 values of the 4 column from Table 2.16 correspond to the ions in the oxidation state (II), wherefrom the values corresponding to the oxidation state (III) can be obtained by the simple one-to-one multiplication with the factor 3/2, thus obtaining the 5th column in Table 2.16.

TABLE 2.15 Examples of Jahn-Teller Distorted Complexes; After (Lancashire, 2002)

Species	Inter-Distances (in pico-meters: pm)
$CuBr_2$	4Br at 240pm/2Br at 318pm
$CuCl_2$	4Cl at 230pm/2Cl at 295pm
$CuCl_2.2H_2O$	2O at 193pm/2Cl at 228pm/2Cl at 295pm
$CsCuCl_3$	4Cl at 230pm/2Cl at 265pm
CuF_2	4F at 193pm/2F at 227pm
$CuSO_4.4NH_3.H_2O$	4N at 205pm/1O at 259pm/1O at 337pm
K_2CuF_4	4F at 191 pm/2F at 237pm
CrF_2	4F at 200pm/2F at 243pm
$KCrF_3$	4F at 214pm/2F for 200pm
MnF_3	2F at 209pm/2F at 191pm/2F at 179pm

TABLE 2.16 The Energies of the Ligands Field Stabilization Energies (LFSE) in Relation With the Width of the Orbital Separation Δ_o in Octahedral Field for the d^n Configurations of High Spin for the O_h and T_d Symmetries; After (Lancashire, 2002)

Electronic configuration	O_h				T_d		Examples of ions
	actual	uniform	LFSE/ $\Delta_o(B^{2+})$	LFSE/ $\Delta_o(B^{3+})$	LFSE/ $\Delta_o(A^{2+})$	LFSE/ $\Delta_o(A^{3+})$	
d^0	t_{2g}^{0}	(0)	0	0	0	0	Fe^{3+}, Mn^{2+}
d^1	t_{2g}^{1}	1(3/5)	2/5	3/5	12/45	2/5	Fe^{2+}, Co^{3+}
d^2	t_{2g}^{2}	2(3/5)	4/5	6/5	8/15	4/5	Co^{2+}
d^3	t_{2g}^{3}	3(3/5)	6/5	9/5	16/45	8/15	Ni^{2+}
d^4	$t_{2g}^{3}e_g^{1}$	4(2/5)	3/5	9/10	8/45	4/15	Mn^{3+}
d^5	$t_{2g}^{3}e_g^{2}$	5(2/5)	0	0	0	0	Fe^{3+}, Mn^{2+}
d^6	$t_{2g}^{4}e_g^{2}$	6(3/5)	2/5	3/5	12/45	2/5	Fe^{2+}, Co^{3+}
d^7	$t_{2g}^{5}e_g^{2}$	7(3/5)	4/5	6/5	8/15	4/5	Co^{2+}
d^8	$t_{2g}^{6}e_g^{2}$	8(3/5)	6/5	9/5	16/45	8/15	Ni^{2+}
d^9	$t_{2g}^{6}e_g^{3}$	9(2/5)	3/5	9/10	8/45	4/15	Mn^{3+}
d^{10}	$t_{2g}^{6}e_g^{4}$	10(2/5)	0	0	0	0	Fe^{3+}, Mn^{2+}

The transposition to the tetrahedral coordination symmetry (T_d) can be also directly obtained from the 4th column (in Table 2.16) for the metallic ions in the oxidation state (II) by the multiplication with the factor 4/9, according to the relations (2.130) and (2.164), yet with the inversed correspondences, because the T_d energies corresponding to minus of the O_h energies, that is the 4th column of Table 2.16 will be one-to-one multiplied with the factor 4/9 and the results will be written in reverse order in the 6th column of Table 2.16.

The transposition from the $LFSE(T_d)/\Delta_0$ values for the ions in the valence state (II) to those corresponding to the ions in the valence state (III), for the T_d symmetry, will be done, again, through the one-to-one multiplication of the 6th column with the factor 3/2 thus obtaining the 7th column of Table 2.16.

In this way the $ESCL/\Delta_0$ variations can be represented for the O_h and T_d coordinative types, and also for their difference, see Figure 2.44-left, which also indicates the degree of occupancy's preference of the central ions in a octahedral coordinated location paralleling its increasing stabilization energy (on a negative scale).

The utility of the presented LFSE consists precisely in the prediction power by which a particular electronic configuration of a metallic ion will be predominantly occupied for a octahedral or otherwise coordinated location.

Remarkably, this type of analysis proves to be the only rationalization mean (on quantum grounds) on the way of assignment of the spinel structure, for instance, see Figure 2.44-right, as being prioritized in the "normal" or "reverse" state. This because, the rule of rationalization by

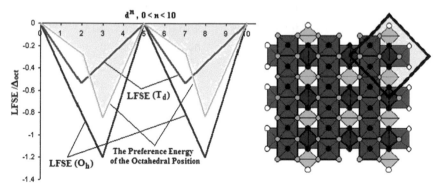

FIGURE 2.44 Left: the O_h and T_d variations for the ligand field stabilization energies (LFSE) and the preference energies for the octahedral location; right: the spinel structure representation by the O_h and T_d locations' types; after (Lancashire, 2002).

the ionic radii ratio predicts how the cations with a small-radius prefer the tetrahedral locations (such as $MgAl_2O_4$), wherefrom the most spinels would result as "reversed"; instead, the analysis of lattice's energy (by the calculation of the network's Madelung constant) predicts how the cations with high charge would preferentially occupy the octahedral locations, such as $TiMg_2O_4$, which would classify the most spinels as being "normal". Certainly, in the presence of these two criteria with opposite effects, one needs for a decisive argument in order to rationalize the spinels' structure as being the ones observed.

At this point, one calls for the energetic stabilization process of the ligand field to clarify the type of preference, being such criteria consistent with the quantum base of the coordination in a given (O_h or T_d) symmetry and with experimental observations too.

Accordingly, the examples of LFSE calculation for a few specific cases aiming to establish the preferential spinel type are in Table 2.17 presented, being the calculation procedure similarly and generally for any other particular situation.

The configuration that produces the biggest LFSE (on a negative scale) is more stable since causing also the structure's main preference as being of the normal $A^{2+}(T_d)[B^{3+}(O_h)]_2O_4$ or of inverse $[A^{3+}(O_h)B^{3+}(O_h)]B^{2+}(Td)O_4$ types, respectively.

The crystalline and the ligands field analysis, interpretation and the prediction were presented on the qualitative and quantitative quantum bases, with applications for the chemical complexes' structures, while reiterating the power of coordination symmetry tools in establishing the streamline of the electronic structures.

TABLE 2.17   The "Normal/Inverse" Spinel Structure Decision by the Calculation of the Ligand Field Stabilization Energy, As Based on the Data of Table 2.16; After (Lancashire, 2002)

	Mn_3O_4	Fe_3O_4	$NiFe_2O_4$
Normal	$(d^5A^{2+})(d^4B^{3+})_2$: $0+2(9/10)=1.8$	$(d^6A^{2+})(d^5B^{3+})_2$: $(12/45)+2(0)=0.27$	$(d^8A^{2+})(d^5B^{3+})_2$: $(16/45)+2(0)=0.36$
Inverse	$(d^4A^{3+})(d^4B^{3+})(d^5B^{2+})$: $(4/15)+(9/10)+0=1.17$	$(d^5A^{3+})(d^5B^{3+})(d^6B^{2+})$: $0+0+(2/5)=0.4$	$(d^5A^{3+})(d^7B^{3+})(d^8B^{2+})$: $0+(6/5)+(6/5)=2.4$
Observed	**Normal**	**Inverse**	**Inverse**

However to the apparently remained opened question of "who precedes whom"? refereeing to the (quantum) paternity concepts of (group) symmetry and molecular orbitals, a possible answer is given by the development of the quantum picture of the matter electronic structure by the primate of the electronic symmetry concept towards that one of eigen-wave function, which becomes the Molecular Orbital Theory as was essentially exposed in the preceding Section 2.2. Yet, the reactivity of the molecular structure will be next discussed, while changing the wave-function based molecular orbital paradigm with the density functional one, in the next Sections.

2.6 VSEPR MODEL OF CHEMICAL BOND

There is history on how Newton G. Lewis had introduced the doublet and octet electronic rule in the valence layer by the cubic phenomenology, see Chapter 3 of the Volume IV of the present five-volume work (Putz, 2016b). However, even if the quantum theory that follows had invalidated the "cubic atom" paradigm, the rationalization of the chemical bond through the electrons doublet or of the electrons pair (with associated spins) survived all the orbital approaches, being definitively consecrated by the Pauli Exclusion Principle.

On the other side, as abundantly shown in crystal chemistry, see Volume IV of the present five-volume work (Putz, 2016b), the constitutive particles representation of a crystal is taking the form of a rigid spheres with a characteristic and constant dimension (at least for a certain structure under concern) directly leads to the conclusion of their arrangement way in crystalline lattice, resulting that the symmetry of elementary cell and the crystalline reason will be conditioned by geometric factors.

Thus, the positional relations established in a structure, between a particle and all the other generically define the "*neighborhood paradigm*" of the considered particle, geometrizing the notion of chemical bond.

The picture of complete neighborhood of a particle is given by the totality of bonding vectors which move from a particle towards other particles, identical and not identical. The symmetry of this "vector beam" is in concordance with the symmetry of the position of the particle in focus. The neighboring particles, towards which lead the bonding vectors of the same length, form a "sphere of coordination".

The closest neighbors, i.e., the particles towards lead the shortest vectors form the particles of the first sphere of coordination, while the following neighbors – of the second sphere and so on.

If the n sphere includes m particles (i.e., on the sphere with radius equal with the module of the nth bonding vector are found m particles) is considered that the particle in cause has inside the n sphere the coordination number m.

The common description of a structure does not take into account the pictures of complete neighborhood, but it limits only to some spheres.

The structure character is, in general, defined, if is going until the 4th sphere; often there are studied only the first and the second spheres.

By the union of m particles of the same spheres, the coordination polyhedron of n-th order arises. In general, its symmetry is in concordance with the proper symmetry to the particle from its center.

But there is always the possibility that the polyhedron coordination symmetry is superior to the symmetry of the particle associated with the center of coordination.

For this reason, polyhedra coordination bodies may arise with impossible crystallographic symmetry, for example with axis of 5th order.

The equivalence of charge distribution, as of the electrostatic attraction intensity in the vertices direction of the coordination polyhedron, indicates the possibility of application of the notion related to the valence orbital hybridization and to the ionic compounds, thus extending the applicability of the paradigm of formal ions, see Volume IV of the present five-volume work (Putz, 2016b).

Actually, this means that, as the ratio of combination is determinant for the assessment of a certain coordination number, the deduced type of hybridization based on the electronic configuration of cation allows the choice of the shape of coordination polyhedron.

After about 50 years since the cubic atom paradigm of Lewis, (Gillespiele, 1966), accumulating all the ideas above listed, reused the idea of geometric rationalization of geometric bonds, in terms of pairs of electrons in valence layers of chemical compounds in general (not necessarily in solid state) asserting that "the geometry (to read structure) of a molecule is determined by the pairs repulsion of electrons in valence layer (*Valence Shell Electron Pair Repulsion*-VSEPR) of the central atom".

Thus the VSEPR chemical bond paradigm appears. There is clearly how, from now on, the Lewis theory is extended to the level of any geometric symmetry, not necessarily cubic, being related to the so-called central atom.

Therefore, from this view, the central atom has the role of "interstitial", and the atoms of coordination being called ligands. Therefore, the VSEPR generalizes the polyhedral coordination (in a crystal) at the level of (molecular) ligands.

TABLE 2.18 The Relation Between the Hybridization Type, the Coordination Number (CN), the Number of Bonding Electron Pairs and the Form of Polyhedra With the Ligands of Coordination; After Chiriac-Putz-Chiriac (2005)

CN	*#2e⁻*	*Hybridization*	*Coordination Polyhedron*	*Examples*
1	*1*	s, p, d	 *line*	H_2, F_2, HCl
2	*2*	sp, ds, dp	 *line*	$HgCl_2$, HF_2^-
			 *broken line*	H_2O
3	*3*	sp^2, dp^2, d^2s	 *triangles*	BF_3
		d^2p	 *trigonal pyramid*	NH_3
4	*4*	dsp^2	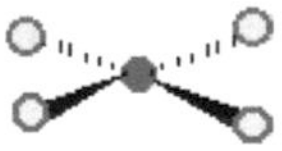 *square*	$[PtCl_4]^{2-}$
		sp^3, d^3s	 *tetrahedron*	CF_4, $[SiO_4]^{4-}$
5	*5*	dsp^3, d^3sp, d^3sp	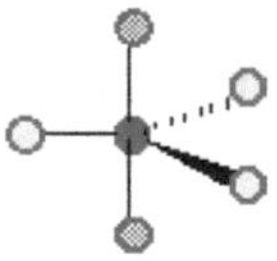 *trigonal bipyramid*	PCl_5, PF_5 (3F at 120^0, 2F at 90^0)

TABLE 2.18   Continued

CN	#2e⁻	Hybridization	Coordination Polyhedron	Examples
		d^4s	*quadratic pyramid*	IF_5
6	6	d^2sp, sp^3d^2	*octahedron*	SF_6, NaCl
		d^4sp	*trigonal prism*	MoS_2, NiAs
7	7	d^3sp^3	*pentagonal bipyrami*	$[UF_7]^{3-}$, IF_7 (5F at 72⁰, 2F at 90⁰)
		d^4sp^2	*trigonal pyramidal prism (on a quadratic face)*	$[ZrF_7]^{3-}$
		d^5sp^2	*cube*	CsI
8	8	d^4sp^3 $(d_\varepsilon + d_x^2 - d_y^2)$	*quadratic antiprism*	$[TaF_8]^{3-}$
		d^4sp^3 $(d_\varepsilon + d_z^2)$	*trigonal dodecahedron (doubled bisfenoid)*	$[Mo(CN)_8]^{4-}$

Table 2.18 shows the coordination polyhedra corresponding to the ligands in the inorganic combinations (lattices), according to the VSEPR theory.

The way of geometrical correlation with the molecular structure, by the virtue of the separation between central atom-ligands, is based on the identification of coordination number and the number of electron pairs (2e⁻) involved in the chemical bond.

As noted, the VSEPR model includes also the pairs of non-participants electrons and may include also multiple bonds (with the rigorous distortions towards the ideal geometries, due to the directional effects of hybrid orbital).

In any case, the molecular geometry is adjusted so that to maximize the separation of electron pairs.

VSEPR theory is, in general, accurate and very useful for the explanation of the halide structures, oxides, hybrids, but less valid for ligands of larger dimensions than those of the central atom (i.e., when the ligand-ligand interaction becomes important, thus becoming more and more important the interaction between the non-participant pairs of electrons).

Yet, the correspondences as central atom = *interstitial* & *ligands* = *coordination polyhedron* generates a functional paradigm which qualitatively rationalizes the chemical bond, emphasizing on the electronic pairs distributed in geometric-symmetrical way.

As an application, the Figure 2.45-bottom is illustrated the famous example of the $\boxed{XeF_6}$ molecule for which the VSEPR model correctly

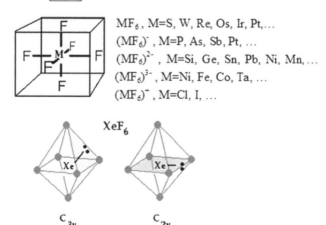

FIGURE 2.45 The VSEPR geometric rationalization (with central atom = interstitial & ligands = coordination polyhedron) for the set of [MF_6]⁻ compounds (and ions) (up), with the application for the [XeF_6]⁻ compound (down); after (Chiriac-Putz-Chiriac, 2005).

prescribed the distortion of the coordination octahedron around of Xe due to the existence of a non-participant pair of electrons, thus allowing, instead of symmetry O_h (initially proposed, as in examples of Figure 2.45-up), one of the two C_{3v} or C_{2v} symmetries, according to the experimental studies.

But does this VSEPR geometric rationalization of chemical bond having also a physical fundament, in terms of electrons quantum nature? The answer is affirmative and its argumentation will be exposed in the following (Chiriac-Putz-Chiriac, 2005):

- *The geometric distribution of the electron pairs in a bond is related to the density of electrons $\rho(r)$;*
- *Molecular stability (and of the structures in general) is associated with the minimization of the total energy of the interaction of the electrons inside that structure, by which the structure is geometrically optimized.*

Moreover, for the bonding stable states, the electrons' total energy, seen as an electronic gas, is equivalent to "$-T$" (minus kinetic energy of electrons), by the virtue of the virial theorem.

This fact can be easily demonstrated if the operator of total energy is written

$$-E_{tot} = T + V = -\frac{h^2}{8\pi^2 m_0}\sum_i \Delta_i \bullet + \sum_{i,j\neq i} V\left(\rho(r_i),\rho(r_j),\frac{1}{|r_i-r_j|}\right) \quad (2.195)$$

through the quantum correspondence involving the momentum transformation

$$p\bullet \rightarrow -h/(2\pi i)(\partial/\partial r)\bullet \quad (2.196)$$

with h, m_0 corresponding to the Planck constant and respectively to the electron's mass.

From Eq. (2.195), at any electronic redistribution of the electrons, $r_i \rightarrow \lambda \tilde{r}_i, \forall i$, the total energy will be:

$$-\tilde{E}_{tot} = \tilde{T} + \tilde{V} = -\frac{h^2}{8\pi^2 m_0}\sum_i \tilde{\Delta}_i \bullet + \sum_{i,j\neq i} V\left(\rho(\tilde{r}_i),\rho(\tilde{r}_j),\frac{1}{|\tilde{r}_i-\tilde{r}_j|}\right) = \lambda^2 T + \lambda V$$

$$(2.197)$$

and the parametric minimum when the following condition is satisfied will be reached:

$$\left(\frac{\partial \tilde{E}_{tot}}{\partial \lambda}\right)_{\lambda=1} = 0 \Rightarrow 2T + V = 0 \overset{(2.195)}{\Rightarrow} -E_{tot} = -T = \frac{h^2}{8\pi^2 m_0}\sum_i \Delta_i \bullet (2.195)$$

(2.198)

Then, since the electronic total energy is a negative measure (of minimum) of the stability of a structure, the relation (2.198) for electronic density will be controlled, eventually, by the topology (space distribution) of Laplacian (Δ) of electronic density, with minus sign, i.e., $-\Delta\rho(r)$. On the other side, the fact that the VSEPR paradigm takes in consideration the electron pairs and not the individual electrons should be quantified, for which reason a particular measure for the inter-separation of the electron pairs will be introduced.

The electrons in their spatial distribution in pairs are considered between the limit of non-interaction ($\lambda=0$) and the limit of interaction ($\lambda=1$) they carry, at any level of parameterized interaction λ, by their "double goer" (*doppelgänger*), modeling the virtual ability to inverse its spin, thus defining the so-called *Fermi hole*, see the Volume II of the present five-volume work (Putz, 2016c)

This way, the Fermi hole is the quantum measure of the Pauli Exclusion principle, at the level of exchange energy (because it takes into account the reciprocal orientation of spin) and correlation (because it generalizes the Coulombian classical interaction),

$$E_{XC} = \frac{1}{2}\int_0^1 \left[\iint \frac{\rho(\mathbf{r}_1)}{|\mathbf{r}_1 - \mathbf{r}_2|} h_\lambda^\alpha (\mathbf{r}_1, \mathbf{r}_2) d\mathbf{r}_1 d\mathbf{r}_2\right] d\lambda$$

(2.199)

being described by the distribution function $h^\alpha(r_1, r_2)$ which measures the decreasing in the probability to find an electron with the same spin (α), at a certain distance (r_2), respecting the one considered as a reference electron (r_1).

In this quantum context the VSEPR model can be interpreted such as: the Fermi holes of the reference electrons, properly placed in the maximum of the concentrations of valence charge of the participant and non-participant pairs (of the central atom), are located such as the relative maximum of concentration to be maximum separated (participant pairs vs. non-participant pairs), resulting the effect corresponding to the energetic minimization and geometric optimization, in conjunction with Pauli exclusion principle.

The case of molecule ClF_3O is in Figure 2.46 illustrated as treated in the VSEPR-quantum-topological context for deciding the optimum structures between the possible geometries: C_s with 2F axial atoms (F_a) and C_{3v} with all the F atoms in equatorial plane (F_e).

The non-participant electrons pair (n) is, in the first case equatorially placed, and axially in the second case, so being represented by the solid triangles, Figure 2.46-bottom side.

There is noted that both in the topological analyze of Laplacian electronic density (Figure 2.46-middle) as in the case of Femi holes distribution (the shaded areas of Figure 2.46-bottom), the distribution $-\Delta\rho(r)$ results as the "largest", while the overlapping of the Fermi holes $h^\alpha(r_1, r_2)$ is the "smallest" for the C_s geometry thus corresponding this symmetry to the most favorable structure at the equilibrium of the ClF_3O molecule.

Therefore, there is clearly how the geometric rationalization of the chemical bond in the central atom & ligands paradigm (VSEPR), founds a rigorous quantum substrate in terms of electrons pairs, repulsion and spin.

Accordingly, as based on such valid foundation, the geometric model of chemical bonds can be extended beyond the case of ionic or formal ionic bonds, for example to those of the hydrogen bonds and complex biological combinations; whose celebrated example is how the pairs of nucleotides bases are preferentially combined in the pairs A:T and G:C (thus forming the so-called *Watson-Crick pairs*).

However, in the presence of a metallic cation also the oligonucleotides can be formed from non-preferential pairs and thus, the detected arrangement appears under a symmetrical geometric form with the cation as interstice (!) and the coordination bases as ligands, Figure 2.47.

Moreover, in the super-molecular chemistry and through the modern techniques of experimental investigation (for example the "electro-spray-ionization" ESI technique) there was undoubtedly proved how the dendritic increases and the ligands form the molecular and poly-metallic polygons, polyhedra and self-organized cages of coordination so that to forming geometric periodic structures, with as higher as possible symmetry, so

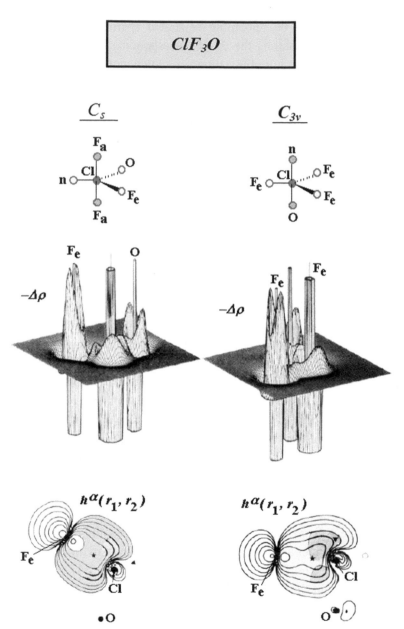

FIGURE 2.46 the quantum rationalization through the Fermi holes (down) and Laplacian of electronic distribution (middle) for the stabilization of optimal geometry of the structure of the ClF_3O compound between the C_s and C_{3v} symmetries, respectively on left-up and on right-up; after (Bader, 1988).

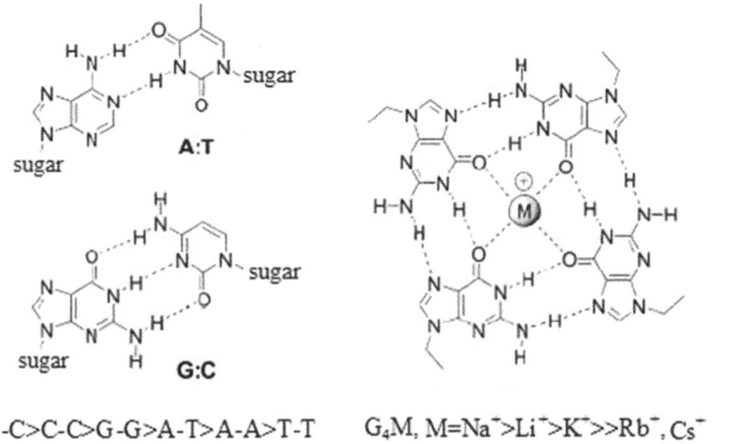

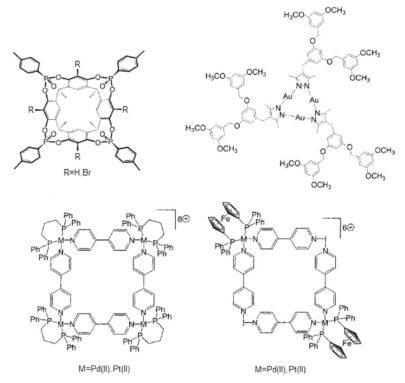

G-C>C-C>G-G>A-T>A-A>T-T G_4M, $M=Na^+>Li^+>K^+>>Rb^+, Cs^+$

FIGURE 2.47   The Bonding for nucleotides (left) and for oligonucleotides (right); after (Schalley, 2001).

R=H,Br

M=Pd(II), Pt(II)   M=Pd(II), Pt(II)

FIGURE 2.48   Geometric rationalization examples for the super-molecular compounds); after (Schalley, 2001).

reflecting the optimal structure at equilibrium, for which in Figure 2.48 only some examples are illustrated.

Finally, by all these considerations, there appears that the idea that the structures analysis of the possible symmetries and of the properties abstracted from the model of interaction following the line of the interstitial-cage (polyhedron, sphere) of coordination paradigms is justified, and can fundamentally contribute at the deep understanding and actually towards the controlled guidance (by growth and synthesis) of the chemical matter's compounds, in a physical framework (quantum, by energetic means), mathematical (by topological and geometric tools) and finally chemical (by the nature of the bonds in terms of electronic density).

2.7 CONCLUSION

The main lessons to be kept for the further theoretical and practical investigations of the quantum chemical modeling of molecular structure that are presented in the present chapter pertain to the following:

- Identifying the main quantum mechanical issues of electrons chemical bonding: electronic vibration, electronic localization, and inter-electronic dipole interaction (van der Waals bonding);
- Employing the symmetry concept and transformation to parallel the quantum eigen-problem so opening the door for further symmetry-quantum symmetry modeling;
- Writing the symmetry operation through recognizing the symmetry elements as of point, line and mirroring plane;
- Dealing with symmetry chart of transformation;
- Characterizing the symmetry charts by associate matrices also viewed as eigen-solution for certain symmetry types;
- Understanding the symmetry operations through the vectorial algebraic properties of groups and of their transformation thus certifying the self-consistent feature of symmetry elements and operations as carrying the quantum nature in general and in special when applies on molecular structure;
- Describing the molecular wave function by superposition of wavefunctions associated with eigen-symmetry transformations for a given symmetry group and molecule;

- Learning he great orthogonal theorem of decomposing of a given molecular symmetry and group on irreducible transformation, as corresponding with the eigen- thus of quantum nature transformations of the symmetry properties of molecule inside that molecule;
- Treating the symmetry adapted linear combination procedure towards assessing the bonding (along with non- and anti-bonding) orbitals for allyl molecule as an illustrative example of the C_{2v} as a freshmen example for organic molecular treatment by quantum mechanical means
- Solving the quantum structure of a given molecule by the linear combination of the atomic orbitals involved in bonding, within the Hückel approximation of the contingent neighbors counting in quantum chemical bonding;
- Formulating the connection between the symmetry adapted orbitals and linear combination of atomic orbitals towards understanding the symmetry eigen-terms as corresponding with eigen-energies of symmetrical modes of a given molecule at their turn corresponding with quantum eigen-energies f the molecular structure itself;
- Interpreting atomic in molecules' coordination as generating the crystalline symmetry and of the associated potential/field in which the central ion is evolving, especially with eth effect in lifting the inherent quantum degeneracies of its eigen-energies, from where the specific properties in inorganic molecules;
- Connecting multi-electronic effects with spin interaction driving the chemical bonding in crystal-field (inorganic) molecules: understanding the Hund rules, and the generating of the molecular states by molecular terms rooting in spin-spin interactions,, multiplicities and correlation (Tanabe-Sugano) diagrams of these terms while lifting or inter-crossing from weak to strong applied field interaction as driven by the molecular symmetry under concern;
- Developing the ligand field theory of chemical bonding through completing the crystal field (based on repulsive forces) with eth attractive one, from where the additional Jahn-Teller distortions associated with high and low spin configurations raised as main "vectors" in forming chemical bonding by such complex hybridization scheme;
- Finding applications of the symmetry based paradigms of quantum modeling of molecules by employing the geometrical structure

of atoms-in-molecules at the valence level of either atoms and the formed molecules, in what was consecrated as VSEPR (Valence Shell Electron Pair Repulsion) of chemical bond, despite missing the bosonic character of the electronic pairs in bonding – as the bosonic-bonding eventually sub-quantum nature of the chemical bonding ultimately prescribes (see Chapter 1).

KEYWORDS

- **atoms-in-molecule**
- **biatomic molecule**
- **coordinative complexes**
- **crystal field analysis**
- **electronic localization in bonding**
- **Hückel approximation**
- **inter-atomic vibration**
- **ligand field analysis**
- **linear combinations of atomic orbitals (LCAO)**
- **molecular symmetry**
- **multi-electronic orbitals**
- **operations and groups**
- **symmetry adapted linear combinations (SALC)**
- **symmetry elements**
- **VSEPR model**

REFERENCES

AUTHOR'S MAIN REFERENCES

Putz, M. V. (2016a). *Quantum Nanochemistry. A Fully Integrated Approach: Vol. I. Quantum Theory and Observability*. Apple Academic Press & CRC Press, Toronto-New Jersey, Canada-USA.

Putz, M. V. (2016b). *Quantum Nanochemistry. A Fully Integrated Approach: Vol. IV. Quantum Solids and Orderability*. Apple Academic Press & CRC Press, Toronto-New Jersey, Canada-USA.

Putz, M. V. (2016c). *Quantum Nanochemistry. A Fully Integrated Approach: Vol. II. Quantum Atoms and Periodicity*. Apple Academic Press & CRC Press, Toronto-New Jersey, Canada-USA.

Putz, M. V., Chiriac, A. (2008). Quantum Perspectives on the Nature of the Chemical Bond. In: *Advances in Quantum Chemical Bonding Structures*, Putz, M. V. (Ed.), Transworld Research Network, Kerala, Chapter 1, pp. 1–43.

Putz, M. V. (2006). *The Structure of Quantum Nanosystems* (in Romanian), West University of Timişoara Publishing House, Timişoara.

Chiriac, V., Putz, M. V., Chiriac, A. (2005). *Crystalography* (in Romanian), West University of Timişoara Publishing House, Timişoara.

SPECIFIC REFERENCES

Allendoerfer, R. D. (1990). Teaching the shapes of the hydrogen like and hybrid atomic orbitals. *J. Chem. Educ.* 67(1), 37–39.

Anderson, P. W. (1968). Self-consistent pseudopotentials and ultralocalized functions for energy bands. *Phys. Rev. Lett.* 21, 13; ibid. (1969). Localized Orbitals for Molecular Quantum Theory. I. The Hückel Theory. *Phys. Rev.* 181, 25–32.

Bader, R. F. W. (1990). *Atoms in Molecules*, Clarendon Press, Oxford.

Bader, R. F. W. (2001). The zero-flux surface and the topological and quantum definitions of an atom in a molecule. *Theor. Chem. Acc.* 105, 276–283.

Bader, R. F. W. (2003). Letter to the editor: Quantum mechanics, or orbitals? *Int. J. Quantum Chem.* 94(3), 173–177.

Bader, R. F. W., Bayles, D., Heard, G. L. (2000). Properties of atoms in molecules: Transition probabilities. *J. Chem. Phys.* 112, 10095–10105.

Bader, R. F. W., Becker, P. (1988). Transferability of atomic properties and the theorem of Hohenberg and Kohn. *Chem. Phys. Lett.* 148, 452–458.

Bader, R. F. W., Carroll, M. T., Cheeseman, J. R., Chang, C. (1987). Properties of atoms in molecules: atomic volumes. *J. Am. Chem. Soc.* 109(26), 7968–7979.

Bader, R. F. W., Essén, H. (1984). The characterization of atomic interactions. *J. Chem. Phys.* 80, 1943–1960.

Bader, R. F. W., Gillespie, R. J., MacDougall, P. J. (1988). A physical basis for the VSEPR model of molecular geometry. *J. Am. Chem. Soc.* 110(22), 7329–7336.

Bader, R. F. W., Heard, G. L. (1999). The mapping of the conditional pair density onto the electron density. *J. Chem. Phys.* 111, 8789–8797.

Bader, R. F. W., Nguyen-Dang, T. T., Tal, Y. (1979). Quantum topology of molecular charge distributions. II. Molecular structure and its change. *J. Chem. Phys.* 70(9), 4316–4329.

Baders, R. F. W. (2002). A comment on "Some fundamental problems with zero-flux partitioning of electron densities". *Theor. Chem. Acc.*107, 381–382.

Barrow, G. M. (1963). *The Structure of Molecules*, Benjamin, New York.

Becke, A. D., Edgecombe, K. E. (1990). A simple measure of electron localization in atomic and molecular systems. *J. Chem. Phys.* 92, 5397–5403.

Bendazzoli, G. L. (1993). The variational principle illustrated by simple examples. *J. Chem. Educ.* 70(11), 912–9113.

Bentley, W. A. (1962). *Snow Crystals*, Dover, New York.

Berlin, T. (1951). Binding Regions in Diatomic Molecules. *J. Chem. Phys.* 19, 208–213.

Berski, S., Andres, J., Silvi, B., Domingo, L. R., (2003). The Joint Use of Catastrophe Theory and Electron Localization Function to Characterize Molecular Mechanisms. A Density Functional Study of the Diels–Alder Reaction between Ethylene and 1,3-Butadiene. *J. Phys. Chem. A* 107, 6014–6021.

Boeyens, J. C. A. (1995). Understanding electron spin. *J. Chem. Educ.* 72(5), 412–415.

Breneman, G. L. (1988). Order out of chaos: shapes of hydrogen orbitals. *J. Chem. Educ.* 65, 31–33.

Buckingham, A. D., Rowlands, T. W. (1991). Can addition of a bonding electron weaken a bond? *J. Chem. Educ.* 68(4), 282.

Campbell, J., Moyers, B. (1988). *The Power of Myth*, Doubleday, New York, pp. 226.

Casida, M. E., Harriman, J. E. (1986) Geometry of density matrices. VI. Superoperators and unitary invariance. *Int. J. Quantum Chem.* 30(2), 161–212.

Cassam-Chenaï, P., Jayatilaka, D. (2002). A complement to "Some fundamental problems with zero flux partitioning of electron densities". *Theor. Chem. Acc.* 107, 383–384.

Chakraborty, A. (2007). Thesis, University of Kalyani, West Bengal, India.

Clinton, W. L., Galli, A. J., Henderson, G. A., Lamers, G. B., Massa, L. J., Zarur, J. (1969). Direct determination of pure-state density matrices. V. Constrained eigenvalue problems. *Phys. Rev.* 177, 27–33.

Clinton, W. L., Massa, L. J. (1972). Determination of the electron density matrix from X-ray diffraction data. *Phys. Rev. Lett.* 29, 1363–1366.

Clinton, W. L., Frishberg, C. A., Massa, L. J., Oldfield, P. A. (1973). Methods for obtaining an electron-density matrix from X-ray diffraction data. *Int. J. Quantum Chem. Symp.* 7, 505–514.

Cotton, A. F. (1990). *Chemical Applications of Group Theory,* 3rd ed., Wiley, New York.

Cotton, F. A., Wilkinson, G., Gaus, P. L. (1995). *Basic Inorganic Chemistry*, 3rd edition, John Wiley and Sons, Inc. New York.

Coulson, C. A. (1939). The electronic structure of some polyenes and aromatic molecules. VII. Bonds of fractional order by the molecular orbital method. *Proc. Roy. Soc. (London)* A169, 413–428.

Coulson, C. A. and Longuet-Higgins, H. C. (1947). The electronic structure of conjugated systems. I. General theory., Proc. Roy. Soc. (London), A191, 39–60.

Daudel, R., Leroy, G., Peeters, D., Sana, M. (1983). *Quantum Chemistry*, John Wiley & Sons, New York.

Delle Sie, L. (2002). Bader's interatomic surfaces are unique. *Theor. Chem. Acc.* 107, 378–380.

Demus, O. (1988). *The Mosaic Decoration of San Marco, Venice.* U. Chicago, Chicago plate 60.

Dias, J. R. (1989). A facile Huckel molecular orbital solution of buckminsterfullerene using chemical graph theory. *J. Chem. Educ.* 66(12), 1012–1015.

Dicke, R. H., Wittke, J. P. (1960). *Introduction to Quantum Mechanics*, Addison Wesley.

Dou, Y. (1990). Equations for calculating Dq and, B. *J. Chem. Educ.* 67(2), 134–135.

Douglas, B. E., McDaniel, D. H., Alexander, J. J. (1983). *Concepts and Models of Inorganic Chemistry*, 2nd edition, John Wiley & Sons, New York.

Dreizler, R. M., Gross, E. K. U. (1990). *Density Functional Theory*; Springer Verlag, Heidelberg.

Duch, S. (2002). *Lectures Notes*, Natural and Applied Sciences, University of Wisconsin, Green Bay.

Edmiston, C., Rüdenberg, K. (1963). Localized atomic and molecular orbitals *Rev. Mod. Phys.* 35, 457–465.

Eisberg, R., Resnick, R. (1985). *Quantum Physics*, 2nd Ed., Wiley.

Emmer, M. (Ed.) (1993). *The Visual Mind: Art and Mathematics*. MIT Press, Cambridge, folia J2.

Feynman, R. P. (1939). Forces in molecules. *Phys. Rev.* 56, 340–343.

Figgis, B. N. (1966). *Introduction to Ligand Fields*, Wiley, New York.

Fiolhais, C., Nogueira, F., Marques, M., Eds., (2003). *A Primer in Density Functional Theory*; Springer-Verlag, Berlin.

Foresman, J. B. (1997). *Using Computers in Chemistry and Chemical Educations,* In: Zielinski, T. J., Swift, M. L. (Eds.), American Chemical Society, Washington, p.243.

Frishberg, C., Massa, L. J. (1981). Idempotent density matrices for correlated systems from x-ray-diffraction structure factors. *Phys. Rev. B* 24, 7018–7024.

Gaine, A. F., Page, F. M. (1980). J. Chem. Res. (S) 200

Gallup, G. A. (1988). The Lewis electron-pair model, spectroscopy, and the role of the orbital picture in describing the electronic structure of molecules. *J. Chem. Educ.* 65(8), 671–674.

Gangi, R. A., Bader, R. F. W. (1971). Study of the Potential Surfaces of the Ground and First Excited Singlet States of H_2O. *J. Chem. Phys.* 55, 5369–5377.

Giancoli, D. C. (1995). *Physics*, 4th Ed., Prentice-Hall.

Gillespie, R. J. (1966). The Structures of PF_5, CH_3PF_4, and $(CH_3)_2PF_3$. *Inorg. Chem.* 5(9), 1634–1635.

Greene, B. (1999). *The Elegant Universe*, W. W. Norton.

Greenwood, N. N., Earnshaw, A. (1984). *The Chemistry of the Elements*, Pergamon Press, Oxford.

Haken, H., Wolf, H. C. (1996)*The Physics of Atoms and Quanta*, 5th Ed., Springer-Verlag.

Hales, T. (2012). *Dense Sphere Packings. A Blueprint for Formal Proofs* (London Mathematical Society Lecture Note Series: 400). Cambridge University Press, Cambridge.

Harriman, J. E. (1979). Geometry of density matrices. III. Spin component. *Int. J. Quantum Chem.* 15, 611–643.

Harriman, J. E. (1978). Geometry of density matrices. I. Definitions, N matrices and 1 matrices. Phys. Rev. A 17, 1249–1257.

Harriman, J. E. (1983). Geometry of density matrices. IV. The relationship between density matrices and densities. *Phys. Rev. A* 27, 632–645; ibid. (1984). Geometry of density matrices. V. Eigenstates. *Phys. Rev. A* 30, 19–29.

Harriman, J. E. (1986). Densities, operators, and basis sets. *Phys. Rev. A* 34, 29–39.

Henderson, G. A., Zimmerman, R. K. Jr. (1976). One-electron properties as variational parameters. *J. Chem. Phys.* 65, 619–622.

Herzberg, G. (1944). *Atomic Spectra and Atomic Structure*, 2nd Ed., Dover.

Heyes, S. J. (1999). Lectures' Notes, *Four Lectures in the 1st Year Inorganic Chemistry Course*, Oxford University.

Hirshfeld, F. L., Rzotkiewics, S. (1974). Electrostatic binding in the first-row AH and A2 diatomic molecules. *Mol. Phys.* 27, 1319–1343.

Huheey, J. A. (1983). *Inorganic Chemistry*, 3rd edition, Harper & Row, New York.

HyperPhysics (2010): http://hyperphysics.phy-astr.gsu.edu/Hbase/hframe.html

Jones, E. R., Childers, R. L. (1990). *Contemporary College Physics*, Addison-Wesley.

Julg, A. (1967). *Chimie Quantique*, Dunod, Paris.

Kappraff, J. (1990). *Connections: The geometric Bridge between Art & Science*. McGraw, New York, pp. 195.

Kaufmann, W. J. III. (1991). *Universe*, 3rd Ed., W. H. Freeman.

Keller, A. (1983). *Infancy of Atomic Physics, Hercules in His Cradle*, Oxford, Clarendon Press, pp. 215.

Kettle, S. F. A. (1998). *Physical Inorganic Chemistry*, Oxford University Press, New York.

Klein, D. J., Trinajstić, N. (1990). Valence-bond theory and chemical structure. *J. Chem. Educ.* 67(8), 633–637.

Koch, W., Holthausen, M. C. (2002). *A Chemist's Guide to Density Functional Theory*, 2nd ed., Wiley-VCH, Weinheim.

Koga, T., Umeyama, T. (1986). Approximate interaction energy in terms of overlap integral. *J. Chem. Phys.* 85, 1433–1437.

Kohout, M., Pernal, K., Wagner, F. R., Grin, Y. (2004). Electron localizability indicator for correlated wavefunctions. I. Parallel-spin pairs. *Theor. Chem. Acc.* 112, 453–459.

Kolos, W., Roothaan, C. C. J., Sack, R. A. (1960). Ground state of systems of three particles with coulomb interaction. *Rev. Mod. Phys.* 32, 178–179.

Konig, E. (1971). The nephelauxetic effect calculation and accuracy of the interelectronic repulsion parameters, I. Cubic high-spin d^2, d^3, d^7, and d^8 systems *Structure and Bonding* 9, 175–212.

Kryachko, E. S. (2002). Comments on "Some fundamental problem with zero flux partitioning of electron densities". *Theor. Chem. Acc.* 107, 375–377.

Kryachko, E. S., Petkov, I. Zh., Stoitsov, M. V. (1987). Method of local-scaling transformations and density functional theory in quantum chemistry. II. The procedure for reproducing a many-electron wave function from x-ray diffraction data on one-electron density. *Int. J. Quantum Chem.* 32, 467–472.

Lancashire, R. J. (2002). *Lectures Notes* on Chemistry of Transition Metal Complexes, The Department of Chemistry, University of the West Indies.

Lawlor, R. (1982). *Sacred Geometry*. Thames & Hudson, New York, pp. 106.

Lehner, E. (1950). *Symbols, Signs & Signets*, Dover, New York, pp. 77; 85.

Leighton, R. B. (1959). *Principles of Modern Physics*, McGraw-Hill.

Lever, A. B. P. (1984). *Inorganic Electronic Spectroscopy*, 2nd Edition, Elsevier Publishing Co., Amsterdam.

Levine, I. N. (1970). *Quantum chemistry*, Allyn and Bacon, Boston.

Löwdin, P.-O. (1955a). Quantum theory of many-particle systems. I. Physical interpretations by means of density matrices, natural spin-orbitals, and convergence problems in the method of configurational interaction. *Phys. Rev.* 97, 1474–1489; ibid. (1955b). Quantum theory of many-particle systems. II. Study of the ordinary hartree-fock approximation. *Phys. Rev.* 97, 1490–1508; ibid. (1955). Quantum theory of many-particle systems. III. Extension of the Hartree-Fock scheme to include degenerate systems and correlation effects. *Phys. Rev.* 97, 1509–1520; ibid. (1960). Expansion theorems for the total wave function and extended Hartree-Fock schemes. *Rev. Mod. Phys.* 32, 328–334.

Mackworth-Praed, B. (1993). *The Book of Kells*, Studio, London Plate XIII.

Malmqvist, P. Å. (1986). Calculation of transition density matrices by nonunitary orbital transformations. *Int. J. Quantum. Chem.* 30(4), 479–494.

March, N. H. (1991). *Electron Density Theory of Many-Electron Systems*; Academic Press, New York.

Massa, L., Goldberg, M., Frishberg, C., Boehme, R. F., La Placa, S. J. (1985). Wave functions derived by quantum modeling of the electron density from coherent X-ray diffraction: beryllium metal. *Phys. Rev. Lett.* 55, 622–625.

Mebane, R. C., Schanley, S. A., Rybolt, T. R., Bruce, C. D. (1999). The correlation of physical properties of organic molecules with computed molecular surface areas. *J. Chem. Educ.* 76(5), 688–693.

Meissler, G. L., Tarr, D. A. (1998). *Inorganic Chemistry*, 2nd edition, Prentice Hall, New Jersey.

Merzbacher, E. (1998). *Quantum Mechanics*, 3rd Ed., Wiley.

Mohallem, J. R. (2002). Molecular structure and Bader's theory. *Theor. Chem. Acc.* 107, 372–374.

Morrison, J. C., Weiss, A. W., Kirby, K., Cooper, D. (1993). *Encyclopedia of Applied Physics*; In: G. L. Trigg (Ed.), VCH, New York, Vol. 6, p. 45.

Mulliken, R. S. (1955). Electronic population analysis on LCAO–MO molecular wave functions. I. *J. Chem. Phys.* 23, 1833–1840; ibid. (1955). Electronic population analysis on LCAO–MO molecular wave functions. II. Overlap populations, bond orders, and covalent bond energies. *J. Chem. Phys.* 23, 1841–1846; ibid. (1955). Electronic population analysis on LCAO – MO molecular wave functions. III. Effects of hybridization on overlap and gross AO populations. *J. Chem. Phys.* 23, 2338–2342; ibid. (1955). Electronic population analysis on LCAO – MO molecular wave functions. IV. Bonding and antibonding in LCAO and valence – bond theories. *J. Chem. Phys.* 23, 2343–2346.

Nesbet, R. K. (2002). Orbital functional theory of linear response and excitation. *Int. J. Quantum Chem.* 86, 342–346.

Nicholls, D. (1971). *Complexes and First-Row Transition Elements*, Macmillan Press Ltd, London.

Nordholm, S. (1988). Delocalization the key concept of covalent bonding. *J. Chem. Educ.* 65(7), 581–584.

Novak, I. (1995). Molecular isomorphism. *Eur. J. Phys.* 16(4), 151.

Novak, I. (1999). Electronic states and configurations: visualizing the difference. *J. Chem. Educ.* 76(1), 135–137.

Ohanian, H. (1989). *Physics*, 2nd Ed Expanded, W. W. Norton.

Page, C. C., Moser, C. C., Chen, X., Dutton, P. L. (1999). Natural engineering principles of electron tunneling in biological oxidation-reduction. *Nature* 402(6757), 47–52.

Parson, R. (1993). Visualizing the variation principle: an intuitive approach to interpreting the theorem in geometric terms. *J. Chem. Educ.* 70, 115–119.

Pauling, L., Wilson, E. B. (1935). *Introduction to Quantum Mechanics*, McGraw-Hill.

Phillips, J. C., Kleinman, L. (1959). New method for calculating wave functions in crystals and molecules. *Phys. Rev.* 116, 287–294; ibid. (1960). Crystal potential and energy bands of semiconductors. III. Self-consistent calculations for silicon. *Phys. Rev.* 118, 1153–1167.

Pickett, W. E. (1989). *Pseudopotential Methods in Condensed Matter Applications*, North-Holland.

Ponec, R. (1982). On the accuracy of a localized description of chemical bonding. *J. Mol. Str. (Theochem)* 86, 285–290.

Prospero (1944). *The Book of Symbols: Magic*, Chronicle, San Francisco, pp. 22.

Purcell, K. F., Kotz, J. C. (1977). *Inorganic Chemistry*, W. B. Saunders Company, Philadelphia.

Richtmyer, F. K., Kennard, E. H., Cooper, J.N (1969). *Introduction to Modern Physics*, 6th Ed, McGraw-Hill.

Ringe, D., Petsko, G. (1999). Quantum enzymology: tunnel vision. *Nature* 399, 417–418.

Rohlf, J. W. (1994). *Modern Physics from A to Z0*, Wiley.

Roothaan, C. C. J. (1951). A study of two-center integrals useful in calculations on molecular structure. I. *J. Chem. Phys.* 19, 1445–1458.

Roothan, C. C. J. (1960). Self-consistent field theory for open shells of electronic systems. *Rev. Mod. Phys.* 32, 179–185.

Rüdenberg, K. (1951). A study of the two-center exchange integrals in molecular problems. *J. Chem. Phys.* 19, 1459–1477.

Sannigrahi, A. B., Kar, T. (1988). Molecular orbital theory of bond order and valency. *J. Chem. Educ.* 65(8), 674–676.

Savin, A., Dolg, M., Stoll, H., Preuss, H., Flesch, J. (1983). The correlated electron density of alkil atoms: pseudopotential and density functional results. *Chem. Phys. Lett.* 100(5), 455–460.

Schalley, C. A. (2001). Molecular recognition and supramolecular chemistry in the gas phase. *Mass Spectrom. Rev.* 20(5), 253–309.

Schlafer, H. L., Gliemann, G. (1969). *Basic Principles of Ligand Field Theory*, Wiley-Interscience, New York.

Schmidera, H. L., Becke, A. D. (2002). Two functions of the density matrix and their relation to the chemical bond. *J. Chem. Phys.* 116, 3184–3193.

Schwinger, J. (1951). The theory of quantized fields. I. *Phys. Rev.* 82, 914–927.

Schwinger, J. (1951). The theory of quantized fields. I. *Phys. Rev.* 82, 914–926.

Sen, K. D., Slamet, M., Sahni, V. (1993). Atomic shell structure in Hartree—Fock theory. *Chem. Phys. Lett.* 205, 313–316.

Serway, R. A. (1990). *Physics for Scientists and Engineers with Modern Physics*, 3rd Ed., Saunders College.

Serway, R. A., Moses, C. J., Moyer, C. A. (1997). *Modern Physics*, 2nd Ed., Saunders College.

Sholl, D., Steckel, J. A. (2009). *Density Functional Theory: A Practical Introduction*; Wiley-Interscience: Hoboken.

Shriver, D. F., Atkins, P. W. (1999). *Inorganic Chemistry*, 3rd edition, W. H. Freeman, New York.

Shusterman, G. P., Shusterman, A. J. (1997). Teaching chemistry with electron density models. *J. Chem. Educ.* 74(7), 771.

Simons, J. (1991). An experimental chemist's guide to ab initio quantum chemistry. *J. Phys. Chem.* 95(4), 1017–1029.

Smith, D. W. (2000). The Antibonding Effect. *J. Chem. Educ.* 77(6), 780–784.

Smith, R. A. (1969). *Wave mechanics of crystalline solids*, second edition, Chapman and Hall, London.

Srebrenik, S., Bader, R. F. W. (1975). Towards the development of the quantum mechanics of a subspace. *J. Chem. Phys.* 63, 3945–3961.

Srebrenik, S., Bader, R. F. W., Nguyen-Dang, T. T. (1978). Subspace quantum mechanics and the variational principle. *J. Chem. Phys.* 68, 3667–6679; Bader, R. F. W. Srebrenik, S., Nguyen-Dang, T. T. ibid., Subspace quantum dynamics and the quantum action principle. 68, 3680–3691.

Summerfield, J. H., Beltrame, G. S., and Loeser, J. G. (1999). A Simple Model for Understanding Electron Correlation Methods. *J. Chem. Educ.* 76(10), 1430–1438.

Tanabe, Y., Sugano, S. (1954). On the absorption spectra of complex ions. I. *J. Phys. Soc. Japan* 9, 753–766; ibid. (1954). On the absorption spectra of complex ions. II. *J. Phys. Soc. Japan* 9, 766–779.

Thom, R. (1975). *Structural Stability and Morphology (An Outline of a General Theory of the Models)*, Benjamin Inc., Reading, Massachusetts.

Thornton, S. T., Rex, A. (1993). *Modern Physics for Scientists and Engineers*, Saunders College Publishing.

Tipler, P. A. (1992). *Elementary Modern Physics*, Worth.

Tipler, P. A., Llewellyn, R. A. (1999). *Modern Physics*, 3rd Ed., W. H. Freeman.

Velders, G. J. M., Feil, D. (1993). Comparison of the Hartree-Fock, Møller-Plesset, and Hartree-Fock-Slater method with respect to electrostatic properties of small molecules. *Theor. Chim. Acta* 86, 391–416.

White, H. E. (1972). *Modern College Physics*, 6th Ed., van Nostrand Reinhold.

Young, H. D. (1968). *Optics and Modern Physics*, McGraw-Hill.

CHAPTER 3

QUANTUM CHEMICAL REACTIVITY OF ATOMS-IN-MOLECULES

CONTENTS

ABSTRACT

Electronegativity and hardness stand within the minimum dimensioned set of global indices that characterize bonding and reactivity as the electronic density and the effective applied potential function closely relate with the inner structure of atoms and molecules. This chapter advocates that a proper combination between these two sets of global and local indices can generate a whole plethora of density functionals with a role in quantifying the many-electronic structures and their transformation at the conceptual rather computational quantum level of comprehension. This is proved though applying the obtained electronegativity and hardness atomic scaled to selected problematical chemical reaction to provide the prediction of reactivity and stabilization of bonds in accordance with the main principles of chemistry: equalization and inequality of electronegativity and hardness, known as the electronegativity equalization, inequality of chemical potential, hard and soft acids and base, and maximum hardness principles, respectively. In this context, a novel reactivity index for quantifying the maximum of hardness realization was proposed with reliability proved throughout providing the hierarchy for a series of hard and soft Lewis bases. In all these, once again, the chemical action influence appears in playing the role of averaged quantum fluctuations that stabilize the molecules at the end of bonding process. There is also for the first time indicated the appropriate complete bonding scenario based exclusively on the correlated quantum quantities and principles of the electronegativity and hardness. This way, the complete set of global electronegativity-hardness indicators of reactivity of atoms and molecules for various physical-chemical conditions is formulated in an elegant analytical manner within the conceptual density functional theory. Therefore, there is still hope that the present scenario will be accompanied by some advanced ultra fast frozen movie of atomic encountering in bonding.

3.1   INTRODUCTION

Quoting Roald Hoffmann: "There is nothing more fundamental to chemistry than the chemical bond" and still, according with Charles A. Coulson: "It does not exist. No one has ever seen one. No one ever can. It is a figment of our own imagination". Just like the millenary search for the Holy Grail, the revelation of the engines that promote, hold and activate a molecular structure remains a permanent challenge for the human intelligence. Shortly, it is worth noting the seminal contributions of the *dualist theory* of Berzelius (1819) advancing for the first time the idea of electrostatic interaction between two opposite charged atoms in defining chemical bonding. However, without taking into account the causes of the charges involved, the theory fails to explain the bonding between two identical atoms, as well as the plethora or organic compounds. It was the *unitary theory* of Dumas (1834) that solves the dichotomy by assuming the bonding forces to be of the same kind whatever the component atoms considered may be. Nevertheless, each of these theories assesses, in fact, a specific type of the chemical bond, the ionic and covalent ones, respectively. Still, Pandora's

Box was opened when the very connection between these two extremes was hidden under the inorganic and organic roughly classification of the chemical compounds. Despite the efforts of star chemist as Kekulé (1857, 1858), Couper (1858), Butlerov (1861), van't Hoff (1874), Le Bel (1874) or Werner (1893) in the second part of the nineteenth century to elucidate the structural constitution of molecules on conceptual grounds, the history of chemical bonding remains with the concept of valence as another mysterious benchmark of the nature's mode of action.

Then, wile the first half of the twentieth century brings to light the *quantum theory* of matter, the subsequent searches of accommodating the valence concepts with the quantum principles dominate the conceptual chemistry through the cornerstone works of Lewis (1916), Kossel, Heitler and London (1927; Heitler, 1931), Pauling (1928, 1931), Mulliken (1942), Hund (1931), Hückel (1930; Fox & Matsen, 1985), Herzberg (1929), Schrödinger (1926), Dirac (1930) and Slater (1930, 1931, 1934). It follows that the chemical bond widescreen can be summarized as the inter-connections between the four fundamental types of bonds: covalent, ionic, metallic, and van der Waals. At this point, it is worth noting the seminal contribution of Lewis (1916) through his "The Atom and the Molecule" (1916) work, where the chemical intuition overwhelms the already three-year old Bohr Theory of hydrogenic atoms (Bohr, 1913, 1921) by introducing the surreal concept of "cubical atom". Although, at first sight, such a paradigm may seem strange now, there it was the first affirmation of the necessity that the atom itself has to be assumed with an inherent structure, viz. orbitals, of symmetry types different even circumvented by the spherical one. Such intuition was, more than ten years later, confirmed when the Schrödinger equation was analytically solved for the hydrogenic atoms and recovering the Bohr's energy in addition to the celebrated orbital functions (Schrödinger, 1926).

Moreover Lewis' lone and bond pair or electrons, abstracted from its "cubical atomic" combinations through connecting of their edges with electronic occupancy between 0 and 2, becomes the main "lingua franca" of chemical bonding analysis leading to the disputed concepts of bonding localizations both at the orbital (intensive) and functional (global) approaching levels. However, the atomic structure was afterwards found as the key of both explaining the atomic periodicities, i.e., recovering and definitely certifying the Mendeleyev systematic arrangement of the elements in its Table, and providing the quantitative tools, i.e., atomic orbitals, with the help of which the entire molecular panorama seems to be on the way of unfolding. The fundamental works of Hartree (1928), Fock (1930), Roothaan (1951), again Slater (1951), and Kohn (Hohenberg & Kohn, 1964; Kohn & Sham, 19657) further enlightened the quantum nature of the chemical bond at the intensive level of electronic spin-orbitals.

Consequently, from the second part of the twentieth century nowadays the first rate scientifically research has been focused mainly on the synergistic quantum approaches to the structure and properties of the natural complex systems, i.e., the polyatomic and biomolecular ones (Journal of Computational Chemistry, 2007).

While pure physics struggled on the great unification paradigm through the fundamental forces in nature, "being subject, in the last decade, to a continuous reform, a similar attitude is now emerging in chemistry, at the quantum level of representation, related to the existing natural chemical bonds. However, because the types of the chemical bonds coexist in various degrees and combinations in the organization of the matter, only a unitary quantum treatment, based on the first physical-chemical principles, can release an estimation of the structure-properties correlations across the complex natural nano-systems: metals, clusters, fullerenes, liquid crystals, polymers, ceramics, biomaterials, metaloenzymes (Levin & Krüger, 1977; Bochicchio et al., 1989; Richard et al., 1993; Gao, 1997; Calatayud et al., 2001; Brinkmann et al., 2003; Sato, 2003; Putz, 2007a).

This way, a unitary picture to link and flexibly adapt the quantum mechanical formalisms on the chemical bonding problem were intensively studied (Journal of Computational Chemistry, 2007). Still, with the belief that the unification of the chemical bonds can be achieved through a single equation or force (Putz, 2007a) we advance in this chapter the iterative link between the intensive, local and global levels of chemical bond in a unitary presentation (Putz, 2003, 2007a, 2009a-b, 2011a-f, 2012a-c).

3.2 COEXISTENCE OF ATOMS-IN-MOLECULES

3.2.1 DENSITY DESCRIPTION OF AIM

A natural approach of chemical bonding is the integration of atoms in molecule (AIM), i.e., understanding the atom respecting to the rest of the molecule (especially) by the electronic contribution which is brought by it to the global electronic distribution in molecule; in this regard, the following "program" of AIM can be unfold by means of a series of definitions and principles (Rychlewski & Parr, 1986; Li & Parr, 1986).

DEFINITION AIM1: One introduces the partition function of the atom A in a molecule, the positive spatially dependent measure $\alpha_A(x)$, $0 \le \alpha_A(x) \le 1$, $x \in \mathfrak{R}^3$, which multiplied with the total electronic density (positive) from the molecule gives the electronic density associated with the atom in question:

$$\alpha_A(x) = \frac{\rho_A(x)}{\rho(x)}, \ 0 \le \alpha_A(x) \le 1, \ x \in \mathfrak{R}^3 \tag{3.1}$$

With this definition, as specified also in the introductive chapter of this Volume, we can consider the following definition of physical-topological field that may be associated to the atom in molecule.

DEFINITION AIM2 (of Bader): The physical-topological space available for an atom (A) in a molecule can be defined by a special choice of the partition function, namely

$$\alpha_A(x) = \begin{cases} 1, x \in \Omega_A \left\| [\nabla \rho_A(x) \cdot \vec{n}(x)]_{x \in \partial\Omega_A} \right\|_{x \in \partial\Omega_A} = 0 \\ 0, in \ rest \end{cases} \tag{3.2}$$

The advantage of this model is the atoms correlation in molecule with the geometric shape of the distribution of its own electrons, in addition to the advantage of a topological treatment on its own subspaces associated to atoms.

However, the disadvantages are more consistent. Firstly, the physical and intuitive image suffers, especially about to the atomic "choice" to reach the inter-atomic bonds in the molecule. In addition, for the covering of the physical-topological spaces of the atoms in chemical bonds one can imagine the inability to distinctly treat the ionic bond from the covalent one.

Another approach, much more physically, is based on the optimization criterion of the energy of promoting the atoms from their isolated state in the bonded one in molecules.

DEFINITION AIM3 (of Parr): For a given atoms set $\{A_1, ..., A_k\}$ with the characteristics from the isolated state given by the densities sets $\{\rho(x)_1^0, ..., \rho(x)_k^0\}$, and of the fundamental (or of valence) states with the energies, $\left\{E\left[\rho_1^0\right],\ldots,E\left[\rho_k^0\right]\right\}$, the connection state in a molecule is characterized by the sets of densities, $\left\{\rho(x)_1^*,\ldots,\rho(x)_k^*\right\}$ of the energies, $\left\{E\left[\rho_1^*\right],\ldots,E\left[\rho_k^*\right]\right\}$ and of the potentials, $\left\{\mu_1^*,\ldots,\mu_k^*\right\}$ satisfying the conditions:

$$\sum_{i=1}^{k}\alpha_{A_i}(x)=1 \tag{3.3}$$

$$\mu_1^* = \mu_2^* = \ldots = \mu_k^* = \mu \tag{3.4}$$

$$E_p = \sum_{i=1}^{k}E\left[\rho_i^*\right]-\sum_{i=1}^{k}E\left[\rho_i^0\right]= \min \tag{3.5}$$

Here μ stays for the equalized potential of the atoms in molecule (corresponding to the electronegativity equalization) and E_p represents the promotion energy of the isolated atoms in the chemical bond within a molecule.

There is interesting to study the atomic properties in molecule, based on the definition AIM3, with the great advantage of the energy-minimizing criterion, which takes out the atoms from the isolated state and promotes them in the molecular bonds.

3.2.2 ENERGY DESCRIPTION OF AIM

To write the energy expression, which characterizes an atom in a molecule, for simplicity, a diatomic molecule AB is considered for which the molecular wave function Ψ is known. This way, we can write the first-order density matrix (Rychlewski & Parr, 1986; Li & Parr, 1986):

$$\rho(x,x')= N\int \Psi^*\Psi ds_1 d\tau_2\ldots d\tau_N \tag{3.6}$$

where the molecule was considered with N electrons, and the elementary space-spin volume $d\tau = dsdx$.

Under these conditions, the molecular kinetic energy will be written (for simplicity will be skipped the above "*" notation for optimized state, being implicitly if not otherwise specified):

$$T[\rho]= T[\rho_A]+T[\rho_B] \tag{3.7}$$

where the functional kinetic energies has the form:

$$T[\rho_A] = \int t[\rho_A(x)]dx \qquad (3.8)$$

$$t[\rho_A(x)] = -\frac{1}{2}\nabla^2 \rho_A(x,x')\big|_{x=x'} = \alpha(x)t[\rho(x)] \qquad (3.9)$$

The corresponding functional energy of molecular electronic repulsion will be written as:

$$V_{ee}[\rho] = V_{ee}[\rho_A] + V_{ee}[\rho_B] + V_{ee}[\rho_A, \rho_B] \qquad (3.10)$$

where

$$V_{ee}[\rho_A] = \int\int G_A(x,x')dxdx' = \int\int \frac{\alpha(x)\alpha(x')\rho^2(x,x')}{|x-x'|}dxdx' \qquad (3.11)$$

$$V_{ee}[\rho_A,\rho_B] = \int\int G_{AB}(x,x')dxdx' = \int\int \frac{\{[\alpha(x)\beta(x') + \alpha(x')\beta(x)]\rho^2(x,x')\}}{|x-x'|}dxdx'$$

$$(3.12)$$

and with $\beta(x)$ as *the partition function* of electronic density for the atom B.

In the same manner, the total nucleus-electron molecular energy will be formed by the terms:

$$V_{en}[\rho] = V_A[\rho_A] + V_B[\rho_A] + V_A[\rho_B] + V_B[\rho_B] \qquad (3.13)$$

where the constituent terms have the form:

$$V_A[\rho_A] = \int V_{AA}(x)dx = -\int \frac{\rho_A(x)Z_A}{|x-R_A|}dx = -\int \frac{\alpha(x)\rho(x)Z_A}{|x-R_A|}dx \qquad (3.14)$$

$$V_B[\rho_A] = \int V_{BA}(x)dx = -\int \frac{\rho_A(x)Z_B}{|x-R_B|}dx = -\int \frac{\alpha(x)\rho(x)Z_B}{|x-R_B|}dx \qquad (3.15)$$

This way the density functional character for the terms involved in the molecular energies is emphasized. For example, the energy of the atom A in the considered molecule associates with the density functional energy

$$E[\rho_A] = \int [t_A(x) + V_{AA}(x) + \int G_A(x,x')dx']dx \qquad (3.16)$$

with gives the immediate generalization for the k atoms from the Definition AIM3:

$$E[\rho_k] = \int [t_k(x) + V_{kk}(x) + \int G_k(x,x')dx']dx \qquad (3.17)$$

Taking into account the expression of the promotion energy previously defined, together with the minimizing condition in the molecular bond – which can be seen as a constraint, the functional can be formed:

$$\Pi = E_p - \int \pi(x) \left[\sum_{i=1}^{k} \alpha_i(x) - 1 \right] dx \qquad (3.18)$$

where $\pi(x)$ is the Lagrange multiplier. From the effective functional minimization respecting the atomic partition functions one yields:

$$t(x) - \frac{Z_i\rho(x)}{|x-x_i|} + 2\int \frac{\alpha_i(x')\rho^2(x,x')}{|x-x'|}dx' = \pi(x) \qquad (3.19)$$

The k-equations of the last relation type form together with the energy minimization condition a closed system leaving with determination of the k-atomic partition functions in molecule, $\alpha_i(x)$, and the Lagrange multiplier $\pi(x)$. The phenomenological relationship between these two last AIM measures can be deduced, if the last relation (3.19) is multiplied with $\alpha_i(x)$ to get the form:

$$\alpha_i(x)t(x) - \frac{Z_i\alpha_i(x)\rho(x)}{|x-x_i|} + \int \alpha_i(x)\frac{\alpha_i(x')\rho^2(x,x')}{|x-x'|}dx'$$

$$= \alpha_i(x) \left[\pi(x) - \int \alpha_i(x')\frac{\rho^2(x,x')}{|x-x'|}dx' \right] \qquad (3.20)$$

where the energy density of the i-atom at the coordinate x in molecule is now easily recognized. Comparing the last two relations, one can observe that the energy density of the representative atom "i" is identically null if its partition function in the molecule is zero, even if the Lagrange multiplier $\pi(x)$ is not vanishing.

In the diatomic case, the molecular promotion energy will be written as:

$$E_p[\alpha] = \int \left\{ t(x) + V_{AA}(x) + V_{BB}(x) + \int [G_A(x,x') + G_B(x,x')]dx' \right\} dx$$
$$- E[\rho_A^0] - E[\rho_B^0] \tag{3.21}$$

where the specific dependence in terms of the partition function for the forming energy has been highlighted by specifying it occurs only for the atoms in the molecule, i.e., for the initial energetic terms; in isolated state, this dependence does not count.

The minimization condition of the promotion energy can be manifested by a variational equation, in this case obtained from (3.21) relation, by taking the derivative in relation with the molecular partition function in molecule, and so it is obtained:

$$\left(\frac{Z_B}{|x - R_B|} - \frac{Z_A}{|x - R_A|} \right) \rho(x) - 2\int \frac{\rho^2(x,x')}{|x - x'|} dx' + 4\int \alpha(x') \frac{\rho^2(x,x')}{|x - x'|} dx' = 0 \tag{3.22}$$

The analytical solution of this equation towards finding the explicit form of the atomic partition function in molecule involves a procedure for solving the Fredholm integral equations. However, such equation can be formally solved, by considering it as a infinite linear set of algebraic equations, and after matrix-formalization one can resuming it as following (Rychlewski & Parr, 1986; Li & Parr, 1986)

$$4hGa = 2hg + V \tag{3.23}$$

with h the integral volume element, with the other matrices having the elements:

$$G_{kl} = \frac{\rho^2(x_k, x_l)}{|x_k - x_l|} \tag{3.24}$$

$$g_k = \sum_l G_{kl} \tag{3.25}$$

$$V_k = \left(\frac{Z_A}{|x_k - R_A|} - \frac{Z_B}{|x_k - R_B|} \right) \rho(x_k) \tag{3.26}$$

$$\alpha_l = \alpha(x_l) \tag{3.27}$$

Then, immediately the matrix for *atomic partition functions in molecule* results as:

$$\alpha = (1/4)G^{-1}\left[2g + h^{-1}V\right] \tag{3.28}$$

from which for $V(x) = 0$ the constant partition function, $\alpha(x) = 1/2$, is obtained; it means that the deviation from this value in partitioning the atomic influences in molecule is due to the attraction between the electrons and the atomic nuclei.

In this competition of the atoms influence in the molecule, there should be noted that the promotion energy of atoms in a molecule, E_p, involves both the molecular formation, the binding energy E_b, in respecting the energies of the isolated atoms, as well as by the interaction energy, E_I, in the molecule, relating the total electronic energy of the molecule,

$$E[\rho] = T[\rho] + V_{ee}[\rho] + V_{en}[\rho] \tag{3.29}$$

and the repulsion energy between the molecule nuclei,

$$V_{nn} := V_{AB}^N = \frac{Z_A Z_B}{R_{AB}} \tag{3.30}$$

The overall relations between these energies are represented in Figure 3.1, and written accordingly:

$$E_p = -E_I + E_b \tag{3.31}$$

$$E_b = E[\rho] + V_{AB}^N - \left(E[\rho_A^0] + E[\rho_B^0]\right) \tag{3.32}$$

$$
\begin{aligned}
E_I &= E[\rho] + V_{AB}^N - \left(E[\rho_A] + E[\rho_B]\right) \\
&= V_B[\rho_A] + V_A[\rho_B] + V_{ee}^{int}[\rho_A, \rho_B] + V_{AB}^N \\
&= \int\left[\left[V_{BA}(x) + V_{AB}(x) + \int G_{AB}(x, x')dx'\right]dx + V_{AB}^N\right. \\
&= -\int\left[\frac{Z_B\rho_A(x)}{|x - R_B|} + \frac{Z_A\rho_B(x)}{|x - R_A|}\right]dx + \iint\frac{\rho_A(x)\rho_B(x')}{|x - x'|}dxdx' + V_{AB}^N + E_{xc}[\rho_A, \rho_B]
\end{aligned}
\tag{3.33}
$$

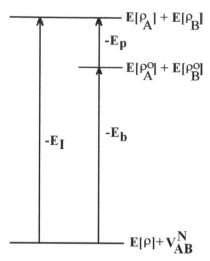

FIGURE 3.1 The energies of "binding (b)", "promotion (p)" and "interaction (I)" for the diatomic molecule AB; after (Rychlewski & Parr, 1986; Li & Parr, 1986).

Here, $E_{xc}[\rho_A, \rho_B]$ represents the exchange and correlation energy between the atoms A and B in molecule.

The interaction energy has a very consistent physical-chemical meaning as this can be seen also from the involved terms' analysis. Moreover, the interaction energy can be written as the sum of energetic contributions for the two atoms involved in the ionic and covalent characters of the chemical bonds in the molecule:

$$E_I = \Delta E_{ion} + \Delta E_{cov} \tag{3.34}$$

If not considering the overlapping of the charged spheres $(Z_A - N_A)$ and $(Z_B - N_B)$ the expressions associated to the two contributions at coordination can be written as:

$$\Delta E_{ion} = \frac{(Z_A - N_A)(Z_B - N_B)}{R_{AB}} \tag{3.35}$$

$$\Delta E_{cov} = E_I - \Delta E_{ion} = V_B[\rho_A] + V_A[\rho_B] + V_{ee}^{int}[\rho_A, \rho_B]$$

$$+ \frac{Z_A Z_B - (Z_A - N_A)(Z_B - N_B)}{R_{AB}} \tag{3.36}$$

with the covalent percentage of the bond given by:

$$\theta_{cov}\% = \frac{\Delta E_{cov}}{E_I}\%$$ (3.37)

The whole process of interaction promotion can be intuitively represented by the densities transformations associated to the atomic and molecular states:

$$\rho_A^0 + \rho_B^0 \xrightarrow{E_p} \rho_A^* + \rho_B^* \xrightarrow{E_I} \rho_{AB}(x)$$ (3.38)

In conjunction with the fact that the interaction energy from the promotion energy contains the information of the percent ionic and covalent character, there following can be stated (Rychlewski & Parr, 1986; Li & Parr, 1986): PRINCIPLE AIM1: In the molecular formation, the charge accumulated in the region of the chemical bonds is distributed so that to maintain the minimum promotion energy.

From above, we can see that the minimization of the promotion energy is equivalent to the minimizing of the interaction energy respecting to the atomic partition function in molecule.

In writing the total energies factorized form in kinetic energies in the promoted state associated to atoms A and B forming a molecule, the proper energies of interaction and energies of interaction result from the condition:

$$\frac{\delta E_p}{\delta \alpha(x)} = 0 = \frac{\delta(E[\rho_A] + E[\rho_B])}{\delta \alpha(x)}$$ (3.39)

that for each atom in molecule is validated by the relation of type (representatively written only for the atom A):

$$\frac{\delta E[\rho_A]}{\delta \alpha(x)} = 0$$ (3.40)

$$E[\rho_A(x)] = T[\rho_A(x)] + V_A[\rho_A(x)] + E_{I(A)}[\rho_A(x)]$$ (3.41)

When a scaled the molecular wave function is considered

$$\Psi(x) = \zeta^{3N/2}\Psi(\xi, x)$$ (3.42)

then the relation (3.41) becomes:

$$E[\rho_A(\xi,x)] = \xi^2 T[\rho_A(1,t)] + \xi V_A[\rho_A(1,t)] + \xi E_{I(A)}[\rho_A(1,t)] , t = \xi x$$

(3.43)

for which, the allied minimization condition as

$$\frac{\partial E[\rho_A]}{\partial \xi}\bigg|_{\xi=1} = \int \frac{\delta E[\rho_A]}{\delta \alpha(x)} \frac{\partial \alpha(x)}{\partial \xi}\bigg|_{\xi=1} dx = 0$$

(3.44)

so providing the analytical form of the virial theorem for an atom in molecule:

$$-2T[\rho_A] = V_A[\rho_A] + E_{I(A)}[\rho_A] + \sum_{i=A,B} R_i \cdot \nabla_i E[\rho_A]$$

(3.45)

and respectively for the whole molecule, when the two relation of this kind are added to give the general form:

$$-2T[\rho] = V_A[\rho_A] + V_B[\rho_B] + E_{I(A)}[\rho_A] + E_{I(B)}[\rho_B]$$

$$+ \sum_{i=A,B} R_i \cdot \nabla_i (E[\rho_A] + E[\rho_B])$$

$$= V_A[\rho_A] + V_B[\rho_B] + E_I + \frac{Z_A Z_B}{R_{AB}} + \sum_{i=A,B} R_i \cdot \nabla_i E[\rho] \quad (3.46)$$

3.2.3 POPULATION ANALYSIS OF AIM

Related with the chemical bonding description, there is for first interest also a method for determining the atomic partition function in molecule; here the Mulliken picture of the population analysis, α_{ij}, will be presented yet modified, in the way of assuming the overlapping of charge unequally distributed between the atoms involved in the bond. Considering the linear combination of atomic orbitals (LCAO) expansion of the molecular function, the density associated to the atomic orbitals $\{\Psi_{ij}\}$ is written (Rychlewski & Parr, 1986; Li & Parr, 1986):

$$\rho(x,x') = \sum_{ij} \rho_{ij} \psi_i(x)\psi_j(x')$$

(3.47)

where

$$\rho_{ij} = \sum_{\upsilon}^{occup} c_{i\upsilon} c_{j\upsilon}$$

(3.48)

represents the population density associated to the atomic orbitals set. Within these conditions, we can write the density associated to the atom A:

$$\rho_A(x,x') = \sum_{i,j}^{A} \rho_{ij}\psi_i(x)\psi_j(x') + \sum_{i}^{A}\sum_{j}^{B}\alpha_{ij}\rho_{ij}[\psi_i(x)\psi_j(x') + \psi_j(x)\psi_i(x')]$$

(3.49)

with which the expression of the total electronic energy for the atom A is becoming:

$$E[\rho_A] = \sum_{i,j}^{A}\rho_{ij}I_{ji}^{A} + \sum_{i}^{A}\sum_{j}^{B}2\alpha_{ij}\rho_{ij}I_{ji}^{A}$$

$$+ \frac{1}{2}\{[\sum_{i,j,k,l}^{A}\rho_{ij}\rho_{kl}(ij|kl) + \sum_{i}^{A}\sum_{j}^{B}\sum_{k}^{A}\sum_{l}^{B}4\alpha_{ij}\alpha_{kl}\rho_{ij}\rho_{kl}(ij|kl)$$

$$+ \sum_{i}^{A}\sum_{j}^{B}\sum_{k}^{A}\sum_{l}^{A}4\alpha_{ij}\rho_{ij}\rho_{kl}(ij|kl)]$$

$$- \frac{1}{2}[\sum_{i,j,k,l}^{A}\rho_{ij}\rho_{kl}(il|jk) + \sum_{i}^{A}\sum_{j}^{B}\sum_{k}^{A}\sum_{l}^{B}4\alpha_{ij}\alpha_{kl}\rho_{ij}\rho_{kl}(il|jk)$$

$$+ \sum_{i}^{A}\sum_{j}^{B}\sum_{k}^{A}\sum_{l}^{A}4\alpha_{ij}\rho_{ij}\rho_{kl}(il|jk)]\}$$

(3.50)

and where:

$$I_{ji}^{A} = \int \psi_j(x)\left(-\frac{1}{2}\nabla^2 - \frac{Z_A}{|x - R_A|}\right)\psi_i dx$$

(3.51)

$$(ij|kl) = \int \psi_i(x_1)\psi_j(x_1)\frac{1}{|x_1 - x_2|}\psi_k(x_2)\psi_l(x_2)dx_1 dx_2$$

(3.52)

Also in this case the optimization condition can be imposed, i.e.

$$\frac{\delta E[\rho_A]}{\delta \alpha_{ij}} = 0$$

(3.53)

which implies an equation of electronic molecular existence with the form

$$\sum_k^A \sum_l^B 2(2\alpha_{ij} - 1)\rho_{kl}\left[(ij|kl) - \frac{1}{2}(il|jk)\right] + \sum_k^A \sum_l^A \rho_{kl}\left[(ij|kl) - \frac{1}{2}(il|jk)\right]$$

$$-\sum_k^B \sum_l^B \rho_{kl}\left[(ij|kl) - \frac{1}{2}(il|jk)\right] - \frac{1}{2}(j|V_A - V_B|i) = 0, \ i \in A, j \in B$$

(3.54)

where one has noted

$$(j|V_A - V_B|i) = I_{ji}^A - I_{ji}^B \tag{3.55}$$

Noting, for simplicity, that $\alpha_{ij} = \alpha$ for all the indices i and j, one obtains (Rychlewski & Parr, 1986; Li & Parr, 1986):

$$\alpha = \frac{1}{2} + \frac{\sum_i^A \sum_j^B \rho_{ij}\left\{\begin{array}{c} \sum_k^B \sum_l^B \rho_{kl}\left[(ij|kl) - \frac{1}{2}(il|jk)\right] \\ -\sum_k^A \sum_l^A \rho_{kl}\left[(ij|kl) - \frac{1}{2}(il|jk)\right] + \frac{1}{2}(j|V_A - V_B|i) \end{array}\right\}}{\sum_i^A \sum_j^B \sum_k^A \sum_l^B 4\rho_{ij}\rho_{kl}\left[(ij|kl) - \frac{1}{2}(il|jk)\right]}$$

(3.56)

Note that when A and B are atoms of the same species, form the (3.56) relation remains only $\alpha_{kl} = 1/2$, for all the indices k and l. The presented conclusion corresponds to the covalent description of the atoms-in- molecule.

From all of the analyzed properties of the atoms in molecule, while taking into account the Definition of AIM2, another fundamental principle of the AIM approach can be formulated:

PRINCIPLE AIM2: The atoms in molecule are coordinated so that the bonding charges to concentrate near the nuclei while the charge transfer between them to be as small as possible.

3.3   ELECTRONEGATIVITY OF ATOMS-IN-MOLECULES

3.3.1   ELECTRONEGATIVITY AS CONNECTIVITY OF ATOMS IN MOLECULES

One of the special cases of atoms in molecule, yet important, is that when the total molecular energy can be partitioned into a sum of the atomic contributions:

$$E(N_A, N_B, ...; Z_A, Z_B, ...; R_{AB}, ...) = \sum_A \langle \hat{T}^A + \hat{V}_{ne}^A + \hat{V}_{ee}^A + \hat{V}_{nn}^A \rangle \quad (3.57)$$

It basically unfolds the sum over the expected values of kinetic energies, of nuclei-electrons, of electrons-electrons and of nuclei-nuclei interactions.

In this framework, we can consider a spherical distribution of the atoms clouds charge in molecule, so that the molecular energy can be rewritten accordingly (Genechten et al., 1987):

$$E = \sum_A E_A = \sum_A \left(E_A^{INTRA} + E_A^{INTER} \right)$$

$$= \sum_A \left[T^A + V_{ne}^A + V_{ee}^A \right] + \sum_A \left[-N_A \sum_{B \neq A} \frac{Z_B}{R_{AB}} + \frac{1}{2} \sum_{B \neq A} \frac{N_A N_B}{R_{AB}} \right] + \frac{1}{2} \sum_A \sum_{B \neq A} \frac{Z_A Z_B}{R_{AB}}$$

$$= \sum_A \left[E_A^* + \left(\frac{\partial E_A^*}{\partial N} \right) \Delta N_A + \frac{1}{2} \left(\frac{\partial^2 E_A^*}{\partial N^2} \right) (\Delta N_A)^2 \right]$$

$$+ \sum_A \left[-N_A \sum_{B \neq A} \frac{Z_B}{R_{AB}} + \frac{1}{2} \sum_{B \neq A} \frac{N_A N_B}{R_{AB}} \right] + \frac{1}{2} \sum_A \sum_{B \neq A} \frac{Z_A Z_B}{R_{AB}}$$

$$= \sum_A \left[E_A^* + \mu_A^* \Delta N_A + \eta_A^* (\Delta N_A)^2 \right]$$

$$+ \sum_A \left[-N_A \sum_{B \neq A} \frac{Z_B}{R_{AB}} + \frac{1}{2} \sum_{B \neq A} \frac{N_A N_B}{R_{AB}} \right] + \frac{1}{2} \sum_A \sum_{B \neq A} \frac{Z_A Z_B}{R_{AB}}$$

$$(3.58)$$

with the notations:

$$\Delta N_A = N_A - N_A^0 \quad (3.59)$$

$$\mu_A^* = \mu_A^0 + \Delta\mu_A \tag{3.60}$$

$$\eta_A^* = \eta_A^0 + \Delta\eta_A \tag{3.61}$$

From the atoms-in-molecules (AIM) energy expression (3.58), the chemical potential, respectively the effective electronegativity with changed sign, for the atom A in molecule automatically results:

$$-\chi_A = \mu_A = \left(\frac{\partial E}{\partial N_A}\right)_{N_B,\dots,R_{AB},\dots}$$

$$= \left(\mu_A^0 + \Delta\mu_A\right) + 2\left(\eta_A^0 + \Delta\eta_A\right)\Delta N_A - \sum_{B \neq A}\frac{Z_B - N_B}{R_{AB}} \tag{3.62}$$

Worth noting the dependency of the electronegativity of the A-atom in molecule by the chemical potential and force of the neutral atom, μ_A^0, η_A^0, along the induced influence by the modification in size $\Delta\mu_A$, by $\Delta\eta_A$ of the atom when integrated in molecule as driven by the transferred charge ΔN_A in molecule, and also by the applied external potential on the A-atom, $\sum_{B \neq A}\frac{Z_B - N_B}{R_{AB}}$, contains, the atom connectivity information in molecule.

Further on, one need to work the general case of the polyatomic molecules where the bond energy of each atom is subject to a parabolic shape (Red, 1981):

$$-E(N) = aN + bN^2 \tag{3.63}$$

From now the electronegativity of every atom will be given as defined by the partial derivative of the decreasing energy in respect with the charges' change, by the expression:

$$\chi_i(N_i) = a_i + 2b_i N_i \tag{3.64}$$

Admitting that there are K atoms in molecule, the electronegativity equalization principle will be applied as:

$$\chi_1(N_1) = \dots = \chi_i(N_i) = \dots = \chi_K(N_K) \tag{3.65}$$

From this equality one gets for example, the number of charges associated to the i-atom in molecule:

$$N_i = \frac{a_1 + 2b_1 N_1 - a_i}{2b_i} \qquad (3.66)$$

The conservation of charge is assured by performing the summary after all the atoms (K) from the molecule

$$\sum_{i=1}^{K} N_i = \Theta \qquad (3.67)$$

whereas by further dividing it to their numbers one get the equation:

$$\frac{\Theta}{K} = \frac{a_1}{2}\left(\frac{1}{K}\sum_{i=1}^{K}\frac{1}{b_i}\right) + b_1 N_1\left(\frac{1}{K}\sum_{i=1}^{K}\frac{1}{b_i}\right) - \frac{1}{2}\left(\frac{1}{K}\sum_{i=1}^{K}\frac{a_i}{b_i}\right)$$

$$:= \frac{a_1}{2}\left(\overline{\frac{1}{b}}\right) + b_1 N_1\left(\overline{\frac{1}{b}}\right) - \frac{1}{2}\left(\overline{\frac{a}{b}}\right) \qquad (3.68)$$

whose solution for b_1, while back substituted in the electronegativity expression of the atom "1" in molecule, as the average electronegativity of the molecule actually results:

$$\chi_1(N_1) = \chi_{molec}(\Theta) = \frac{2\left(\frac{\Theta}{K}\right) + \left(\overline{\frac{a}{b}}\right)}{\left(\overline{\frac{1}{b}}\right)} = a_{molec} + 2\frac{\Theta}{K}b_{molec} \qquad (3.69)$$

and where there were made the notations:

$$b_{molec} = \left(\overline{\frac{1}{b}}\right)^{-1} \quad , \quad \left(\overline{\frac{a}{b}}\right)b_{molec} = a_{molec} \qquad (3.70)$$

Moreover, from the molecular electronegativity expression there can be determined, by direct integration, also the molecular total energy, if it considered the condition $E(\Theta = 0) = 0$, and will have the form

$$-E(\Theta) = a_{molec}\Theta + \frac{1}{K}b_{molec}\Theta^2 \qquad (3.71)$$

with Θ, as the total number of the electrons in molecule, as before stated.

Thus, by the two situations presented one realizes how close they are: the bond, the connectivity, and the atomic influence respecting the rest of the molecule, however driven by the external potential (or nuclei potential for isolated molecules); all these influences are reflected in electronegativity, which by its nature resumes the averaging contributions of the atoms in the molecule (Bader et al., 1987). Therefore, this section can be concluded with the connectivity of AIM principle (Mullay, 1986, 1988a,b; Boyd & Markus, 1981; Edgecombe & Boyd, 1987; Labanowski et al., 1989; del Re, 1983; Gesteiger & Marsili, 1980; Magnusson, 1986; Magnusson, 1988; Balbás et al., 1983; Komorowski, 1987; Böhm et al., 1981; Nalewajski, 1989, 1990): PRINCIPLE AIM-EN1 (of connectivity): The electronegativity of the atom in molecule depends on the electronegativities and the chemical hardness of all the other atoms in molecule.

3.3.2 FRACTIONAL CHARGE ELECTRONEGATIVITY

The conceptual (chemically ontological) problem of N- and V-representability of a chemical system with its N-electrons evolving within applied V-potential have been solved by a construction of statistical density operators in the Hohenberg-Kohn-Sham formulation (see the Volume I of this five-volume book dedicated to quantum nanochemistry (Putz, 2016a). However, there still remains the problem regarding the continuity of the functional energy and of the chemical potential with the parametrical dependence of the number of electrons and respectively of the applied potential. The continuity is a fundamental problem as long as the basic relations between the main concepts, i.e., the electronegativity (the chemical potential with changed sign), the chemical hardness, the Fukui functions, and the chemical softness (the reverse of the chemical hardness) – involve derivatives in relation to the electronic density as functional measure of the total electrons number in system (Garza & Robles, 1993).

With the aim of treating the fractional charged chemical systems, one may consider the accordingly modified particle conservation law for the total fractional number of electrons (Parr & Bartolotti, 1983)

$$\int \rho(x)dx = N + \omega, \ N \in Z , 0 \le \omega \le 1 \tag{3.72}$$

with

$$E_v[\rho] = \int \rho(x)V(x)dx + F_{frac}[\rho] \qquad (3.73)$$

where the fractional functional Hohenberg-Kohn is expressed by the minimization of the statistical density operator:

$$F_{frac}[\rho] = \min_{\substack{\hat{D} \to \rho \\ \hat{D} = \alpha_N|\Psi_N\rangle\langle\Psi_N| + \alpha_{N+1}|\Psi_{N+1}\rangle\langle\Psi_{N+1}| \\ \alpha_N + \alpha_{N+1} = 1}} tr\{\hat{D}(\hat{T} + \hat{V}_{ee})\} \qquad (3.74)$$

Accordingly, a working expression for the Hohenberg-Kohn functional can be written if the expression of the electronic density is written through the statistical density operator as it is next considered

$$\rho(x) = tr(\hat{D}\hat{\rho}(x)) = \alpha_N\langle\Psi_N|\hat{\rho}(x)|\Psi_N\rangle + \alpha_{N+1}\langle\Psi_{N+1}|\hat{\rho}(x)|\Psi_{N+1}\rangle \qquad (3.75)$$

Upon integration over the space, it provides the recursive relationship:

$$N + \omega = \alpha_N N + \alpha_{N+1}(N+1) = N + \alpha_{N+1} \qquad (3.76)$$

The identity (3.76) can be solved by the identifications: $\alpha_{N+1} = \omega$, $\alpha_N = 1 - \omega$, allowing for the rewriting of the HK functional (3.74) for its minimization condition:

$$F_{frac}[\rho] = \min_{\substack{\hat{D} \to \rho \\ \hat{D} = (1-\omega)|\Psi_N\rangle\langle\Psi_N| + \omega|\Psi_{N+1}\rangle\langle\Psi_{N+1}|}} Tr\{\hat{D}(\hat{T} + \hat{V}_{ee})\} \qquad (3.77)$$

With this expression the energy, which corresponds to the fundamental state characterized by the fractional number $N + \omega$ electrons can be accordingly calculated:

$$E_{N+\omega}[\rho] = \min_{\substack{\rho(x) \\ \int \rho(x)dx = N+\omega}} \left[\int \rho(x)V(x)dx + \min_{\hat{D} \to \rho} Tr\{\hat{D}(\hat{T} + \hat{V}_{ee})\} \right]$$

$$= \min_{\substack{\rho(x) \\ \int \rho(x)dx = N+\omega}} \min_{\substack{\Psi_N, \Psi_{N+1} \\ \rho(x) = (1-\omega)\langle\Psi_N|\hat{\rho}(x)|\Psi_N\rangle + \omega\langle\Psi_{N+1}|\hat{\rho}(x)|\Psi_{N+1}\rangle}}$$

$$\{(1-\omega)\langle\Psi_N|\hat{H}|\Psi_N\rangle + \omega\langle\Psi_{N+1}|\hat{H}|\Psi_{N+1}\rangle\} \qquad (3.78)$$

wherefrom the density for the correct fundamental state will be given as:

$$\rho(x) = (1-\omega)\rho_N(x) + \omega\rho_{N+1}(x) \qquad (3.79)$$

while the minimum energy takes the form

$$E_{N+\omega} = (1-\omega)E_N + \omega E_{N+1} \tag{3.80}$$

From Eq. (3.80) there is immediately that for an open system, the fractional change of electrons between a chemical system and a reservoir will have as an effect a chemical potential of equilibrium through the limit:

$$\mu(N) = \lim_{\omega \to 0} \frac{\partial E_{N+\omega}}{\partial N} = -E_N + E_{N+1} = \begin{cases} -I(Z), & Z-1 < N < Z \\ -A(Z), & Z < N < Z+1 \end{cases} \tag{3.81}$$

which corresponds to the two situations of accepting or disposing electrons toward the reservoir (Z being the nuclear charge of the system).

It remains to show the continuity of the chemical potential between the limits of Eq. (3.81).

For this, we consider the differential expression of electronic density (3.79), which is then converted to finite differences for the electronic density variation:

$$\Delta\rho = (1-\omega)\Delta\rho_N + \omega\Delta\rho_{N+1} + (\rho_{N+1} - \rho_N)\Delta\omega \tag{3.82}$$

and similarly for the variation of the energy:

$$\Delta E = (1-\omega)\int \frac{\delta E_N}{\delta\rho_N}\Delta\rho_N dx + \omega\int \frac{\delta E_{N+1}}{\delta\rho_{N+1}}\Delta\rho_{N+1} dx + \mu\Delta\omega \tag{3.83}$$

with μ given by the above limits.

On the other side, between the last two expressions there is the formal integral identity:

$$\Delta E = \int \left(\frac{\delta E}{\delta\rho}\right)\Delta\rho dx \tag{3.84}$$

where, by explicit replacement of the variation of the electronic density (3.82), one further gets:

$$\Delta E = (1-\omega)\int \frac{\delta E}{\delta\rho}\Delta\rho_N dx + \omega\int \frac{\delta E}{\delta\rho}\Delta\rho_{N+1} dx + \left[\int \left(\frac{\delta E}{\delta\rho}\right)(\rho_{N+1} - \rho_N)dx\right]\Delta\omega \tag{3.85}$$

By identification of the two forms of the energy variation, Eqs. (3.83) and (3.85), the relevant identities are obtained, namely:

$$\int \left[\frac{\delta E}{\delta \rho} - \frac{\delta E_N}{\delta \rho_N} \right] \Delta \rho_N dx = 0 \tag{3.86}$$

$$\int \left[\frac{\delta E}{\delta \rho} - \frac{\delta E_{N+1}}{\delta \rho_{N+1}} \right] \Delta \rho_{N+1} dx = 0 \tag{3.87}$$

Next, remembering that the relative states with the densities ρ_N and ρ_{N+1} are properly normalized to the N and $N+1$ numbers of electrons, they also satisfy the conditions:

$$\int \Delta \rho_N dx = 0 , \; \int \Delta \rho_{N+1} dx = 0 \tag{3.88}$$

so leaving the two energetic identities (3.86) and (3.87) with the successive identities:

$$\mu = \frac{\delta E_{N+1}}{\delta \rho_{N+1}} + CT. = \frac{\delta E_N}{\delta \rho_N} + CT. \tag{3.89}$$

The result (3.89) says that for the fractional number exchanging of the electrons between a given chemical system and a particles reservoir (in fact very natural situation), the chemical potential (or the electronegativity with changed sign) has a continuous behavior on segments and features a jump which corresponds to a constant between these portions. The constant is naturally associated to the neutral system and corresponds to the situation when the fractional charge is canceled. However, and most important, the derivatives of the chemical potential at the opening of an electronic system toward a reservoir exist and are continuous, i.e., correspond to a continuous exchange until the regaining of the potential equilibrium in the new state.

Adding these conceptual-analytical results, one concludes that the chemical potential with changed sign, i.e., the electronegativity of an (open) system places itself within the electronic affinity (EA) and ionization potential (IP) energetic realm (Parr & Bartolotti, 1983):

$$EA < \chi = -\mu < IP \tag{3.90}$$

This fact has a very interesting result for the electronegativity: observe that the energetic interval in which the electronegativity of an electronic open system can take values corresponds to the $IP-EA$ measure; yet, as previously demonstrated, this interval corresponds/represents the chemical hardness η. Therefore, a large interval of values models higher chemical hardness and in fact represents a large system of reaction channels forbidden to the electronic transfer for this system (and increases the system stability). This fact substantiates the chemical hardness maximization in relation to the HOMO-LUMO band, which actually corresponds to the *IP–EA* difference.

This way, there is apparent that the density functional theory has the ability to illuminate new aspects of the chemical reactivity, for different levels of analysis for the electronic transfer in close and open quantum systems, in accordance with the fundamental principles in chemical bonding formation.

3.3.3 DENSITY FUNCTIONAL ELECTRONEGATIVITY

The essence of the density functional theory consists in writing the total energy of an *N*-electronic system as a function of electronic density of the fundamental state associated to the system

$$E[\rho] = \int \rho(1)V(1)dx_1 + F[\rho] \qquad (3.91)$$

where V is the external potential applied to the electrons' system (for example the nuclei potential), and $F(\rho)$ is the universal functional Hohenberg-Kohn.

The univocity of the relation between the external potential applied to the electronic system and the electronic density is provided by the Hohenberg-Kohn Theorems, see the Volume I of this five-volume book dedicated to quantum nanochemistry (Putz, 2016a). Moreover, one of the theorems also states the inequality relation between a density functional energy of any electronic state, $E[\rho]$, and the true functional energy of the fundamental electronic state of the system, $E[\rho']$, namely as

$$E_V[\rho'] \geq E_V[\rho] \qquad (3.92)$$

which is equivalent to the existence of a stationary equation:

$$\delta\{E_V[\rho'] - \mu N[\rho']\} = 0 \qquad (3.93)$$

where μ is the Lagrange multiplier associated with the total electrons number in system.

For the true/optimized density of the fundamental state, the multiplier μ works as the chemical potential associated to the system and is provided by the functional derivative (Harbola et al., 1991; Chattaraj & Parr, 1993):

$$\mu = \left(\frac{\delta E[\rho]}{\delta \rho(1)} \right)_{\rho=\rho(V)} \qquad (3.94)$$

which is reduced, from the stationary principle, to a partial derivative, in relation with the electrons number from the fundamental state:

$$\mu = \left(\frac{\partial E}{\partial N} \right)_V \qquad (3.95)$$

from where it is associated by Parr with the electronegativity with changed sign $\mu = -\chi$.

Applying the finite-difference approach for the system states variation by an electron acquirement (i.e., producing $E_0 - EA$ with EA the electronic affinity) and respectively towards the ionized one (having $E_0 + IP$ with IP the ionization potential) respecting the neutral system energy (E_0), the so called Mulliken electronegativity expression is obtained for the chemical neutral (in equilibrium) systems

$$-\mu = \chi_M = \frac{IP + EA}{2} \qquad (3.96)$$

Proceeding to calculating the functional derivative in relation with the external potential $V(1)$, one forms the functional derivative (Bartolotti, 1987):

$$\rho(1) = \left(\frac{\delta E[\rho]}{\delta V(1)} \right)_{\rho=\rho(V)} \qquad (3.97)$$

wherefrom, together with the analogous expression of the chemical potential, one can unitarily expresses the total differential of the functional energy, $E = E(N,V)$, under the form of the first differential equation of the chemical reactivity (Kohn et al., 1996):

$$dE = \mu dN + \int \rho(x) dV(x) dx \qquad (3.98)$$

Applying the Maxwell relations to this equation, the following series of identities, which define the Fukui functions $f(x)$ is formed

$$\left(\frac{\delta \mu}{\delta V(x)} \right)_N = \left(\frac{\partial \rho(x)}{\partial N} \right)_V = f(x) \qquad (3.99)$$

Similarly, one expresses the total differential of the chemical potential as a function of the system's electronic number and the external potential applied:

$$d\mu = \left(\frac{\partial \mu}{\partial N}\right)_V dN + \int \left(\frac{\delta \mu}{\delta V(x)}\right)_N \delta V(x) dx \qquad (3.100)$$

where under the integral term the Fukui function is recognized by the Eq. (3.99) identity, while the first term from the right site of Eq. (3.100) corresponds to the chemical force associated to the electronic system relating chemical hardness through the basic definitions (Berkowitz et al., 1985):

$$2\eta = \left(\frac{\partial^2 E}{\partial N^2}\right)_V = \left(\frac{\partial \mu}{\partial N}\right)_V \qquad (3.101)$$

Employing the same finite difference methodology as used for Mulliken electronegativity deduction, one can approximately express the (Pearson) chemical hardness (global):

$$\eta \cong \frac{IP - EA}{2} \qquad (3.102)$$

Then, the total difference of the chemical potential is rewritten such as the second differential equation of the chemical reactivity is unfolded (Parr et al., 1978; Parr & Gázquez, 1993):

$$d\mu = 2\eta dN + \int f(x)\delta V(x) dx \qquad (3.103)$$

In these conditions, the Euler-Lagrange equation of the electronic system is abstracted from (3.93) under the form of the chemical potential general expression

$$\mu = \left(\frac{\delta E}{\delta \rho(x)}\right)_V = V(x) + \frac{\delta F[\rho]}{\delta \rho(x)} \qquad (3.104)$$

which implicitly satisfies the differential form (3.103).

Overall, Eqs. (3.98) and (3.103) are the fundamental relations in terms of functional densities, which will be processed, and transformed aiming in modeling the chemical reactivity description with the aid of electronegativity and chemical hardness reactivity principles (Berkowitz & Parr, 1988).

3.3.4 FROM CHEMICAL POTENTIAL TO ELECTRONEGATIVITY RELATIONSHIPS

As based on the chemical potential measure within the density functional formalism, further the various implications of the electronegativity concept will be present, since the direct inter-relationship by the sign changing.

For example, with the aid of the chemical potential, a new form for expressing the total energies of the electronic system as density functional can be written (Parr et al., 1978):

$$E[\rho] = N\mu - Q[\rho] \tag{3.105}$$

$$-Q[\rho] \equiv F[\rho] - \int \frac{\delta F}{\delta \rho(x)} \rho(x) dx \tag{3.106}$$

Expression (3.105) may be formalized under the total differential form:

$$dE = Nd\mu + \mu dN - dQ \tag{3.107}$$

which, when compared with the first differential equation of the chemical reactivity, Eq. (3.98), leads with the identity:

$$Nd\mu = dQ + \int \rho(x) dV(x) dx \tag{3.108}$$

so recovering the celebrated Gibbs-Duhem equation for the concerned electronic system, within the density functional theory.

Moreover, with unfolding the total differential for the expression $Q[\rho]$

$$dQ = -dF + d\left[\int \frac{\delta F}{\delta \rho(x)} \rho(x) dx \right]$$

$$= -\int \frac{\delta F}{\delta \rho} d\rho dx + \int \frac{\delta F}{\delta \rho} d\rho dx + \int \rho(x) \left[d\left(\frac{\delta F}{\delta \rho} \right) \right] dx$$

$$= \iint \frac{\delta^2 F}{\delta \rho(x) \delta \rho(x')} \rho(x) d\rho(x') dx dx' \qquad (3.109)$$

the Gibbs-Duhem equation can take the more consistent form:

$$N d\mu = \int dx \rho(x) \left\{ dV(x') + \int \frac{\delta^2 F}{\delta \rho(x) \delta \rho(x')} d\rho(x') dx' \right\} \qquad (3.110)$$

Further chemical modeling reactivity may be performed with the homogeneous functionals assumption, as in next formalized.

DEFINITION AIM-EN1: A functional $U[\rho]$ is considered as homogeneous of k-degree/kind/order in ρ if satisfying the identity:

$$\int \frac{\delta U}{\delta \rho(x)} \rho(x) dx = kU \qquad (3.111)$$

In these conditions, if the functionals $Q[\rho]$ and $E[\rho]$ are homogeneous with the functional Hohenberg-Kohn, at its turn orbital-linearly expanded as:

$$F[\rho] = \sum_i f_i[\rho] \qquad (3.112)$$

then the functional $Q[\rho]$ will have the form

$$Q[\rho] = \sum_i (k_i - 1) f_i[\rho] \qquad (3.113)$$

and the total energy will be accordingly expressed from (3.105)

$$E[\rho] = N\mu - \sum_i (k_i - 1) f_i[\rho] \qquad (3.114)$$

Further assuming the fact that the homogeneous energy functional has also the orbital-linearized form:

$$E[\rho] = \sum_i e_i[\rho] \qquad (3.115)$$

the equation $N\mu$ can be equally resumed as:

$$N\mu = \sum_i k_i e_i[\rho] \qquad (3.116)$$

This result is readily analytically representable, for instance for the (homogeneous) *Thomas-Fermi* (TF) theory for atoms, where the energy functional is homogeneously expressed by the form:

$$E_{TF}[\rho] = T + V_{ne} + J = C_F \int \rho^{5/3}(x)dx - Z\int \frac{\rho(x)}{x}dx + \frac{1}{2}\iint \frac{\rho(x)\rho(x')}{x_{12}}dxdx'$$

(3.117)

for which the correspondent equation $N\mu$ will be given by:

$$N\mu = \frac{5}{3}T + V_{ne} + 2J$$

(3.118)

When Eqs. (3.117) and (3.118) are combined with the virial theorem

$$0 = 2T + V_{ne} + J$$

(3.119)

the Thomas-Fermi energy functional follows the string identities:

$$E_{TF}[\rho] = -T = \frac{3}{7}(V_{ne} + N\mu) = 3(N\mu - J)$$

(3.120)

Going to the level of chemical potential, i.e., when performing the functional derivative of the TF-energy respecting the electronic density, one obtains:

$$\mu = \frac{5}{3}C_F\rho^{2/3} - \left(\frac{Z}{x} - \int \frac{\rho(x')}{|x-x'|}dx'\right) = \frac{5}{3}C_F\rho^{2/3} - \phi(x) \quad (3.121)$$

where the second term represents the electrostatic potential in the point x produced by the system with the electronic density ρ and the nuclear charge Z. For the atom or neutral molecule ($N–Z$) one implements the condition $\phi(x) \to 0$, when $x \to \infty$, and the electronic density becomes asymptotically so readily uniformly; under these conditions, the chemical potential becomes:

$$\mu_{TF}(N = Z) = 0$$

(3.122)

and the TF-energy is analytically reduced to:

$$E_{TF}[\rho] = \frac{3}{7}V_{ne} = -3J$$

(3.123)

For a positively ionized molecule (or atom) the chemical potential associated has the form:

$$\mu_{TF} = -\frac{Z - N}{R} \qquad (3.124)$$

where R is the radius (towards the molecule mass center, for example) beyond which the electron density is canceled, i.e. the energy is properly rewritten from (3.120). Obviously, for other models of functional energy expressions the above relations are going to be accordingly modified. However, there is interesting to note the relation between the chemical potential, i.e., the electronegativity with the changed sign, and the charge or variation of nuclear charge from the system. This way, the total differential of the energy system $E=E(N,Z)$ will be considered as

$$dE = \left(\frac{\partial E}{\partial N}\right)_Z dN + \left(\frac{\partial E}{\partial Z}\right)_N dZ \qquad (3.125)$$

wherefrom, by taking into account the relation between N and Z given by the charge parameter $q = Z - N$, the relation (3.125) can be rewritten as:

$$\left(\frac{\partial E}{\partial N}\right)_{N=Z-q} = \left(\frac{\partial E}{\partial N}\right)_Z + \left(\frac{\partial E}{\partial Z}\right)_N \qquad (3.126)$$

or equivalently

$$\left(\frac{\partial E}{\partial N}\right)_{N=Z-q} = -\chi + \frac{V_{ne}}{Z} \qquad (3.127)$$

or in terms of Mulliken electronegativity (3.96):

$$\left(\frac{\partial E}{\partial N}\right)_{N=Z-q} = -\frac{IP + EA}{2} + \frac{V_{ne}}{Z} \qquad (3.128)$$

For the isolated atoms – or neutral systems, the electronegativity is much smaller than the electrons-nuclei attraction potential, and therefore, the relation (3.127) is reduced to:

$$V_{ne} = Z\left(\frac{\partial E}{\partial Z}\right)_{N=Z} \qquad (3.129)$$

wherefrom, when the general expression of the energy $E=CZ$ is considered, it simply becomes

$$V_{ne} = \beta E \tag{3.130}$$

known as the general relation of Fraga, representing a peculiar representation for the above Thomas-Fermi case.

Returning to the electronegativity equalization principle within density functional formalism, it can be extended to the orbital equalization of electronegativity, as starting from the stationary equation of the atoms-in-molecule energy; this will be in next exposed, by starting with a new definition of AIM.

DEFINITION AIM-EN2: The density of a fermionic system can be expressed by its natural orbitals ψ_k and by the associated occupancy numbers n_k:

$$\rho(x) = \sum_k n_k |\psi_k(x)|^2 \tag{3.131}$$

satisfying the conservation conditions:

$$0 \le n_k \le 2 \ , \ \sum_k n_k = N, \quad \int \psi_k^*(x) \psi_l(x) dx = \delta_{kl} \tag{3.132}$$

Therefore, the electronic density with occupancy numbers includes all the properties of the fundamental (ground) state and allows the construction of the density functional of the fundamental state energy in terms of *natural orbitals* (NO), namely:

$$E^{NO}[n_k, \psi_k] = \sum_k n_k \int |\psi_k(x)|^2 V(x) dx + F^{NO}[n_k, \psi_k] \tag{3.133}$$

for which the stationary equation Euler-Lagrange now takes the form:

$$\delta \left\{ E_V^{NO}[n_k', \psi_k'] - \mu N[n_k'] - \sum_{k,l} \lambda_{kl} \langle \psi_k' | \psi_k' \rangle \right\} = 0 \tag{3.134}$$

where the Lagrange multipliers λ_{kl} were also incorporated as associated with the orthonormalization condition of the natural orbitals.

The Lagrange equation (3.134) can be rewritten in a more explicit form:

$$0 = \sum_k \delta n_k' \left[\left(\frac{\partial E_V^{NO}}{\partial n_k'} \right)_{n_j',\psi'} - \mu \right] + \sum_k \int \delta \psi_k' \left[\left(\frac{\delta E_V^{NO}}{\delta \psi_k'} \right)_{n',\psi_j'} - 2 \sum_l \lambda_{kl} \psi_l \right] dx$$

(3.135)

leaving with the orbital electronegativity result:

$$- \chi = \mu = \left(\frac{\partial E}{\partial n_k} \right)_{n_j,\psi} , \forall k$$

(3.136)

Another important result in conjunction with the electronegativity of the atom and molecule regards the bonding/anti-bonding states (closely related with the valence states); it can nevertheless be naturally and elegantly deduced in the context of density functionals theory as in next exposed (Gázquez et al., 1987, 1990; Galván et al., 1988; Berkowitz, 1987; Parr & Yang, 1984; Yang et al., 1984; Gázquez & Ortiz, 1984; Vela & Gázquez, 1988; Ohwada, 1984; De Amorim & Ferreira, 1981):

PRINCIPLE AIM-EN2: A molecular orbital will have a bonding (anti-bonding) character if its electronegativity is bigger (smaller) than the electronegativity corresponding to the atomic orbitals.

The proof of this principle starts from the Euler-Lagrange relation for the chemical potential, applied to the atom and respectively to the molecule and then writing the difference, so obtaining the atoms-in-molecule chemical potential difference:

$$\mu_M - \mu_{at} = V_M(x) - V_{at}(x) + \frac{\delta F[\rho_M]}{\delta \rho_M} - \frac{\delta F[\rho_{at}]}{\delta \rho_{at}}$$

(3.137)

Using now the virial theorems for the atom and molecule, namely:

$$T[\rho_{at}] = -E[\rho_{at}]$$

(3.138)

$$T[\rho_M] = -E[\rho_M] - R \frac{dE[\rho_M]}{dR}$$

(3.139)

the expressions corresponding to the Hohenberg-Kohn functionals in both cases result as:

$$F[\rho_{at}] = -\int V_{at}(x) \rho_{at}(x) dx - T[\rho_{at}]$$

(3.140)

$$F[\rho_M] = -\int V_M(x)\rho_M(x)dx - T[\rho_M] - R\frac{dE[\rho_M]}{dR} \qquad (3.141)$$

wherefrom, by taking into account the integral identity:

$$\frac{\partial}{\partial\rho(x)}\int V(x)\rho(x)dx = V(x) \qquad (3.142)$$

the expression of the atoms-in-molecule chemical potential difference (3.137) becomes:

$$\mu_M - \mu_{at} = -\frac{\delta T[\rho_M]}{\delta\rho_M} + \frac{\delta T[\rho_{at}]}{\delta\rho_{at}} - \frac{\delta}{\delta\rho_M}\left(R\frac{dE[\rho_M]}{dR}\right) \qquad (3.143)$$

This relation can be still processed if the Hellmann-Feynman theorem, in terms of the electrostatic force that the atoms exert to the nuclei, is applied, i.e.

$$F_e = -\frac{dE[\rho_M]}{dR} \qquad (3.144)$$

thanks to which Eq. (3.143) further becomes (De Amorim & Ferreira, 1981):

$$\mu_M - \mu_{at} = -\frac{\delta T[\rho_M]}{\delta\rho_M} + \frac{\delta T[\rho_{at}]}{\delta\rho_{at}} + \frac{\delta}{\delta\rho_M}\left(\vec{R}\cdot\vec{F_e}\right) \qquad (3.145)$$

Moreover, if the kinetic-energy functional is written in the local Thomas – Fermi (TF) approximation:

$$T_L[\rho] = CT.\int \rho^{5/3}(x)dx \qquad (3.146)$$

we arrive to the *TF* chemical potential difference *of the atoms-in-molecule*

$$\mu_M - \mu_{at} = CT.\left[\rho_M^{2/3} - \rho_{at}^{2/3}\right] + \frac{\delta}{\delta\rho_M}\left(\vec{R}\cdot\vec{F_e}\right) \qquad (3.147)$$

or respectively in terms of electronegativities:

$$\chi_M - \chi_{at} = CT.\left[\rho_M^{2/3} - \rho_{at}^{2/3}\right] - \frac{\delta}{\delta\rho_M}\left(\vec{R}\cdot\vec{F_e}\right) \qquad (3.148)$$

The relation (3.148) and the AIM-EN2 Principle open the next discussion. If, for example, one considers that an electron is moving from an atomic orbital to a molecular one, its energy becomes more negative, and its kinetic energy increases; therefore, also the functional derivatives of the kinetic energy will increase, so that

$$\left[\rho_M^{2/3} - \rho_{at}^{2/3}\right] > 0 \tag{3.149}$$

On the other hand, the term $\left(\vec{R} \cdot \vec{F}_e\right)$ corresponds to the nuclei movements' energy when $\rho_{at} \to \rho_M$ and for a stable state corresponds to a negative value:

$$\left(\vec{R} \cdot \vec{F}_e\right) < 0 \tag{3.150}$$

Therefore, by the enlargement of the molecular density ρ_M also the functional derivative becomes as more negative

$$\frac{\delta}{\delta \rho_M}\left(\vec{R} \cdot \vec{F}_e\right) < 0 \tag{3.151}$$

There is clear now that when considering together the two contributions, (3.149) and (3.151), the assertion of the AIM-EN2 Principle appears as true.

Of the first interest is also the relation between the effective electronegativity calculated in the density functionals theory and the Pauling electronegativity.

For developing on this issue, an expression of electronegativity in accordance with the potential (electrostatically) picture will be firstly considered (Ohwada, 1984):

$$\chi = \int \frac{x_m}{e} V\left(x_m + \Delta x\right) dx \tag{3.152}$$

where x_m is the radial distance (of the electronic system towards the center of mass-CM of the associated nuclei system) where the radial electronic density $4\pi x^2 \rho(x)$ takes the smallest value (from where the electronic region of valence is arguable to begin) and e is the elementary electric charge.

In this framework, one next considers the potential expansion only until the first order

$$V(x_m + \Delta x) = V(x_m) + \left(\frac{\partial V}{\partial x}\right)_{x_m} (\Delta x) \qquad (3.153)$$

while replacing the expressions found by Politzer (1980) for the relations of the density with the electrostatic electronic potential

$$\left(\frac{\partial V}{\partial x}\right)_{x_m} = -\frac{2V(x_m)}{x_m} \qquad (3.154)$$

$$V(x_m) = \int_{x_m}^{\infty} \frac{\rho(x')}{|x_m - x'|} dx' = \frac{N_m e}{2x_m} \qquad (3.155)$$

with N_m being the numbers of electrons located beyond of the radial distance x_m; so the electronegativity (3.152) takes the working form:

$$\chi = \int_{-\Delta x_m}^{\Delta x_m} \frac{N_m}{2}\left[1 - \frac{2(\Delta x)}{x_m}\right] dx \qquad (3.156)$$

To retrieve the Pauling scale the integration domain is considered as bordered by the value:

$$\Delta x_m = \frac{(N_m + 1)}{2N_m} f(n) \qquad (3.157)$$

with the empirical value accounting for the periodicity by the function $f(n)$ given by (Ohwada, 1984):

$$f(n) = \left[1 - \frac{2}{9}(n-1)^2\right], n \geq 1 \qquad (3.158)$$

where n-counting the quantum principal (periodic shell) number. While denoting x_c by x_m (for the electronic core radius) and N_m by N_v (for the number of valence electrons), the Pauling electronegativity is revealed now with the expression:

$$\chi_P = \frac{N_v + 1}{2} f(n) \qquad (3.159)$$

On the other side, in the density functional theory in order to derive the electronegativity one firstly has to written the expression of the energy functionals as a function of electronic density and then performed the functional derivation.

Here, for the energy density functional the expression of Thomas-Fermi-Weizsäcker theory with the modified term Weizsäcker will be considered (having, in the Wang-Parr approach, the coefficient 1/72 instead of the classical 1/8) (Ohwada, 1984; Bartolotti et al., 1980; Robles & Bartolotti, 1984; Parr & Zhou, 1993; Harbola, 1992):

$$E[\rho] = T_o[\rho] + T_W[\rho] + E_{ne}[\rho] + E_{ee}[\rho] + E_{xc}[\rho]$$

$$= \frac{3(3\pi^2)^{2/3}}{10} \int \rho^{5/3}(x)dx + \frac{1}{72} \int \frac{[\nabla\rho(x)]^2}{\rho} dx$$

$$+ \int \left(V_{ne} + \frac{1}{2}V_{ee}\right)\rho(x)dx + \frac{3}{4}\left(\frac{3}{\pi}\right)^{1/3} \int \rho^{4/3}(x)dx \qquad (3.160)$$

from where automatically electronegativity is extracted through the functional derivative in relation to the electronic density:

$$-\chi = \mu = \frac{\delta E[\rho]}{\delta \rho(x)} = \frac{1}{2}(3\pi^2)^{2/3}\rho^{2/3} + \frac{1}{72}\left(\frac{\nabla\rho}{\rho}\right)^2$$

$$-\frac{1}{36}\frac{\nabla^2\rho}{\rho} + \left(V_{ne} + \frac{1}{2}V_{ee}\right) + \frac{3}{\pi}\rho^{1/3} \qquad (3.161)$$

where only the gradient terms were considered as having non-zero contributions at the domain frontier, the rest being vanishing as $\rho \to 0$ when $x \to \infty$.

Now there remains to choose the only form of the electronic density to obtain an analytical formula for electronegativity. Taking into account the quantum wave nature of valence/peripheral electrons behaving asymptotically on the long-range interaction, in accordance with the basic postulate of quantum mechanics, see the Volume I of the present five-volume book (Putz, 2016a), there is natural to consider the exponential form on the electronic levels for the fermionic system/electronic density in question, that is:

$$\rho(x) = \begin{cases} A_1 \exp[-2\lambda_1 x], & 0 \leq x \leq X_1 (level\ K) \\ A_2 \exp[-2\lambda_2 x], & X_1 \leq x \leq X_2 (level\ L) \\ A_3 \exp[-2\lambda_3 x], & X_2 \leq x \leq X_3 (level\ M) \\ \dots \end{cases} \qquad (3.162)$$

By combining the exponential form of the electronic normalized density for the asymptotic density form in the valence electrons region with the electronegativity main domain frontier contribution above, the effective Wang-Parr (WP) electronegativity expression may be written as:

$$\chi_{WP} = \frac{1}{72}(2\lambda_n)^2 \tag{3.163}$$

The connection between this relation and the Pauling electronegativity (3.159) is obtained for a properly choice of the parameter λ_n, for instance

$$\lambda_n = C\sqrt{\chi_P} + D = C\sqrt{\frac{N_v+1}{2}f(n)} + D \tag{3.164}$$

by which the Wang-Parr electronegativity becomes (Ohwada, 1984):

$$\chi_{WP} = \frac{1}{18}\left(C^2\chi_P + 2CD\sqrt{\chi_P} + D^2\right) \tag{3.165}$$

By adjusting the constants in the last expression, through various semi-empirical methods, one can interplay the electronegativities calculated by Pauling with those calculated by the density functional theory, as practical (applications) need demand.

3.3.5 ELECTRONEGATIVITY ADDITIVE EQUALIZATION PRINCIPLE

Aiming to analytically expressing an atoms-in-molecule expression of electronegativity and chemical hardness, the starting point is to consider the expansion of the atomic energy of an atom, around its N-electronic isolated ("0") status, up to the second order, when attempts to a molecular coordination throughout the charge transfer (ΔN):

$$E_{\langle\rangle}(N_{\langle\rangle}) \cong E_{\langle\rangle}(N) + \left(\frac{\partial E_{\langle\rangle}}{\partial N_{\langle\rangle}}\right)_0 (N_{\langle\rangle} - N) + \frac{1}{2}\left(\frac{\partial^2 E_{\langle\rangle}}{\partial N_{\langle\rangle}^2}\right)_0 (N_{\langle\rangle} - N)^2$$

$$\equiv E_{\langle\rangle}(N) - \chi\Delta N + \eta(\Delta N)^2$$

$$\tag{3.166}$$

where the second order derivation of the total energy respecting with the number of electrons was identified as the first order electronegativity derivation respecting the number of electrons defining the *chemical hardness* (Parr & Pearson, 1983).

Following above convention, the relation between electronegativity of an atom in coordination and its isolated electronegativity and chemical hardness looks like:

$$\chi_{\langle\,\rangle} = -\frac{\partial E_{\langle\,\rangle}}{\partial N_{\langle\,\rangle}} = \chi - 2\eta\Delta N \qquad (3.167)$$

However, to pass from the atomic electronegativity to the molecular one a working principle is required. Due to the identification of electronegativity with the chemical potential, at all organization levels of electronic systems, it can be used the thermodynamically feature of the chemical potential. When systems with different chemical potentials are combined, they exchange the particles (charges, electrons) between them until their chemical potentials will equalize. In terms of electronegativity, the corresponding principle is noted as the Sanderson equalization principle of electronegativities and sounds like follows (Sanderson, 1976): *for the molecules in their fundamental state, the electronegativities of different electronic regions in molecule – are equal.*

Nevertheless, at this point, a shortly revelatory discussion has to take place (Tachibana, 1987; Tachibana & Parr, 1992; Tachibana et al., 1999). Let's consider that the molecule is seen as the total system <*A*>+<*B*> of the two binding regions: the dynamic acceptor region <*A*> and the dynamic donor region <*B*>. Obviously, the conservation in the total energy and number of electrons is preserved, throughout any kind of molecular partition in dynamic acceptor and donor regions:

$$E = E_{\langle A \rangle} + E_{\langle B \rangle} \qquad (3.168)$$

$$N = N_{\langle A \rangle} + N_{\langle B \rangle} \qquad (3.169)$$

The regional electronegativities will respectively be given as:

$$\chi_{\langle A \rangle} = -\left(\frac{\partial E_{\langle A \rangle}}{\partial N_{\langle A \rangle}}\right)_{N_{\langle B \rangle}}, \quad \chi_{\langle B \rangle} = -\left(\frac{\partial E_{\langle B \rangle}}{\partial N_{\langle B \rangle}}\right)_{N_{\langle A \rangle}} \qquad (3.170)$$

Now, the electronegativity of the total system (molecule) is representable, according with the Sanderson equalization principle of regional electronegativities, consecutively as:

$$\chi = -\left(\frac{\partial E}{\partial N}\right) = -\left(\frac{\partial E}{\partial N_{\langle A \rangle}}\right)_{N_{\langle B \rangle}} = -\left(\frac{\partial E}{\partial N_{\langle B \rangle}}\right)_{N_{\langle A \rangle}}$$

$$= \chi_{\langle A \rangle} - \left(\frac{\partial E_{\langle B \rangle}}{\partial N_{\langle A \rangle}}\right)_{N_{\langle B \rangle}} = \chi_{\langle B \rangle} - \left(\frac{\partial E_{\langle A \rangle}}{\partial N_{\langle B \rangle}}\right)_{N_{\langle A \rangle}} \tag{3.171}$$

This result, however, seems to prohibit the equality between global and regional electronegativities:

$$\chi \neq \chi_{\langle A \rangle} \neq \chi_{\langle B \rangle} \tag{3.172}$$

Beyond of the perplexity given by such inequalities, this paradox is simple solved observing that:

$$\Delta\chi_{\langle B \rangle}^{\langle A \rangle} = \left|\chi_{\langle B \rangle} - \chi_{\langle A \rangle}\right| = \left|\chi_{\langle A \rangle} - \chi_{\langle B \rangle}\right| = \Delta\chi_{\langle A \rangle}^{\langle B \rangle}$$

$$= \left|\left(\frac{\partial E_{\langle A \rangle}}{\partial N_{\langle B \rangle}}\right)_{N_{\langle A \rangle}} - \left(\frac{\partial E_{\langle B \rangle}}{\partial N_{\langle A \rangle}}\right)_{N_{\langle B \rangle}}\right|$$

$$= \left|V_{\langle A \rangle} - V_{\langle B \rangle}\right| \equiv \Delta\chi \tag{3.173}$$

where the *transfer potentials* have been introduced:

$$V_{\langle A \rangle} = \left(\frac{\partial E_{\langle A \rangle}}{\partial N_{\langle B \rangle}}\right)_{N_{\langle A \rangle}} , \; V_{\langle B \rangle} = \left(\frac{\partial E_{\langle B \rangle}}{\partial N_{\langle A \rangle}}\right)_{N_{\langle B \rangle}} \tag{3.174}$$

There is clear now that the Sanderson's principle of regional electronegativities equalization takes place when the transfer potentials cancel each other or, in other words, when the zero minimum difference between regional electronegativities, within the bond, is achieved:

$$\Delta\chi \rightarrow \min \rightarrow 0 \qquad (3.175)$$

Even more, the last condition leads just with the variational principle:

$$d\chi = 0 \qquad (3.176)$$

If this condition is regarded inversely, so to speak when $\Delta\chi$ is non-zero, the binding process is promoted, i.e., the *Principle of Frontier Electron Theory* follows (Parr & Yang, 1984, 1989; Yang et al., 1984; Berkowitz, 1987): *out of two different sites with generally similar disposition for reacting with a given reagent, the reagent prefers the one which on the reagent's approach is associated with the maximum response of the system's electronegativity* (n. a.). *In short, $\Delta\chi$ big is good for reactivity* (n. a.) (Parr & Yang, 1984, 1989). An illustrative example on how the electronegativity differences promote the binding and molecular formation regards the next discussion.

Let considering the formation of a diatomic molecule *AB* or *a bond* with constant atomic nuclear charges at the equilibrium separating distance R_{AB}.

For an infinitesimal transfer of electronic charges between the molecule's atoms,

$$N_{\langle A\rangle} = N_A + dN_{\langle A\rangle} \qquad (3.177)$$

and

$$N_{\langle B\rangle} = N_B - dN_{\langle B\rangle} \qquad (3.178)$$

the variation in the total energy

$$E = E_{\langle A\rangle} + E_{\langle B\rangle} \qquad (3.179)$$

can be written as:

$$dE = \left(\frac{\partial E}{\partial N_{\langle A\rangle}}\right)_{N_{\langle B\rangle},R_{AB}} (N_{\langle A\rangle} - N_A) - \left(\frac{\partial E}{\partial N_{\langle B\rangle}}\right)_{N_{\langle A\rangle},R_{AB}} (N_{\langle B\rangle} - N_B)$$

$$+ \left(\frac{\partial E}{\partial R_{AB}}\right)_{N_{\langle A\rangle},N_{\langle B\rangle}} dR_{AB} \qquad (3.180)$$

Since, in the fundamental equilibrium state, we have

$$\partial E / \partial R_{AB} = 0, \, dE = 0 \qquad (3.181)$$

the last equation can be reduced as:

$$\left(\frac{\partial E_{\langle A \rangle}}{\partial N_{\langle A \rangle}} \right)_{N_{\langle B \rangle}, R_{AB}} = \left(\frac{\partial E_{\langle B \rangle}}{\partial N_{\langle B \rangle}} \right)_{N_{\langle A \rangle}, R_{AB}} \qquad (3.182)$$

from which, the principle of equalization of the atomic (bond) electro-negativities in the formed molecule is assured.

Next, by employing this principle on the atoms A and B in AB it yields:

$$\chi_{\langle A \rangle} = \chi_A - 2\eta_A \Delta N = \chi_{\langle B \rangle} = \chi_B + 2\eta_B \Delta N = \chi_{AB} \qquad (3.183)$$

from which, immediately follows the number of the transferred charges – with the expression:

$$\Delta N = \frac{\chi_A - \chi_B}{2(\eta_A + \eta_B)} \qquad (3.184)$$

The released energy due to the previous charge transfer will be therefore successively written as:

$$\Delta E = (E_{\langle A \rangle} - E_A) + (E_{\langle B \rangle} - E_B)$$

$$= [-\chi_A + \eta_A \Delta N + \chi_B + \eta_B \Delta N] \Delta N$$

$$= [-\chi_B - 2\eta_B \Delta N - \eta_A \Delta N + \chi_B + \eta_B \Delta N] \Delta N$$

$$= [-\eta_B \Delta N - \eta_A \Delta N] \Delta N = -\frac{1}{2} [\chi_A - \chi_B] \Delta N$$

$$= -\frac{1}{4} \frac{(\chi_A - \chi_B)^2}{(\eta_A + \eta_B)} \qquad (3.185)$$

It is clear now that the electronegativity difference is crucial for binding promotion and bonding formation, as well in terms of exchanged number of electrons as for the released total energy.

The average value of the equalized electronegativity of atoms in molecule can be obtained as:

$$\chi_{AB} \equiv \overline{\chi} = \frac{\eta_A \chi_B + \eta_B \chi_A}{\eta_A + \eta_B} \qquad (3.186)$$

However, the present (often called Parr-Pearson) approach does not provide a direct evaluation of molecular hardness from the atomic ones, even a formula can be laid down if assumed the same proportionality between electronegativity and hardness at both atomic and molecular levels:

$$\chi_A = \theta\, \eta_A \qquad (3.187a)$$

$$\chi_B = \theta\, \eta_B \qquad (3.187b)$$

$$\chi_{AB} = \theta\, \eta_{AB} \qquad (3.187c)$$

being θ a proportional constant.

Therefore, it turns out for the average molecular hardness the result:

$$\eta_{AB} \equiv \overline{\eta} = 2\frac{\eta_A \eta_B}{\eta_A + \eta_B} \qquad (3.188)$$

However, one just notes that the assumptions made claims that all pairs electronegativity-hardness, either for atoms or molecules, are correlated by the same factor θ. Therefore, this treatment seems that doesn't properly correlate the average electronegativity $\overline{\chi}$ with the average of the chemical hardness $\overline{\eta}$ without such "universal" proportionality. Further expansions of this Parr-Pearson model were made (Nalewajski, 1998), but the difficulties concerning the inherent correlation between $\overline{\chi}$ and $\overline{\eta}$ still remain.

3.3.6   OBSERVABLE ELECTRONEGATIVITY BY AVERAGING OF ATOMS IN MOLECULES

Beside the electronegativity aspects already presented, those regarding the relationship between the electronegativity and the parabolic energy of the atomic bonds in molecules (or of the electrons-nuclei bonded systems in general) will be here discussed through employing the derivation of a special electronegativity expression including the electrostatic dependence.

The present approach is based on the functional energy formulation in terms of the occupancy numbers and (natural) molecular orbitals (Gáspár & Nagy, 1988a,b):

$$
E[n_k, \Psi_k] = \sum_i n_i \int \Psi_i^*(x) \left(-\frac{1}{2} \nabla^2 \right) \Psi_i(x) dx - Z \int \frac{\rho(x)}{x} dx
$$

$$
+ \frac{1}{2} \iint \frac{\rho(x_m)\rho(x')}{|x - x'|} dx dx'
$$

$$
- \frac{3}{4} \alpha C_x \int [\rho(x)]^{4/3} dx \qquad (3.189)
$$

The last term in (3.189) corresponds to the exchange and correlation term written in a generalized way with the aid of the parameter α, thus recovering the functional expression in the $X\alpha$ version of the density functional theory as anticipated by Slater in 1974.

The total energy minimization procedure respecting with the molecular orbitals, while respecting the custom density *orthonormalization condition*, provides the set of stationary equations:

$$
\left(-\frac{1}{2} \nabla^2 + V_{\text{eff}} \right) \Psi_i = \varepsilon_i \Psi_i \qquad (3.190)
$$

$$
V_{\text{eff}} = -\frac{Z}{x} + \int \frac{\rho(x')}{|x - x'|} dx' - \alpha C_x \rho^{1/3} \qquad (3.191)
$$

Now, by using the *Janak's Theorem* one can employ the total energy variation respecting the occupancy numbers, so allowed the orbital relaxation conditions, so that recovering the eigen-states (necessary the stationary states) from the $(i-)$ orbital electronegativity with changed sign, with the working definitions:

$$
-\chi_i = \mu_i = \left(\frac{\partial E}{\partial n_i} \right)_{n_j \neq n_i} = \varepsilon_i \qquad (3.192)
$$

On the next level of chemical reactivity, aiming to evaluate of the orbital chemical hardness (for an open chemical state and with unoccupied orbitals in neutral state) the derivative of the orbital electronegativity relation (3.192) respecting the occupancy orbital numbers will be considered while taking into account the total energy expression (3.189), so obtaining:

$$\eta_i = \frac{1}{2}\left(\frac{\partial^2 E}{\partial n_i^2}\right)_{n_j \neq n_i} = \frac{1}{2}[J(i) + K(i)] \tag{3.193}$$

$$J(i) = \int \frac{|\Psi_i(x)|^2 |\Psi_i(x')|^2}{|x - x'|} dx dx' \tag{3.194}$$

$$K(i) = -\frac{1}{3}\alpha C_x \int \frac{|\Psi_i(x)|^2 |\Psi_i(x)|^2}{[\rho(x)]^{2/3}} dx \tag{3.195}$$

Worth noting that the ionization potential and the electronic affinity can be accordingly expressed in terms of orbital electronegativity and chemical hardness functions

$$IP(i) = \chi_i + \eta_i = -\varepsilon_i + \frac{1}{2}[J(i) + K(i)] \tag{3.196}$$

$$EA(i) = \chi_i - \eta_i = -\varepsilon_i - \frac{1}{2}[J(i) + K(i)] \tag{3.197}$$

Although the last two relations are accurate, they require a considerable computational effort. To avoid as much possible the numerical complications, one can advance a model of the orbital relaxation, considering in fact the potential above as being approximately given by an electrostatic term with a screening S-constant (Gázquez et al., 1984, 1987, 1990; Galván et al., 1988; Vela & Gázquez, 1988):

$$\left[-\frac{1}{2}\nabla^2 - \frac{Z - S(q)}{x}\right]\Psi_i = \varepsilon_i \Psi_i \tag{3.198}$$

with $q = Z-N$ having the charge parametrical role. The Eq. (3.198) resembles in a quantum manner the analogous classical problem of proper values for an electrostatic potential, so working out the eigen-orbital energies' solution:

$$\varepsilon_i(q) = -\frac{1}{2}\frac{[Z - S(q)]^2}{v_i^2} \qquad (3.199)$$

with v_i the main quantum number associated to the i-orbital.

Considering the difference between the shielding constants associated to the neutral orbital states respecting a current one, it goes proportionally to the net charge,

$$S(q) - S(0) = kq \qquad (3.200)$$

so that the allied orbital energetic difference cast as:

$$\varepsilon_i(q) - \varepsilon_i(0) = \frac{Z - S(0)}{v_i^2}kq - \frac{k^2 q^2}{2v_i^2} = \langle x^{-1}\rangle_i kq - \frac{k^2 q^2}{2v_i^2} \qquad (3.201)$$

where $\langle x^{-1}\rangle$ associates with the average over a hydrogenic orbital.

To determine a value of k *parameter*, two coordinate – extreme behaviors for the applied potential on chemical matter will be considered: $-Z/x$ for $x \to 0$, and $-(q+1)/x$ for $x \to \infty$; in terms of shielding constant this means $S(q) \to 0$ for $x \to 0$, and $S(q) \to (Z-q-1)$ for $x \to \infty$. However, when the arithmetic average is considered for shielding constant, $S(q) \cong (Z-q-1)/2$, the actual difference $S(q)-S(0)$ is compared with the analogous difference in (3.200) resulting in the constant $k \cong -1/2$. Therefore, the energetic relation (3.201) also becomes:

$$\varepsilon_i(q) - \varepsilon_i(0) \cong -\frac{1}{2}\langle x^{-1}\rangle_i q \qquad (3.202)$$

where the last term from the previous expression was omitted (since actually cancelling for the valence electrons with the main orbital high and very high quantum numbers).

Next, one seeks the relation between the difference (3.202) and the chemical hardness computed by considered the Taylor expansion for the total energy around the neutral system ($N=Z$) restrained to the second order of charge influence:

$$E(Z,q) = E(Z,0) + q \left.\frac{\partial E}{\partial q}\right|_{q=0} + \frac{q^2}{2!} \left.\frac{\partial^2 E}{\partial q^2}\right|_{q=0} + \dots \qquad (3.203)$$

Upon differentiation of the relation (3.203) respecting the q – net charge

$$\frac{\partial E(Z,q)}{\partial q} = \left.\frac{\partial E}{\partial q}\right|_{q=0} + q \left.\frac{\partial^2 E}{\partial q^2}\right|_{q=0} \qquad (3.204)$$

one arrives to a form, which allows the immediate application of the Janak's Theorem towards the searched energetics:

$$\varepsilon_i(q) - \varepsilon_i(0) = -q \frac{\partial^2 E}{\partial q^2} \stackrel{(3.193)}{=} - q(2\eta)_i \qquad (3.205)$$

Now, comparing the expressions (3.202) and (3.205) the chemical hardness expression (orbital) in the shielding effect approximation simply yields:

$$\eta_i = \frac{1}{4}\langle x^{-1}\rangle_i \qquad (3.206)$$

With this working form of orbital chemical hardness the ionization potential and the electronic affinities expressions (3.196) and (3.197) respectively become:

$$IP(i) = \chi_i + \eta_i = -\varepsilon_i + \frac{1}{4}\langle x^{-1}\rangle_i \qquad (3.207)$$

$$EA(i) = \chi_i - \eta_i = -\varepsilon_i - \frac{1}{4}\langle x^{-1}\rangle_i \qquad (3.208)$$

which have the virtue of being easier to be numerically implemented.

Worth to give also a conceptual-modeling example of the present orbital-electrostatic picture of reactivity: for instance we just seen that the electronegativity evaluation of an AB molecule arrives to its averaged (atoms-in-molecule) expression recalling (3.186)

$$\chi_{AB} = \overline{\chi} = \frac{\eta_A \chi_B^0 + \eta_B \chi_A^0}{\eta_A + \eta_B} \qquad (3.209)$$

This expression can be immediately transposed into the actual electrostatic – density functionals formalism by identifying the atomic electronegativities with the proper atomic HOMO energy and respectively the chemical hardness with the calculated $\langle x^{-1} \rangle$ average, both calculated for the same orbital states. Accordingly, the electrostatic atoms-in-molecule electronegativity of the diatomic molecule AB writes as:

$$\chi_{AB} = \overline{\chi} = -\frac{\varepsilon_A \langle x^{-1} \rangle_B + \varepsilon_B \langle x^{-1} \rangle_A}{\langle x^{-1} \rangle_A + \langle x^{-1} \rangle_B} \qquad (3.210)$$

Therefore, there is noted that, within density functionals theory the electronegativity does not change its original meaning of "attracting electrons", but instead deepens this significance with a functional definition featuring a large degree of implementation and interpretation, that of electrostatic orbital included in an average manner (consistent with quantum mechanical description).

3.3.7 MODELING ELECTRONIC (DE)LOCALIZATION IN CHEMICAL BONDING

The characterization of chemical bonding has gone a long way since Lewis' first attempt to put it on a firm basis (Lewis, 1916). In the past years, several tools have been designed to this aim, many of them using in one-way or another the exchange-correlation density (Ruedenberg, 1962). In particular, the exchange correlation density has been used to characterize the amount of electron sharing between two atoms A and B through (Fradera et al., 2002; Matito et al., 2007)

$$D(A, B) = 2 \int_A \int_B \rho(\mathbf{r}, \mathbf{r}') d\mathbf{r} d\mathbf{r}' \qquad (3.211)$$

as was first suggested in the seminal work of Bader (Bader & Stephens, 1974, 1975; Bader, 1994, 1998). $D(A;B)$ corresponds to the covariance of the electron atomic populations of A and B (Matito et al., 2007; Angyaan et al., 1999), thus giving a measure between the correlation of these two populations but, unlike quantities in conceptual DFT, it has no predictive power about how the electron sharing may change upon external perturbation. From this analysis we deduce that the *new softness kernel* (Matito & Putz, 2011)

$$s(\mathbf{r},\mathbf{r}') = \left(\frac{\partial \rho_{xc}(\mathbf{r},\mathbf{r}')}{\partial \mu} \right)_V \tag{3.212}$$

corresponds to the change in the electron fluctuation between two regions of the space upon modification of the chemical potential. In particular, one may calculate the new kernel integrated into two atomic regions, A and B,

$$s(A,B) = \frac{1}{2} \frac{\partial D(A,B)}{\partial \mu} = \int_A \int_B \frac{\rho_{xc}(\mathbf{r},\mathbf{r}')}{\partial \mu} d\mathbf{r} d\mathbf{r}' \tag{3.213}$$

Note that the new softness kernel definition is related with the *exchange-correlation density* (XCD) (Ruedenberg, 1962)

$$\rho_{xc}(\mathbf{r},\mathbf{r}') = \rho(\mathbf{r})\rho(\mathbf{r}') - \rho_2(\mathbf{r},\mathbf{r}') \tag{3.214}$$

where $\rho_2(\mathbf{r},\mathbf{r}')$ is the second-order density matrix, is the workhorse for the methods that account for electron localization. The XCD has been used to define many popular tools in bonding analysis, such as the electron-sharing indices (or bond-orders) (Fradera et al., 1998), the domain-averaged Fermi holes (Ponec, 1997) and the electron localization function (Becke & Edgecombe, 1990; Ponec, 1998). Upon integration of the XCD we obtain the density

$$\int \rho_{xc}(\mathbf{r},\mathbf{r}') d\mathbf{r}' = \rho(\mathbf{r}) \tag{3.215}$$

taking the derivative with respect to μ at each side we obtain

$$\int \left(\frac{\partial \rho_{xc}(\mathbf{r},\mathbf{r}')}{\partial \mu} \right)_V d\mathbf{r}' = \left(\frac{\partial \rho(\mathbf{r})}{\partial \mu} \right)_V = s(\mathbf{r}) \tag{3.216}$$

which, naturally, gives the definition of the softness kernel above.

However, in order to further illustrate the significance of the new softness kernel we will provide an example based on a simple model. We take an *AB* system, where electrons flow between *A* and *B*. Following (Nalewajski, 1984, 1985) let the charge transfer be described to first-order by, see also (3.184)

$$\frac{1}{2}D(A,B) \equiv \Delta N = \frac{|\mu_B - \mu_A|}{2(\eta_A + \eta_B)} \qquad (3.217)$$

and the chemical potential be given by (3.186) here under the chemical potential form

$$\mu = \frac{\mu_A \eta_B + \mu_B \eta_A}{2(\eta_A + \eta_B)} \qquad (3.218)$$

then

$$\frac{\partial \mu}{\partial \mu_A} = \frac{\eta_B}{2(\eta_A + \eta_B)} \qquad (3.219)$$

which, applied into Eq. (3.217) gives

$$\left[\frac{\partial D(A,B)}{\partial \mu}\right]_A = \frac{\partial D(A,B)}{\partial \mu_A}\frac{2(\eta_A + \eta_B)}{\eta_B} = 2s_B \qquad (3.220)$$

i.e., the change of the electron sharing between A and B upon modification of the chemical potential in A, depends on the softness of B. Since the chemical potential is conceptually equivalent to the electronegativity (Parr et al., 1978), we reach the intuitive condition that the change of electronegativity in A will bring a change in the electron sharing between A and B which is proportional to the softness of B. The present example puts forwards the significance of this new softness kernel definition.

3.3.8 ELECTRONEGATIVITY GEOMETRIC EQUALIZATION PRINCIPLE

Alternatively, one may consider the so called geometrical mean of atoms-in-molecule electronegativity, namely through the relationships (Parr & Bartolotti, 1982)

$$\chi_{AB} = \chi_{\langle A \rangle} = \chi_{\langle A \rangle}(N_A + \Delta N) \qquad (3.221)$$

$$\chi_{AB} = \chi_{\langle B \rangle} = \chi_{\langle B \rangle}(N_B - \Delta N) \qquad (3.222)$$

Together, Eqs. (3.221) and (3.222) can equivalently be written like a geometric mean

$$\chi^2_{AB} = \chi_{\langle A\rangle}\chi_{\langle B\rangle} = \chi_{\langle A\rangle}(N_A + \Delta N)\chi_{\langle B\rangle}(N_B - \Delta N) \quad (3.223)$$

Because of the energy conservation reasons, a similar relation has to be obeyed also by the isolated atoms

$$\chi^2_{AB} = \chi_A\chi_B \quad\quad\quad (3.224)$$

By equating the right hand sides of last two equations the geometrical general form for the dependence of the atomic electronegativity in molecule results

$$\chi_{\langle\rangle} = \chi\exp[-\gamma\,\Delta N] = \chi\exp[-\gamma(N_{\langle\rangle} - N)] \quad (3.225)$$

with

$$\Delta N = \begin{cases} +\Delta N...\,for\ A \\ -\Delta N...\,for\ B \end{cases} \quad\quad (3.226)$$

being γ an exponential scaling parameter, with the working expression

$$\gamma = \frac{2\eta_{\langle\rangle}}{\chi_{\langle\rangle}} \quad\quad\quad (3.227)$$

since

$$2\eta_{\langle\rangle} = -\frac{\partial\chi_{\langle\rangle}}{\partial N_{\langle\rangle}} = \gamma\chi\exp[-\gamma\,\Delta N] = \gamma\chi_{\langle\rangle} \quad\quad (3.228)$$

In these conditions, applying the equalization principle between the atomic electronegativities in molecules, $\chi_{\langle A\rangle} = \chi_{\langle B\rangle}$, the transferred charge that promotes the binding process of the AB molecule casts as

$$\Delta N = -\frac{1}{2\gamma}\ln\left(\frac{\chi_B}{\chi_A}\right) \quad\quad (3.229)$$

For completeness, the corresponding released energy is now calculated as

$$- \Delta E = \int_{N_A=Z_A^*}^{N_A+\Delta N} \chi_{\langle A \rangle} dN_{\langle \rangle} + \int_{N_B=Z_B^*}^{N_B-\Delta N} \chi_{\langle B \rangle} dN_{\langle \rangle}$$

$$= \chi_A \frac{1}{\gamma} [1 - \exp(-\gamma \Delta N)] + \chi_B \frac{1}{\gamma} [1 - \exp(+\gamma \Delta N)]$$

$$\cong -(\chi_B - \chi_A)\Delta N - \frac{1}{2}(\chi_A + \chi_B)\gamma \, (\Delta N)^2 \qquad (3.230)$$

As may be seen, the charge transfer form by geometrical mean Eq. (3.229) is richer than that provided by the additive model equation (3.185) for atoms-in-molecule because containing also the equilibrium information through the parameter (3.227). Consequently, the electronic sharing index of atoms-in-molecule by geometric mean (3.229) becomes

$$\frac{1}{2}D(A, B) \equiv \Delta N = \frac{1}{2\gamma} \ln\left(\frac{\chi_A}{\chi_B}\right) \qquad (3.231)$$

leading with the successive softness type relationships (Putz, 2011b, 2012b)

$$-\frac{1}{2}\left[\frac{D(A, B)}{\partial \chi_{\langle \rangle}}\right]_A = -\frac{1}{2}\left[\frac{\partial D(A, B)}{\partial \chi_A}\right]\left(\frac{\partial \chi_A}{\partial \chi_{\langle \rangle}}\right)$$

$$= -\frac{1}{2\gamma\chi_A \exp[-\gamma \Delta N]} = -\frac{1}{2\gamma\chi_{\langle \rangle}} = -\frac{1}{4\eta_{\langle \rangle}} = -\frac{1}{2}S_{\langle \rangle} \quad (3.232)$$

and in a similar way

$$-\frac{1}{2}\left[\frac{D(A, B)}{\partial \chi_{\langle \rangle}}\right]_B = +\frac{1}{2}S_{\langle \rangle} \qquad (3.233)$$

The results of Eqs. (3.232) and (3.233) are nevertheless a quarter of the electronic sharing indices within the additive model of atoms-in-molecules, (3.217) and (3.220), respectively, modeling therefore weaker bonds. Nevertheless, the geometrical mean model still prescribes the sharing index as behaving like the bonding softness, within equilibrated atoms-in-molecule, i.e., the states $\langle \, \rangle$, being this way superior to the additive model that involve only the softness of isolated atoms; it is therefore more adapted to the binding reality, since more consistent with the molecular realm either in its structural (sharing) index or in its reactivity counterpart (i.e., chemical softness).

3.3.9 CONNECTING ABSOLUTE AND MULLIKEN ELECTRONEGATIVITIES BY REACTIVITY

The electronegativity of an atom in a molecule ($\overline{\chi}$) can be evaluated starting from the atomic electronegativity (χ_0) and hardness (η_0) through the relations (Sanderson, 1976; Mortier, 1987; Parr & Bartolotti, 1982; Parr et al., 1978):

$$\overline{\chi} = \chi_0 - 2\eta_0 \Delta N \tag{3.234}$$

$$\overline{\chi} = \chi_0 \exp(-\gamma \Delta N) \cong \chi_0 - \chi_0 \gamma \Delta N \tag{3.235}$$

being γ a fall-off parameter (Parr & Bartolotti, 1982).

The reliability of the recently proposed electronegativity expressions, see Putz (2003, 2008b) and Volume II of the present five-volume work (Putz, 2016b),

$$\chi(N) = -\frac{1}{\sqrt{a}} \arctan\left[\frac{N}{\sqrt{a}}\right] - \frac{b}{a+N^2} - NC_A \frac{1}{a+N^2} \tag{3.236}$$

$$\chi_M(N) = \frac{b+N-1}{2\sqrt{a}} \arctan\left(\frac{N-1}{\sqrt{a}}\right) - \frac{b+N+1}{2\sqrt{a}} \arctan\left(\frac{N+1}{\sqrt{a}}\right)$$

$$+ \frac{C_A - 1}{4} \ln\left[\frac{a+(N-1)^2}{a+(N+1)^2}\right] \tag{3.237}$$

can be checked showing that they are consistent with (3.234) and (3.235) equations.

To this aim, one can consider the atom placed at the center of a sphere of infinite radius in which has to be evaluated the influence of the electronegativity of the same atom, up to where the probability to find electrons is very low. Therefore, if we apply the limit $N(r \to \infty) \to 0$, on the actual

(3.236) and (3.237) absolute and Mulliken electronegativity formulations, the following equations are obtained:

$$\lim_{N(r\to\infty)\to 0} \chi(N) = -\frac{b}{a} \tag{3.238a}$$

$$\lim_{N(r\to\infty)\to 0} \chi_M(N) = -\frac{b}{\sqrt{a}} \arctan\left(\frac{1}{\sqrt{a}}\right) \tag{3.238b}$$

If the Poisson equation within the long-range condition is considered:

$$\nabla V(r) \cong -4\pi\rho(r)\Delta r \tag{3.239a}$$

$$V(r) \cong 4\pi\rho(r)[\Delta r]^2 \tag{3.239b}$$

the components a and b in Eq. (3.238) can be re-arranged, respectively, as (Putz et al., 2005):

$$a \cong \sum_{i=1}^{N} \frac{\nabla\rho_i}{4\pi\rho_i\Delta r_i}\Delta r_i = \frac{1}{4\pi}\sum_{i=1}^{N}\frac{\nabla\rho_i}{\rho_i} \cong \frac{1}{4\pi}\sum_{i=1}^{N}\frac{\Delta\rho_i}{\rho_i\Delta r_i} \cong \frac{1}{4\pi}\sum_{i=1}^{N}\frac{\Delta\rho_i}{\Delta\rho_i} = \frac{N}{4\pi} \tag{3.240a}$$

$$b \cong \int\frac{\nabla\rho}{4\pi\rho\Delta r}4\pi\rho[\Delta r]^2\,dr = \int\nabla\rho\,\Delta r\,dr \cong \int\frac{\Delta\rho}{\Delta r}\Delta r\,dr = \int\Delta\rho(r)\,dr = \Delta N \tag{3.240b}$$

Introducing (3.240) in (3.238), they become:

$$\lim_{r\to\infty}\chi(N) = -\frac{4\pi}{N}\Delta N \equiv \overline{\chi} - \chi_0 \tag{3.241a}$$

$$\lim_{r\to\infty}\chi_M(N) = -\sqrt{\frac{4\pi}{N}}\arctan\left(\sqrt{\frac{4\pi}{N}}\right)\Delta N \equiv \overline{\chi} - \chi_0 \tag{3.241b}$$

that are formally identical to:

$$\overline{\chi} - \chi_0 = -2\eta_0\Delta N \cong -\gamma\chi_0\Delta N \tag{3.242}$$

so proving the reliability of the analytical expressions (3.236) and (3.237).

3.3.10 POLY-ATOMIC ELECTRONEGATIVITY BY CHEMICAL HARDNESS AND FUKUI FUNCTION

The molecule can be seen as a collection of atoms (Putz & Chiriac, 1998), thus, its global quantum properties – in general, and that of chemical hardness and softness – in particular, can be computed as additive contributions of atomic components. In this respect, the molecular softness is written as:

$$S_M = \sum_A S_A \tag{3.243}$$

since it progressively accounts for the electronic cloud deformations of atoms in molecules (Torrent-Sucarrat et al., 2005). To describe the molecular hardness, also a frontier partition correction factor is necessary, e.g., the so-called atomic Fukui function (Li & Evans, 1995), see also (3.99), is introduced

$$f_A(\mathbf{r}) = \left(\frac{\partial \rho_A(\mathbf{r})}{\partial N_A} \right)_{V_A(\mathbf{r})} = -\left(\frac{\delta \chi_A}{\delta V_A(\mathbf{r})} \right)_{N_A} \tag{3.244}$$

which account here for the atomic density ρ_A or electronegativity χ_A variation respecting the total number of electrons N_A or that of the bare potential V_A in atoms, respectively; consequently it can be re-written as:

$$\eta_M = \sum_A f_A \eta_A \tag{3.245}$$

Therefore, an operational atoms-in-molecules chemical hardness expression is to be achieved once the Fukui function is determined. This goal may be elegantly achieved since the Fukui function also applies at the softness level in a reciprocal manner than it links the atoms in molecules for chemical hardness, namely:

$$S_A = f_A S_M \tag{3.246}$$

Now, combining these expressions the atomic Fukui function gets out as:

$$f_A = \frac{S_A}{\sum_A S_A} \qquad (3.247)$$

while by further accounting of the chemical hardness-softness relationship it provides the global atomic containing molecular chemical hardness:

$$\eta_M = \frac{n^{AIM}}{\sum_A \dfrac{n^A}{\eta_A}} \qquad (3.248)$$

which depends on the total number of atoms in molecule and the number of identically atoms, n^{AIM} and n^A, respectively.

By generalizing this expression it leads with the polyatomic hardness in the same spirit:

$$\eta_{M^{\pm\delta}} = \frac{n^{AIM} \pm \delta}{\sum_A \dfrac{n^A}{\eta_A}} \qquad (3.249)$$

Therefore, employing the universal invariant θ for their proportion it results also for electronegativity that

$$\chi_{M^{\pm\delta}} = \frac{n^{AIM} \pm \delta}{\sum_A \dfrac{n^A}{\chi_A}} \qquad (3.250)$$

where the sum of the n^A atoms of each species present in the molecule recovers the total atoms in molecule n^{AIM},

$$\sum_A n^A = n^{AIM} \qquad (3.251)$$

being also the overall molecular charge $\pm\delta$ included in the formula, for the shake of completeness.

Now, unfolding the AIM and MOL schemes for electronegativity and chemical hardness implementation in absolute of compactness aromaticity definitions above, one recalls the AIM electronegativity generalization to the polyatomic molecules (Bratsch, 1985)

$$\chi_{AIM} = \frac{n_{AIM}}{\sum_A \frac{n_A}{\chi_A}} \tag{3.252}$$

where the total atoms in molecule n_{AIM} is the sum of the n_A atoms of each A-species present in the molecule

$$\sum_A n_A = n_{AIM} \tag{3.253}$$

Analogously, although not in the same manner, the AIM chemical hardness writes, beside the contributions of the atomic chemical hardnesses η_A, in terms of atomic Fukui functions f_A (Parr & Yang, 1984; Yang et al., 1984), and of atomic global softness values S_A, in equivalent forms

$$\eta_{AIM} = \sum_A f_A \eta_A = \frac{\sum_A S_A \eta_A}{\sum_A S_A} \tag{3.254}$$

since the atomic Fukui function alternative unfolding (Putz, 2011a)

$$f_A = \frac{1}{n_{AIM}} \frac{\eta_{AIM}}{\eta_A} = \frac{S_A}{\sum_A S_A} \tag{3.255}$$

It is worth remarking that with the help of formulation (3.255) for the Fukui function the local reactivity on a specific atom in a molecule may be evaluated according to its values, namely:

$$f_A = \begin{cases} >1...nucleophilc\ attack \\ <1...electrophlic\ attack \\ =1...radicalic\ attack \end{cases} \tag{3.256}$$

without the need of the fashioned derivative of the density to the total number of electrons in molecule – see Eq. (3.244); actually, the Fukui function recipe of Eqs. (3.255) & (3.256) indicates the degree to which the product $n_{AIM} \times \eta_A$ that involves local chemical hardness is underneath, is overcoming or equals the overall AIM chemical hardness for nucleophilic, electrophilic, and radicalic attacks, respectively. This is nevertheless a newer result to be applied in further communications.

3.4   CHEMICAL HARDNESS OF ATOMS IN MOLECULES

3.4.1   ITERATIVE CHEMICAL HARDNESS OF AIM

The chemical reactivity is a concept with a physical reality which contains the compromise between the tendency of electronegativity to promote the coordination and the inertia of the chemical species manifested by the chemical hardness (Ray et al., 1979; Parr & Bartolotti, 1982; Pearson, 1985).

The electronegativity was already discussed; however, the link between the electronegativity, respectively the chemical potential with changed sign and the chemical hardness consists in a differential connection. Based on such relationship, the study and respectively the information about the chemical hardness can be naturally extended by appealing to the principle of the chemical potential equalization (also in differential form) within or between molecule(s), by taking into account also the secondary effects of coordination, meaning that an initial flow of charges between two groups of atoms in molecule generates also the reaction as manifested by celebrated Le Châtelier-Braun Principle (Nalewajski, 1990).

PRINCIPLE AIM-HD1 (of Le Châtelier-Braun): A primary change of the charges number of an atom A in a molecule, say dN, induces a (secondary) redistribution δN of the charges number between the atoms from the rest of the molecule, in order to decrease the primary "hardness" which produced the change in the atom A charge. The chemical potentials equilibrium and of chemical hardness in molecule will be ultimately accomplished by a readjustment of the net charges between the atoms (molecular fragments) involved in the primary and secondary processes.

Therefore, by considering the secondary effects of the charges redistribution between the atoms in molecule, there is expected to obtain information about the global chemical strength of the molecule, by the hardness of the atoms or of the fragments of the molecule; the forthcoming discussion follows the references (Nalewajski, 1988; Nalewajski &

Korchewiec, 1989a-b, Nalewajski et al., 1988; Nalewajski & Koniński, 1988).

For the beginning, the case of the diatomic molecule will be considered. In the primary movement of the charges in molecule one considers $dN_A = dN > 0, dN_B = 0$, so that the chemical potentials associated to the atoms as a result of this primary process will become:

$$d\mu_A(dN_A) = \eta_{A,A}dN \qquad (3.257)$$

$$d\mu_B(dN_A) = \eta_{B,A}dN \qquad (3.258)$$

In the secondary movement, the atoms potentials in molecule will be properly modified:

$$\delta\mu_A(\delta N_{AB}) = (\eta_{A,B} - \eta_{A,A})\delta N \qquad (3.259)$$

$$\delta\mu_B(\delta N_{AB}) = (\eta_{B,B} - \eta_{B,A})\delta N \qquad (3.260)$$

The value of the secondary movement will be determined by the principle of the atoms' potential equalization in molecule:

$$d\mu_{AB}^*(dN) = d\mu_A^*(dN) = d\mu_B^*(dN)$$
$$\Leftrightarrow d\mu_A(dN) + \delta\mu_A(dN) = d\mu_B(dN) + \delta\mu_B(dN)$$
$$\Rightarrow \delta N = \frac{\eta_{A,A} - \eta_{A,B}}{\eta_{A,A} + \eta_{B,B} - 2\eta_{A,B}}dN \qquad (3.261)$$

from where the global chemical hardness of the molecule AB results:

$$\eta_{AB} = \frac{d\mu_{AB}^*(dN)}{dN} = \frac{\eta_{A,A}\eta_{B,B} - \eta_{A,B}^2}{\eta_{A,A} + \eta_{B,B} - 2\eta_{A,B}} \qquad (3.262)$$

The chemical hardness of the coordinated atoms in molecule can also be expressed by taking into account the net charge that the atoms had changed during the primary and secondary processes of the charge movement:

$$dN_B^*(dN) = \delta N \qquad (3.263)$$

$$dN_A^*(dN) = dN - \delta N = \frac{\eta_{B,B} - \eta_{A,B}}{\eta_{A,A} + \eta_{B,B} - 2\eta_{A,B}} dN \qquad (3.264)$$

$$\Rightarrow \eta_A = \frac{d\mu_A^*(dN)}{dN_A^*(dN)} = \frac{\eta_{A,A}\eta_{B,B} - \eta_{A,B}^2}{\eta_{B,B} - \eta_{A,B}} \qquad (3.265)$$

$$\Rightarrow \eta_B = \frac{d\mu_B^*(dN)}{dN_B^*(dN)} = \frac{\eta_{A,A}\eta_{B,B} - \eta_{A,B}^2}{\eta_{A,A} - \eta_{A,B}} \qquad (3.266)$$

With these relations, the global molecular chemical hardness can be expressed with the chemical atomic hardness aid:

$$\eta_{AB}dN = \eta_A dN_A^*(dN) = \eta_B dN_B^*(dN)$$
$$= \eta_{A,A}dN_A^*(dN) + \eta_{A,B}dN_B^*(dN)$$
$$= \eta_{B,A}dN_A^*(dN) + \eta_{B,B}dN_B^*(dN) \qquad (3.267)$$

Now, we notice the chemical hardness of the atoms in molecule actually depends on the chemical strengths (the mixed chemical hardnesses) of the coordination corresponding to the primary and secondary transfers that is taking place between the atoms in molecule.

These relationships, deduced in the diatomic case, can be found also at the level of more complex molecules, as in sequel.

In considering the case of the triatomic molecules $M=(ABC)$ there are three types of partitions. Supposing the partition $M_1=(A|BC)$ and the primary movement of charge $dN=(dN_A = dN,0,0)$ upon which the potentials AIM will be analogically given at the diatomic cases:

$$d\mu_A(dN_A)=\eta_{A,A}dN \qquad (3.268)$$

$$d\mu_B(dN_A)=\eta_{B,A}dN \qquad (3.269)$$

$$d\mu_C(dN_A)=\eta_{C,A}dN \qquad (3.270)$$

Considering the secondary charge movement inside the fragment (BC) from B to C, it will affect each chemical potential AIM, as based on the Principle AIM-HD1:

$$\delta\mu_A(dN_{BC}) = (\eta_{A,C} - \eta_{A,B})\delta N_{BC} \qquad (3.271)$$

$$\delta\mu_B(dN_{BC}) = (\eta_{B,C} - \eta_{B,B})\delta N_{BC} \qquad (3.272)$$

$$\delta\mu_C(dN_{BC}) = (\eta_{C,C} - \eta_{C,B})\delta N_{BC} \qquad (3.273)$$

As a result of this movement, the atoms B and C are brought to the same internal equilibrium, ("+"), but not yet also with the atom A.

So, until now, from the potentials equalization, we have

$$d\mu_{BC}^+(dN_A) = d\mu_B^+(dN_A) = d\mu_C^+(dN_A)$$

$$\Leftrightarrow d\mu_B(dN_A) + \delta\mu_B(dN_{BC}) = d\mu_C(dN_A) + \delta\mu_C(dN_{BC}) \qquad (3.274)$$

while the internal charge movement within the fragment (BC) will be determined upon above procedure to be:

$$\delta N_{BC} = \frac{\eta_{B,A} - \eta_{C,A}}{\eta_{B,B} + \eta_{C,C} - 2\eta_{B,C}} dN_A \qquad (3.275)$$

wherefrom the non-diagonal hardness corresponding to the partition M_1 results in terms of the chemical hardness of AIM correlated by primary and secondary charges:

$$\eta_{A,BC} = \frac{d\mu_{BC}^+(dN_A)}{dN_A}\bigg|_{\substack{A-PRIMARY \\ BC-SECONDARY}} = \frac{d\mu_A^+(dN_{BC})}{dN_{BC}}\bigg|_{\substack{BC-PRIMARY \\ A-SECONDARY}}$$

$$= \eta_{BC,A} = \frac{\eta_{A,B}(\eta_{C,C} - \eta_{B,C}) + \eta_{A,C}(\eta_{B,B} - \eta_{B,C})}{\eta_{B,B} + \eta_{C,C} - 2\eta_{B,C}} \qquad (3.276)$$

The diagonal term of the BC fragment can be obtained as previously illustrated, considering that this fragment is the subject of the primary movement dN_{BC}, while maintaining $dN_A = 0$:

$$\eta_{BC,BC}^+ = \frac{d\mu_{BC}^+(dN_{BC})}{dN_{BC}} = \frac{\eta_{B,B}\eta_{C,C} - \eta_{B,C}^2}{\eta_{B,B} + \eta_{C,C} - 2\eta_{B,C}} \tag{3.277}$$

The diagonal term of the chemical hardness of the atom A in partition M_I has to take into account the effect of the Principle AIM-HD1 so becoming

$$\eta_{A,A}^+ = \frac{d\mu_A^+(dN_A)}{dN_A} = \frac{d\mu_A(dN_A)}{dN_A} + (\eta_{A,C} - \eta_{A,B})\frac{\delta N_{BC}(dN_A)}{dN_A}$$

$$= \eta_{A,A} - \frac{(\eta_{A,B} - \eta_{A,C})^2}{\eta_{B,B} + \eta_{C,C} - 2\eta_{B,C}} \tag{3.278}$$

The last step in completing the equalization of the AIM potentials is the charge transfer between the BC fragment reached to the internal equilibrium and the atom A found in the last state described, by reiterating to another level the relationships in the chemical potentials of the molecular fragments considered:

$$\delta\mu_A\left(\delta N_{A|BC}\right) = (\eta_{A,BC} - \eta_{A,A})\delta N_{A|BC} \tag{3.279}$$

$$\delta\mu_{BC}\left(\delta N_{A|BC}\right) = (\eta_{BC,BC} - \eta_{BC,A})\delta N_{A|BC} \tag{3.280}$$

Again, following the equalization of the fragments potentials in the final molecular equilibrium:

$$d\mu_{ABC}^*(dN) = d\mu_A^*(dN) = d\mu_{BC}^*(dN)$$
$$\Leftrightarrow d\mu_A^+(dN) + \delta\mu_A\left(\delta N_{A|BC}\right) = d\mu_{BC}^+(dN) + \delta\mu_{BC}\left(\delta N_{A|BC}\right) \tag{3.281}$$

the charge exchanged by the two fragments, (A) and (BC), assures the internal (molecular) equilibrium – yet analytically distinguishable, thus resulting at the global equilibrium to be:

$$\delta N_{A|BC} = \frac{\eta_{A,A}^+ - \eta_{A,BC}}{\eta_{A,A}^+ + \eta_{BC,BC} - 2\eta_{A,BC}} dN \tag{3.282}$$

With the aid of this relation, the molecular global hardness immediately yields:

$$\eta_{ABC} = \frac{d\mu^*_{ABC}(dN)}{dN} = \frac{\eta^+_{A,A}\eta_{BC,BC} - \eta^2_{A,BC}}{\eta^+_{A,A} + \eta_{BC,BC} - 2\eta_{A,BC}} \qquad (3.283)$$

Then, the net charges transferred between the atomic fragments in molecule appear as:

$$dN^*_{BC}(dN) = \delta N_{A|BC} \qquad (3.284)$$

$$dN^*_A(dN) = dN - \delta N_{A|BC} = \frac{\eta_{BC,BC} - \eta_{A,BC}}{\eta^+_{A,A} + \eta_{BC,BC} - 2\eta_{A,BC}} dN \qquad (3.285)$$

while the individual chemical hardnesses of the atoms accounting for global equilibrium in molecule are respectively given by:

$$\eta_A = \frac{d\mu^*_A(dN)}{dN^*_A(dN)} = \frac{\eta^+_{A,A}\eta_{BC,BC} - \eta^2_{A,BC}}{\eta_{BC,BC} - \eta_{A,BC}} \qquad (3.286)$$

$$\eta_{BC} = \frac{d\mu^*_{BC}(dN)}{dN^*_{BC}(dN)} = \frac{\eta^+_{A,A}\eta_{BC,BC} - \eta^2_{A,BC}}{\eta^+_{A,A} - \eta_{A,BC}} \qquad (3.287)$$

Thus, the various expressions of the global molecular chemical hardness at global equilibrium can be written, respectively according to the initial individual atomic hardnesses, correlated with multiple net charge transfer, until attaining the final equilibrium, in all the possible variants of partition:

$$\eta_{ABC}dN = \eta_A dN^*_A(dN) = \eta_B dN^*_B(dN) = \eta_C dN^*_C(dN)$$

$$= \eta_{BC}dN^*_{BC}(dN) = \eta_{AB}dN^*_{AB}(dN) = \eta_{AC}dN^*_{AC}(dN)$$

$$= \eta_{A,A}dN^*_A(dN) + \eta_{A,B}dN^*_B(dN) + \eta_{A,C}dN^*_C(dN)$$

$$= \eta_{C,A}dN^*_A(dN) + \eta_{C,B}dN^*_B(dN) + \eta_{C,C}dN^*_C(dN)$$

$$= \eta_{B,A}dN^*_A(dN) + \eta_{B,B}dN^*_B(dN) + \eta_{B,C}dN^*_C(dN) \qquad (3.288)$$

Now, by comparing the global formulas in case of diatomic and triatomic molecules, one can generalize (by induction) similar results for a working molecule, M, randomly partitioned in molecular fragments, say $M^\pi = (X|Y)$.

In this case, the global molecular hardness will be given by a relation abstracted by induction, with the form:

$$\eta_{M^{II}} = \frac{\eta_{X,X}^{\pi}\eta_{Y,Y}^{\pi} - \left(\eta_{X,Y}^{\pi}\right)^2}{\eta_{X,X}^{\pi} + \eta_{Y,Y}^{\pi} - 2\eta_{X,Y}^{\pi}} \qquad (3.289)$$

where the diagonal and non-diagonal components are recursively given if one considers, for example, a new partition of the fragment $Y^{\sigma} = (I|J)$; then, by induction, the general relations are inferred:

$$\eta_{X,Y}^{\pi} = \frac{\eta_{X,I}^{\sigma}\left(\eta_{J,J}^{\sigma} - \eta_{I,J}^{\sigma}\right) + \eta_{X,J}^{\sigma}\left(\eta_{I,I}^{\sigma} - \eta_{I,J}^{\sigma}\right)}{\eta_{I,I}^{\sigma} + \eta_{J,J}^{\sigma} - 2\eta_{I,J}^{\sigma}} \qquad (3.290)$$

$$\eta_{X,X}^{\pi} = \eta_{X,X} - \frac{\left(\eta_{X,I}^{\sigma} - \eta_{X,J}^{\sigma}\right)^2}{\eta_{I,I}^{\sigma} + \eta_{J,J}^{\sigma} - 2\eta_{I,J}^{\sigma}} \qquad (3.291)$$

From the relations above another principle can be established, namely:

PRINCIPLE AIM-HD2: For a given molecule, composed by an arbitrary atoms' collection, the global molecular chemical hardness is recursively expressed in terms of the chemical hardness associated to the various molecular partitions, between which the charge transfers are manifested in order to reach the coordination equilibrium.

However, the molecular equilibrium condition requires also the following principle:

PRINCIPLE AIM-HD3: For a formed molecule to be stable there is required that at equilibrium both the global chemical hardness and the internal one (for molecular fragments) have to be positive toward the molecular partitions (basins), meaning to provide a physical-chemical inertia at equilibrium.

The equivalent condition for the internal stability criterion can be deduced from the last relations above, where the second term (with the sign), being a secondary effect (since Principle AIM-HD1), has appeared from the tendency to reduce the primordial "action", so having to be negative, which implies the fact that the denominator of the last relations above must be positive. Therefore, generalizing at the partitioning level, the condition of internal stability results as

$$\frac{\eta_{X,X}^{\pi} + \eta_{Y,Y}^{\pi}}{2} > \eta_{X,Y}^{\pi} \qquad (3.292)$$

Corroborating this result with the requirement of positivity of the global hardness, one finds that for assuring the molecular stability there is required that also the numerator of the expression (3.291) of the molecular chemical hardness to be positive, leaving with the condition:

$$\eta_{X,X}^{\pi}\eta_{Y,Y}^{\pi} > \left(\eta_{X,Y}^{\pi}\right)^2 \qquad (3.293)$$

Therefore, one notes how important and even decisive role the chemical hardness has in general and the atomic partitioning for fragments' hardness in a molecule, in particular, for the stabilization of the molecular coordination from the charge exchanges between the atoms that compose it.

3.4.2 SOFTNESS AND FUKUI INDICES RELATING CHEMICAL HARDNESS

Next, a partitioning of a random molecule will be considered in the general form $M=(X|Y|R)$. Firstly, the fragment A is considered as rigid, frozen, and so not participating in coordination. In these conditions, the diagonal and non-diagonal terms of the molecular hardness will be determined in a similar way as in the previous section, yet by introducing some new parameters for defining the molecular reactivity; the forthcoming discussion follows the references (Nalewajski, 1988, 1990; Nalewajski & Korchowiec, 1989a).

Now, by $\eta_{ik}^{X,Y}$ will be indicated the chemical hardness associated to the coupling between the atom i from fragment X and the atom k from fragment Y and, similarly, η_{ij}^{X} denotes the chemical hardness of the coupling of the atoms i and j from the same molecular fragment X (Nalewajski, 1989b, 1990). In these conditions one may formulate the followings:

DEFINITION AIM-HD1: Consider a molecule with a fragment X characterized by the chemical hardness set of parameters $\left\{\eta_{ij}^{X}\right\} = \eta^{X}$; then one defines: (a) the chemical availability matrix (=softness) associated to the molecular fragment X writing as

$$s^X = (h^X)^{-1} \tag{3.294}$$

with the property that the matrix elements are satisfying the closure relationship:

$$\sum_j^X \sigma_{ij}^X \eta_{jk}^X = \delta_{ik} \tag{3.295}$$

(b) the softness of molecular fragment is noted as S_X and is expressed by the sum of atomic softnesses associated to the atoms from the generic fragment X, successively unfolded by the sum over the matrix elements of chemical softnesses over the fragment X:

$$S_X = (\eta_X)^{-1} = \sum_i^X s_i^X = \sum_i^X \sum_j^X \sigma_{ij}^X \tag{3.296}$$

(c) the atomic Fukui indices for atoms in the molecular X fragment is defined by the ratio between the atomic softness and the encompassing molecular fragment softness:

$$f_i^X = \frac{s_i^X}{S_X} = \frac{\partial N_i^X}{\partial N_X} \tag{3.297}$$

where N_i^X and N_X are the electronic populations of the i-atoms in molecule and of the X-fragment in molecule, respectively.

Then, one considers that at the initial moment the molecular X-fragment and also the other fragments, say X,Y and R, are at an internal equilibrium:

$$\mu_j^X = \mu_i^X = ... = \mu_X \tag{3.298}$$

$$\mu_k^Y = \mu_l^Y = ... = \mu_Y \tag{3.299}$$

$$\mu_s^R = \mu_t^R = ... = \mu_R \tag{3.300}$$

When a primordial charge movement dN_X is considered, with the other fragments considered as frozen

$$d\mu_X = \eta_X dN_X \tag{3.301}$$

then the X-fragment will react with a secondary charge variation in order to reestablish the internal equilibrium:

$$\mu_X - \mu_i^X = \sum_j^X \eta_{ij}^X \delta N_j^X \qquad (3.302)$$

from where the secondary charge results as:

$$\delta N_i^X (dN_X) = \sum_j^X \sigma_{ij}^X \left(\mu_X - \mu_j^X \right) = \sum_j^X \sigma_{ij}^X \eta_X dN_X = f_i^X dN_X \quad (3.303)$$

This secondary charge will affect also the potentials of the Y-fragment, even if it is still maintained as rigid, i.e., without being free to regain its internal equilibrium:

$$d\mu_k^Y = \sum_i^X \eta_{ki}^{Y,X} \delta N_i^X (dN_X) = \left(\sum_i^X \eta_{ki}^{Y,X} f_i^X \right) dN_X = \eta_{k,X}^M dN_X \quad (3.304)$$

If the Y-fragment is relaxed from the rigidity constraint at coordination, the atoms of this fragment will spontaneously transfer a charge as above in the X-fragment case, evolving to reach the equilibrium, both internal and in the coordination with X-fragment:

$$\delta N_k^Y (dN_X) = \sum_l^Y \sigma_{kl}^Y \left(\eta_{Y,X}^M - \eta_{l,X}^M \right) dN_X \qquad (3.305)$$

However, by employing the charge conservation relation for the charge transfers specific to the Y-fragment,

$$\sum_k^Y \delta N_k^Y = 0 \qquad (3.306)$$

the equivalent successive expressions results (thanks to the Definition AIM-HD1):

$$\eta_{Y,X}^M = \frac{\partial \mu_Y}{\partial N_X} = \sum_k^Y f_k^Y \eta_{k,X}^M = \sum_i^X \eta_{Y,i}^M f_i^X = \sum_i^X \sum_k^Y f_k^Y \eta_{ki}^{Y,X} f_i^X$$

$$(3.307)$$

where for the last equality the form provided by the last equality of $d\mu_k^Y$ in (3.304) have been used.

Worth noting that for the $\eta^M_{Y,X}$ relation (3.307) expressing the non-diagonal hardness of the molecular fragments involved in coordination, the atomic Fukui indices have the purpose to connect the atoms' composing the molecular fragments with their global behavior.

As a result of the Y-fragment relaxation towards the internal equilibrium also with the X-fragment, the charge transfer may spontaneously occur in Y, which in turn will influence the hardness (the chemical potential) of the X-fragment in the sense stated by the principle AIM-HD1:

$$\eta^M_{X,X} = \frac{\partial \mu_X}{\partial N_X} = \eta_X + \frac{d\mu_X \left\{ \delta N^Y_k (dN_X) \right\}}{dN_X} = \eta_X + \frac{\sum^Y_k \eta^M_{X,k} \delta N^Y_k (dN_X)}{dN_X}$$

$$\Rightarrow \eta^M_{X,X} = \eta_X + \delta \eta^M_{X,X}(dN_Y) = \eta_X + \left[S_Y \left(\eta^M_{X,Y} \right)^2 - \sum^Y_{k,l} \eta^M_{X,k} \sigma^Y_{k,l} \eta^M_{l,X} \right]$$

$$\tag{3.308}$$

If next also the R-fragment is totally relaxed, a charge transfer internal to the R-fragment will spontaneously appear, aiming to restore the internal equilibrium also with the rest of the molecule:

$$\delta N^R_s (dN_X) = \sum^R_t \sigma^R_{st} \left(\eta^M_{R,X} - \eta^M_{t,X} \right) dN_X \tag{3.309}$$

Then, the diagonal and non-diagonal expressions of the molecular chemical hardnesses above will be modified as a result of the influence exercised by the charges redistribution from R-fragment in the rest of the molecule, as based on the Principle AIM-HD1, so that at the final equilibrium between all the molecular fragments the averaged expressions cast as:

$$\overline{\eta}^{-M}_{X,X} = \eta^M_{X,X} + \delta \eta^M_{X,X}(\delta N_R) = \eta^M_{X,X} + \left[S_R \left(\eta^M_{X,R} \right)^2 - \sum^R_{s,t} \eta^M_{X,s} \sigma^R_{s,t} \eta^M_{t,X} \right]$$

$$\tag{3.310}$$

$$\overline{\eta}^{-M}_{X,Y} = \eta^M_{X,Y} + \delta \eta^M_{X,Y}(\delta N_R) = \eta^M_{X,Y} + \left[S_R \eta^M_{X,R} \eta^M_{R,Y} - \sum^R_{s,t} \eta^M_{X,s} \sigma^R_{s,t} \eta^M_{t,Y} \right]$$

$$\tag{3.311}$$

These relations can be interpreted as a generalization of the expressions in the previous section, now in terms of atomic based molecular fragments available, with the intervention of the frontier Fukui functions.

3.4.3 SENSITIVITY OF ATOMS-IN-MOLECULES

The sensitivity of a system composed of interacting atoms (or more specific for atoms in molecules) should take into account the way of reaction (also in the feedback sense) of the system response to the variations of the specific parameters under the influence of the external factors.

The definition, the modeling and the application of the sensitivities as a working reactive information of an atomic system, eventually as a subsystems of a molecule, is the purpose of this section; the forthcoming discussion follows the references (Nalewajski, 1988, 1990; Nalewajski & Korchowiec, 1989a).

There is known that while a closed atomic system is characterized by its N-number of electrons, once it will became an open system, say by having contact with a reservoir of electrons (coming from catalytic substances, surfaces, for instance) the conjugated parameter with N will characterize the system, namely the potential chemical:

$$\mu = \frac{\partial E(N,Z)}{\partial N}\Big|_Z \qquad (3.312)$$

where Z is the molecule nuclear charge, and $E(N,Z)$, the equilibrium state energy. Also the Z parameter is characteristic of the closed system, having at its turn a conjugate parameter for the open system (by relaxation in relation to a reservoir of nuclei), namely the electron-nucleus attraction energy per unit of nuclear charge:

$$\varsigma = \frac{\partial E(N,Z)}{\partial Z}\Big|_N = \frac{V_{en}}{Z} \qquad (3.313)$$

where the last equality follows from the application of the Hellmann-Feynman theorem. However, a formal picture about the sensitivity of a molecule in question can be provided by considering the whole set of parameters:

$$(X,Y) = \{(N,Z),(\mu,Z),(N,\varsigma),(\mu,\varsigma)\} \qquad (3.314)$$

with the associated potentials' set, respectively:

$$P(X,Y) = \{E(N,Z),Q(\mu,Z),K(N,\varsigma),R(\mu,\varsigma)\} \qquad (3.315)$$

Next, one may consider the forces associated with these parameters' variations; this way, in analogy with the model of the elastic forces, the bonds in the various potentials representations $P(X,Y)$ can be written considering the adjacent (matrix) variation of the system's parameters so that the action of the conjugated forces having the forms:

$$[d\mu, d\zeta]^T = \begin{bmatrix} E_{NN} & E_{NZ} \\ E_{ZN} & E_{ZZ} \end{bmatrix} \begin{bmatrix} dN \\ dZ \end{bmatrix} \tag{3.316}$$

$$[-dN, -dZ]^T = \begin{bmatrix} R_{\mu\mu} & R_{\mu\zeta} \\ R_{\zeta\mu} & R_{\zeta\zeta} \end{bmatrix} \begin{bmatrix} d\mu \\ d\zeta \end{bmatrix} \tag{3.317}$$

and, in the same way, one writes the respective relations partly inverted between the forces (or potentials) and charge movements:

$$[-dN, d\zeta]^T = \begin{bmatrix} Q_{\mu\mu} & Q_{\mu Z} \\ Q_{Z\mu} & Q_{ZZ} \end{bmatrix} \begin{bmatrix} d\mu \\ dZ \end{bmatrix} \tag{3.318}$$

$$[d\mu, -dZ]^T = \begin{bmatrix} K_{NN} & K_{N\zeta} \\ K_{\zeta N} & K_{\zeta\zeta} \end{bmatrix} \begin{bmatrix} dN \\ d\zeta \end{bmatrix} \tag{3.319}$$

where the elements of the potential matrices indicate the "constants" of these *chemical force – charge movement* relations, characterizing in fact the molecular sensitivities.

These sensitivities as well as the relations between them will be next determined by considering as basic elements in these determinations the components of the matrix associated to the energy $E(N,Z)$:

$$dE = \mu dN + \zeta dZ \; ; \tag{3.320}$$

$$\Rightarrow E_{NN} = \frac{\partial^2 E}{\partial N^2} = \frac{\partial \mu}{\partial N}\Big|_Z \equiv a \tag{3.321}$$

$$\Rightarrow E_{NZ} = \frac{\partial^2 E}{\partial N \partial Z} = \frac{\partial \mu}{\partial Z}\Big|_N = \frac{\partial \xi}{\partial N}\Big|_Z \equiv b \tag{3.322}$$

$$\Rightarrow E_{ZZ} = \frac{\partial^2 E}{\partial Z^2} = \frac{\partial \zeta}{\partial Z}\Big|_N \equiv c \tag{3.323}$$

where a corresponds to the chemical hardness of the molecule, b measures the strength of the electrons-nuclei coupling, c measures the hardness of the inter-nuclei coupling; note that for the relation of b the Maxwell identities were taken into account (equivalent to the Cauchy condition according which the partially mixed derivatives are equal, regardless the derivation order, such that the basic function, the $E -$ energy in this case, to maintain the exact total differential, i.e., to correspond to a state function, so integrable).

The other elements of the matrices above will be determined in terms of a, b, and c, parameters considering the appropriate Legendre transformations and employing the partial derivatives properties in relation with the associated Jacobians' determinants.

This way, for the potential $Q(\mu, Z)$ we have the transformation:

$$Q(\mu, Z) = E - \mu N$$
$$\Rightarrow dQ = -N d\mu + \zeta dZ \qquad (3.324)$$

wherefrom the elements (molecular sensitivities for the matrix action Q) are obtained as:

$$Q_{\mu\mu} = \frac{\partial^2 Q}{\partial \mu^2} = -\frac{\partial N}{\partial \mu}\Big|_Z = -\frac{1}{a} \qquad (3.325)$$

$$Q_{\mu Z} = \frac{\partial^2 Q}{\partial \mu \partial Z} = -\frac{\partial N}{\partial Z}\Big|_N = \frac{\partial \xi}{\partial \mu}\Big|_Z = \frac{b}{a} \qquad (3.326)$$

$$Q_{ZZ} = \frac{\partial^2 Q}{\partial Z^2} = \frac{\partial \zeta}{\partial Z}\Big|_\mu = \frac{\partial[\zeta, \mu]}{\partial[Z, \mu]} = \frac{\partial[\zeta, \mu]}{\partial[[Z, N]} \frac{\partial[Z, N]}{\partial[Z, \mu]} = \frac{\partial[\zeta, \mu]}{\partial[Z, N]}\left(\frac{\partial N}{\partial \mu}\right)_Z$$

$$= \begin{vmatrix} \dfrac{\partial \zeta}{\partial Z}\Big|_N & \dfrac{\partial \zeta}{\partial N}\Big|_Z \\ \dfrac{\partial \mu}{\partial Z}\Big|_N & \dfrac{\partial \mu}{\partial N}\Big|_Z \end{vmatrix} \times \left(\frac{\partial N}{\partial \mu}\right)_Z = \left[\left(\frac{\partial \zeta}{\partial Z}\right)_N \left(\frac{\partial \mu}{\partial N}\right)_Z - \left(\frac{\partial \mu}{\partial Z}\right)_N \left(\frac{\partial \zeta}{\partial N}\right)_Z\right]\left(\frac{\partial N}{\partial \mu}\right)_Z$$

$$= \left(\frac{\partial \zeta}{\partial Z}\right)_N \left(\frac{\partial \mu}{\partial N}\right)_Z \left(\frac{\partial N}{\partial \mu}\right)_Z - \left(\frac{\partial \mu}{\partial Z}\right)_N \left(\frac{\partial \zeta}{\partial N}\right)_Z \left(\frac{\partial N}{\partial \mu}\right)_Z = \frac{ca - b^2}{a}$$

$$(3.327)$$

Analogously, for the potential $K(N,\zeta)$ we have the transformation:

$$K(N,\zeta) = E - \zeta Z$$
$$\Rightarrow dK = \mu dN - Z d\zeta \tag{3.328}$$

and the elements of the molecular sensitivity for the K action matrix look like:

$$K_{NN} = \frac{\partial^2 K}{\partial N^2} = \frac{\partial \mu}{\partial N}\Big|_\zeta = \frac{ac - b^2}{c} \tag{3.329}$$

$$K_{N\zeta} = \frac{\partial^2 K}{\partial N \partial \zeta} = \frac{\partial \mu}{\partial \zeta}\Big|_N = -\frac{\partial Z}{\partial N}\Big|_\zeta = \frac{b}{c} \tag{3.330}$$

$$K_{\zeta\zeta} = \frac{\partial^2 K}{\partial \zeta \partial \zeta} = -\frac{\partial Z}{\partial \zeta}\Big|_N = -\frac{1}{c} \tag{3.331}$$

Finally, for the potential $R(\mu,\zeta)$ the next transformation is considered:

$$R(\mu,\zeta) = E - \mu N - \zeta Z$$
$$\Rightarrow dR = -N d\mu - Z d\zeta \tag{3.332}$$

wherefrom the elements of the molecular sensitivity for the R- the action matrix are obtained:

$$R_{\mu\mu} = \frac{\partial^2 R}{\partial \mu^2} = -\frac{\partial N}{\partial \mu}\Big|_\zeta = \frac{c}{b^2 - ac} \tag{3.333}$$

$$R_{\mu\zeta} = \frac{\partial^2 R}{\partial \mu \partial \zeta} = -\frac{\partial N}{\partial \zeta}\Big|_\mu = -\frac{\partial Z}{\partial \mu}\Big|_\zeta = \frac{b}{ac - b^2} \tag{3.334}$$

$$R_{\zeta\zeta} = \frac{\partial^2 R}{\partial \zeta \partial \zeta} = -\frac{\partial Z}{\partial \zeta}\Big|_\mu = \frac{a}{b^2 - ac} \tag{3.335}$$

To determine the hierarchy of these sensitivities along with their displacements applied of the system by the registered conjugated forces, the assumptions such as $a < 0, b < 0, c < 0$ are considered as based on the

(Le Châtelier–Braun) Principle AIM-HD1 and the Principle of the molecular stability (AIM-HD3), The inertia, the positive molecular chemical hardness was discussed in the previous section.

A negative value of b implies, for an increase of the electrons number in the system, $dN > 0$, the electrons-nuclei coupling decreasing, for Z (the number of nuclei) kept constant, i.e., $d\zeta < 0$.

Analogously, a negative value of c, implies that for an increase in the nuclear charges of a system, $dZ > 0$, the electrons-nuclei coupling also decreases, for the number of electrons (N) kept constant, $d\zeta < 0$.

Based on these natural, physically intuitive, criteria, the relations giving the (sensitivities) elements of the chemical potentials associated to the molecule provide the set of inequalities

$$0 < a = \frac{\partial \mu}{\partial N}\Big|_{z} < \frac{\partial \mu}{\partial N}\Big|_{\varsigma} \tag{3.336}$$

which will be properly commented.

The successive inequalities that need to be commented are only in the second part of (3.336). An increase of the electrons number in molecule, $dN > 0$, has as a consequence in the plane of the conjugated forces, i.e., for an increase of the chemical potential (electron-electron coupling) $d\mu > 0$, and for a decrease of the electrons-nuclei coupling, $d\zeta < 0$. But to maintain ζ as a constant a force surplus $d\zeta > 0$ should be involved; it nevertheless involves from the definition of c (above) a decrease of the nuclear charges, $dZ < 0$, which implies an additional (arguable) increase of the electron-electron coupling, $d\mu > 0$.

In conclusion, the chemical potential modification, when a variation of electrons occurs in the system, is greater when the force ζ has to be kept constant, than when we have to keep constant the number of nuclear charges.

The same reasoning is valid also for explaining the inequalities:

$$\frac{\partial \xi}{\partial Z}\Big|_{\mu} < \frac{\partial \zeta}{\partial Z}\Big|_{N} = c < 0 \tag{3.337}$$

$$\frac{\partial \xi}{\partial N}\Big|_{\mu} < \frac{\partial \zeta}{\partial N}\Big|_{z} = b < 0 \tag{3.338}$$

$$\left.\frac{\partial \mu}{\partial Z}\right|_{\varsigma} < \left.\frac{\partial \mu}{\partial Z}\right|_{N} = b < 0 \qquad (3.339)$$

A very important observation is required for the relations above, i.e., for a and c: both assure the variations of the conjugated forces, either as the intensive parameters, respecting the associated charge movements, and the extensive parameters, under the constraint of the extensive/intensive opposite parameters. From the analysis of these relations worth observing that the greatest (positive or negative) variation belongs to the rationale when the opposite potential must be kept constantly. This means that a maximal reaction (effort, or feedback) is required in order to remove the effect generated by the primary (charge or chemical) movement toward the equilibrium state. On the contrary, in a maximum relaxation condition, the system opens the potential equalization with the associated reservoir, and consequently the effort for reaching the equilibrium is minimal.

Finally, these relations and considerations allow for the generalization of the Le Châtelier Principle (Nalewajski, 1989, 1990):

PRINCIPLE AIM-HD1* (generalized Le Châtelier): one considers a system (molecule) departed from a stable equilibrium by a chemical movement δx_k of one of its extensive parameters. Then, the system reacts through a feedback modification in its intensive parameters, δP_k, for decreasing the initial displacement and restoring the equilibrium. The system response is maximal when all the other parameters, extensive and intensive, are kept frozen and decreases with their relaxation, partial or total; in analytical terms:

$$\left(\frac{\partial P_k}{\partial x_k}\right)_{x(\neq k)} \geq \left(\frac{\partial P_k}{\partial x_k}\right)_{P_1, x[\neq(1,k)]} \geq \dots \geq \left(\frac{\partial P_k}{\partial x_k}\right)_{P_1, \dots, P_{k-1}, x_{k+1}, \dots, x_m} > 0 , \; k = \overline{1, m}$$

$$(3.340)$$

This way, the chemical reactivity analysis in terms of molecular sensitivities allows a thermodynamic approach of the fundamental problem in Chemistry, i.e., the stability characterization of the various atomic coordination in molecule and of its reactivity in relation with the environment as well; in applied chemistry a reservoir may associate with a catalyst, an electrode, or another molecular fragment, much larger, or by the contact with various active centers on an extended chemical space (e.g., the local coupling, in which some centers may manifest as donors and others as acceptor of electrons, respectively nuclear charges etc.).

3.4.4 NOTE ON CHEMICAL HARDNESS COEFFICIENTS

The hardness may posses the same characteristics as its originator, the electronegativity, having the role of accompanying the reactivity global description from the force perspective so completing the driving reactive influence prescribed by the potential nature of electronegativity.

The present issue here has to be dedicated to necessity of the factors (1/2) included in the definition of electronegativity and hardness in employed expressions so far.

One way is to consider that both electronegativity and hardness are regulated by the same occupation number q:

$$\chi = \tilde{q}(IP + EA) \tag{3.341}$$

$$\eta = \tilde{q}(IP - EA) \tag{3.342}$$

due to the concerting effects that both indices assume, as approaching and establishing bonding, respectively. Now, in order that indeed χ and η to be affordable as parametrical minimal bond description one notes that the occupation number has to achieve the value $\tilde{q} = 0.5$ as each atom within a bond contribute with one electron to its covalence. The non-integer value of q is in complete accordance with the generic quantum nature of the (sharing) electrons in bonding.

The second way is based on the equivalent gauge reactions, in which an acid-base complex is formed:

$$A^+ + {}^{\bullet}_{\bullet}B^- \leftrightarrow A{}^{\bullet}_{\bullet}B \leftrightarrow {}^{\bullet}A^- + {}^{\bullet}B^+ \tag{3.343}$$

Eventually, the left reaction implies that partial charge transfer through Lewis acids and bases occurs, while the right reaction states for the complete charge transfer redox process. In these conditions, the exchange of electrons between the radicalic extremes of *Lewis reaction* arises through the ionic/covalent – acid/base involved structures by means of general charge path:

$$\frac{1}{\widetilde{q}} = \int_{N-1}^{N+1} dq = \int_{N-1}^{N} dq + \int_{N}^{N+1} dq = 1+1 = 2 \qquad (3.344)$$

The integration path records at once the acidic (electron-accepting: $N \leq q \leq N+1$) and basic (electron-donating $N-1 \leq q \leq N$) behavior of the species, making it an inherent part of the description of their chemical reactivity (Putz, 2008b, 2012b). Therefore, the electronegativity and hardness indices should be averaged by the charge factor $\widetilde{q} = 0.5$ along their reactive paths leaving with the working semi-sum and semi-difference of the electron-releasing and electron-attaching energies, IP and EA, respectively (Guo & Whitehead, 1989a; Nesbet, 1997; Orsky & Whitehead, 1987).

Being about modeling reactivity, an elegant way of describing it calls the perturbation of the ground or valence state energy of an isolated system when engaging to an interaction (Gordy, 1946; Iczkowski & Margrave, 1961). This is to write the energy as its expansion respecting to the changed charge ΔN into a specific reaction. At this point, even formerly such series expansion was considered up to the fourth order in ΔN (Hinze & Jaffé, 1962, 1963; Hinze et al., 1963; Zhang, 1982; Tachibana, 1987), we consider that a simple and in principle complete scheme of global reactivity can be achieved on the electronegativity and hardness basis only, due to their character as the chemical potential and force, driving the changing in the total energy of the system.

In this context, the interaction energy $E_{\Delta N}$ of an electronic system that changes the charge ΔN with environment assumes the paradigmatic parabolic analytical form, see also Eq. (3.166):

$$E_{\Delta N} = E_{0/v} + \mu_1 \Delta N + \eta_1 (\Delta N)^2 \qquad (3.345)$$

standing for the total ground (subscript "0") or the valence (subscript "v") perturbed energy $E_{0,v}$ in the course of reaction through the chemical potential μ_1 and force η_1. Worth again noting here the virtual nature of electronegativity and hardness as they are proper to a certain system in the absolute sense but becoming manifested chemical ones since the reaction flows, i.e., when $\Delta N \neq 0$.

Then, while considering the finite-difference approximations to the electronegativity and chemical hardness definitions, in a Koopmans' theorem environment (Koopmans, 1934) they take the working forms (Parr & Yang, 1989; Chermette, 1999; Geerlings et al., 2003; Putz, 2012b):

$$\chi \equiv -\left(\frac{\partial E_N}{\partial N}\right)_{V(r)} \cong -\frac{E_{N+1}-E_{N-1}}{2} = \frac{(E_{N-1}-E_N)+(E_N-E_{N+1})}{2}$$

$$= \frac{IP+EA}{2} \cong -\frac{HOMO+LUMO}{2} \qquad (3.346)$$

recovering (3.98), and respectively

$$\eta = \left(\frac{\partial^2 E_N}{\partial N^2}\right)_{V(r)} \cong E_{N+1} - 2E_N + E_{N-1}$$

$$= IP - EA \cong LUMO - HOMO \qquad (3.347)$$

specializing (3.101), in terms of lowest unoccupied molecular orbital (*LUMO*) and highest occupied molecular orbital (*HOMO*) energies, although similar relations (in terms of ionization potential *IP* and electron affinity *EA*) hold for atomic systems as well, see Figure 3.2.

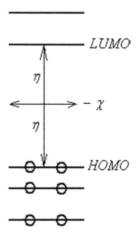

FIGURE 3.2 Orbital energy diagram for a molecule (Pearson, 1987), showing the electronegativity and hardness, on which basis the equalization of electronegativity and hardness principles, EE and HSAB respectively follows.

3.4.5 HSAB PRINCIPLE AND MAXIMUM HARDNESS INDEX

However, EE-energy based relation permits the immediate recovery of the "Hard and Soft Acids and Bases" (HSAB) principle (Pearson, 1985; Parr & Pearson, 1983; Klopman, 1968; Chandrakumar & Pal, 2002a; Chattaraj & Maiti, 2003), an important conceptual principle to treat the molecular binding and reactive processes. Writing the energy variation at transfer under the following forms (Chattaraj & Parr, 1993):

$$\Delta E = \Delta\Omega_A + \Delta\Omega_B \qquad (3.348)$$

$$\Delta\Omega_A = -\frac{\eta_A}{4(\eta_A + \eta_B)^2}(\Delta\chi)^2 \qquad (3.349)$$

$$\Delta\Omega_B = -\frac{\eta_B}{4(\eta_A + \eta_B)^2}(\Delta\chi)^2 \qquad (3.350)$$

$$\Delta\chi = \chi_A - \chi_B \qquad (3.351)$$

the optimal energetic transfer will imply, for instance, the minimization of $\Delta\Omega_A$ respecting η_A, in conditions in which $\Delta\chi$ and η_B are maintained constant:

$$\left(\frac{\partial}{\partial\eta_A}\Delta\Omega_A\right)_{\Delta\chi,\eta_B} = 0 \qquad (3.352)$$

Therefore, the condition to achieve the optimum transfer results to be:

$$\eta_A = \eta_B \qquad (3.353)$$

thus implying the fact that *the species with a high chemical hardness prefer the coordination with species that are high in their chemical hardness, and respectively the species with low softness (the inverse of the chemical hardness) will prefer reactions with species that are low in their softness.* This way, the HSAB – Pearson principle, was established (Pearson, 1985; Parr & Pearson, 1983).

The above relations were inferred for the general case, for which the form of the chemical potential (respectively the electronegativity with changed sign) is not specified.

From now on, the molecular systems are recognized as hard and soft acids and bases (HSAB), in the sense that each molecule can be seen as hard-hard, soft-soft, hard-soft or soft-hard bonding combinations between acids and bases. The associate HSAB principle of chemical reactivity was formulated as well, providing that "hard acids prefer hard bases and soft acids prefer soft bases" (Drago & Kabler, 1972; Pearson, 1972; Chattaraj et al., 1991):

$$h_1 - s_1 + s_2 - h_2 \leftrightarrow h_1 - h_2 + s_2 - s_1 \tag{3.354}$$

Despite the qualitative character (Nalewajski, 1984; Chattaraj & Sengupta, 1996; Komorowski et al., 1996; Ponti, 2000; Chandrakumar & Pal, 2002a-c; Chattaraj & Maiti, 2003; Torrent-Sucarrat et al., 2005; Putz et al., 2004.) of the HSAB principle, an appropriate quantum index to smoothly distinguish between the soft and hard character of acids, bases, and their bonding, would switch HSAB towards a quantitative theory.

In quantum mechanical characterization of bonding, the chemical hardness appears to play the inner stabilization role, behaving as the main quantum influence (or force) (Putz, 2007a). However, its involvement takes place in two correlated stages:

(i)   one stage regards the fulfillment of the HSAB principle, where adducts react according to their reciprocal strengths, that is to optimize themselves in bonding such that approaching Eq. (3.353) prescription;

(ii) the second stage in bonding accounts for the minimization of the quantum fluctuations around the energetic equilibrium of bond that corresponds to the so-called maximization of hardness (MH), from where the associated MH principle (Parr & Chattaraj, 1991; Chattaraj et al., 1995).

Since the two chemical hardness stages and principles drive the quantum chemical bond, the quest for their linked description, both at the phenomenological and analytical levels, appears as a natural endeavor. In this respect, aiming to quantify "in one shoot" the chemical HSAB and MH principles, the so called *maximum hardness index (MHI)* (Putz, 2008b,c) was recently proposed:

$$Y = 1 - \frac{1}{2\eta^2} \tag{3.355}$$

This expression was derived through considering also the chemical softness index as the inverse of the global chemical hardness (Pearson, 1997), quantifying the degree of electronic cloud polarizability (propensity to deformation), in competition with the chemical hardness, under the normalized form:

$$Y = \frac{\eta - S}{\eta} = \frac{\eta}{\eta} - \frac{S}{\eta} \qquad (3.356)$$

It is clear that the MHI definition emphases on how the chemical bond stability is related to the difference between the hard-hard (η/η) and soft-hard (S/η) ratios, transposing in an analytical manner the two equilibrium sides of bonding equilibrium.

In the next, let's comment on some faces of the meaning of the maximum hardness index Y.

First, on the associate symbol, one could remark that the electronegativity index was historically assigned by "X" while the chemical hardness, which in above chemical bonding phenomenology follows the electronegativity equalization principle, should be identified by letter "Y". Other literal argument was offered elsewhere (Putz, 2008b).

Second, the MHI definition incorporates the hard-hard (or soft-soft) and hard-soft (or soft-hard) contest of the chemical bond by means of η/η and S/η terms, respectively. As such, the difference implied measures the degree by which the equilibrium in h-s reaction is broken to favor or not the stabilization of hard-hard and soft-soft bonds.

From the point of view of the values that Y acquire, two main states of bonding may be revealed. One is quantified by the values $Y \in [0,1]$, in which case the equilibrium in h-s reaction is departed to its right side; however, as Y closely tends to 1, as the hard-hard bond is more favorable to the soft-hard state of adducts ($h - h \gg s - h$, or $\eta/\eta \gg S/\eta$).

The other accounts for the values of Y bellow to zero, that indicates the stabilization process is not yet completed, according to the HSAB principle; in other words, the equilibrium in h-s reaction is shifted to its left side as $h - h$ bonding is less favorable, respecting $s - h$ one ($\eta/\eta < S/\eta$). Actually, the negative of Y means a sort of violation of maximum hardness requirement for chemical bond stabilization, that is achieved between 0 and 1 and is completed when $Y \rightarrow 1$.

Nevertheless, another interesting feature of above MHI gets out when rewriting them in the equivalent form (Putz, 2008c):

$$1 = \frac{S}{\eta} + Y \tag{3.357}$$

This form may be easily assimilated to both a conservation principle (of chemical bond from adducts) and a *probabilistic equation*. In fact, one can interpret the chemical bond formation by the competition between the hard-soft and maximum hardness (hard-hard) bonding probabilities. The probability character is crucial to certify the quantum character of the chemical hardness involved in bonding, not only phenomenological but also analytically. Moreover, since the hard-soft term (S/η) basically express the emerging HSAB principle and Y values associates with MH quantum effects, the unified chemical hardness of bonding equation may be formulated as (Putz, 2008c):

$$1 = HSAB + MH \tag{3.358}$$

This equation may constitute the foreground relation for future quantum chemical kinetics of the chemical bond.

Going to the physical-chemical meaning of the maximum hardness, its probabilistic nature will be first justified. Although linear, the definition may be immediately rearranged under exponential form, with the associate limiting points within the [0, 1] realm (Putz, 2008c):

$$Y_e = \exp\left(-\frac{1}{2\eta^2}\right) = \begin{cases} 0 & \dots \eta \to 0 \\ \frac{1}{\sqrt{e}} \cong 0.606531 & \dots \eta = 1 \\ 1 & \dots \eta \to \infty \end{cases} \tag{3.359}$$

Worth noting, the above exponential maximum hardness index resembles the original hyperbolic one on the chemical hardness domain where $\eta > 1$, whereas the unfolded hyperbolic version covers in more detail the chemical bonding regions, as shown by the limits (Putz, 2008c):

$$Y_h = 1 - \frac{1}{2\eta^2} = \begin{cases} -\infty & \dots \eta \to 0 \\ 0 & \dots \eta = \frac{1}{\sqrt{2}} \cong 0.707107 \\ 0.5 & \dots \eta = 1 \\ 1 & \dots \eta \to \infty \end{cases} \tag{3.360}$$

and by the graphical representation in Figure 3.3.

From the Figure 3.3, the almost complete superposition between the exponential and hyperbolic MH indices is evident for values of chemical hardness exceeding unity, but the hyperbolic function covers considerably more cases than the always-positive exponential form, for values below unity. In fact, the hyperbolic MH index allows the quantum characterization of the chemical bonds as appears in the HSAB reaction. This is due to the dual positive and negative values of Y_h, a behavior that provides maximum structural information on the concerned bond.

Remarkably, the $Y_h(h)$ graph offers the chemical bonding partition in three correlated regions.

The first one corresponds to values of the chemical hardness higher than unity, a case in which Y_h holds values over 0.5 probability for the equilibrium in *h-s* reaction flowing to its right side; as higher chemical hardness values of the bond are assessed as hard-hard binding frame is preferred.

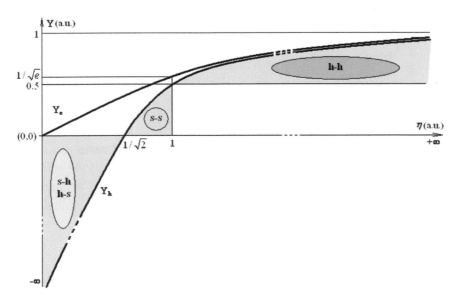

FIGURE 3.3 Comparison between the hyperbolic maximum hardness index (3.360), with marked regions of the hard-hard (h-h), soft-soft (s-s), soft-hard (s-h), and hard-soft (h-s) natures of the chemical bonding in reaction (3.354), and the associate exponential form (3.359). Both scales are set in arbitrary units (a.u.) (Putz, 2008c).

The second region rises within the interval $1/\ddot{O}2<h<1$ in which the U_h probability stands bellow 0.5 values, however, with positive nonzero figures, such that the chemical bond can be still formed as the soft-soft combination in the right side of *h-s* reaction.

The last region provides negative values for Y_h indicating the "hole" or anti-bonding character that can further be associated to an anti-binding entropy S of states with negative statistical probability $\exp(S/k_B)$, k_B being the Boltzmann constant; since the bonding states are restricted through the induced negative potential barrier, the equilibrium in *h-s* reaction is driven to its left side and, consequently, assigned to the soft-hard (or hard-soft) bonding character of concerned molecules. Moreover, due to the negative probabilities assumed, the hard-soft (or soft-hard) bonding situation may link with the back-scattering framework of adducts in a velocity limiting step within the virtual (transition state) channels of *h-s* reaction respecting hard-hard and soft-soft rearrangements (Zewail, 1992).

Resuming, the maximum hardness index helps in prescribing the hard and soft nature of the acid-base chemical bonding against the domains of chemical hardness as follows (Putz, 2008c):

$$\left.\begin{array}{l} hard-hard \\ soft-soft \\ hard-soft\,(soft-hard) \end{array}\right\} \begin{array}{l} ACID-BASE \\ BONDING \end{array} \left\{\begin{array}{l} Y_h \geq 0.5 \quad\quad ...\ \eta \geq 1 \\ 0 \leq Y_h < 0.5 \ ...\ 1/\sqrt{2} \leq \eta < 1 \\ Y_h < 0 \quad\quad\quad ...\ 0 < \eta < 1/\sqrt{2} \end{array}\right.$$

$$(3.361)$$

Additionally, the soft-to-hard classification of acids and bases can in any case be established through identifying on the Y(h) diagrams the soft, borderline, and hard detached "islands" of Y values respecting η range. All these aspects should leave with quantum elucidation of the chemical bond and bonding nature via *maximum hardness index* Y.

3.4.6 MAXIMUM HARDNESS INDEX OF LEWIS ACIDS AND BASES

According with Pearson, classification of acids and bases as hard and soft needs the recourse to the concept of strength although little insight this way is given since further appeal to the experimental enthalpy of reaction is involved (Pearson, 1985, 1997).

Even the consideration of the ionic and covalent bonding contributions as appeared in the well know four-parameter equation of Drago and Wayland (1965), helps not so much in quantifying HSAB principle, albeit a famous scientifically debate was produced on the issues whether or not they may constitute viable quantum measures for the strength and softness, respectively (Drago & Kabler, 1972; Pearson, 1972). In short, the main problem states that: having a given molecule or a bond to which the chemical hardness may be eventually evaluated by some experimentally or computationally based method to establish the hard and soft nature of the bond itself, and even more, to precise, if possible, the hard and soft nature of the bonding components.

There was further conjectured that if a sort of universal soft-to-hard scale of molecular strength is produced, and each time confirmed or never invalidated, then the HSAB principle itself will be consecrated among the chemical universal principles. The fact that such universal classification was still not produced it does not mean that the HSAB principle is not applicable. Contrary, in our opinion, by both epistemological (the postulates' need) and structural energetically arguments, see the last part of introduction, we are conducted to proceed in a reverse way, namely to assume the HSAB as valid principle in any circumstances and to apply it in order to establish the specific hard and soft character of a particular chemical bond or molecular strength.

The present maximum hardness index Y is introduced on the ground of assumed HSAB as true principle resulting in consistent picture of the HSAB-MH chemical bonding. Within this picture the specific chemical hardness ranges were produced to assess the type of hard and soft bond (as global character) and bonding (as interaction character) without any other artifacts.

In this context, the actual hard and soft classification scheme will be in next compared and discussed against the traditional Pearson classification for a limited, however representative, series of molecular Lewis acids and bases as displayed in the Tables 3.1 and 3.2, respectively.

In order to compute the associate chemical hardness for the molecules of Tables 3.1 and 3.2, the previous exposed atoms-in-molecules

TABLE 3.1 Qualitative Pearson Classification of Lewis Acids Tested (Pearson, 1963, 1997)

Soft				Borderline			Hard	
CH_2	$GaCl_3$	BH_3	SO_2	GaH_3	$B(CH_3)_3$	SO_3	$AlCl_3$	BCl_3
A1	*A2*	*A3*	*A4*	*A5*	*A6*	*A7*	*A8*	*A9*

TABLES 3.2 Qualitative Pearson Classification of Lewis Bases Tested (Pearson, 1963, 1997)

Soft				Borderline			Hard	
CH_3SH	CO	C_6H_6	N_2	C_5H_5N	$C_6H_5NH_2$	H_2O	N_2H_4	NH_3
B1	*B2*	*B3*	*B4*	*B5*	*B6*	*B7*	*B8*	*B9*

methodology is here implemented. However, for complete discussions also various theoretical ways of computing atomic hardness are considered. Since, ultimately, the atomic chemical hardness depends on the atomic radii two different set of values will be here implemented, namely those based on the Putz-DFT electronegativity, see Putz et al. (2003) and Volume II of the present five-volume work and on the Ghosh & Biswas (2002) (GB) scales, respectively. Moreover, for the benchmark considerations, the atomic hardness will be implemented in relation with the vertical ionization potential (IP) and the electronic affinity (EA) as well. That is further made in two distinct ways. One is to simply use the finite difference (FD) approximation to the standard differential energetically definition of hardness (Kohn et al., 1996; Parr & Pearson, 1983; Putz, 2008b), here going as half of the prescription (3.347) since accounting for the acid-base averaging as indicated by charge factor (3.344), so regaining (3.102)

$$\eta = \frac{1}{2}\left(\frac{\partial^2 E_N}{\partial N^2}\right)_{V(r)} \cong \frac{E_{N+1} - 2E_N + E_{N-1}}{2} = \frac{IP - EA}{2} \qquad (3.362)$$

when the IP and EA experimental atomic scales are employed (Lackner & Zweig, 1983).

The other approach is based on the Pearson charge conducting sphere model of atomic systems in which, by considering the classical electrostatic expression for the total energy, $q^2 Z_{eff}/(4\pi\varepsilon_0 R)$, the specialization of the above chemical hardness relation will yield another practical expression for the chemical hardness:

$$\eta = \frac{Z_{eff}}{2R}(a.u.)$$ (3.363)

This index will be recognized as softness based chemical hardness (η^S) on the ground of close similitude with the reciprocal relation according which the softness will appear directly proportional with atomic radius. Now, for the Putz-DFT chemical hardness together with its FD and softness based forms above both the considered atomic (Putz-)DFT and GB radii scales are employed to produce the chemical hardness computed for all atoms involved in the molecules of Tables 3.1 and 3.2, in all possible variants, in Table 3.3, while for the molecules themselves the respective hardness and maximum hardness η and Y values are reported in the Tables 3.4–3.8, respectively.

For facilitating the discussion respecting the prototype picture of Figure 3.4, the results of Tables 3.6 and 3.7 are represented in Figures 3.4–3.6 for

TABLE 3.3 Hardness Values for the Atomic Species Used Computed Both By Considering (Putz) Density Functional Electronegativity (DFT) (Putz et al., 2003) and Ghosh & Biswas (2002) (GB) Atomic Radii for the Putz-DFT Chemical Hardness Considered, i.e., Based on Density Functional Electronegativity (DFE)

Atomic Species	η_{DFE}^{Putz}	η_{GB}^{Putz}	η_{DFE}^{S}	η_{GB}^{S}	η_{FD}
H	6.45	6.45	6.45	6.45	6.45
B	53.06	122.72	14.34	10.97	4.06
C	174.22	372.6	22.67	17.1	4.99
N	593.97	958.9	32.52	24.69	7.59
O	3157.6	2521.7	44.44	33.81	6.14
Al	35.51	87.26	14.08	8.8	2.81
S	920.54	520.65	33.87	21.41	4.16
Cl	17059.7	934.87	42.56	26.73	4.68
Ga	86.49	116.12	19.21	10.89	2.81

*The finite-difference (FD) definition based on vertical ionization energy (IP) and electron affinity (EA) scales was added for experimental assessment (Lackner & Zweig, 1983). For comparison, the softness based chemical hardness values based on sphere-charged model of Pearson was also employed (Pearson, 1997). In all cases the atomic values were computed upon hydrogen calibration to its experimental 6.45 eV value. All values are in electron-volts (Putz, 2008c).

TABLE 3.4 Hardness Values for the Molecular Lewis Acids Tested in this Work Computed Upon Atomic Values of Table 3.3 in Molecular Chemical Hardness Recursive Approach

Lewis Acids	η_{DFE}^{Putz}	η_{GB}^{Putz}	η_{DFE}^{S}	η_{GB}^{S}	η_{FD}
BCl_3	210.28	352.19	28.53	19.67	4.51
$AlCl_3$	141.16	272.68	28.27	17.71	4.01
SO_3	1964.2	1286.03	41.22	29.53	5.49
$B(CH_3)_3$	9.08	9.21	8.14	7.82	5.8
GaH_3	8.39	8.44	7.73	7.18	4.87
SO_2	1744.5	1105.5	40.25	28.34	5.3
BH_3	8.27	8.45	7.48	7.19	5.62
$GaCl_3$	340.78	338.39	32.64	19.6	4.01
CH_2	9.5	9.59	8.47	8.14	5.88

*All values are in electron-volts (Putz, 2008c).

TABLE 3.5 Hardness Values for the Molecular Lewis Bases Tested in this Work Computed Upon Atomic Values of Table 3.3 in Molecular Chemical Hardness Formula

Lewis Bases	η_{DFE}^{Putz}	η_{GB}^{Putz}	η_{DFE}^{S}	η_{GB}^{S}	η_{FD}
NH_3	8.57	8.58	8.07	7.91	6.7
N_2H_4	9.62	9.64	8.8	8.56	6.79
H_2O	9.67	9.66	9.02	8.83	6.34
$C_6H_5NH_2$	12.48	12.7	10.14	9.48	5.79
C_5H_5N	13.65	13.93	10.72	9.93	5.76
N_2	593.97	958.9	32.52	24.69	7.59
C_6H_6	12.44	12.68	10.04	9.37	5.63
CO	330.22	649.27	30.02	22.7	5.51
CH_3SH	9.57	9.6	8.65	8.27	5.66

*All values are in electron-volts (Putz, 2008c).

the Lewis acids and bases of Tables 3.1 and 3.2, respectively. In Figure 3.4 only the finite-difference based results were depicted. They are grounded on the atomic chemical hardness with the experimental atomic IP and EA (Lackner & Zweig, 1983). Despite this "experimentally" assumed picture the predicted chemical hardness ordering looks like (Putz, 2008c):

TABLE 3.6   Values of the Maximum Hardness Index for the Lewis Acids Tested in this Work Computed Upon Molecular Chemical Hardness of Table 3.4 in Maximum Hardness Definition

Lewis Acids	Y_{DFE}^{Putz}	Y_{GB}^{Putz}	Y_{DFE}^{S}	Y_{GB}^{S}	Y_{FD}
BCl_3	.999989	.999996	.999386	.998707	.975395
$AlCl_3$	.999975	.999993	.999374	.998406	.968944
SO_3	1.	1.	.999706	.999427	.983393
$B(CH_3)_3$	.993938	.994105	.99245	.991828	.985117
GaH_3	.992899	.992987	.991642	.990307	.978937
SO_2	1.	1.	.999691	.999377	.982195
BH_3	.992681	.993001	.99106	.99033	.984184
$GaCl_3$	.999996	.999996	.999531	.998699	.968944
CH_2	.994459	.994566	.993031	.992454	.985523

*All values are in arbitrary units (Putz, 2008c).

TABLE 3.7   Values of the Maximum Hardness Index for the Lewis Bases Tested in this Work Computed Upon Molecular Chemical Hardness of Table 3.5 in Maximum Hardness Definition

Lewis Bases	Y_{DFE}^{Putz}	Y_{GB}^{Putz}	Y_{DFE}^{S}	Y_{GB}^{S}	Y_{FD}
NH_3	.993191	.993209	.992316	.992011	.988867
N_2H_4	.9946	.994622	.993546	.993172	.989155
H_2O	.994648	.994645	.993855	.993591	.987574
$C_6H_5NH_2$	.996792	.9969	.995137	.994438	.985067
C_5H_5N	.997318	.997423	.995646	.994926	.984942
N_2	.999999	.999999	.999527	.99918	.991321
C_6H_6	.996769	.99689	.995042	.994301	.984208
CO	.999995	.999999	.999445	.999031	.983505
CH_3SH	.99454	.994579	.993315	.992693	.984367

*All values are in arbitrary units (Putz, 2008c).

$$\text{FD: } (A8, \mathbf{A2}) < (A9, \mathbf{A5}) < (A4, \mathbf{A7}, A3, A6, A1) \qquad (3.364)$$

$$\text{FD: } (\mathbf{B2}, \mathbf{B3}, \mathbf{B1}, B5, B6) < (B7, B9, B8) < B4 \qquad (3.365)$$

providing the *soft<borderline<hard* classifications of acids and bases of Tables 3.1 and 3.2, respectively.

TABLE 3.8 Numerical Parameters for the Compact Finite Second (2C)-, Fourth (4C)- and Sixth (6C)-Order Central Differences; Standard Padé (SP) Schemes; Sixth (6T)- and Eight (8T)-Order Tridiagonal Schemes; Eighth (8P)- and Tenth (10P)-Order Pentadiagonal Schemes up to Spectral-Like Resolution (SLR) Schemes for the Electronegativity and Chemical Hardness of Eqs. (3.375) and (3.376) (Rubin & Khosla, 1977; Putz et al., 2004; Putz, 2011a,d; 2012b; Putz et al., 2013)

Scheme	Electronegativity					Chemical Hardness				
	a1	b1	c1	α1	β1	a2	b2	c2	α2	β2
2C	1	0	0	0	0	1	0	0	0	0
4C	$\frac{4}{3}$	$-\frac{1}{3}$	0	0	0	$\frac{4}{3}$	$-\frac{1}{3}$	0	0	0
6C	$\frac{3}{2}$	$-\frac{3}{5}$	$\frac{1}{10}$	0	0	$\frac{12}{11}$	$\frac{3}{11}$	0	$\frac{2}{11}$	0
SP	$\frac{5}{3}$	$\frac{1}{3}$	0	$\frac{1}{2}$	0	$\frac{6}{5}$	0	0	$\frac{1}{10}$	0
6T	$\frac{14}{9}$	$\frac{1}{9}$	0	$\frac{1}{3}$	0	$\frac{3}{2}$	$-\frac{3}{5}$	$\frac{1}{5}$	0	0
8T	$\frac{19}{12}$	$\frac{1}{6}$	0	$\frac{3}{8}$	0	$\frac{147}{152}$	$\frac{51}{95}$	$-\frac{23}{760}$	$\frac{9}{38}$	0
8P	$\frac{40}{27}$	$\frac{25}{54}$	0	$\frac{4}{9}$	$\frac{1}{36}$	$\frac{320}{393}$	$\frac{310}{393}$	0	$\frac{344}{1179}$	$\frac{23}{2358}$
10P	$\frac{17}{12}$	$\frac{101}{150}$	$\frac{1}{100}$	$\frac{1}{2}$	$\frac{1}{20}$	$\frac{1065}{1798}$	$\frac{1038}{899}$	$\frac{79}{1798}$	$\frac{334}{899}$	$\frac{43}{1798}$
SLR	1.303	0.994	0.038	0.577	0.09	0.216	1.723	0.177	0.502	0.056

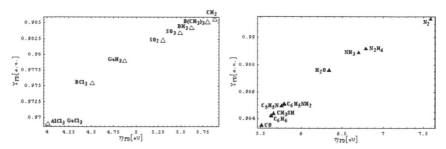

FIGURE 3.4 Graphical correlations between the values of the molecular chemical hardness η and the maximum hardness index Y by employing the Tables 3.4–3.7 for the Lewis acids and bases of Tables 3.1 and 3.2, in the case of experimental finite-difference (FD) based definition of atomic chemical hardnesses of Table 3.3, in left and right pictures, respectively (Putz, 2008c).

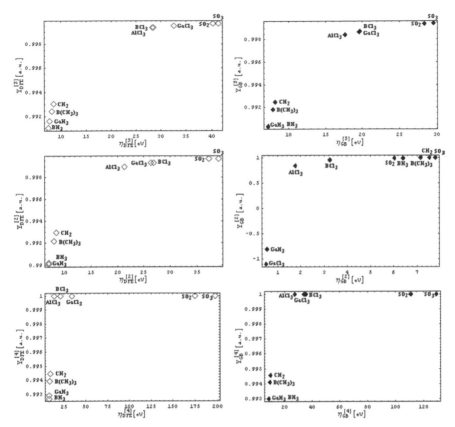

FIGURE 3.5 Graphical correlation between the values of the molecular chemical hardness η and the maximum hardness index Y for the Lewis acids of Table 3.1 employing the softness-, the second- and the fourth- order density functional electronegativity -DFE and Ghosh-Biswas-GB based atomic radii values of chemical hardness of Tables 3.4–3.7, from the top to bottom, in left and right sides on the draws, respectively (Putz, 2008c).

The FD chain relationships confirm that the Pearson classification is only partly fulfilled: by marking in bold the cases when the actual analysis fits with Pearson one, clearly appears that in the case of acids, in each Pearson classes (soft, borderline, and hard) only one acid from the computed set is recovered; for bases only those classified as soft are here recovered as such although in an enlarged set. Therefore, the percentage of actual/Pearson approaches goes to 33% for both acids and bases considered apart of some internal ordering relativity.

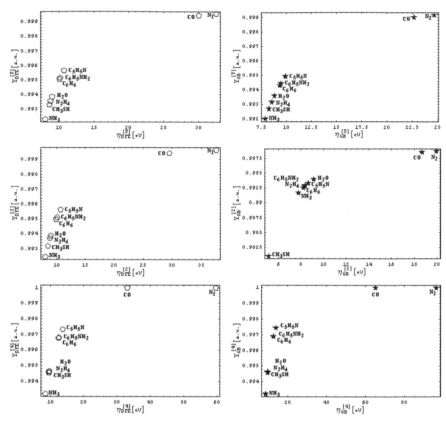

FIGURE 3.6 Graphical correlation between the values of the molecular chemical hardness η and the maximum hardness index Y for the Lewis bases of Table 3.2 employing the softness-, the second and the fourth order density functional electronegativity (DFE) and Ghosh-Biswas (GB) based atomic radii values of chemical hardness of Tables 3.4–3.7, from the top to bottom, in left and right sides on the draws, respectively (Putz, 2008c).

On the other hand, the actual endeavor gives insight also into the type of chemical bonding: in accordance with the acid-base bonding characterization all considered FD computed molecules are of hard-hard acid-base interaction type, although with different resulting maximum hardness values.

In Figure 3.5 all other ways for chemical hardness computation are collected for the acids of Table 3.1.

As before, the bond and bonding issues are addressed: the soft-to-hard classification for chemical bond nature and the soft-hard types of chemical bonding.

Concerning the first issue, several ordering combinations are obtained (Putz, 2008c):

$$DFE^S: (\mathbf{A3}, A5, A6, \mathbf{A1}) < (A8, A9) < (A2, A4, \mathbf{A7}) \qquad (3.366)$$

$$GB^S: (A5, \mathbf{A3}, A6, \mathbf{A1}) < (A8, A2, A9) < (A4, \mathbf{A7}) \qquad (3.367)$$

$$DFE^{Putz}: (\mathbf{A3}, A5, A6, \mathbf{A1}) < (A8, A9, A2) < (A4, \mathbf{A7}) \qquad (3.368)$$

$$GB^{Putz}: (A5, \mathbf{A3}, A6) < (A1, A8, A2) < (\mathbf{A9}, A4, \mathbf{A7}) \qquad (3.369)$$

The results are complex and deserve a close inspection. At a glance, there is again observed the case of drastically discrepancy of the actual orderings respecting the Pearson scheme of Table 3.1. However, as before, the fit with Pearson classification do not exceeds 33% of cases – and this by assuming the comparison between classes while neglecting the exact relative orderings. Moreover, the actual maximum hardness approach offers the perspective of hard and soft classification of Lewis acids, beyond the simple assessment of global chemical hardness values. That is the case, for instance, of the DFE^{Putz} and GB^{Putz} frames of computations when, at almost equal values among the hardness values of different species, the soft and borderline classification was decided by the graphical splitting in Y groups providing that as bigger it is as more stable bond is associated.
Actually, the graph $Y(\eta)$ furnishes the global soft-to-hard classification respecting the displacements of the Y values within "islands" on its (interpolated or virtual) curve of the type of that represented in Figure 3.3. Nevertheless, the inner quantum chemical bonding is described as hard-hard type in all cases.

The last remark is nothing else than the confirmation of the fact that a more complex way of atomic radii involvement in chemical hardness definition, in the sense of atomic potential and of chemical action influences, may lead with better results. Following this line we may conclude the analysis of Lewis acids of Table 3.1 with the recommendation of the grouping DFE^{Putz} as the best soft-to-hard ordering; this is also the most complex computational approach with the most higher frequency of ordering appearance among the compared models. Moreover, the hard-hard acid-base interaction stands as the dominant mechanism of chemical bonding.

In the same manner, the investigation of Figure 3.6 provides the chemical hardness orderings of Lewis bases of Table 3.2 along the computational scheme implemented (Putz, 2008c):

$$DF^S: (B9, \mathbf{B1}, B8, B7){<}(B3, \mathbf{B6}, \mathbf{B5}){<}(B2, B4) \qquad (3.370)$$

$$GB^S: (B9, \mathbf{B1}, B8, B7){<}(B3, \mathbf{B6}, \mathbf{B5}){<}(B2, B4) \qquad (3.371)$$

$$DFE^{Putz}: (B9, \mathbf{B1}, B8, B7){<}(B3, \mathbf{B6}, \mathbf{B5}){<}(B2, B4) \qquad (3.372)$$

$$GB^{Putz}: (B9, \mathbf{B1}, B8, B7){<}(B3, \mathbf{B6}, \mathbf{B5}){<}(B2, B4) \qquad (3.373)$$

Although the ordering chemical hardness percentage respecting Pearson classification of Table 3.2 records no sensible modification, few notable differences have now appeared: there is noted quite an inversion between the Pearson recommended hard bases B7-B9 which are now classified as soft, while the previous soft and borderline bases B2 and B4 are here situated as hard in almost all ordering schemes above. This strongly suggests that the Pearson classification of the chemical compounds against their enthalpy of formation is rather relative to chemical reaction considered and not to inner structural atoms-in-molecule information.

The second observation is that unlike the acids previous cases the actual bases ordering looks quite similar for all employed finite difference, Ghosh-Biswas and density functional electronegativity recipes. Therefore, the present suggested soft-to-hard Lewis bases classification is that recommended by DFE^{Putz} hierarchy – rooting on the most complex conceptual-computational containing algorithm.

From the bonding perspective all discussed bases originate in hard-hard acid-base interactions in accordance with the introduced maximum hardness criteria.

Future studies on different molecular, atomic and ionic compounds are in future envisaged for further clarifying the maximum hardness index role in elucidation of quantum nature of the chemical bond bonding as driven by inner electronic structures.

3.4.7 APPLICATION LIMITS OF HSAB PRINCIPLE

Here, the exchange reactions in gas-phase between two acid-base com-plexes have been considered and the reaction energy has assumed as a measure of the stabilization of hard-hard and soft-soft adducts in com-parison to the corresponding hard-soft and soft-hard counterparts. The used general scheme for the acid-base reactions during which a base (X), bonded to the hard acid (H^+), removes a hard base (OH^-) from a soft acid (HO^+) is:

$$H\text{-}X^+ + HO\text{-}OH \leftrightarrow HO\text{-}X^+ + H\text{-}OH \tag{3.374}$$

and to determine which reaction is more favorable the corresponding reac-tion energies are compared. Indeed, the more negative the reaction energy the softer the base X would be. For computational implementation, one considers the general working formulae for electronegativity and for the chemical hardness within the "spectral" like resolution, see Volume I/ Section 4.5.4.4 of the present five-volume set, with the respective forms (Putz et al., 2004; Putz, 2011a,d; 2012b; Putz et al., 2013)

$$
\begin{aligned}
\chi_{CFD} = &-\left[a_1\left(1-\alpha_1\right) + \frac{1}{2}b_1 + \frac{1}{3}c_1 \right]\frac{\varepsilon_{HOMO(1)} + \varepsilon_{LUMO(1)}}{2} \\
&-\left[b_1 + \frac{2}{3}c_1 - 2a_1\left(\alpha_1 + \beta_1\right) \right]\frac{\varepsilon_{HOMO(2)} + \varepsilon_{LUMO(2)}}{4} \\
&-\left(c_1 - 3a_1\beta_1\right)\frac{\varepsilon_{HOMO(3)} + \varepsilon_{LUMO(3)}}{6}
\end{aligned}
\tag{3.375}
$$

along with the respective chemical hardness

$$
\begin{aligned}
\eta_{CFD} = &\left[a_2\left(1-\alpha_2 + 2\beta_2\right) + \frac{1}{4}b_2 + \frac{1}{9}c_2 \right]\frac{\varepsilon_{LUMO(1)} - \varepsilon_{HOMO(1)}}{2} \\
&+\left[\frac{1}{2}b_2 + \frac{2}{9}c_2 + 2a_2\left(\beta_2 - \alpha_2\right) \right]\frac{\varepsilon_{LUMO(2)} - \varepsilon_{HOMO(2)}}{4} \\
&+\left[\frac{1}{3}c_2 - 3a_2\beta_2 \right]\frac{\varepsilon_{LUMO(3)} - \varepsilon_{HOMO(3)}}{6}
\end{aligned}
\tag{3.376}
$$

where the involved parameters discriminate between various schemes of computations and the spectral-like resolution (SLR), see Table 3.8 (Rubin & Khosla, 1977; Putz et al., 2004; Putz, 2011a,d; 2012b; Putz et al., 2013).

Quantum chemical computations have been performed at *ab initio* Hartree-Fock and local-DF VWN (Vosko et al., 1980), gradient corrected-DF BP86 (Becke, 1988; Perdew, 1986) and hybrid-DF B3LYP (Becke, 1988; Perdew, 1986; Lee et al., 1988) levels of theory together with the standard 6–31+G* basis set. Geometries have been fully optimized to evaluate global hardness values for all the bases X that are: NH_3, CH_3O^-, H_2O, CN^-, CH_3S^- and CH_3SH. The same levels of theory have been used to carry out geometry optimization for all adducts involved into the considered acid-base reactions to evaluate reaction energies and zero point energy (ZPE) corrections have been included. All the computations have been performed using the Gaussian (1998) program. Global hardness values have been calculated using both general expression and appropriate coefficients for each approximation level.

With the aim to test the above expanded second derivative compact difference schemes for chemical hardness calculation, in Tables 3.9–3.14 are reported the hardness values for the bases (X= NH_3, CH_3O^-, H_2O, CN^-,

TABLE 3.9 The Hartree-Fock, B3LYP, SVWN, and BP86/6–31G Chemical Hardness Computations for the $[CH_3O]^-$ Molecule by Using Orbital Energies and Vertical Total Energies for Ionization Potential and Electron Affinity Calculation, Respectively; IRHT=internally resolved hardness tensor; H/L=HOMO/LUMO (Putz et al., 2004)

Schemes	HF		B3LYP		SVWN		BP86	
η (eV)	*orbital*	*vertical*	*orbital*	*vertical*	*orbital*	*vertical*	*orbital*	*vertical*
2C	7.21	5.67	3.21	5.97	2.31	6.09	2.31	5.95
4C	8.31	5.86	3.65	6.26	2.62	6.43	2.61	6.26
6C	5.83	3.53	2.52	3.86	1.79	4.01	1.78	3.87
SP	6.78	4.36	2.95	4.72	2.1	4.88	2.08	4.73
6T	9.05	6.45	4.	6.89	2.88	7.09	2.86	6.89
8T	5.4	3.4	2.34	3.69	1.67	3.82	1.66	3.7
8P	5.34	3.81	2.34	4.05	1.68	4.16	1.67	4.05
10P	5.65	4.99	2.54	5.17	1.85	5.24	1.84	5.14
SLR	7.27	8.69	3.43	8.75	2.57	8.73	2.55	8.66
Other	5.29 from IRHT method; 3.18 from H/L method; 6 from EA/IP							

TABLE 3.10   The Hartree-Fock, B3LYP, SVWN, and BP86/6–31G Electronegativity and Chemical Hardness Computations for the H_2O Molecule By Using Orbital Energies and Vertical Total Energies for Ionization Potential and Electron Affinity Calculation, Respectively; IRHT=internally resolved hardness tensor; H/L=HOMO/LUMO (Putz et al., 2004)

Schemes	HF		B3LYP		SVWN		BP86	
η (eV)	orbital	vertical	orbital	vertical	orbital	vertical	orbital	vertical
2C	9.59	8.16	4.66	8.33	3.47	8.43	3.53	8.26
4C	11.03	8.54	5.27	8.73	3.89	8.87	3.95	8.66
6C	7.71	5.25	3.61	5.38	2.63	5.49	2.66	5.33
SP	8.97	6.43	4.23	6.58	3.1	6.7	3.14	6.52
6T	12.24	9.41	5.95	9.61	4.45	9.76	4.5	9.54
8T	7.12	5.02	3.33	5.15	2.42	5.24	2.46	5.1
8P	7.	5.53	3.32	5.65	2.43	5.74	2.47	5.6
10P	7.44	7.09	3.65	7.23	2.73	7.29	2.78	7.17
SLR	9.87	12.04	5.22	12.22	4.08	12.25	4.16	12.15
Other	9.5 from Pearson; 8.49 from IRHT method; 4.01 from H/L method; 8.35 from EA/IP							

TABLE 3.11   The Hartree-Fock, B3LYP, SVWN, and BP86/6–31G Electronegativity and Chemical Hardness Computations for the NH_3 Molecule by Using Orbital Energies and Vertical Total Energies for Ionization Potential and Electron Affinity Calculation, Respectively; IRHT=internally resolved hardness tensor; H/L=HOMO/LUMO (Putz et al., 2004)

Schemes	HF		B3LYP		SVWN		BP86	
η (eV)	orbital	vertical	orbital	vertical	orbital	vertical	orbital	vertical
2C	8.44	7.29	4.15	7.36	3.16	7.4	3.18	7.26
4C	9.45	7.56	4.56	7.67	3.31	7.75	3.33	7.56
6C	6.39	4.58	2.85	4.69	2.03	4.76	2.05	4.61
SP	7.53	5.64	3.43	5.75	2.49	5.83	2.51	5.66
6T	10.18	8.31	4.77	8.46	3.52	8.54	3.55	8.33
8T	5.99	4.4	2.72	4.49	1.96	4.55	1.98	4.42
8P	6.13	4.91	2.92	4.97	2.19	5.01	2.2	4.9
10P	6.89	6.4	3.56	6.42	2.8	6.43	2.81	6.34
SLR	9.68	11.06	5.58	10.99	4.64	10.94	4.64	10.9
Other	8.2 from Pearson; 7.6 from IRHT method; 3.72 from H/L method; 7.52 from EA/IP							

TABLE 3.12 The Hartree-Fock, B3LYP, SVWN, and BP86/6–31G Electronegativity and Chemical Hardness Computations for the [CN]⁻ Molecule by Using Orbital Energies and Vertical Total Energies for Ionization Potential and Electron Affinity Calculation, Respectively; IRHT=internally resolved hardness tensor; H/L=HOMO/LUMO (Putz et al., 2004)

Schemes	HF		B3LYP		SVWN		BP86	
η (eV)	orbital	vertical	orbital	vertical	orbital	vertical	orbital	vertical
2C	9.01	7.65	4.41	7.88	3.18	7.83	3.21	7.72
4C	10.5	8.12	5.09	8.37	3.64	8.29	3.68	8.19
6C	7.46	5.1	3.58	5.26	2.53	5.19	2.57	5.12
SP	8.63	6.18	4.16	6.38	2.95	6.31	2.99	6.22
6T	11.46	8.9	5.59	9.17	4.02	9.08	4.07	8.96
8T	6.89	4.85	3.31	5.01	2.34	4.95	2.38	4.88
8P	6.72	5.26	3.25	5.42	2.32	5.38	2.35	5.31
10P	6.94	6.57	3.42	6.76	2.49	6.74	2.5	6.65
SLR	8.56	10.78	4.4	11.07	3.33	11.1	3.31	10.96
Other	4.91 from IRHT method; 2.64 from H/L method; 6.73 from EA/IP							

TABLE 3.13 The Hartree-Fock, B3LYP, SVWN, and BP86/6–31G Electronegativity and Chemical Hardness Computations for the [CH₃S]⁻ Molecule by Using Orbital Energies and Vertical Total Energies for Ionization Potential and Electron Affinity Calculation, Respectively; IRHT=internally resolved hardness tensor; H/L=HOMO/LUMO (Putz et al., 2004)

Schemes	HF		B3LYP		SVWN		BP86	
η (eV)	orbital	vertical	orbital	vertical	orbital	vertical	orbital	vertical
2C	6.5	5.68	3.03	5.54	2.22	5.54	2.2	5.43
4C	7.51	6.04	3.47	5.88	2.54	5.9	2.52	5.76
6C	5.28	3.81	2.42	3.69	1.76	3.71	1.74	3.61
SP	6.13	4.61	2.82	4.48	2.06	4.5	2.04	4.38
6T	8.19	6.65	3.81	6.47	2.79	6.49	2.77	6.34
8T	4.89	3.62	2.24	3.51	1.63	3.53	1.62	3.44
8P	4.82	3.9	2.22	3.8	1.62	3.81	1.61	3.72
10P	5.07	4.85	2.37	4.74	1.75	4.74	1.73	4.65
SLR	6.48	7.94	3.13	7.81	2.35	7.77	2.34	7.69
Other	4.57 from IRHT method; 2.04 from H/L method; 4.91 from EA/IP							

TABLE 3.14   The Hartree-Fock, B3LYP, SVWN, and BP86/6–31G Electronegativity and Chemical Hardness Computations for the **CH₃SH** Molecule by Using Orbital Energies and Vertical Total Energies for Ionization Potential and Electron Affinity Calculation, Respectively; IRHT=internally resolved hardness tensor; H/L=HOMO/ LUMO (Putz et al., 2004)

Schemes	HF		B3LYP		SVWN		BP86	
η (eV)	*orbital*	*vertical*	*orbital*	*vertical*	*orbital*	*vertical*	*orbital*	*vertical*
2C	7.33	6.39	3.63	6.31	2.72	6.29	2.7	6.19
4C	8.42	6.77	4.11	6.69	3.05	6.69	3.02	6.56
6C	5.89	4.23	2.82	4.2	2.06	4.21	2.04	4.11
SP	6.85	5.14	3.3	5.09	2.43	5.11	2.41	4.99
6T	9.16	7.38	4.48	7.31	3.33	7.31	3.3	7.17
8T	5.46	4.04	2.62	4.	1.93	4.01	1.91	3.92
8P	5.41	4.4	2.64	4.35	1.96	4.31	1.95	4.26
10P	5.76	5.52	2.9	5.44	2.2	5.39	2.18	5.33
SLR	7.49	9.11	3.98	8.93	3.12	8.79	3.1	8.75
Other	4.70 from IRHT method; 2.66 from H/L method; 5.94 from EA/IP							

CH₃S⁻ and CH₃SH) involved in the acid-base reaction scheme (3.374). For each level of theory are reported two columns that are labeled *orbital* and *vertical* whether orbital energies or calculated vertical IP and EA are used in working formulas.

Some general conclusions can be drawn from the reported results. For each scheme o*rbital* hardness values are about two times smaller than the *vertical* ones in all cases, except when the HF/6–31+G* computational protocol is adopted. In this latter case the differences are less pronounced and, on the contrary, the *vertical* values are smaller than the *orbital* ones. Results obtained at local VWN and non-local BP86 levels are coincident and in some cases differences are negligible also with respect to the B3LYP values. The HF values are completely different when the *orbital* operational definition of hardness is used, whereas *vertical* values are very close to those obtained at all the other levels of theory.

Concerning the trend of values along each column, as the second derivative computational scheme is varied no regular increase or decreasing can be observed. It is noteworthy that when the *orbital* operational definition is used the first, 2C, and the last, SLR, values are close to each other, whereas a difference of at least two eV exists between the same *vertical* values.

The bases have been chosen according to their general classification as hard (CH_3O^-, NH_3, H_2O) and soft (CN^-, CH_3S^-, CH_3SH) and the reported values of reaction free energies correspond to their expected behavior. Since the more negative the DG the softer the base X, i.e., the more able to remove the hard base OH^- from the soft acid HO^+, in Table 3.15 bases are arranged from the softest, CN^-, to the hardest, H_2O, ones.

Consequently, the calculated hardness values would satisfy this requirement and according to ΔG values the hardness order would be (Putz et al., 2004):

$$CN^- < CH_3S^- < CH_3SH < CH_3O^- < NH_3 < H_2O.$$

In Figure 3.7 ΔG values are reported *versus* hardness computed using all the other schemes. At a first glance appears that none of the applied scheme is able to provide hardness values in agreement with the trend of reaction energies. The classification of the chosen bases as hard, for CH_3O^-, NH_3, H_2O, and soft, for CN^-, CH_3S^-, CH_3SH, does not correspond to the calculated hardness values, whatever is the used level of theory and the computational scheme. It is noteworthy that the hardness for the hardest base, H_2O, is comparable to that of the softest one, CN^-.

In Table 3.15, the free reaction energies (ΔG) are given for each of the probe base used in the exchange paradigmatic reaction considered. The DF values of free reaction energies calculated employing different functional are similar among them and only B3YP ones have been reported. Although values of ΔG calculated at HF level differ with respect to the DF ones the trend remains unaltered.

All the effects due to (Putz et al., 2004):

1. the different values of orbital energies calculated at HF and DF levels,
2. correlation,

TABLE 3.15 Qualitative Classification of Lewis Bases Tested in this Work According With Free Reaction Energies ΔG_{HF}(kcal/mol) Computed in B3LYP Mode (Putz et al., 2004)

Soft			Hard		
CN^- <	CH_3S^- <	CH_3SH <	CH_3O^- <	NH_3 <	H_2O
-45.65	-35.93	-30.62	-6.51	1.6	17.01

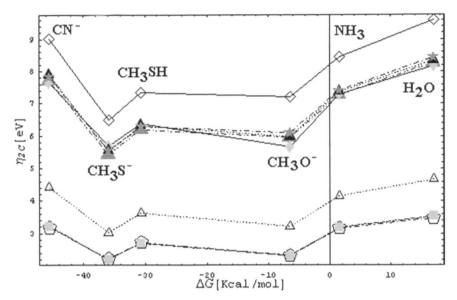

FIGURE 3.7 Second central-2C (ordinary, or simple first, based on Eq. (3.362)) order in compact finite difference representation of the chemical hardness for the molecules presented in Tables 3.9–3.14 as a function of the free energies of their reactions with water molecules. The adopted notations for the computational methods are: HF orbital: $\diamondsuit$-, B3LYP orbital: $\cdot\triangle\cdot$, VWN orbital: $\heartsuit$-, BP86 orbital: $-\blacksquare-$, BP86 vertical: $\cdot\blacktriangle\cdot$, HF vertical: $-\blacklozenge-$, B3LYP vertical: $\cdot\blacktriangle\cdot$, and VWN vertical: $\cdot\bigstar-$, respectively (Putz et al., 2004).

3. improved computational schemes for the second order derivatives,

4. use of *orbital* and *vertical* energies seem do not influence at all the final result. In spite of the differences in hardness values due the adoption of computational schemes which improve the approximate calculation of second order derivatives, the final trend is practically the same The computational scheme corresponding to lowest level of approximation as well as the highest one give hardness values which do not support the "preference" for the formation of hard-hard and soft-soft adducts.

Finally, the results of this work have been used to check the parabolic behavior of the energy as a function of the electron number, $E(N)$, which is the limiting assumption for the use of the operational definition. Figure 3.8 reported the plots of the energy as a function of N for each probe base for all the employed levels of theory in the range from

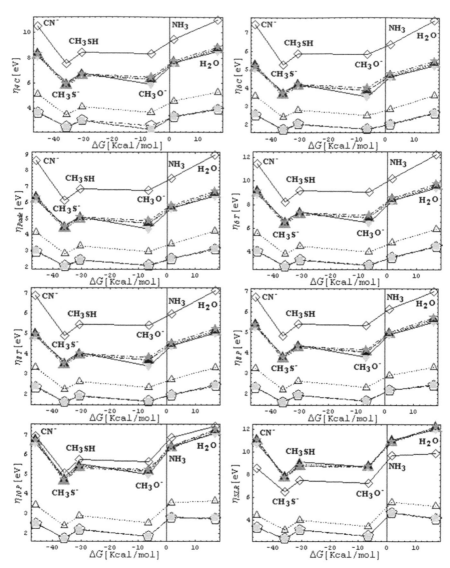

FIGURE 3.8 Superior orders (4C-, 6C-, SP-, 6T, 8T, 8P, 10P up to spectral-like resolution- SLR) in the compact finite difference representations of the chemical hardness for the molecules presented in Table 3.9 to Table 3.14 as a function of the free energies of their reactions with water molecules. The adopted notations for the computational methods are: HF orbital: ◇-, B3LYP orbital: ·△·, VWN orbital: ·⟡·, BP86 orbital: -■-, BP86 vertical: ·▲·, HF vertical: -◆-, B3LYP vertical: ▲, and VWN vertical: ·★-, respectively (Putz et al., 2004).

N–3 to N+3. It clearly appears that the energy behavior is parabolic and hardness correctly represents the curvature of the plot. The use of rough approximations for the hardness computation appears to be justified and the origin of the discrepancies, more and more numerous, between HSAB predictions and experimental and theoretical results has to be found elsewhere.

3.5 CHEMICAL REACTIVITY BY FRONTIER FUKUI INDEX PRINCIPLES

3.5.1 CHEMICAL HARDNESS PRINCIPLES RELOADED

The molecules with the electronic layers occupied such as those in Figure 3.9 left allow the introduction of very useful chemical reactivity (CR) definitions in analyzing the reactions between acids and bases; the forthcoming discussion follows the references (Purser, 1988; Tykodi, 1988; Wu & Sheng, 1994; van Hooydonk, 1986).

DEFINITION CR1: (a) A molecule (ion) is defined as hard if it has an energetic *broad* energetic gap between the orbitals HOMO and LUMO. (b) One defines a molecule (ion) as being soft if it has a *narrow* energetic gap between the orbitals HOMO and LUMO.

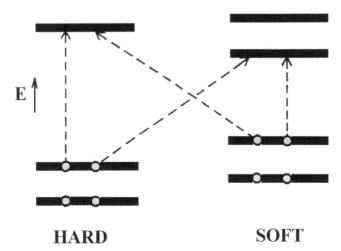

FIGURE 3.9 Delocalization and polarization in the chemical reaction process of a hard molecule with a soft one, having the same electronegativity.

This presentation allows the definition of the acids and bases as weak and strong, and also of the principle that governs their interactions.

DEFINITION CR2: (a) A *weak* base: is a donor system with high polarizability, low electronegativity, slightly oxidable, and associated to the orbitals with low energy. (b) A *strong* base is a donor system with weak polarizability, high electronegativity, hardly to reduce, and associated to the orbitals unoccupied with high energy and therefore inaccessible. (c) A *weak* acid is a donor system, polarizable, of low positive charge, with large dimension, and with some valence electrons slightly excitable. (d) A *strong* acid is a donor system, un-polarizable, with high positive charge, small dimension, and without valence electrons slightly excitable.

Therefore, these defining properties are reasonable in the light of the so far presented concepts of chemical reactivity. The weak acids and bases have properties, which ensure them that the HOMO levels are relatively high in energy and the LUMO levels are relatively at low energies. The strong acids and bases have opposite properties.

Regarding the polarization, the quantum theory proves that in the presence of an external electric field, the excited states are mixed with the fundamental one, so that the resulted energy to be as low as possible. For the weak acids and bases, i.e., characterized by a narrow energetic band $LUMO–HOMO = I - A$, the polarizability is high because all the excited states in spectrum are enlarged near the fundamental state. Therefore, also a principle regarding the reactivity (as an interaction) between the acids and bases can be expressed:

PRINCIPLE CR1 (of "Hard and Soft Acids and Bases" – HSAB): Hard acids prefer coordination with the hard bases, and the soft acids prefer coordination with the soft bases.

In relation with the LUMO – HOMO band size the two natural tendencies can be provided, under the following principles (Pearson, 1986, 1987, 1988a-b, 1989; Parr & Pearson, 1983):

PRINCIPLE CR2 (of second order Jahn–Teller effect): An energetic narrow LUMO – HOMO band predisposes in general to an increase in chemical reactivity.

PRINCIPLE CR3 (of Maximum Chemical Harness): It is a rule of Nature that the molecular systems tend to combine them as it is harder to further destroy them.

The principle may be seen in terms of stability at the formation, so that a new molecular complex created with a large LUMO – HOMO band gives a stability increase, after formation.

There is instructive to perform also the reactivity analysis of two molecular systems, one hard and one weak, but with the same electronegativity as in Figure 3.9, placed so that the corresponding energy to the middle of the LUMO – HOMO band coincides for both systems. In this case three effects may occur:

(i) A partial electrons transfer from the HOMO level of each molecule to the LUMO level of the other one takes place, a mechanism called delocalization, by which the old molecular bonds are broken and a new one is formed. If only the four frontier orbitals are considered as involved, then an approximate value for the energy lowered due to delocalization will be:

$$\Delta E = \frac{2\beta_1^2}{A_1 - I_2} + \frac{2\beta_2^2}{A_2 - I_1} \tag{3.377}$$

where the β values correspond to the exchange integrals between the involved molecular orbitals and can correspond both to the σ or π bonds. From the Eq. (3.377) there is noted that the stability of the new bond increases if the affinities of the two initial systems are high and respectively the ionization potentials are small, which corresponds to reality if the LUMO – HOMO bands for both systems are both narrow or soft.

(ii) The second effect is due to the mixing between the occupied molecular orbitals with the empty ones for each molecule, producing an effect called *polarization*, with the consequence of decreasing the inter-electronic repulsion, as the molecules are brought nearby. As a polarization result, the two molecules interaction will be more favorable if both are weak.

(iii) The electrostatic effect also occurs due to the interaction of reactants' charges and dipoles.

All these effects are able to certify the validity of the Principles CR1, i.e., the HSAB principle. Therefore, allowing for a more nuanced redefinition of what it is meant by strong molecule, or with chemical inertia at reactivity, given in qualitative formulation:

DEFINITION CR3: One calls a hard molecule or a molecule with chemical inertia at reactivity, beside a resistance at electronic transfer with another electronic system, also the resistance at the exchange of the space distribution of the electric charge.

Once defined the global measure characterizing molecular systems, i.e., the absolute electronegativity, there is realized that the chemical hardness is not constant over the entire molecule. Therefore, also the local sizes can be defined, which have been called as Fukui functions by Parr and Yang (1984):

DEFINITION CR4: For a molecular system the sizes associated to the electrons densities for the associated HOMO and LUMO states can be defined in terms of their electronic densities:

$$f^- = \rho_{HOMO} \tag{3.378}$$

$$f^+ = \rho_{LUMO} \tag{3.379}$$

$$f^0 = \frac{1}{2}\left(\rho_{HOMO} + \rho_{LUMO}\right) \tag{3.380}$$

so that these functions may correspond with the molecular electronegativity through the manifested electrophilic f^-, nucleophilic f^+ or radicalic f^0 characters, respectively.

PRINCIPLE CR4: The chemical reaction occurs from that part of molecule where one of the Fukui function has a maximum value, which corresponds to an electronic transfer imported by that region from the molecule.

As a working example, the thiocyanate ion (NCS^-) may react as a weak electrophilic molecule at its sulfur side and as a strong electrophilic (by reaction electrostatically controlled) by its nitrogen with the Lewis strong acids. Therefore, the electronegativity is actually locally complemented with its companion, the chemical hardness, or with the chemical softness, by the Fukui functions so providing a complete molecular picture of the chemical reactivity, i.e., of the tendency of the complex electronic systems to chemically react.

3.5.2 FRONTIER CONTROLLED CHEMICAL REACTIVITY

As noted, the primary role for the electrons transfer in the molecular formation and in the development of the chemical reactions is, in broad sense, assured by the farthest electrons from the nuclei, or of those for whose action is the weakest felt.

In other words, the *valence* electrons and their way of distribution have the dominant role in charge transfers.

More specifically, the principal electronic levels are those named by HOMO (the highest-energetic-occupied molecular orbital) and LUMO (the lowest-energetic-unoccupied molecular orbital). As already shown, the combination between these two energy levels is representative for molecular systems behavior (i.e., for electrons interacting with a collection of nuclear charges) in their various combinations.

However, the necessity for unification of concepts and of their submission to a Principle is natural, and this can be formulated in the context of the density functionals theory; the forthcoming discussion follows the references (Parr & Bartolotti, 1983; Parr et al., 1978; Parr & Gázquez, 1993; Parr & Yang, 1984; Yang et al., 1984):

PRINCIPLE CR5 (of Parr and Yang):

(a) The Fukui Function (3.99) is a local measure, which allows various values in different locations in a chemical species and can be represented as a (density) contour map.

(b) The Fukui function has a border/frontier function role for the relative chemical species and allows the identification in the chemical species of locations, which are favorable for the chemical reactivity.

(c) The direction preferred by a chemical species in reactivity is defined by the species' location/localization where the Fukui function has the maximum value; it corresponds to a maximum initial value of the chemical potential $|d\mu|$ for the discussed species.

(d) The Fukui function being expressed by a functional derivative of the chemical potential, allows discontinuities, or continuities on portions, as the chemical potential does.

(e) The continuous values on portions for the Fukui functions (in relation to the electronic number values respecting the total nuclear charge of the species) define the attack type to which the discussed species is the subject and correspond to the analytic expressions:

- for the electrophilic attack:

$$f^-(x)=\left(\frac{\partial \rho(x)}{\partial N}\right)_V^- \cong \rho_{HOMO}(x) \qquad (3.381)$$

- for the nucleophilic attack:

$$f^+(x)=\left(\frac{\partial \rho(x)}{\partial N}\right)_V^+ \cong \rho_{LUMO}(x) \qquad (3.382)$$

- for the radical attack:

$$f^0(x)=\left(\frac{\partial \rho(x)}{\partial N}\right)_V^0 \cong \frac{1}{2}[\rho_{HOMO}(x)+\rho_{LUMO}(x)] \qquad (3.383)$$

where the last approximations are made assuming the frozen electronic core hypothesis, i.e., the equivalency of the electronic density variation for the entire system with the valence electron density variation.

The CR4 principle affirms an essential aspect, beyond its sequential presentation, namely the prediction for the evolution of a chemical system (viewed as a collection of electrons and nuclei subject to the mutual interaction and with the environment too).

With this principle, *Chemistry* affirms a new stage of the comprehensibility of the reactivity phenomena under review, namely the capacity of prediction and the analytical way in which this is made, representing the cornerstone of any theory or even science. Note that there is very interesting how the density functionals theory, through its thermodynamics and statistical sides, naturally provides the instruments of predictability in Chemistry.

Next, we will analytical deepen this principle and it will be extended to include the spin variable for determining the reactivity direction (chemical stereo-selection "vector").

According to the Hohenberg-Kohn-Sham theory, an electronic system with a continuous number of orbital occupancy admits the stationary problem by Kohn-Sham (KS) self-consistency:

$$\left(-\frac{1}{2}\nabla^2 + \frac{\delta U[\rho]}{\delta \rho} + \frac{\delta E_{xc}[\rho]}{\delta \rho}\right)\Psi_i = \varepsilon_i \Psi_i \qquad (3.384)$$

$$\rho(x) = \sum_i n_i |\Psi_i(x)|^2 \qquad (3.385)$$

$$E[\rho] = T_s[\rho] + U[\rho] + E_{xc}[\rho] \qquad (3.386)$$

where $T_s[\rho]$ represents the density functional of kinetic energy of the system with electrons without interaction, $U[\rho]$ is the density functional of the energy developed by the classical electrostatic forces, and $E_{xc}[\rho]$ is the density functional of the exchange and correlation energy in the Kohn-Sham formalism.

For a system "slightly" negative ionized, i.e., with $N + \delta$ electrons, $\delta > 0$, in the fundamental state, the KS equation becomes, for the LUMO orbital:

$$\left[-\frac{1}{2}\nabla^2 + V_{II}^+ + V_{xc}^+\right]\Psi_{N+1}^+ = \varepsilon_{N+1}^+ \Psi_{N+1}^+ \qquad (3.387)$$

where the classical electrostatic potential and the exchange potential are appropriately recognized.

In the same time, the associated Euler equation, with the normalization condition for the density for the system slightly ionized, with $N + \delta$ electrons, so becomes:

$$\mu^+ = V_{II}^+(x) + V_{xc}^+(x) + \frac{\delta T_s[\rho^+]}{\delta \rho(x)} \qquad (3.388)$$

Now, through considering the Janak theorem, or the equivalent, by combining the last two relations, one is left with the simple result:

$$\frac{\delta T_s[\rho^+]}{\delta \rho(x)} = -\frac{1}{2}\frac{\nabla^2 \Psi_{N+1}^+}{\Psi_{N+1}^+} \qquad (3.389)$$

from where immediately follows the conclusion that $\rho^+(x)$ is an orbital density functional equivalently with the form $\rho[\Psi_{N+1}^-]$. Therefore, for the $N + \delta$ electrons in fundamental state, with the occupation numbers satisfying the occupancy $n_i = 1$ for $i \leq N$ and $n_{N+1} = \delta$, it can be expressed, recalling (3.131), as next:

$$\rho_{N+\delta}(x) = \sum_{i=1}^{N} |\Psi_i^+(x)|^2 + \delta |\Psi_{N+1}^+(x)|^2 \qquad (3.390)$$

wherefrom the nucleophilic Fukui function can be rewritten such as:

$$f^+(x) = \lim_{\delta \to +0} \frac{\partial}{\partial N}[\rho_{N+\delta}(x)] = \sum_{i=1}^{N} \frac{\partial}{\partial N} |\Psi_i^-(x)|^2 + |\Psi_{N+1}(x)|^2 \qquad (3.391)$$

The result tells that for a nucleophilic attack of the system, the frontier (Fukui) function is separated in the uniformly occupied orbitals response and the one where the attack occurred.

Similarly, for a system slightly positively ionized, with $N + \delta$ electrons, $\delta < 0$, in the fundamental state. The KS equation for the HOMO orbital becomes:

$$\left[-\frac{1}{2}\nabla^2 + V_{II}^- + V_{xc}^-\right]\Psi_N^- = \varepsilon_N^- \Psi_N^- \qquad (3.392)$$

while the Euler equation associated with the density normalization condition for the system "slightly" ionized to the $N + \delta$ electrons, will be in this case:

$$\mu^- = V_{II}^-(x) + V_{xc}^-(x) + \frac{\delta T_s[\rho^-]}{\delta \rho(x)} \qquad (3.393)$$

Again, taking into account the Janak theorem, by combining the last relations for the system "slightly" positive ionized one yields with the next identity results:

$$\frac{\delta T_s[\rho^-]}{\delta \rho(x)} = -\frac{1}{2}\frac{\nabla^2 \Psi_N^-}{\Psi_N^-} \qquad (3.394)$$

wherefrom the immediate conclusion that $\rho^-(x)$ is an orbital functional which has the form $\rho[\Psi_N^-]$.

As above, one can express the electronic density and the frontier function by the orbital partitioning, for example, for an electrophilic attack:

$$\rho_{N+\delta}(x) = \sum_{i=1}^{N} |\Psi_i^-(x)|^2 + \delta |\Psi_N^-(x)|^2 \qquad (3.395)$$

wherefrom the associated Fukui function can be rewritten as:

$$f^-(x) = \lim_{\delta \to -0} \frac{\partial}{\partial N}[\rho_{N+\delta}(x)] = \sum_{i=1}^{N-1} \frac{\partial}{\partial N}|\Psi_i(x)|^2 + |\Psi_N(x)|^2 \qquad (3.396)$$

From the limiting relations for f^+ and f^- there results that in the approximation of the frozen electronic core, the first derivatives in relation with the electronic total number of core electronic orbitals distribution are canceled, thus naturally resulting the approximations expressions from the Principle CR5.

An immediate application with major implications for the frontier functions can be rendered in connection with the electrical charge transfer occurring in the molecular formation of chemical reactions.

If the formation of an electronic and nuclei system is considered as a result of a chemical process of charge transfer between two initial systems A and B, the *electronegativity equalization principle* or of the *chemical potential* (electronegativity with changed sign) can be formulated in one of the formulation from the above dedicated Section 3.3, and also in the light of the definition relation of the Fukui function within the density functional theory, recall (3.103), with the total differential unfolding (Berkowitz, 1987; Berkowitz & Parr, 1988):

$$d\mu_A = 2\eta_A dN_A + \int f_A(x)dV_A(x)dx$$

$$= 2\eta_B dN_B + \int f_B(x)dV_B(x)dx = d\mu_B \qquad (3.397)$$

In this equalization, the external potentials which act on the systems A and B have two sources in a first approximation: the potential of nuclei movement and the potential due to the movements of the electrons between A and B; so it can be modeled, for example, for the system A, by the form:

$$dV_A(x) = dV^n(x) + \int \frac{\delta\rho_B(x')}{|x-x'|}dx'$$

$$= dV^n(x) + \int \frac{f_B(x)dN_B}{|x-x'|}dx' \qquad (3.398)$$

and similarly for the system B.

Under these circumstances, the equalization equation of the chemical potentials in molecule becomes:

$$2\eta_A dN_A + \int f_A(x) dV''(x) dx + \left[\iint \frac{f_A(x) f_B(x')}{|x-x'|} dx dx' \right] dN_B$$

$$= 2\eta_B dN_B + \int f_B(x) dV''(x) dx + \left[\iint \frac{f_A(x) f_B(x')}{|x-x'|} dx dx' \right] dN_A \quad (3.399)$$

wherefrom the charge conservation, $dN_A = -dN_B = dN$, once assuming, for example, the charge transfer from the system B to the system A, leads with the charge transfer expression, such as:

$$\Delta N = -\frac{\int [f_A(x) - f_B(x')] dV''(x) dx}{(2\eta_A + 2\eta_B) - 2J_f} \quad (3.400)$$

$$J_f = \iint \frac{f_A(x) f_B(x')}{|x-x'|} dx dx' \quad (3.401)$$

Further on, the charge transfer in the approximation of the frozen electronic core for both systems is considered, so that the difference of the frontier functions can be accordingly considered (Bartolotti, 1980; Robles & Bartolotti, 1984; Parr, Zhou, 1993; Harbola, 1992):

$$f_A(x) - f_B(x) \cong \rho_{LUMO}^A - \rho_{HOMO}^B = \Delta\rho(x) \quad (3.402)$$

through which the charge transfer will be simplified by rewritten it as:

$$\Delta N = -\frac{\int \Delta\rho(x) dV''(x) dx}{2(\eta_A + \eta_B - J_f)} \quad (3.403)$$

The analysis of these relations shows how much the charge transfer depends on the frontier functions. In this case, an integral J_f of the Fukui functions with a large orbital overlapping, has the role to reduce the inhibitory effects induced by the chemical strengths of the two chemical systems and to promote the charges transfer.

Observe that the denominator of the transferred charge (3.403) is correlated with the covalent character of the chemical bond, which is formed between the systems A and B through an electronic redistribution as a result of the exchange (including the charge transfer), which is more efficient as the covalent bond is much stronger.

By all these aspects, the chemical bond gets in the density functionals picture a local-global self-consistent understanding and interpretation.

3.5.3 SPIN EFFECTS ON CHEMICAL REACTIVITY

The more refined approach of chemical bond in general and of the orbital overlapping and electronic exchange distribution naturally appeals the necessity to integrate and to specify the spin distributions, i.e., by counting the up and down electronic spin behavior, as reflected by the main equations of the density functional theory; the forthcoming discussion follows the references (Manoli & Whitehead, 1984, 1988a-b; Orski & Whitehead, 1987; Guo & Whitehead, 1989a-b, 1991).

Accordingly, the following constructions of the electronic density will be considered

$$\rho(x) = \rho_\uparrow(x) + \rho_\downarrow(x) \tag{3.404}$$

$$\rho_s(x) = \rho_\uparrow(x) - \rho_\downarrow(x) \tag{3.405}$$

which satisfy the conditions:

$$\int \rho_\uparrow(x)dx = N_\uparrow \tag{3.406}$$

$$\int \rho_\downarrow(x)dx = N_\downarrow \tag{3.407}$$

$$\int \rho(x)dx = N_\uparrow + N_\downarrow = N \tag{3.408}$$

$$\int \rho_s(x)dx = N_\uparrow - N_\downarrow = N_s \tag{3.409}$$

so providing a density way for defining the electronic number with up or down spin, the total number of electrons in the system and the spin number, respectively. Especially the last relation introduces an explicit dependence of effective spin distribution, which has a contribution in the system.

Therefore, the density functional of the total energy of a system placed in an external potential, with an oriented magnetic field which "activate" the orientation ordered by spin, is written as a generalization of the total-energy functional (Gázquez et al., 1984, 1987, 1990; Galván et al., 1988; Vela & Gázquez, 1988)

$$E[\rho, \rho_S, V, B] = F[\rho, \rho_S] + \int \rho(x)V(x)dx - \mu_B \int B(x)\rho_S(x)dx \qquad (3.410)$$

with the associated Euler generalized equations, recall (3.104):

$$\mu_N = \left(\frac{\delta E}{\delta \rho(x)}\right)_{V, \rho_S, B} = V(x) + \frac{\delta F[\rho]}{\delta \rho(x)} \qquad (3.411)$$

$$\mu_S = \left(\frac{\delta E}{\delta \rho_S(x)}\right)_{V, \rho, B} = -\mu_B B(x) + \frac{\delta F[\rho]}{\delta \rho_S(x)} \qquad (3.412)$$

where μ_B stands here for the Bohr magneton.

Next, one expresses the total differential of the total functional energy, extending (3.98), as:

$$dE = \int \left(\frac{\delta E}{\delta \rho(x)}\right)_{\rho_S, V, B} \delta \rho(x)dx + \int \left(\frac{\delta E}{\delta \rho_S(x)}\right)_{\rho, V, B} \delta \rho_S(x)dx$$

$$+ \int \left(\frac{\delta E}{\delta V(x)}\right)_{\rho_S, \rho, B} \delta V(x)dx + \int \left(\frac{\delta E}{\delta B(x)}\right)_{\rho, \rho_S, V} \delta B(x)dx \qquad (3.413)$$

By evaluating the last two functional derivatives under the integrals of (3.413) with the aid of the total spin energy and by using the Euler relations (3.412), one can re-rewrite the last total differential as:

$$dE = \mu_N dN + \mu_S dN_S + \int \rho(x)\delta V(x)dx - \mu_B \int \rho_S(x)\delta B(x)dx \qquad (3.414)$$

On the other hand, another form of the total differential (3.414) can be obtained through considering the total energy as a function of variables (N, N_S, B, V):

$$dE = \left(\frac{\partial E}{\partial N}\right)_{N_S, V, B} dN + \left(\frac{\partial E}{\partial N_S}\right)_{N, V, B} dN_S$$

$$+ \int \rho(x)\delta V(x)dx - \mu_B \int \rho_S(x)\delta B(x)dx \qquad (3.415)$$

By comparing the equivalent last two relations, the electronic and spin potentials above reshape now, recalling (3.95):

$$\mu_N = \left(\frac{\partial E}{\partial N}\right)_{V,N_S,B} \tag{3.416}$$

$$\mu_S = \left(\frac{\partial E}{\partial N_S}\right)_{V,N,B} \tag{3.417}$$

A very important result drawn from these last two relations regards the electronic potential and respectively that of electronic spin as functions of the natural variables (N, N_S, B, V), respectively. Therefore, the total differentials associated to these potentials, as generalized expressions having incorporated the Fukui functions can be also written, extending (3.103):

$$d\mu_N = 2\eta_{NN}dN + 2\eta_{NS}dN_S + \int f_{NN}(x)\delta V(x)dx - \mu_B \int f_{SN}(x)\delta B(x)dx \tag{3.418}$$

$$d\mu_S = 2\eta_{SN}dN + 2\eta_{SS}dN_S + \int f_{NS}(x)\delta V(x)dx - \mu_B \int f_{SS}(x)\delta B(x)dx \tag{3.419}$$

From the unfolds (3.418) and (3.419) the definitions of the new electronic measures appear from the variable's coefficients' identification in the total differential and respectively by the Maxwell relations for the mixed equal derivatives in the total differential of the total energy, as one may formulated.

As such, the chemical hardness is enriched with a *mixed hardness* feature so measuring the chemical potential variation with the spin number variation, or by commensurate the spin potential variation with the system's electronic number variation, so extending (3.101) to:

$$\eta_{NS} = \left(\frac{\partial \mu_N}{\partial N_S}\right)_{N,V,B} = \left(\frac{\partial \mu_S}{\partial N}\right)_{N_S,V,B} = \left(\frac{\partial^2 E}{\partial N_S \partial N}\right)_{V,B} = \eta_{SN} \tag{3.420}$$

Equally, the spin chemical hardness new definition:

$$\eta_{SS} = \left(\frac{\partial \mu_S}{\partial N_S}\right)_{N,V,B} = \left(\frac{\partial^2 E}{\partial N_S^2}\right)_{N,V,B} \tag{3.421}$$

enriches the electronic chemical hardness with the significance of inertia at spin modification for the chemical coordination between the two systems characterized by electronic spin.

In terms of chemical reactivity analysis, as the initial value of the *electronic potential* $|d\mu_N|$ is big the system's reactivity will be established with high values of the frontier functions, extending (3.99) to become

$$f_{NN}(x) = \left(\frac{\partial \mu_N}{\partial V(x)} \right)_{N,N_S,B} = \left(\frac{\partial \rho(x)}{\partial N} \right)_{N_S,V,B} \tag{3.422}$$

$$f_{SN}(x) = -\frac{1}{\mu_B} \left(\frac{\partial \mu_N}{\partial B(x)} \right)_{N,N_S,B} = \left(\frac{\partial \rho_S(x)}{\partial N} \right)_{N_S,V,B} \tag{3.423}$$

according to the Principle CR4 above.

On the other hand, in the case for that the initial value of the *spin potential* $|d\mu_N|$ is big the system reactivity will be established with high values of the respective spin related frontier functions, this time extending (3.99) as:

$$f_{NS}(x) = \left(\frac{\partial \mu_S}{\partial V(x)} \right)_{N,N_S,B} = \left(\frac{\partial \rho(x)}{\partial N_S} \right)_{N,V,B} \tag{3.424}$$

$$f_{SS}(x) = -\frac{1}{\mu_B} \left(\frac{\partial \mu_S}{\partial B(x)} \right)_{N,N_S,B} = \left(\frac{\partial \rho_S(x)}{\partial N_S} \right)_{N,V,B} \tag{3.425}$$

wherefrom the way in which the spin potential, or of its distribution, determines the considered system reactivity so revealed in accordance with the Principle CR4 above.

These equivalent relations for the mixed Fukui functions represent the local criterion of selection of the chemical reactivity in a system where also the spin effects are considered.

Remarkably, when considering the spin distribution, the coordination analysis will involve a double equalization or a chemical reactivity unfold in two steps: one refers to the charges exchange, respectively to the electronic chemical potentials equalization, yet accompanied by the spin potentials equalization, respectively corresponding to a "spin exchange" after which the up and down spins are rearranged in the given coordination

(Filippetti, 1998; Ray & Parr, 1980; Koga & Umeyama, 1986; Schmidt & Böhm, 1983; Böhm & Schmidt, 1986; Harrison, 1987).

Another spin contribution can be identified in the frozen-core approximation, in order to express the frontier functions from the Principle CR5above, as in next exposed.

To this aim, one starts from the limiting relations of the Fukui functions where the densities associated to the up and down spins with the HOMO and LUMO orbitals can be expressed for a system under various conditions.

Then, for a system with $N + \delta$ electrons and with the constant spin number N_S the functional dependencies for the various ionization states' densities can be written as follow (Gázquez et al., 1987, 1990; Galván et al., 1988):

(i) $$\delta > 0 : \rho_\uparrow^+ = \rho_\uparrow^+ \left[\Psi_{LUMO\uparrow} \right] \quad \rho_\downarrow^+ = \rho_\downarrow^+ \left[\Psi_{LUMO\downarrow} \right] \tag{3.426}$$

(ii) $$\delta < 0 : \rho_\uparrow^- = \rho_\uparrow^- \left[\Psi_{HOMO\uparrow} \right] \quad \rho_\downarrow^- = \rho_\downarrow^- \left[\Psi_{HOMO\downarrow} \right] \tag{3.427}$$

(iii) $$\delta = 0 : \rho_\uparrow^0 = \rho_\uparrow^0 \left[\Psi_{LUMO\uparrow}, \Psi_{HOMO\uparrow} \right] \quad \rho_\downarrow^0 = \rho_\downarrow^0 \left[\Psi_{LUMO\downarrow}, \Psi_{HOMO\downarrow} \right] \tag{3.428}$$

Instead, for a system with a fixed number of N-electrons and with the spin number $N_S + \delta_S$ the functional dependencies for the various ionization states' densities cast as:

(i-s) $$\delta_S > 0 : \rho_\uparrow^+ = \rho_\uparrow^+ \left[\Psi_{LUMO\uparrow} \right] \quad \rho_\downarrow^+ = \rho_\downarrow^+ \left[\Psi_{HOMO\downarrow} \right] \tag{3.429}$$

(ii-s) $$\delta_S < 0 : \rho_\uparrow^- = \rho_\uparrow^- \left[\Psi_{HOMO\uparrow} \right] \quad \rho_\downarrow^- = \rho_\downarrow^- \left[\Psi_{LUMO\downarrow} \right] \tag{3.430}$$

(iii-s) $$\delta_S = 0 : \rho_\uparrow^0 = \rho_\uparrow^0 \left[\Psi_{LUMO\uparrow}, \Psi_{HOMO\uparrow} \right] \quad \rho_\downarrow^0 = \rho_\downarrow^0 \left[\Psi_{LUMO\downarrow}, \Psi_{HOMO\downarrow} \right] \tag{3.431}$$

where one notes the LUMO and HOMO orbital combinations appearing for each description of the system in discussion, also when the spin number is considered.

Therefore, in the electronic frozen-core approximation (with spins too) the Fukui frontier functions become in each of the presented ionization cases accordingly specializing (3.381)–(3.383) as (Gázquez & Ortiz, 1984; Vela & Gázquez, 1988):

(a) for the frontier $f_{NN}(x)$ function one has the actual realizations:

$$\delta > 0 : f_{NN}^+(x) \cong \frac{1}{2}\left[\left|\Psi_{LUMO\uparrow}(x)\right|^2 + \left|\Psi_{LUMO\downarrow}(x)\right|^2\right] \tag{3.432}$$

$$\delta < 0 : f_{NN}^-(x) \cong \frac{1}{2}\left[\left|\Psi_{HOMO\uparrow}(x)\right|^2 + \left|\Psi_{HOMO\downarrow}(x)\right|^2\right] \tag{3.433}$$

$$\delta = 0 : f_{NN}^0(x) \cong \frac{1}{2}\left[f_{NN}^+(x) + f_{NN}^-(x)\right] \tag{3.434}$$

(b) for the frontier $f_{SN}(x)$ function one has the respective realizations:

$$\delta > 0 : f_{SN}^+(x) \cong \frac{1}{2}\left[\left|\Psi_{LUMO\uparrow}(x)\right|^2 - \left|\Psi_{LUMO\downarrow}(x)\right|^2\right] \tag{3.435}$$

$$\delta < 0 : f_{SN}^-(x) \cong \frac{1}{2}\left[\left|\Psi_{HOMO\uparrow}(x)\right|^2 - \left|\Psi_{HOMO\downarrow}(x)\right|^2\right] \tag{3.436}$$

$$\delta = 0 : f_{SN}^0(x) \cong \frac{1}{2}\left[f_{SN}^+(x) + f_{SN}^-(x)\right] \tag{3.437}$$

For these local functions a conservation of the HOMO or LUMO orbital type involved in the calculation of a given ionization case is noted, with the specification that in the situation (b) the difference rather than the summation of probabilities for the orbital frontier functions is involved.

However, for the other set of frontier functions, one continues with specializations:

(c) for the frontier $f_{NS}(x)$ function the respective valence orbitalic formulations are:

$$\delta_S > 0 : f_{NS}^+(x) \cong \frac{1}{2}\left[\left|\Psi_{LUMO\uparrow}(x)\right|^2 - \left|\Psi_{HOMO\downarrow}(x)\right|^2\right] \tag{3.438}$$

$$\delta_S < 0 : f_{NS}^-(x) \cong \frac{1}{2}\left[\left|\Psi_{HOMO\uparrow}(x)\right|^2 - \left|\Psi_{LUMO\downarrow}(x)\right|^2\right] \tag{3.439}$$

$$\delta_S = 0 : f_{NS}^0(x) \cong \frac{1}{2}\left[f_{NS}^+(x) + f_{NS}^-(x)\right] = f_{SN}^0(x) \tag{3.440}$$

(d) for the frontier function $f_{SS}(x)$ the working expressions are:

$$\delta_S > 0 : f_{SS}^+(x) \cong \frac{1}{2}\left[\left|\Psi_{LUMO\uparrow}(x)\right|^2 + \left|\Psi_{HOMO\downarrow}(x)\right|^2\right] \qquad (3.441)$$

$$\delta_S < 0 : f_{SS}^-(x) \cong \frac{1}{2}\left[\left|\Psi_{HOMO\uparrow}(x)\right|^2 + \left|\Psi_{LUMO\downarrow}(x)\right|^2\right] \qquad (3.442)$$

$$\delta_S = 0 : f_{SS}^0(x) \cong \frac{1}{2}\left[f_{SS}^+(x) + f_{SS}^-(x)\right] = f_{NN}^0(x) \qquad (3.443)$$

Analyzing the (c) and (d) expressions one notes the resembling character is sign combinations with those provided by (a) and (b) expressions, still with a mixed character in the distribution of electrons with spin in the frontier's orbitals.

These expressions can be numerically implemented for a set of coefficients for the initial atomic orbitals in the system, as well as for other basis functions (e.g., of hydrogenic, Gaussian, or Slater type). An alternative method for computational implementation is to self-consistently solve the equations from the Hohenberg-Kohn-Sham density functional theory, properly modified in order to include the extension of the spin characterization, wherefrom the molecular orbitals corresponding to the electronic distribution and of spin may directly result, hence, retaining only the HOMO and LUMO orbitals in the electronic frozen-core approximation with the help of which one can calculate and represent the contours of the frontier functions in any of the above (a) to (d) variants.

By such procedure the global chemical measures defined by the electronic chemical potential and of electronic spin potential, along the various chemical harnesses, with the electronic, mixed or of spin specializations can also be evaluated.

Overall, the density functional theory allows a natural and immediate extension of the chemical reactivity analysis, as stated by the Principle CR2, to the systems characterized also by the spin number, allowing a very complex analysis of the atomic and molecular combinations in the light of allied global and local reactivity indices, eventually numerically or self-consistently evaluated.

3.6   MOLECULAR DFT PARABOLIC DEPENDENCY E=E(N) ON REACTIVITY INDICES

Since the special energetic (accuracy about 1kJ/mol) and space of electronic manifestation of chemical reactivity (the valence state, frontier orbitals) appropriate physical-chemical indicators were advanced in order to better quantify, model, and finally control specific systems (atoms, molecules, atoms in molecules, etc.) and of their properties. Basically, once a system is computed for its energy in various states of donating and accepting electronic the total energy vs. number of electrons $E = E(N)$ is provided to represent a sort of dynamical mark of the electronic sample under given external potential influence $V(\mathbf{r})$ (Putz, 2011a). Now, under the assumption that such a curve is always convex for Coulomb interacting systems (Ayers & Parr, 2001), one can extract the physical-chemical information from it by considering the first and second derivative on the curve point associated with the neutral system, i.e., the tangent and the curvature's values on the neutral point eventually provides the prediction for further systems evolution outside of it. This way, the two basic indices of reactivity are at once advanced, namely the electronegativity (EL or χ) or the negative of the chemical potential (Iczkowski & Margrave, 1961; Parr et al., 1978) and the chemical hardness (HD or η) (Parr & Pearson, 1983; Putz, 2011a).

However, while being recently reviewed the electronegativity-chemical hardness orthogonal space of reactivity $\{\chi, \eta | \chi \perp \eta\}$ (Putz, 2011a, 2012c), the present work review at its turn the applicability of such binomial set in assessing general energetic parabolic dependency respecting the total/valence electrons or biological endpoints in the interacting systems, producing either consistent chemical reactivity or biological activity models. There follows that indeed Density Functional Theory parabolic recipe is enough reliable to be assumed as further basis for describing the open systems, due to its universal feature in displaying global minimum/equilibrium alongside the activating system under the perturbation (external) influences. Further work may be therefore advanced for characterizing the chemical-biological systems with the aid of electronegativity and chemical hardness, viewed as the "velocity/slope" and "acceleration/curvature" in an abstract, yet with a natural projective reality, towards assessing the specific reactivity-activity principles.

3.6.1 BIVARIATE ELECTRONEGATIVITY-CHEMICAL HARDNESS RELATIONSHIP

3.6.1.1 Physical Necessity for Parabolic E(N) Dependency

Employing the Kohn-Sham equation for the (reactivity) equalized chemical potential (minus electronegativity) eigenvalue, among atomic or molecular one-electronic orbitals, recalling for instance (3.65) and (3.298)–(3.300),

$$\mu_i = \mu_j = ... = \mu \tag{3.444}$$

in quantum mechanically sense, we firstly get (Putz, 2008a):

$$\mu = \frac{\int \varphi^*(\mathbf{r}) \left[-\frac{1}{2}\nabla^2 + V_{eff}^{DFT} \right] \varphi(\mathbf{r}) d\mathbf{r}}{\int \varphi^*(\mathbf{r})\varphi(\mathbf{r}) d\mathbf{r}} = \frac{\int \varphi^*(\mathbf{r})\hat{H}\varphi(\mathbf{r}) d\mathbf{r}}{\int \varphi^*(\mathbf{r})\varphi(\mathbf{r}) d\mathbf{r}} = \frac{E}{N} \tag{3.445}$$

For proper characterization of the chemical systems the quantum Ehrenfest version of the fundamental Newtonian law, linking the observed force with the minus of the potential gradient $\mathbf{F} = -\nabla V$, is here written for chemical reactions modeled throughout the charge transfer along the reaction path (Putz, 2008a):

$$F_\mu = -\frac{d}{dN}(\mu) \tag{3.446}$$

Now, combining the last two equations projected on the reaction path we successively obtain:

$$F_\mu = -\frac{d}{dN}\left(\frac{E}{N}\right) = \frac{E}{N^2} - \frac{1}{N}\left(\frac{dE}{dN}\right)_V \tag{3.447}$$

At this point, taking for the natural differential the finite correspondence in what regarding the chemical potential formal (absolute) definition,

$$\left(\frac{dE}{dN}\right)_V = \mu = \frac{E}{N} \tag{3.448}$$

the chemical potential energy equation is unfolded as:

$$E_\mu = \mu N + F_\mu N^2 \tag{3.449}$$

The remaining issue is to identify the chemical-potential related force meaning in above equation. In this respect, by considering recognizing the electronegativity-chemical potential relationship, we have that the electronegativity energy equation becomes:

$$E_\chi = -\chi N + F_\chi N^2 \tag{3.450}$$

while the involved electronegativity related force is seen as the reactive force of the chemical potential,

$$F_\chi = -F_\mu \tag{3.451}$$

recovering the chemical hardness index (Putz, 2008a):

$$F_\chi = -\frac{d}{dN}(\chi) = \eta \tag{3.452}$$

With these considerations, the chemical reactivity energy equation is finally displayed as

$$E = -\chi N + \eta N^2 \tag{3.453}$$

equally supporting the differential (statistical, thermodynamically) form

$$dE = -\chi dN + \eta (dN)^2 \tag{3.454}$$

based on the above finite-differential equivalence between the total (E, N) or exchanged (dE, dN) energy and charge, respectively; this way the earlier parabolic formulations, for example (3.166), are quantum mechanically justified.

3.6.1.2 Chemical Realization for Parabolic E(N) Dependency

What there was actually proofed is that the quadratic chemical reactivity equations for total energy in both finite and differential fashions may be derived employing the Kohn-Sham (as Schrödinger reminiscence) equation for chemical potential eigenvalue combined with the chemical quantum version of the Ehrenfest theorem involving the force concept and its active-reactive peculiar property for the chemical potential and electronegativity, respectively. No particular assumptions were considered being all above arguments only on first principles grounded. Thus, the exposed demonstration is of general value cutting much discussion in the last decades on the viability of the second order truncation in the total energy expansion in terms of chemical reactivity indices, viz. electronegativity and chemical hardness concepts. However, a computational test for this behavior is in next addressed.

One of the outstanding *DFT* working tools regards the baseline density (ρ) to the number of electrons (N) relationship that fulfills the integral relationship throughout eventually under the variational form (Hohenberg & Kohn, 1964; Kohn & Sham, 1965; Kohn et al., 1996; Parr & Yang, 1989; Dreizler & Gross, 1990; March, 1991; Koch & Holthausen, 2002; Fiolhais et al., 2003; Sholl & Steckel, 2003; Putz, 2003, 2012a)

$$\int \Delta\rho(\mathbf{r})d\mathbf{r} = \Delta N \tag{3.455}$$

This nevertheless, allows rethinking upon the chemical reactivity principles since invariable they involve the chemical reactions, which involve the charge transfer. Namely, if one considers the fundamental relation as the modification in the ground state energy re-written as the density variation subject to the Taylor series expansion (assuming $\Delta\rho$ small enough to provide truncation to the second order), it looks like (Putz, 2003, 2012a):

$$E[\rho + \Delta\rho] \cong E[\rho] + \int \left(\frac{\delta E[\rho]}{\delta \rho(\mathbf{r})} \right)_V \Delta\rho(\mathbf{r})d\mathbf{r}$$

$$+ \frac{1}{2}\iint \left(\frac{\delta^2 E[\rho]}{\delta \rho(\mathbf{r})\delta \rho(\mathbf{r}')} \right)_V \Delta\rho(\mathbf{r})\Delta\rho(\mathbf{r}')d\mathbf{r}d\mathbf{r}' \tag{3.456}$$

At this point, the stability of the many-body system, even upon being engaged in reactivity (i.e., by charge transfer or total energy variations) requires that the deviation of energy $E[\rho + \Delta\rho]$ to be minimum. What does mean this in terms of reactivity? It means that the active sites of a reactant molecule are usually placed where the addition or loss of electrons is energetically favorable.

Back into mathematical language, the most favorable sites to add or lose electrons further means that the energy have to be minimized by a function that just accounts for the ground state density variations when electrons are exchanged with a bath; this behavior may be identified with the celebrated Fukui function defined as the density variation when the total number of electrons is changed within the constant applied potential, see Eq. (3.99) (Parr & Yang, 1984; Yang et al., 1984). Thus, the presence of the Fukui function will produce the minimization of the changed energy, by accounting for the preferred reactive sites, leading the analytical successive equivalences (Putz, 2003, 2012a):

$$
\min\left(\Delta E\right) \cong \int \left(\frac{\delta E[\rho]}{\delta \rho(\mathbf{r})}\right)_{V(\mathbf{r})} \left(\frac{\partial \rho(\mathbf{r})}{\partial N}\right)_{V(\mathbf{r})} \Delta \rho(\mathbf{r}) d\mathbf{r}
$$

$$
+ \frac{1}{2} \iint \left(\frac{\delta^2 E[\rho]}{\delta \rho(\mathbf{r}) \delta \rho(\mathbf{r}')}\right)_{V(\mathbf{r})} \left(\frac{\partial \rho(\mathbf{r})}{\partial N}\right)_{V(\mathbf{r})} \left(\frac{\partial \rho(\mathbf{r}')}{\partial N}\right)_{V(\mathbf{r})} \Delta \rho(\mathbf{r}) \Delta \rho(\mathbf{r}') d\mathbf{r} d\mathbf{r}'
$$

$$
= \left(\frac{\partial E}{\partial N}\right)_{V(\mathbf{r})} \Delta N + \frac{1}{2}\left(\frac{\partial^2 E}{\partial N^2}\right)_{V(\mathbf{r})} \left(\Delta N\right)^2
$$

$$
\equiv -\chi\left(\Delta N\right) + \eta\left(\Delta N\right)^2 \tag{3.457}
$$

where the functional derivative rules, together with the standard definitions of electronegativity and of the chemical hardness were considered respectively as in Eqs. (3.346) and (3.347) (Iczkowski & Margrave, 1961; Parr et al., 1978; Parr & Pearson, 1983; Sen & Jørgensen, 1987; Sen & Mingos, 1993).

However, the result enlightens two major achievements. The first one states that the electronegativity and chemical hardness are the minimizing global values corresponding to the local Fukui minimizing function for the change in total energy functional, when the system is favorable to exchanging electrons around its ground state. The second result shows that for enough small variation in the ground state density the shape of the change in energy displays a parabolic dependence on the exchanged number of electrons.

So, the parabolic assumption for the total energy in number of electrons, even not demonstrated to be most generally valid, finds however a natural justification, and fits with a wide range of the electronic exchange processes, as a general behavior either for neutral or ionic systems, see for instance Figure 3.10 for the series of bases NH_3, CH_3O^-, H_2O, CN^-, CH_3S^- and CH_3SH, see Section 3.4.7 and Putz et al. (2004).

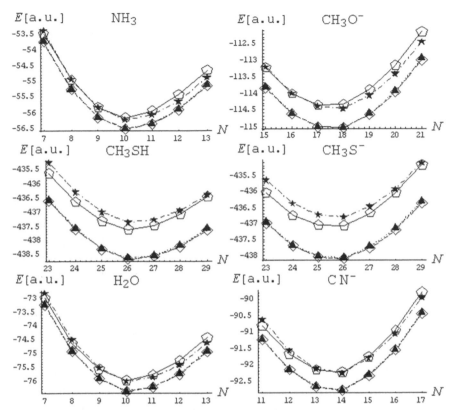

FIGURE 3.10 The total energies joint-plots, within different computational methods abbreviated (see the text) as: HF: ⟨⟩, B3LYP: ····▲····, VWN: ¯⁻ * ¯⁻ , and BP86: ◇), respectively, for the concerned molecular systems as successive ionic states between the N–3 and N+3 total number of electrons, see Section 3.4.7 and Putz et al. (2004).

These are the basic premises for considering the parabolic total energy respecting the variation of the total (or valence) number of electrons as a general framework in assessing either the chemical reactivity principles, atomic and molecular stability as well as regulating the biological activity upon related adapted principles.

3.6.2 THE CHEMICAL BOND: SPONTANEOUS SYMMETRY-BREAKING APPROACH

The method of spontaneous symmetry-breaking originates in the solid state theory (Kleinert, 1989), aiming to quantify the collective modes of excitations with vanishing energy in the long-range wave-lengths with null mass for excitations (e.g., spin waves with magnons in ferromagnets or with phonons in superconductors for Coulombian absence interaction), being established as a viable quantum method for understanding the appearance of new particles with or without mass in unification theories in Physics, especially through electro-weak interaction (Goldstone et al., 1962). It basically employs the concept of Lagrangian of the system identified in terms of a quantum field ϕ to generate a specific equation, while featuring particular invariance condition (in coordinates or in field transformations), thus displaying an original symmetry, to be spontaneously broken by a non-invariant potential (itself as a function of fields as well: $V(\phi)$); the physical-mathematical effect of such broken symmetry relays in springing of a new Lagrangian terms eventually associated with new fields and consequently with new quantum particles; there is commonly say that those particles are created from "quantum vacuum" and intermediate the interaction driven by the field potential $V(\phi)$ (Boeyens, 2005).

On the other side, although the Schrödinger equation was intensively used in assessing essential properties of chemical (many-body) systems it was until now not employed to quantization of the chemical field of chemical bonding in the physical sense. This perhaps because it was not clear whether this is the right (or universal) chemical reactivity potential to be used. However, since the traditional chemical concept of electronegativity χ was reloaded in actual quantum chemistry by many studies in the frame of DFT (Kohn et al., 1996), and especially by its equivalence with negative of the global chemical potential μ of a given system, according with Eq. (3.136) or similar relationships, see Parr & Yang (1998)

$$\chi = -\mu \tag{3.458}$$

the idea of a reactivity potential that contains it along the intriguing chemical hardness η companion (Parr & Pearson, 1983) had appeared while driving the conceptual chemistry by means of associate chemical reactivity principles, i.e., electronegativity equalization, electronegativity inequality principle, hard and soft acids and bases, maximum hardness principle (Putz, 2008b). In this context, the parabolic form of the relationship between the (total) energy and particle exchange modulated by electronegativity and chemical hardness, already proved with the help of classical-chemical correspondence principle (Putz, 2008a), may be assumed as the pattern from the actual (universal) chemical reactivity working field potential (Putz, 2008d)

$$V(\phi) = \begin{pmatrix} \mu \\ or \\ \chi \end{pmatrix} \phi^2 + \frac{1}{2}\eta\phi^4 \qquad (3.459)$$

to generate bonding fields and its particles – the *bondons* by changing from the upper (positive chemical potential) branch to lower (negative chemical potential aka electronegativity) branch of the first order particle ($\sim\phi^2$) potential. In Section 3.6.2.1 the main requisites for assessing chemical bonding fields and associated bondon particles are exposed in agreement within Schrödinger evolution theory; in Sections 3.6.2.2 and 3.6.2.3 the results of symmetry-breaking machinery formalism are presented for the chemical reactivity potential abstracted from Eq. (3.459) worked for both real and complex perturbed scalar fields, while in Section 3.6.2.4 the application for paradigmatic chemical bonds is envisaged for computing the associate quantum bondons' masses (Putz, 2008d).

3.6.2.1   Chemical Bonding Fields and Bondons

The chemical field ϕ carried by electrons of mass m_0 moving under the potential V should be regarded as the wave-function solution of the temporal Schrödinger equation

$$i\hbar\dot{\phi} = \left(-\frac{\hbar^2}{2m_0}\nabla^2 + V\right)\phi \qquad (3.460)$$

Alternatively, it enters in products of direct and conjugated field pairs in the Schrödinger Lagrangian, see the Volume I of the present five volume work

$$\mathcal{L} = i\hbar\phi^*\dot{\phi} - \frac{\hbar^2}{2m_0}\left(\nabla\phi^*\right)\left(\nabla\phi\right) - V\phi^*\phi \qquad (3.461)$$

as one may immediately check through its substitution into Euler-Lagrange equation

$$\frac{\delta\mathcal{L}}{\delta\phi} = \partial_\mu\left(\frac{\delta\mathcal{L}}{\delta\phi_{,\mu}}\right) \qquad (3.462)$$

where was used the derivative 4D notation

$$\phi_{,\mu} = \partial_{,\mu}\phi = \frac{\partial\phi}{\partial x_\mu} \equiv \left(\partial_0 = \frac{\partial}{\partial t}, \partial_1 = \frac{\partial}{\partial x_1}, \partial_2 = \frac{\partial}{\partial x_2}, \partial_3 = \frac{\partial}{\partial x_3} \right) \qquad (3.463)$$

Now, the Schrödinger Lagrangian (3.461) becomes the Lagrangian of the chemical filed once for the potential in Eq. (3.461) that of Eq. (3.459) is considered; however, at this point a subtle discussion regarding the sign of the chemical potential is needed for better emphasizing the symmetry-breaking phenomena in bonding, see Figure 3.11. As such, if one considers the potential (Putz, 2008d)

$$V_\mu(\phi) = \mu\phi^2 + \frac{1}{2}\eta\phi^4 \qquad (3.464)$$

and represent it graphically with $\mu > 0$, $\eta > 0$ the parabolic dependence is obtained (see the upper dashed curve in Figure 3.11) that presents the

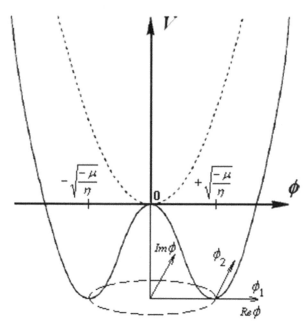

FIGURE 3.11   The potential of Eq. (3.464) with $+\mu$, in dashed curve and with $-\mu = \chi$ on continuous curve, illustrating the symmetry breaking of the parabolic (upper curve) to double well potential (lower curve) that includes also the negative (vacuum) quantum states (Putz, 2008d).

minimum zero potential for the vanishing field $\phi=0$. Instead, a completely different picture is obtained if the same potential is considered with $\mu<0$, $\eta>0$ when two distinct non-zero minimum potential values appears in its negative (vacuum) region for the chemical field acquiring the respective values (Putz, 2008d)

$$\frac{\partial V_\mu}{\partial \phi} = 0 \Rightarrow \phi_{a,b} = \pm\sqrt{\frac{-\mu}{\eta}} = \pm\sqrt{\frac{\chi}{\eta}} \qquad (3.465)$$

There is said that going from the potential driven by chemical potential $+\mu$ to that driven by electronegativity $-\mu=\chi$, see Eq. (3.458), the chemical field becomes the chemical bonding field by shifting the minimum zero potential to its minimum negative range, in quantum vacuum region from where the quantum particles to be spontaneous created, the bondons as the quantum particles of the chemical bonding fields.

For the analytical procedure, the chemical bonding may be described by field potential relating electronegativity and chemical hardness (Putz, 2008d)

$$V_\chi(\phi) = \chi\phi^2 + \frac{1}{2}\eta\phi^4 \qquad (3.466)$$

considered with the Schrödinger Lagrangian of Eq. (3.461).

Now is clear that both chemical bonding potential of Eq. (3.466) and the Lagrangian of Eq. (3.461) are invariant under the field's phase transformation

$$\phi \rightarrow \phi_0 = e^{i\alpha}\phi \qquad (3.467)$$

but not under the chemical field transformations involving the shift given by Eq. (3.465) towards negative minimums of the chemical bonding potential (see Figure 3.11).

However, the shift towards broken symmetry may be realized either along the real axis of the chemical field (Putz, 2008d)

$$\phi \rightarrow \phi' = \phi + \sqrt{\frac{\chi}{\eta}} \qquad (3.468)$$

or when also imaginary contribution to the chemical field in vacuum state is counted as (Putz, 2008d)

$$\phi \to \phi'' = \frac{1}{\sqrt{2}}(\phi_1 + i\phi_2) + \sqrt{\frac{\chi}{\eta}} \tag{3.469}$$

Within the last two situations there will be obtained additional chemical bonding potentials and Lagrangian's creating the new chemical fields and associate quantum particles. Resuming, the present spontaneous symmetry-breaking mechanism we may build up the invariant–noninvariant potential-Lagrangian scheme (Chaichian & Nelipa, 1984; Putz, 2008d):

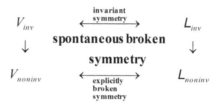

Yet, the chemical bonding fields ϕ_B appeared from the difference between the noninvariant and invariant Lagrangian's are then employed to evaluate the mass of the correspondent chemical bondons; their ansatz formula may look like (Putz, 2008d):

$$m_B = m_0 \frac{1}{2^3} \frac{R_B^3}{a_0^3} |\phi_B|^2 \tag{3.470}$$

based on the fact that both the single electron in Hydrogen atom and the electron of bonding fulfill the phenomenological mass-volume-probability density relationships with the same electronic dimensional constant (Putz, 2008d)

$$m_0 = ct_{[e]} \times \mathcal{V}_0 \times \wp_0, \quad \mathcal{V}_0 = \frac{4\pi}{3} a_0^3, \quad \wp_0 = 1 \tag{3.471}$$

$$m_B = ct_{[e]} \times \mathcal{V}_B \times \wp_B, \quad \mathcal{V}_B = \frac{4\pi}{3}\left(\frac{R_B}{2}\right)^3, \quad \wp_B = |\phi_B|^2 \tag{3.472}$$

This bondons' mass description is in accordance with the assumption that the probability of electron localization within the sphere encompassed by the first Bohr radius a_0 is close to unity while that corresponding to the bonding electron compressed by the sphere designed by the half-bond radius $R_B/2$ equals the squared of the chemical bonding field risen through spontaneous symmetry breaking effect.

3.6.2.2   First Kind of Chemical Bonding Field

Aiming to identify the chemical bonding fields, then used in bondons mass' calculation, one should consider the "reference" invariant Lagrangian obtained under transformation of Eq. (3.467) with chemical bonding potential of Eq. (3.466) replaced in Eq. (3.461) (Putz, 2008d)

$$\mathscr{L} = \mathscr{L}_0 = i\hbar \phi^* \dot{\phi} - \frac{\hbar^2}{2m_0}\left(\nabla \phi^*\right)\left(\nabla \phi\right) + \chi \phi^4 - \frac{1}{2}\eta \phi^6 \tag{3.473}$$

Now, when transformation (3.468) is performed the chemical bonding field is generated by appearance of the enriched Lagrangian respecting that of Eq. (3.473)

$$\mathscr{L}' = \mathscr{L} + \overline{\mathscr{L}}' \tag{3.474}$$

with additional sector (Putz, 2008d)

$$\overline{\mathscr{L}}' = i\hbar\sqrt{\frac{\chi}{\eta}}\dot{\phi} - 3\eta\sqrt{\frac{\chi}{\eta}}\phi^5 - \frac{15}{2}\chi\phi^4 - 6\chi\sqrt{\frac{\chi}{\eta}}\phi^3 - \frac{3\chi^2}{2\eta}\phi^2 + \frac{\chi^2}{\eta}\sqrt{\frac{\chi}{\eta}}\phi + \frac{\chi^3}{2\eta^2} \tag{3.475}$$

Under the condition the Lagrangian (3.474) be invariant for the transformation (3.468), i.e., equal with (3.473), the correction Lagrangian (3.475) should be canceled, that is the higher order differential equation is provided (Putz, 2008d)

$$i\hbar\dot{\phi} = 3\eta\phi^5 + \frac{15}{2}\chi\sqrt{\frac{\eta}{\chi}}\phi^4 + 6\chi\phi^3 + \frac{3\chi^2}{2\eta}\sqrt{\frac{\eta}{\chi}}\phi^2 - \frac{\chi^2}{\eta}\phi - \frac{\chi^3}{2\eta^2}\sqrt{\frac{\eta}{\chi}} \tag{3.476}$$

whose solutions gives the first class of chemical bonding fields. However, since Eq. (3.476) has no analytical solution one may try to solve a more specialized equation derived from it and then employing the solution to a meaningfully physical-chemical picture. Such framework can be achieved when recognizing that transformation (3.468) leaves invariant the Lagrangian (3.473) also under the vanishing electronegativity condition

$$\chi = 0 \tag{3.477}$$

which, when replaced on Eq. (3.476) still preserves its highest order nature for the power field (Putz, 2008d):

$$i\hbar\dot{\phi} = 3\eta\phi^s \tag{3.478}$$

In fact, such condition of electronegativity asymptotic cancellation limit may be assumed with the picture in which the bonding adducts are seen in their first instance as being infinitely separated, from where they are encountering in bonding creation. Moreover, there is noted from Eq. (3.478) that chemical hardness plays the role of driving chemical bonding (out of quantum vacuum) even when electronegativity effect is absent; this may be constitute as another argument in favor of assuming the chemical hardness principles as intrinsic to chemical constitution and reactivity and not only as consequence of those regarding electronegativity; in other words, we may consider chemical hardness not only as the second order (or as a perturbation) accompanying the electronegativity effect in chemical reactivity but also as an independent (or intrinsic) component of it.

Even more, Eq. (3.478) may consecrate the actual *chemical hardness quantum bonding equation*, as rooting in Schrödinger one, however with chemical hardness carrying the Hamiltonian role and acting on a fifth power of chemical field; this may lead with conclusion that chemical reactivity appears in higher order of power fields respecting those of physical quantum (tunneling) phenomena. These special features will be for sure continuing in further related studies.

Integration of Eq. (3.478) within above mentioned asymptotic (ϕ_∞) – to – bonding (ϕ_B) linked regimes can be firstly written under the form (Putz, 2008d):

$$\frac{1}{\phi_\infty^4} - \frac{1}{\phi_B^4} = \frac{12\eta i \Delta t}{\hbar} \tag{3.479}$$

and then, through applying the time-to-temperature Wick transformation (Weiss, 1999)

$$i\Delta t = \hbar\beta, \ \beta = \frac{1}{k_B T} \tag{3.480}$$

with k_B-Boltzmann constant, the first kind spontaneous-breaking-symmetry chemical bonding field takes the (first order) expanded form (Putz, 2008d)

$$\phi_B^I = \phi_\infty \left[1 + 6\beta \left(\frac{1}{2} \eta \phi_\infty^4 \right) \right] \tag{3.481}$$

Note that there was appropriately evidenced the same chemical hardness reactivity correction to asymptotic chemical field as that appeared in chemical reactivity potential (3.466) for better illustration of the analytical consistency of the present treatment. Going to comment upon the chemical bonding solution given by Eq. (3.481) worth remarking that, for infinite separation, i.e., for infinite bonding length, the temperature describing such situation goes infinitely as well while leaving with the limits $\beta \to 0$ and $\phi_B^I \to \phi_\infty$. However, the chemical bonding field correction as the fifth power of the chemical asymptotic field is in close agreement with the fifth order form of the original equation (3.478). On the other side, for practical implementation the asymptotic chemical field has to be nominated; a suitable choice may be abstracted from the asymptotic electronic density behavior as being determined by the ionization potential $IP \approx h$, for finite difference approximation within the framework of Eq. (3.477), within DFT conceptual development (Ayers et al., 1998), to be with the dimensionless form

$$\phi_\infty(R_B, \eta) = \exp\left(-\Xi R_B \sqrt{2\eta}\right) \tag{3.482}$$

involving the calibration constant (Putz, 2008d)

$$\Xi = a_0\left(\mathring{A}\right)^{-1} \cdot 1\,a.u.(eV)^{-1/2} = 0.362 \times 10^{10}\left[\mathring{A}^{-1} \cdot eV^{-1/2}\right] \tag{3.483}$$

Note that the normalization condition is not a compulsory quantum constraint when about creation of particles from quantum vacuum as it is the present case. With these, one may resume that the *first kind of chemical bonding fields* appeared from spontaneous symmetry-broken invariant reactivity chemical field are jointly characterized by chemical hardness, chemical bond length and the sample temperature throughout the working relations (3.481)–(3.483).

3.6.2.3   Second Kind Chemical Bonding Field

When the chemical field transformation (3.469) is undertaken the noninvariant spontaneously broken Lagrangian for the second kind chemical bonding fields yields (Putz, 2008d):

$$\mathscr{L}^{II} = \mathscr{L}^{II}_{10} + \mathscr{L}^{II}_{20} + \mathscr{L}^{II}_{11} + \mathscr{L}^{II}_{22} + \mathscr{L}^{II}_{12} \tag{3.484}$$

with the Lagrangians' sectors corresponding to:
-essentially invariant real component sector

$$\mathscr{L}^{II}_{10} = \frac{i\hbar}{2}\phi_1^*\dot{\phi}_1 - \frac{\hbar^2}{4m_0}(\nabla\phi_1)^2 + \frac{\chi}{4}\phi_1^4 - \frac{1}{16}\eta\phi_1^6 \tag{3.485}$$

-essentially invariant imaginary component sector

$$\mathscr{L}^{II}_{20} = \frac{i\hbar}{2}\phi_2^*\dot{\phi}_2 - \frac{\hbar^2}{4m_0}(\nabla\phi_2)^2 + \frac{\chi}{4}\phi_2^4 - \frac{1}{16}\eta\phi_2^6 \tag{3.486}$$

-the first type chemical bonding field Lagrangian sector for the real component

$$\mathscr{L}^{II}_{11} = \frac{i\hbar}{\sqrt{2}}\sqrt{\frac{\chi}{\eta}}\dot{\phi}_1 - \frac{3\eta}{4\sqrt{2}}\sqrt{\frac{\chi}{\eta}}\phi_1^5 - \frac{15}{8}\chi\phi_1^4 - \frac{3\chi}{\sqrt{2}}\sqrt{\frac{\chi}{\eta}}\phi_1^3 - \frac{3\chi^2}{4\eta}\phi_1^2$$
$$+ \frac{\chi^2}{\sqrt{2\eta}}\sqrt{\frac{\chi}{\eta}}\phi_1 + \frac{\chi^3}{2\eta^2} \tag{3.487}$$

-the perturbation chemical field Lagrangian sector for the imaginary component:

$$\mathscr{L}^{II}_{22} = -\frac{\hbar}{\sqrt{2}}\sqrt{\frac{\chi}{\eta}}\dot{\phi}_2 - \frac{3}{8}\chi\phi_2^4 + \frac{\chi^2}{4\eta}\phi_2^2 \tag{3.488}$$

-and the so called interference chemical fields Lagrangian sector:

$$\mathscr{L}^{II}_{12} = -\frac{\hbar}{2}\dot{\phi}_1\phi_2 - \left(\frac{7}{4}\chi + \frac{3}{16}\eta\phi_1^2 + \frac{3}{16}\eta\phi_2^2 + \frac{3\eta}{2\sqrt{2}}\sqrt{\frac{\chi}{\eta}}\phi_1\right)\phi_1^2\phi_2^2$$
$$- \sqrt{2}\sqrt{\frac{\chi}{\eta}}\left(\frac{\chi}{2}\phi_2 + \frac{3}{8}\eta\phi_2^3\right)\phi_1\phi_2 \tag{3.489}$$

One may easily note that the first two Lagrangians' sectors, Eqs. (3.485) and (3.486), correspond with the invariant Lagrangian (3.473) up to a constant factor $1/\sqrt{2}$ that modifies ϕ into $\phi_1/\sqrt{2}$ and $\phi_2/\sqrt{2}$, respectively; the same type of conversion help in recognizing the Lagrangian (3.487) as being of type (3.475) thus producing no new chemical bonding field out of spontaneous-breaking of quantum vacuum; instead, Lagrangian sectors (3.488) and (3.489) are new and susceptible to produce new type of chemical bonding fields and associate bondonic particles. In this respect the canceling of Lagrangians (3.488) and (3.489) will provide solvable coupled equations under the stationary field conditions:

$$\dot{\phi}_1 = 0, \dot{\phi}_2 = 0 \tag{3.490}$$

With the setting (3.490) the Lagrangian (3.488) leads with the equation (Putz, 2008d)

$$0 = -\frac{3}{8}\chi\phi_2^4 + \frac{\chi^2}{4\eta}\phi_2^2 \tag{3.491}$$

having trivial solution $\phi_2 = 0$ that regains the previous chemical bonding field case, along the non-zero realization (Putz, 2008d)

$$\phi_2 = \pm\sqrt{\frac{2}{3}\frac{\chi}{\eta}} \tag{3.492}$$

However, the field (3.492) cannot be assumed alone as the second type chemical bonding field solution but only along the ϕ_1 component; this last one can be found by substitution of ϕ_2 with (3.492) form on the equation obtained from cancellation of Lagrangian (3.489) under stationary conditions (3.490); it finally produces the chemical bonding field equation (Putz, 2008d)

$$\frac{\eta}{16}\sqrt{\frac{6\chi}{\eta}}\phi_1^3 + \frac{\sqrt{3}}{2}\chi\phi_1^2 + \frac{15}{8}\chi\sqrt{\frac{2\chi}{3\eta}}\phi_1 + \frac{3}{2\sqrt{3}}\frac{\chi^2}{\eta} = 0 \tag{3.493}$$

that is solved for single real triple (equal) solution

$$\phi_2 = -2\sqrt{\frac{2\chi}{\eta}} \tag{w(3.494)}$$

Cumulating both chemical field components (3.492) and (3.494) in the chemical field (3.469) produces the *second kind of chemical bonding fields* in terms of electronegativity and chemical hardness only (Putz, 2008d):

$$\phi_B'' = \frac{1}{\sqrt{2}}(\phi_1 + i\phi_2) = \sqrt{\frac{\chi}{\eta}}\left(-2 \pm \frac{i}{\sqrt{3}}\right) \tag{3.495}$$

Worth noting that although being imaginary in nature, the second kind chemical bonding field (3.495) will produce single real finite electronegativity and chemical hardness contained probability density, namely (Putz, 2008d)

$$\left|\phi_B^{II}\right|^2 = \frac{13}{3}\frac{\chi}{\eta} \qquad (3.496)$$

then used to compute the associate bondonic mass upon Eq. (3.470). Moreover, the second kind chemical bonding field (3.495) resulted to be naturally dimensionless, as well as not necessary normalizable to unity, this way validating both the corresponding dimensionless choice of the first kind chemical bonding field in (3.482) and the particle creation nature of these not normalized fields.

For both types of chemical bonding fields arisen from spontaneous breaking symmetry of the chemical field in Schrödinger Lagrangian the corresponding bondon particles' masses will be quantified for paradigmatic chemical bonds and discussed thereafter in next section.

3.6.2.4 Bondons' Masses For Ordinary Chemical Bonds

Going to compute the present developed chemical bonding fields and bondons' masses for some common chemical AB bonds cases (see Table 3.1) the bonding electronegativity and hardness quantities are firstly evaluated upon the atoms-in-molecules formulas abstracted from (3.249) and (3.250) (Bratsch, 1984)

$$\chi_{Bond} = \frac{2}{\dfrac{1}{\chi_A} + \dfrac{1}{\chi_B}}, \ \eta_{Bond} = \frac{2}{\dfrac{1}{\eta_A} + \dfrac{1}{\eta_B}} \qquad (3.497)$$

entering the chemical bonding fields of the first kind (only η_{Bond}) and of second kind (both χ_{Bond} and η_{Bond}), Eqs. (3.481) and (3.495), once the experimental or otherwise estimated bond length is provided.

The results of Table 3.16 display interesting features: firstly there is noted that bondons have either larger or smaller mass than of electrons, corresponding to localized or delocalized character of bonding; however, there seems that the first kind of chemical bonding fields merely carriers

TABLE 3.16 The Chemical Bond Characteristics: Bond Length R_B (Oelke, 1969), Bond Electronegativity χ_B and Bond Chemical Hardness η_B Computed with Eq. (3.497) Employing the Atomic Experimental Values (Lackner & Zweig, 1983), Entering the Quantum Symmetry-Breaking Chemical Field Quantities: Squared (probability density) of Chemical Bonding Fields ϕ_B of Eqs. (3.481) and (3.495), and Associated Bondons to Electrons Mass' Particles Ratio Given by Eq. (3.470) for the First Kind (I, at the "room temperature" of 300K) and for the Second Kind (II) of Broken Symmetry, Respectively, for Typical Chemical Bonds (Putz, 2008d)

Bond Type	R_B (Å)	χ_{Bond} (eV)	η_{Bond} (eV)	$\left\|\phi_B^I\right\|^2$	$\left\|\phi_B^{II}\right\|^2$	$\dfrac{m_B^I}{m_0}$	$\dfrac{m_B^{II}}{m_0}$
H–H	*0.60*	*7.18*	*6.45*	*243.24*	*4.82*	*44.36*	*0.88*
C–C	*1.54*	*6.24*	*4.99*	*0.07*	*5.42*	*0.21*	*16.71*
C=C	*1.34*	*6.24*	*4.99*	*0.24*	*5.42*	*0.48*	*11.01*
C≡C	*1.20*	*6.24*	*4.99*	*0.74*	*5.42*	*1.08*	*7.91*
N≡N	*1.10*	*6.97*	*7.59*	*0.35*	*3.98*	*0.39*	*4.47*
O=O	*1.10*	*7.59*	*6.14*	*0.83*	*5.36*	*0.94*	*6.02*
F–F	*1.28*	*10.4*	*7.07*	*0.1*	*6.37*	*0.17*	*11.29*
Cl–Cl	*1.98*	*8.32*	*4.68*	*0.01*	*7.7*	*0.1*	*50.49*
I–I	*2.66*	*6.76*	*3.74*	*0.001*	*7.83*	*0.08*	*124.48*
C–H	*1.09*	*6.68*	*5.63*	*1.29*	*5.14*	*1.41*	*5.62*
N–H	*1.02*	*7.07*	*6.97*	*1.15*	*4.40*	*1.03*	*3.94*
O–H	*0.96*	*7.38*	*6.29*	*3.34*	*5.08*	*2.50*	*3.80*
C–O	*1.42*	*6.85*	*5.51*	*0.09*	*5.39*	*0.23*	*13.02*
C=O (in CH2O)	*1.21*	*6.85*	*5.51*	*0.46*	*5.39*	*0.69*	*8.06*
C=O (in O=C=O)	*1.15*	*6.85*	*5.51*	*0.79*	*5.39*	*1.01*	*6.92*
C–Cl	*1.76*	*8.32*	*4.83*	*0.03*	*7.46*	*0.13*	*34.36*
C–Br	*1.91*	*6.85*	*4.60*	*0.02*	*6.45*	*0.11*	*37.97*
C–I	*2.10*	*6.49*	*4.28*	*0.01*	*6.57*	*0.10*	*51.38*

particles with less mass than electrons, thus being more delocalized, in close agreement with asymptotic field dependency of Eq. (3.482) that enters Eq. (3.481); on the other side the second kind chemical bonding fields behave as particles with greater mass than that of electrons due to the electronegativity-chemical hardness ratio appearing in probability density field of Eq. (3.496) that is custom to be over unity since seeing chemical hardness as a second order effect of chemical reactivity, thus with lower value than electronegativity; this is also in agreement with systematically lower mass of bondons of the first kind that depend on chemical hardness but not also on electronegativity, respecting those of the second kind which combines both chemical reactivity indices (Putz, 2008d).

There are also cases when bondons are close with the electron mass, namely for C≡C, N–H, and C=O (in O=C=O) bonds within first kind chemical bonding, and for H–H bond within second kind chemical bonding; most interestingly, the H_2 molecule poses the heaviest bondon in the first kind bonding and the lightest bondon in the second kind bonding pictures; however, noting the first kind of bonding is dependent on bonding temperature as well, there follows that the temperature at which the bondons acquire electronic mass can be calculated for each bonding case; yet, for limiting temperature of 0[°K] the infinite mass bondon is formed, consistent with the fact since created from the quantum vacuum, it may be considered as the quanta of the frozen Universe's wave (Putz, 2008d)!

Nevertheless, Chemistry in general and chemical bonding in special cast at finite temperature; therefore whenever the analysis would demand the temperature factor the first chemical bonding picture is the most suited; instead, when isolated or primarily structural study is envisaged the chemical reactivity indices of electronegativity and chemical hardness drive the chemical bonding field and associate bondon's mass at the optimum bond length. Still, in general, their reciprocal trends are in opposite direction (see the C-C, C-O, and C-X^{VII} series for instance) however preserving the systematic rationale inside of their framework (Putz, 2008d).

Moreover, the first and second kinds of chemical bond may eventually be related (and/or correlated) with a phase-transition phenomena, i.e., to further describe the chemical bonding as a combined process of these two types of bindings, for example, starting with one type at asymptotic encountering adducts and then passing to the other type for the stabilization-optimization of the bonding structure.

3.6.3 TESTS OF BIVARIATE ENERGY

The general bivariate equation linking the energy (various) functionals with electronegativity and chemical hardness, either for atomic and molecular systems, works out with form (Putz, 2008a):

$$E = a + b\chi + c\eta \qquad (3.498)$$

with coefficients a, b, and c being determined throughout consecrated statistical analysis methods. The result of such correlation will lead with two kind of information:

- the degree of correlation itself between the employed energy functional and the couple of electronegativity-chemical hardness structural indices; this is measured by *the standard correlation factor*:

$$r = \sqrt{1 - \frac{\sum_i (y_{iINPUT} - y_{iFIT})^2}{\sum_i (y_{iINPUT} - \overline{y_{iINPUT}})^2}} \qquad (3.499)$$

- the degree of parabolic dependency by checking whether the chemical hardness coefficient (c) is the square of the electronegativity coefficient (b) thus giving the opportunity of introducing the so-called *sigma-pi reactivity index* (Putz, 2008a)

$$\sigma_\pi = sign(b) \frac{c}{b^2} \xrightarrow{\text{parabolic } E=E(\chi,\eta)} -1 \qquad (3.500)$$

At this point worth nothing that as $\sigma_\pi \to -1$ as better the energy fulfills the correlation shape with the established parabolic equation. The b and c signs open further discussion among the allowed combinations for the correlation equation. Firstly, let's observe that the signs of b and c give information about the signs of electronegativity and chemical hardness, respectively. Then, while considering the finite-difference approximations to the electronegativity and chemical hardness definitions, in a Koopmans theorem environment, they take the working forms (3.346) and (3.347) in terms of lowest unoccupied molecular orbital (*LUMO*) and highest occupied molecular orbital (*HOMO*) energies, although similar relations (in terms of ionization potential *IP* and electron affinity *EA*) hold for atomic systems as well. Nevertheless, these expressions help in deciding that only the (−, +) and (+, +) combinations for coefficients (b, c) are allowed, based on the fact that only maximum hardness values (i.e., positive values of c coefficient) are in accordance with the reactivity criteria of maximum hardness principle; in other words, the correlation combination carrying negative values for chemical hardness coefficient are less physically probably since that would imply that $EA>0$ and $LUMO <0$, for atoms and molecules, respectively.

Finally, the sigma-pi index can be used in defining another reactivity index, namely *the efficiency index* (Putz, 2008a):

$$\Sigma_\Pi = |\sigma_\pi| \xrightarrow{\;\;parabolic\;\;E=E(\chi,\eta)\;\;} 1 \qquad (3.501)$$

measuring the power of electronic exchange characterizing the chemical reactivity, which is maximized by the exact parabolic dependence of the energetic functional respecting the electronegativity-chemical hardness jointly influence.

3.6.3.1 Testing Simple Atoms and Molecules

With these, the Tables 3.17–3.21 display the tested energetic functionals against various electronegativity and chemical hardness scales for selected atoms and molecules, while the Tables 3.22 and 3.23 show their quantitative (χ & η) structural–(energy functionals) property relationships (QSPR), respectively; the forthcoming discussion follows (Putz, 2008a).

At a glance, Tables 3.22 and 3.23 reveals that the parabolic energetic dependency on the electronegativity and chemical hardness indices is not exactly recovered since all sign combinations in the coefficients c & b as well as poor correlation itself with the (χ, η) indices is revealed, i.e., widely deviating from the optimum (–,+) and maximum correlation coefficient. However, for atomic analysis, Table 3.22 gives us that the finite-difference scale of electronegativity and chemical hardness provides invariable right (–, +) sign combination in coefficients (b, c), excepting kinetic energy functionals cases, however revealing both scarce reactive efficiency and bilinear correlation in almost all treated cases. Notably, the higher reactivity efficiency on studied atomic system was furnished by the correlation energy merely than the total one, yet with a low correlation factor.

The situation is somehow changed for molecular systems, see Table 3.23, where, albeit not better records are noted in the sense of the maximum reactive efficiency, again excepting the correlation functional case rather than the total energy, a higher correlation factor in semiclassical treatment

TABLE 3.17 Atomic Kinetic, Exchange, and Correlation, Energies (in Hartrees) From Various Schemes of Computations (*)

Atoms	Kinetic energy				Exchange energy			Correlation energy		
	T_{exact}♦	T_0♦	T_0+T_2♦	$T_{Padé}$♦	K_{exact}•	K_0♦	K^{B88}▼	E_c^{exact}♦	$E_c^{(139)}$♠	E_c•
He	2.86168	2.56054	2.87850	2.87639	−1.0260	−0.884	−1.025	−0.0425	−0.0215	−0.0681
Li	7.43273	6.70062	7.50504	7.44941	−1.7812	−1.538	−1.775	−0.0454	−0.0486	−0.0815
Be	14.5730	13.1286	14.6466	14.4223	−2.6669	−2.312	−2.658	−0.0945	−0.0820	−0.1192
B	24.5291	22.0720	24.5228	24.2089	−3.7438	−3.272	−3.728	−0.1247	−0.1197	−0.1625
C	37.6886	34.0144	37.5988	37.2533	−5.0444	−4.459	−5.032	−0.1566	−0.1609	−0.2091
N	54.4009	49.4771	54.3852	54.0643	−6.5971	−5.893	−6.589	−0.1850	−0.2050	−0.2567
O	74.8094	67.8965	74.3573	74.1625	−8.1752	−7.342	−8.169	−0.2579	−0.2512	−0.3035
F	99.4093	90.4598	98.6429	98.6959	−10.003	−9.052	−10.02	−0.332	−0.2996	−0.3510
Ne	128.547	117.761	127.829	128.221	−12.108	−11.03	−12.14	−0.390	−0.3498	−0.3987
Na	161.859	148.809	161.093	161.718	−14.017	−12.79	−14.03	−0.398	−0.3892	−0.4137
Mg	199.614	184.017	198.749	199.578	−15.994	−14.61	−16.00	−0.443	−0.4351	−0.4491
Al	241.877	223.443	240.868	242.008	−18.069	−16.53	−18.06	−0.480	−0.4809	−0.4863
Si	288.854	267.315	287.659	289.139	−20.280	−18.59	−20.27	−0.521	−0.5308	−0.5308
Cl	459.482	426.865	457.321	460.117	−27.512	−25.35	−27.49	−0.714	−0.6901	−0.6710
Ar	526.817	490.017	524.289	527.617	−30.185	−27.86	−30.15	−0.787	−0.7459	−0.7190

♦: from DePristo & Kress (1987) and references therein; ♣: from Lee & Parr (1990) and references therein; ♥: from Becke (1988);
♠: from Eq. (139) of Putz (2008a) – see also the Volume I of the present five-volume series (Putz, 2016a), and Liu et al. (1999);
•: from fitting equation $E_c = -0.04682N + 0.005753<\rho^{2/3}>-0.00096<\rho^{1/3}>$, see Liu et al. (1999) and references therein.
*The exact values are computed with Hartree-Fock densities (Putz, 2008a).

TABLE 3.18 Atomic Exchange-Correlation and Total Energies (in Hartrees) From Various Schemes of Computations (*)

Atoms	Exchange-Correlation energy					Total energy				
	E_{xc}^{exact}♣	$E_{xc}^{I(Xa)}$♠	$E_{xc}^{II(Xa)}$♠	$E_{xc}^{I(Wig)}$♠	$E_{xc}^{II(Wig)}$♠	E_{tot}^{exact}♥	$E_{tot}xc(RG)$♥	$E_{tot}^{xc(LDA)}$▼	E_{tot}^{BLYP}♠	E_{tot}^{PW91}♠
He	-1.0685	-1.0604	-1.0566	-1.0633	-1.0654	-2.9042	-3.0317	-2.8601	-2.9071	-2.9000
Li	-1.8266	-1.8048	-1.8134	-1.8093	-1.8108	-7.4781	-7.6473	-7.3704	-7.4827	-7.4742
Be	-2.7614	-2.7260	-2.7522	-2.7325	-2.7342	-14.6675	-14.8911	-14.4966	-14.6615	-14.6479
B	-3.8685	-3.8126	-3.8415	-3.8215	-3.8177	-24.6538	-24.9158	-24.4097	-24.6458	-24.6299
C	-5.2010	-5.1127	-5.1338	-5.1248	-5.1121	-37.8163	-38.1305	-37.5095	-37.8430	-37.8265
N	-6.7821	-6.6400	-6.6440	-6.6558	-6.6321	-54.4812	-54.8681	-54.1287	-54.5932	-54.5787
O	-8.4331	-8.3599	-8.3405	-8.3796	-8.3450	-75.0271	-75.4597	-74.5979	-75.0786	-75.0543
F	-10.325	-10.327	-10.277	-10.350	-10.305	-99.741	-100.235	-99.247	-99.7581	-99.7316
Ne	-12.498	-12.551	-12.466	-12.579	-12.524	-128.937	-129.522	-128.403	-128.9730	-128.9466
Na	-14.415	-14.462	-14.382	-14.488	-14.445	-162.257	-162.862	-161.624	-162.293	-162.265
Mg	-16.437	-16.482	-16.424	-16.504	-16.484	-200.058	-200.705	-199.340	-200.093	-200.060
Al	-18.549	-18.566	-18.542	-18.583	-18.593	-242.357	-243.028	-241.533	-242.380	-242.350
Si	-20.801	-20.774	-20.791	-20.784	-20.830	-289.356	-290.063	-288.435	-289.388	-289.363
Cl	-28.226	-28.115	-28.272	-28.092	-28.281	-460.196	-461.005	-458.963	-460.165	-460.147
Ar	-30.972	-30.827	-31.037	-30.789	-31.035	-527.605	-528.452	-526.267	-527.551	-527.539

♣: from Lee & Parr (1990); ♠: from Lee & Bartolotti (1991); ♠: from Eqs. (170)–(173) of Putz (2008a) – see also the Volume I of the present five-volume series (Putz, 2016a); and Lee & Parr (1990) ♥: from (Grabo & Gross (1995). ▼: from Hartree-Fock densities (Putz, 2008a).

*The exact values are computed with Hartree-Fock densities (Putz, 2008a).

TABLE 3.19 Values (in Hartrees) of the Structural Indices Electronegativity (χ), Chemical Hardness (η), in Finite-Difference, Putz-DFT (Mulliken, 1934; Lackner & Zweig, 1983) and Putz-SC (Putz, 2007b) (semiclassical) Modes of Computations for Atoms of Tables 3.17 and 3.18 (Putz, 2008a)

Level Atoms	Finite-Difference		Functional		Semiclassical	
	χ_{FD}	η_{FD}	χ_{DFT}	η_{DFT}	χ_{SC}	η_{SC}
He	0.45094	0.45866	1.21132	1.66189	0.57038	0.2172
Li	0.11099	0.16134	0.15105	0.08784	0.00412	0.00334
Be	0.12606	0.21794	0.44248	0.44579	0.00893	0.0047
B	0.15656	0.14921	1.15362	1.34105	0.01526	0.00588
C	0.22933	0.18339	2.76332	2.9695	0.02279	0.00684
N	0.25616	0.27894	5.79566	4.91363	0.03139	0.0076
O	0.27894	0.22566	10.6505	5.91694	0.04072	0.00816
F	0.38221	0.25983	16.9129	4.37707	0.05061	0.00849
Ne	0.39361	0.40132	23.7119	−0.08747	0.06079	0.00864
Na	0.10290	0.10621	0.23153	0.18743	0.00011	0.00005
Mg	0.09555	0.18339	0.49871	0.53142	0.00018	0.00007
Al	0.11834	0.10327	1.04631	1.19882	0.00026	0.00008
Si	0.17200	0.12606	2.10805	2.30724	0.00036	0.00009
Cl	0.30577	0.17120	11.5766	7.7692	0.00074	0.00012
Ar	0.28299	0.29806	17.8831	9.08857	0.00088	0.00013

of the electronegativity and chemical hardness scales is obtained; such behavior was recently validated for hard and soft acids and bases reactions as well. At the same time, the kinetic energy functionals do not fit at all with standard (−,+) correlation scheme for signs of (c, b) coefficients.

Finally, in Figure 3.12 the atomic and molecular parabolic reactivity efficiencies of various energetic functionals are plotted as charts with the widths emphasizing the bilinear electronegativity and chemical hardness correlation values. The comparative behavior reveals many interesting features to be taking into account for further conceptual and computational studies on quantum chemistry:

- the correlation energy appears to provide acceptable parabolic shapes in both atomic and molecular cases, with better bilinear regression

TABLE 3.20 Molecular Kinetic, Exchange, Correlation, Exchange-Correlation, and Total Energies (in Hartrees) From Various Schemes of Computations

Molecules	Kinetic		Exchange		Correlation		Exch.-corr.		Total energy	
	$T_0^{\clubsuit}$	$T_0+T_2^{\clubsuit}$	$K^{exact\clubsuit}$	$K^{PBE\clubsuit}$	$E_c^{VWN\clubsuit}$	$E_c^{GC\clubsuit}$	$E_{xc}^{exact\clubsuit}$	$E_{xc}^{PBE\clubsuit}$	$E_{tot}^{BLYP\heartsuit}$	$E_{tot}^{TH\heartsuit}$
H_2	1.140	1.125	−0.657	−0.648	-95×10^{-3}	-47×10^{-3}	−0.698	−0.691	−1.169	−1.178
LiH	7.978	8.003	−2.125	−2.105	-219×10^{-3}	-93×10^{-3}	−2.212	−2.188	−8.068	−8.070
CH_4	40.050	40.141	−6.576	−6.536	-593×10^{-3}	-328×10^{-3}	−6.883	−6.836	−40.502	−40.515
H_2O	76.150	75.477	−8.910	−8.917	-664×10^{-3}	-365×10^{-3}	−9.292	−9.241	−76.448	−76.433
HF	100.137	99.242	−10.378	−10.385	-704×10^{-3}	-380×10^{-3}	−10.779	−10.720	−100.48	−100.455
N_2	109.115	108.242	−13.094	−13.128	-945×10^{-3}	-506×10^{-3}	−13.665	−13.580	−109.559	−109.54
O_2	149.843	148.369	−16.290	−16.358	-1110×10^{-3}	-599×10^{-3}	−16.958	−16.887	−150.384	−150.337
F_2	198.892	196.729	−19.872	−19.951	-1302×10^{-3}	-697×10^{-3}	−20.661	−20.564	−199.599	−199.533

♣: from Perdew et al. (1998); ♦: from Savin et al. (1986); ♥: from Tozer & Handy (1998).

*The exact values are computed with HF densities (Putz, 2008a).

TABLE 3.21 Values (in Hartrees) of the Structural Indices Electronegativity (χ), Chemical Hardness (η), Computed by Means of the Group Method Within the Finite-Difference (Bratsch, 1985; Putz, 2008b), Putz-DFT and Putz-SC (semiclassical) Modes of Computations for Molecules of Table 3.20 (Putz, 2008a)

Level	Finite-Difference		Functional		Semiclassical	
Molecules	χ_{FD}	η_{FD}	χ_{DFT}	η_{DFT}	χ_{SC}	η_{SC}
H$_2$	0.26387	0.2370	0.26384	0.23704	0.26387	0.23705
LiH	0.15626	0.192	0.19212	0.12818	0.00811	0.006596
CH$_4$	0.25616	0.2239	0.32216	0.29051	0.08468	0.03064
H$_2$O	0.26871	0.2331	0.39097	0.34859	0.09335	0.0229
HF	0.3077	0.2479	0.51964	0.44974	0.08493	0.01639
N$_2$	0.25616	0.2789	5.79566	4.91363	0.03139	0.00761
O$_2$	0.27894	0.2257	10.6505	5.91694	0.04072	0.00816
F$_2$	0.36898	0.2598	16.9129	4.37707	0.05061	0.00849

for molecular analysis, while strongly depending on the electronegativity and chemical hardness used atomic scales;

- the kinetic energy, while displaying poor parabolic shape at atomic level behaves with negative chemical hardness in molecular systems, probably due the positive contribution in bonding that do not stabilize (localize) the electrons within internuclear basin;
- exchange and exchange-correlation functionals reveal similar reactive (parabolic) efficiency as well as close bilinear regression correlation factor for both atomic and molecular cases, leaving with the impression that the exchange contribution is dominant in exchange-correlation functionals and cancelling somehow the good behaving correlation part of the functional.
- overall, the total energy, although with correlation factors in the range of its components' regressions does not fit with parabolic reactive theoretical prescription, at least for present employed set of atoms and molecules.

However, all revealed parabolic features have appeared on the statistical correlation basis and not upon an individual atomic or molecular analysis. At this point we can refer to studies (Putz et al., 2004) that clearly illustrate the individual parabolic behavior of the total energy against the total

TABLE 3.22 Coefficients in Bilinear Correlation of the Energies of Tables 3.17 and 3.18 Against the Electronegativity and Chemical Hardness of Table 3.19 Within Experimental Finite Difference (FD), Putz-DFT and Putz-SC (Semiclassical) Modes of Computations, Respectively (*)

Method of		QSPR results					Method of		QSPR results				
Energy	(χ,η)	a	b	c	σ_π	r	Energy	(χ,η)	a	b	c	σ_π	r
T_{exact}	FD	194.59	741.55	−951.38	−0.0017	0.33	E_{xc}^{exact}	FD	−15.30	−44.80	60.89	−0.0303	0.37
	DFT	51.73	2.09	31.51	7.2	0.62		DFT	−6.56	−0.19	−1.53	40.84	0.58
	SC	186.38	−2315.7	5147.4	−0.001	0.35		SC	−13.76	73.07	−128.0	−0.024	0.38
T_0	FD	179.67	687.26	−881.14	−0.0019	0.33	$E_{xc}^{I(Xa)}$	FD	−15.28	−44.56	60.69	−0.0306	0.37
	DFT	46.99	1.91	29.41	8.07	0.62		DFT	−6.56	−0.2	−1.51	39.11	0.57
	SC	172.42	−2185.3	4874.1	−0.001	0.35		SC	−13.72	72.04	−125.5	−0.024	0.38
T_0+T_2	FD	193.83	736.94	−946.29	−0.0017	0.33	$E_{xc}^{II(Xa)}$	FD	−15.28	−44.81	60.92	−0.0303	0.37
	DFT	51.59	2.07	31.36	7.33	0.62		DFT	−6.52	−0.19	−1.53	40.81	0.57
	SC	185.59	−2312.1	5142	−0.001	0.35		SC	−13.75	74.72	−132.4	−0.024	0.38
T_{Pade}	FD	194.61	741.17	−951.43	−0.0017	0.33	$E_{xc}^{I(Wig)}$	FD	−15.29	−44.53	60.67	−0.0306	0.37
	DFT	51.54	2.08	31.58	7.3	0.62		DFT	−6.58	−0.2	−1.50	38.78	0.57
	SC	186.42	−2334.4	5197.4	−0.001	0.35		SC	−13.72	71.4	−123.8	−0.024	0.38
K_{exact}	FD	−14.91	−43.62	59.38	−0.0312	0.37	$E_{xc}^{II(Wig)}$	FD	−15.30	−44.80	60.95	−0.0304	0.37
	DFT	−6.37	−0.19	−1.49	43.17	0.57		DFT	−6.54	−0.2	−1.52	39.98	0.57
	SC	−13.4	72.74	−128.78	−0.0243	0.38		SC	−13.76	74.1	−130.7	−0.024	0.38
K_0	FD	−13.59	−40.29	54.69	−0.0337	0.37	E_{tot}^{exact}	FD	−194.98	−742.7	952.91	−0.0017	0.33
	DFT	−5.72	−0.17	−1.39	46.96	0.58		DFT	−51.91	−2.10	−31.54	7.15	0.62
	SC	−12.24	68.77	−123.42	−0.026	0.38		SC	−186.73	2316	−5148	−0.001	0.35

TABLE 3.22 Continued

Method of		QSPR results				
Energy	(χ,η)	a	b	c	σ_π	r
K^{B88}	FD	−14.89	−43.61	59.33	−0.0312	0.37
	DFT	−6.37	−0.19	−1.49	42.35	0.57
	SC	−13.39	71.88	−126.54	−0.0245	0.38
E_c^{exact}	FD	−0.39	−1.20	1.52	−1.055	0.38
	DFT	−0.19	−0.0075	−0.03	595.98	0.59
	SC	−0.362	0.232	0.997	18.528	0.38
$E_c^{(139)♣}$	FD	−0.40	−1.13	1.55	−1.207	0.39
	DFT	−0.19	−0.0056	−0.03	1081.3	0.57
	SC	−0.356	0.851	−0.563	−0.778	0.41
$E_c^{•}$	FD	−0.41	−1.10	1.44	−1.1808	0.40
	DFT	−0.22	−0.0063	−0.03	735.5	0.59
	SC	−0.374	−0.224	2.113	−42.24	0.42

Method of		QSPR results				
Energy	(χ,η)	a	b	c	σ_π	r
$E_{tot}^{xc(RG)}$	FD	−195.55	−743.9	954.5	−0.0017	0.33
	DFT	−52.27	−2.11	−31.56	7.1	0.62
	SC	−187.24	2315.	−5143.	−0.001	0.35
$E_{tot}^{xc(LDA)}$	FD	−194.26	−740.7	950.13	−0.0017	0.33
	DFT	−51.59	−2.09	−31.47	7.18	0.62
	SC	−186.1	2313.	−5142.	−0.001	0.35
E_{tot}^{BLYP}	FD	−194.99	−742.7	952.86	−0.0017	0.33
	DFT	−51.94	−2.1	−31.53	7.15	0.62
	SC	−186.74	2315.	−5145	−0.001	0.35
E_{tot}^{PW91}	FD	−194.97	−742.7	952.81	−0.0017	0.33
	DFT	−51.92	−2.1	−31.54	7.15	0.62
	SC	−186.7	2315.	−5145.	−0.001	0.35

♣, •: from associate energetic atomic values of Table 3.17.

*The deviation from the parabolic expansion $E=a+b\chi+c\eta$ in terms of the index $\sigma_\pi=\text{sign}(b)\, c/b^2$ as well as the correlation factor of the QSPR model (r) are both calculated. The color code indicates: the (+, +) for magenta background and (−, +) for green background sign combinations of (b, c), respectively, while the grey background is used in highlighting the representative (σ_π, r) couple for each energy type of employed functionals (Putz, 2008a).

TABLE 3.23 Coefficients in Bilinear Correlation of the Energies of Table 3.20 Against the Respective Electronegativity and Chemical Hardness of the Table 3.21 (*)

Method of		QSPR results				
Energy	χ & η	a	b	c	σ_π	r
T_0	FD	−192.71	821.61	238.59	0.00035	0.77
	DFT	40.446	8.125	4.501	0.068	0.89
	SC	82.14	542.16	−977.9	−0.0033	0.56
K^{exact}	FD	18.18	−70.1	−38.02	0.0077	0.74
	DFT	−5.234	−0.646	−0.805	1.93	0.89
	SC	−9.78	−48.27	94.96	−0.04	0.61
E_c^{VWN}	FD	1.039	−4.06	−2.74	0.167	0.71
	DFT	−0.423	−0.035	−0.06	47.83	0.87
	SC	−0.71	−3.13	6.22	−0.63	0.64
E_{xc}^{exact}	FD	18.95	−72.63	−40.1	0.0076	0.74
	DFT	−5.457	−0.668	−0.846	1.897	0.89
	SC	−10.19	−50.16	98.71	−0.039	0.61
E_{tot}^{BLYP}	FD	193.12	−824.17	−238.96	0.00035	0.77
	DFT	−40.67	−8.149	−4.514	0.068	0.89
	SC	−82.48	−544.35	981.7	−0.003	0.56

Method of		QSPR results				
Energy	χ & η	a	b	c	σ_π	r
T_0+T_2	FD	−190.58	811.48	237.96	0.0004	0.77
	DFT	40.199	8.012	4.497	0.07	0.89
	SC	81.4	537.74	−969.47	−0.003	0.56
K^{PBE}	FD	18.324	−70.48	−38.25	0.0077	0.74
	DFT	−5.219	−0.65	−0.811	1.92	0.89
	SC	−9.81	−48.15	95.	−0.04	0.61
E_c^{GCP}	FD	0.59	−2.27	−1.496	0.289	0.72
	DFT	−0.225	−0.019	−0.033	94.08	0.86
	SC	**−0.36**	**−2.1**	**3.7**	**−0.88**	**0.64**
E_{xc}^{PBE}	FD	18.85	−72.45	−39.62	0.0075	0.74
	DFT	−5.421	−0.666	−0.84	1.898	0.89
	SC	−10.13	−49.97	98.28	−0.04	0.61
E_{tot}^{TH}	FD	193.06	−823.8	−239.04	0.0004	0.77
	DFT	−40.67	−8.145	−4.514	0.068	0.89
	SC	−82.46	−544.3	981.5	−0.003	0.56

*The computational meaning of the output values and the color code are the same as those of Table 3.17 caption (Putz, 2008a).

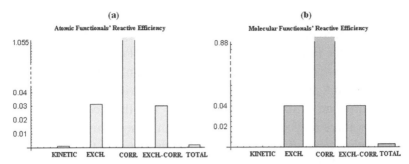

FIGURE 3.12   Charts of energetic functionals showing their reactive efficiency as the height of bars around the parabolic expansion and of their statistical correlation as the width of bars, in terms of electronegativity and chemical hardness, resuming atomic analysis of He, Li, Be, B, C, N, O, F, Ne, Na, Mg, Al, Si, Cl, Ar systems and molecular one for H_2, LiH, CH_4, H_2O, HF, N_2, O_2, F_2 systems (Putz, 2008a).

number of electrons. On the other side, the systematic focus on atomic and molecular large class of systems should provide more or less the same phenomenological results.

The present dichotomy, while confronting the conceptual and computational behavior of the energetic functionals respecting the parabolic bilinear expansion in terms of electronegativity and chemical hardness, gives the indication that the derivation of the energetic functionals (and of the total energy in special) starting from electronegativity and chemical hardness functionals, see Volume II of the present five-volume book and (Putz, 2008b), may eventually conclude with better energy correlation both as parabolic shape and statistical regression with fundamental (χ, η) couple of reactivity indices. Otherwise, we will be forced to admit that the density functionals do not properly model the reactivity through a parabolic fit in total or exchanged electrons of the involved systems at least in a statistical sense.

3.6.3.2   Testing on Anthocyanidins' Reactivity

Anthocyanidins are a class of natural, poly-phenol based colored compounds that are largely responsible for the vibrant colors in most fruits and flowers. They are commonly found in red wine and fruit juices, and have been shown to have properties beneficial to human health, including antioxidant and antitumor activity. Synthetic anthocyanidins have also been used in advanced devices, serving as a basis for organic material in optical data storage; the forthcoming discussion follows (Putz et al., 2008).

The structures of some naturally occurring anthocyanidins studied in the present work are given in Figure 3.13 along with that of flavylium, the parent compound. Also shown in the Figure 3.13 the molecular skeleton is labeled for three rings (A, B, C) and the numbering scheme for the atoms. All the anthocyanidins discovered to date have largely the same chemical structure, yet they are responsible for wide variations in color. This is due in part to the anthocyanidins' equilibrium in aqueous solution between the pigmented benzopyrylium from at low pH and the colorless quinonoidal base mildly acidic to neutral pHs. Other factors, such as co-pigmentation, serve to modulate the color of anthocyanidins in their natural settings.

Chemical structures of anthocyanidin compounds studied in this work, along with their common abbreviations are: *F-Flavilium*: R3=R5=R7=R3'= R4'=R5'=–H; *Ap-Apigeninium*: R3=R3'=R5'=–H; R5=R7=R4=–OH; *Cy-Cyanidin*: R5'=–H; R3=R5=R7=R3'=R4'=–OH; *Dp-Delphinidin*: R3= R5=R7=R3'=R4'=R5'=–OH; *Lt-Luteolinidin*: R3=R5'=–H; R5=R7=R3'= R4'=–OH; *Mv-Malvidin*: R3=R5=R7=R4'=–OH; R3'=R5'=–OHCH3 *Pg-Pelargonidin*: R3'=R5'=–H; R3=R5=R7=R4'=–OH; *Pn-Peonidin*: R5'=H; R3=R5=R7=R4'=–OH; R3'=–OCH3; *Pt-Petunidin*:R3=R5=R7 = R4'=R5'=–OH; R3'=–OCH3, see Figure 3.14.

For the molecules of Figures 3.13 and 3.14 the exited energies ΔE of anthocyanidins for two series of oscillating strengths (I & II) are in Table 3.24 collected along with highest occupied and lowest unoccupied molecular orbitals HOMO, LUMO and with Koopman's approximation

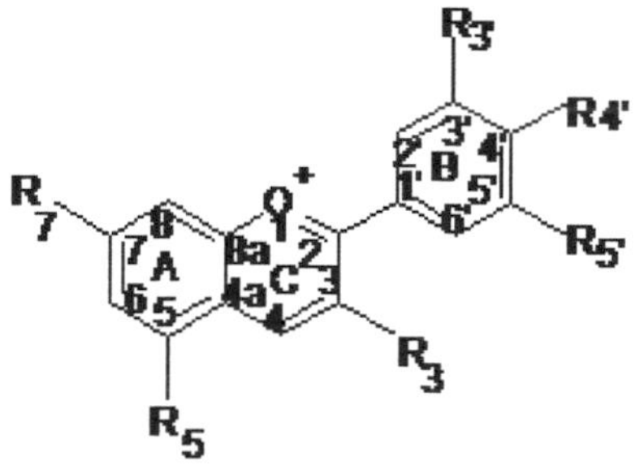

FIGURE 3.13   The generic structure of the studied series of anthocyanidins (Putz et al., 2008).

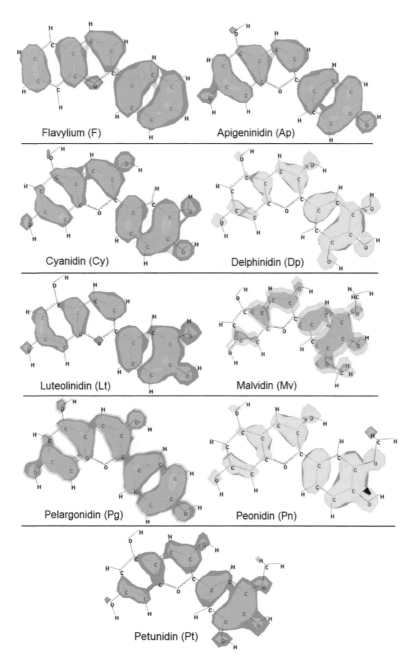

FIGURE 3.14 The orbital structures for working anthocyanidins in HOMO states (Putz et al., 2008).

TABLE 3.24 Synopsis of the HOMO-LUMO Transition Energies Along the HOMO, LUMO, Finite-Difference Electronegativity and Chemical Hardness Values for Each of the Molecule of the Series of Figure 3.14 (Putz et al., 2008)

Crt.	ΔE(I) [eV]	ΔE(II) [eV]	HOMO [eV]	LUMO [eV]	χ [eV]	η [eV]
F	3.11	3.30	−13.05713	−5.889265	9.4732	3.58393
Ap	2.88	3.21	−12.42232	−5.750154	9.08624	3.33608
Cy	2.53	3.17	−12.06511	−5.98948	9.0273	3.03782
Dp	2.53	2.55	−12.05294	−6.041625	9.04728	3.00566
Lt	2.65	3.29	−12.20063	−5.809774	9.0052	3.19543
Mv	2.42	2.60	−11.79443	−5.915446	8.85494	2.93949
Pg	2.64	3.13	−12.23512	−5.936516	9.08582	3.1493
Pn	2.60	3.15	−11.95847	−5.923064	8.94077	3.0177
Pt	2.47	2.61	−11.94764	−5.980243	8.96394	2.9837

electronegativity and chemical hardness while the last reactivity index was here considered by the 0.5 acid-base factor as in (3.103) and (3.362) since the increased complexity of the treated molecular systems.

With these the fitting, reactivity and statistical results are displayed in Table 3.24. From Table 3.25 two cases emerges: for the first series I of molecular strengths there is recorded poor reactivity efficiency while the statistical factor shows a higher correlation; instead, for the second series II of oscillating series in excited states the *reactivity efficiency* Σ_Π approaches the ideal case, however, with a drawback from the correlation statistics.

Such results forced us to admit that when about chemical reactivity the chemical efficiency should be more appropriate index than the quantitative-structure-property correlation factor.

TABLE 3.25 Statistics of the bivariate quantitative structure-transition energy reactivity models computed for the molecules of Figure 3.14 with data of Table 3.24 (Putz et al., 2008).

Crt.	a	b	c	σ_π	r
ΔE(I)	−0.845588	0.0378752	1.00371	699.69	0.984663
ΔE(II)	7.44092	−1.14797	1.89682	−1.43935	0.730148

Finally going to individual molecular analysis, the bilinear $E=E(\chi, \eta)$ equation with computed coefficients from Table 3.25 is employed on electronegativity and chemical hardness data of Table 3.24 to produce the actual computed excitation energies together with the associate residues for both oscillating strengths I & II of Table 3.24, with the results collected in Table 3.26.

From Table 3.26, more exactly, from residues analysis two reactivity hierarchies may be extracted by counting the increasing of the absolute residual values away from ideal no deviation case and compared with the input hierarchy abstracted from Table 3.24 (Putz et al., 2008):

-for series I of oscillating strength we obtain:

$$\text{I (fitted): } F \gg Cy > Dp > Pt > Pg > Mv > Ap > Lt > Pn \quad (3.502)$$

$$\text{I (input): } F > Ap > Lt > Pg > Pn > Cy > Dp > Pt > Mv \quad (3.503)$$

-for series II of oscillating strength we obtain:

$$\text{II (fitted): } F > Lt > Ap > Pg > Pt > Dp > Pn > Mv > Cy \quad (3.504)$$

$$\text{II (input): } F > Lt > Ap > Cy > Pn > Pg > Pt > Mv > Dp \quad (3.505)$$

TABLE 3.26 Computed and Residues of the Transition Energies of Molecules of Figure 3.14 With the Models of Table 3.25 (Putz et al., 2008)

Crt.	Computed		Residues	
	$\Delta E(I)$ (eV)	$\Delta E(II)$ (eV)	$\Delta E(I)$ (eV)	$\Delta E(II)$ (eV)
F	3.11045	3.36407	0.000454106	0.0640665
Ap	2.84703	3.33816	-0.0329728	0.128157
Cy	2.545443	2.84007	0.015427	-0.329927
Dp	2.5139	2.75613	-0.0160958	0.206135
Lt	2.70279	3.1644	0.0527854	-0.125599
Mv	2.4402	2.85142	0.0202035	0.251422
Pg	2.65954	2.98435	0.0195375	-0.145649
Pn	2.52195	2.90125	-0.0780451	-0.248758
Pt	2.48871	2.81015	0.0187062	0.200152

Such ordering helps in further checking the predicted reactivity respecting the initial input data; indeed, again, the series I produces the lowest matching of input-with-fitted ordering reactivity, i.e., only the position "1" with F-molecules being superposed; instead, series II provides fours such matching, namely for the molecules F, Lt, Ap and My in positions "1, 2, 3 and 8" in the hierarchical order of reactivity. Therefore, the series II of strengths in Table 3.24 is expected to furnish the best reactivity results while overall reactive molecule remains the Flavylium (F) structure with its counterpart validated low reactivity for Malvidin (Mv) anthocyanidin. The chemical reactivity of selected anthocyanidins was studied with the help of electronegativity and chemical hardness bivariate parabolic relationship with two series of excited strengths of HOMO-LUMO transitions. There was found that reactivity efficiency prevails on the statistical correlation factor in matching the predicted series of reactive molecules against the oscillating strengths. The Flavylium (F) structure was validated with the highest reactivity while Malvidin (Mv) molecule was found with the lowest validated reactivity among the studied anthocyanidins.

3.6.3.3 Testing on Hydracids HSAB Reactivity

As we go into the analysis of the reactivity principles it is worth involving the molecules in the paradigmatic scheme of HSAB (Pearson, 1973, 1985, 1987) by which a hard acid substitutes a hard base from a soft acid so that the hard-hard and soft-soft acid-bases complexes are formed. Particularization of the HSAB scheme on the hydracids series molecules lead with the associate prototype reactions; the forthcoming discussion follows (Putz & Chiriac, 2008):

$$HF + H_2O_2 \rightleftharpoons FOH + H_2O \quad \text{(HSAB1)} \quad (3.506)$$

$$HCl + H_2O_2 \rightleftharpoons ClOH + H_2O \quad \text{(HSAB2)} \quad (3.507)$$

$$HBr + H_2O_2 \rightleftharpoons BrOH + H_2O \quad \text{(HSAB3)} \quad (3.508)$$

$$HI + H_2O_2 \rightleftharpoons IOH + H_2O \quad \text{(HSAB4)} \quad (3.509)$$

Now, we would like to derive the change in total energy, electronegativity and chemical hardness values among the reactants of (HSAB1)–(HSAB4) reactions and of their inter-correlations, employing different theoretical and computational methods.

As such, for total energy the semiempirical AM1 (Austin Model 1) and PM3 (re-parameterized AM1 with less repulsive nonbonding interactions) were considered among the *ab initio HF* and *DFT* methods. The results are presented in Tables 3.27 and 3.28.

Different models were assessed for the electronegativity and chemical hardness as well, namely the experimental based finite-difference (*FD*) (Mulliken, 1934; Lackner & Zweig, 1983), Putz-DFT (Putz, 2006), and Putz-SC (semiclassical) (Putz, 2007b) ones. The atomic values for atoms involved are displayed in Table 3.29, while the molecular results, based

TABLE 3.27 Values of the Total Energies for the Molecules Considered Through Chemical Reactions (HSAB1)–(HSAB4) Computed with Semi-Empirical (AM1 and MP3) and Self-Consistent Field (HF and DFT) Environments (**)

Method Molecule	−EAM1	−EPM3	−EHF	−EDFT
HF	48.2	43.99	262.33	234.66
HCl	37.42	32.07	1206.75	1134.5
HBr	35.43	35.55	6716.48*	6465.96*
HI	34.83	29.26	18068.2*	17612.6*
FOH	78.7	71.99	458.33	402.71
ClOH	68.13	60.27	1402.9	1278.71
BrOH	65.72	63.26	6911.32*	6575.28*
IOH	65.69	57.6	18263.4*	17709.3*
H_2O_2	64.22	59.41	395.483	334.87
H_2O	33.6	31.32	199.41	171.9

* Computed with small 3–21G orbital base.

**The computations were performed using HyperChem Release 7" (Hypercube, 2002). All values are expressed in MJ/mol, where 1MJ/mole≈10.4eV/atom (Putz & Chiriac, 2008).

TABLE 3.28 Values of the Variation of the Total Energies for the Molecules Considered Through Chemical Reactions (HSAB1)–(HSAB4) in the Computational Cases of the Table 3.27 (*)

Method	ΔEAM1	ΔEPM3	ΔEHF	ΔEDFT
Reaction				
HSAB1	−30.5	0.09	0.073	−5.08
HSAB2	−0.09	−0.11	−0.077	18.76
HSAB3	0.33	0.38	1.233	53.65
HSAB4	−0.24	−0.25	0.873	66.27

*All values are expressed in MJ/mol (Putz & Chiriac, 2008).

TABLE 3.29 Values of the Structural Indices Electronegativity (χ), Chemical Hardness (η), in Finite-Difference (Mulliken, 1934; Lackner & Zweig, 1983), Putz-DFT (Putz, 2006) and Putz-SC (semiclassical) (Putz, 2007b) Modes for the Atoms of the Reactants of Chemical Reactions (HSAB1)–(HSAB4) (*)

Level	Finite-Difference		Functional		Semiclassical	
Atom	χ_{FD}	η_{FD}	χ_{DFT}	η_{DFT}	χ_{SC}	η_{SC}
H	0.69	0.62	0.69	0.62	0.69	0.62
O	0.73	0.59	27.87	15.48	1.07×10^{-1}	0.21×10^{-1}
F	1.00	0.68	44.25	11.45	1.34×10^{-1}	0.22×10^{-1}
Cl	0.80	0.45	30.29	20.33	0.19×10^{-2}	0.32×10^{-3}
Br	0.73	0.41	32.58	24.02	0.32×10^{-4}	0.44×10^{-5}
I	0.65	0.36	21.26	16.56	0.35×10^{-6}	0.5×10^{-7}

*All values are expressed in MJ/mol (Putz & Chiriac, 2008).

on iterative relations, are presented in Table 3.30 for electronegativity and chemical hardness, respectively.

It is worth noting that from Table 3.30 the experimental based finite difference and path integral based semi-classical hardness hierarchy is neatly prescribed as (Putz & Chiriac, 2008):

$$\eta_{HF} > \eta_{HCl} > \eta_{HBr} > \eta_{HI} \tag{3.510}$$

while the OH–X complexes are in general softer than H–OH one, excepting FOH. This situation clearly reverses the empirical expected order thus

TABLE 3.30   Values of the Structural Indices Electronegativity (χ), Chemical Hardness (η), Computed by Means of the Group Method (Bratsch, 1985; Putz, 2008b), Within Finite-Difference, Putz-DFT and Putz-SC (semiclassical) Modes for the Reactants of Chemical Reactions (HSAB1)–(HSAB4) (*)

Level Index	Finite-Difference		Functional		Semiclassical	
Molecule	χ_{FD}	η_{FD}	χ_{DFT}	η_{DFT}	χ_{SC}	η_{SC}
HF	0.82	0.65	1.36	1.18	0.22	0.04
HCl	0.74	0.52	1.35	1.2	0.38×10^{-2}	0.06×10^{-2}
HBr	0.71	0.49	1.35	1.21	0.06×10^{-3}	0.88×10^{-5}
HI	0.67	0.46	1.34	1.2	0.7×10^{-6}	0.1×10^{-6}
FOH	0.79	0.63	1.99	1.70	0.16	0.03
ClOH	0.74	0.54	1.98	1.74	0.56×10^{-2}	0.09×10^{-2}
BrOH	0.72	0.52	1.98	1.75	0.1×10^{-3}	0.01×10^{-3}
IOH	0.69	0.49	1.96	1.73	1.05×10^{-6}	1.5×10^{-7}
H2O2	0.71	0.6	1.35	1.19	0.19	0.04
H2O	0.70	0.61	1.02	0.91	0.25	0.06

*All values are expressed in MJ/mol (Putz & Chiriac, 2008).

proving that the quantum mechanical calculation can overcome the empirical judgments, a situation often met in connection with quantum phenomena. However, the conceptual based *DFT* analysis suggest that while the empirical order is somehow observed, the complex H–OH appears harder than all products OH–X in Table 3.31, according to reactions (HSAB1)–(HSAB2). At this point, it is clear that another cutting criterion has to be

TABLE 3.31   Variations of the Electronegativity (χ) and Chemical Hardness (η) by Employing the Finite-Difference, Putz-DFT and Putz-SC (semiclassical) Methods for Chemical Reactions (HSAB1)–(HSAB4) (*)

Level	Finite-Difference		Putz-Functional		Putz-Semiclassical	
Reaction	$\Delta\chi_{FD}$	$\Delta\eta_{FD}$	$\Delta\chi_{DFT}$	$\Delta\eta_{DFT}$	$\Delta\chi_{SC}$	$\Delta\eta_{SC}$
HSAB1	−0.04	−0.01	0.3	0.24	0.00	0.01
HSAB2	−0.01	0.03	0.3	0.26	0.062	0.02
HSAB3	0.00	0.04	0.3	0.26	0.06	0.02
HSAB4	0.01	0.04	0.29	0.25	0.06	0.02

*All values are in MJ/mol (Putz & Chiriac, 2008).

checked in order to decide which of these approaches should be chosen as most appropriate for the molecular series involved in the present *HSAB* type reaction.

Therefore, the electronegativity and chemical hardness excesses ($\Delta\chi$, $\Delta\eta$) for (HSAB1)–(HSAB2) reactions are reported in Table 3.31 and are finally correlated with total energy differences (ΔE) of Table 3.28 by employing the (spectral) quantitative structure-property relationships (QSPR) analysis (Putz, 2011c).

The QSPR results through all energy and electronegativity and chemical hardness combined methods are presented in Table 3.6, emphasizing both the degree of parabolic dependence, i.e., the closeness $c/b^2 \rightarrow 1$, and the consecrated statistical correlation coefficient (r).

From Table 3.32 there it is now clear that a parabolic form of total energy respecting electronegativity and chemical hardness is possible, it is attained within SC-AM1 approach, and when this is the case the associated correlation coefficient goes asymptotically to unity.

TABLE 3.32 Coefficients of the Correlation of the Variation of the Total Energies of the Table 3.28 With the Finite-Difference Electronegativity and Hardness Variations of the Table 3.31 in Experimental, Density Functional, and Semiclassical Models of Chemical Reactions (HSAB1)–(HSAB4), Respectively (*)

Method of		QSAR results				
$\chi\&\eta$	Energy	a	b	c	c/b2	r
FD	AM1	−35.58	−381.92	965.46	0.007	0.989014
	PM3	−0.93	−30.81	26.15	0.028	0.576147
	HF	1.08	30.35	−9.92	−0.011	0.682117
	DFT	81.55	2451.31	−945.31	−0.0002	0.967383
DFT	AM1	50.56	−1495.0	1531.0	0.0007	0.999937
	PM3	−11.325	36.25	2.25	0.0017	0.681353
	HF	10.438	−54.75	25.25	0.0084	0.527625
	DFT	1020.72	−5070.75	2064.25	0.00008	0.899406
SC	AM1	−65.095	−67.5	3459.5	0.7593	0.999884
	PM3	−5.135	−87.5	522.5	0.068	0.33817
	HF	−34.807	−565.0	3488.0	0.011	0.97237
	DFT	−1306.12	−20600.0	130104	0.00031	0.987422

*The deviation from the parabolic expansion $E=a+b\chi+c\eta$ in terms of the ratio c/b^2 as well as the correlation factor of the QSAR model (r) are also indicated (Putz & Chiriac, 2008).

3.6.4   FROM HÜCKEL- TO PARABOLIC- TO π- ENERGY FORMULATIONS

The Hückel method is simple and has been in use for decades (Hückel, 1931a,b). It is based on the σ-π separation approximation while accounting for the pi-electrons only, i.e., the atomic orbitals involved refer to those $2p_z$ for Carbon atoms as well to the $2p_z$ and $3p_z$ orbitals for the second and third period elements as (N,O, F) and (S, Cl) respectively; further discussion on the d-orbitals involvement may be also undertaken, yet the method essence reside in non explicitly counting on the electronic repulsion with an effective, not-defined, mono-electronic Hamiltonian, as the most simple semi-empirical approximation. In these conditions, for the mono-electronic Hamiltonian matrix elements two basic assumptions are advanced; the forthcoming discussion follows (Putz, 2011d):

- In the case of hydrocarbures (C containing only π-systems) one has:

$$H_{\mu\nu} = \int \phi_\mu H^{eff} \phi_\nu d\tau = \begin{cases} \alpha & \dots \quad \phi_\mu = \phi_\nu \\ 0 & \dots \quad \phi_\mu, \phi_\nu \in non-bonded\ atoms \\ \beta & \dots \quad \phi_\mu, \phi_\nu \in bonded\ atoms \end{cases} \quad (3.511a)$$

where all Columbic integrals are considered equal among them and equal with the quantity α representing the energy of the electron on atomic orbital $(2p_z)_C$; non-diagonal elements are neglected, i.e., for the non-bonding atoms; and the exchange or resonance integral is set equal with the non-definite β integral for neighboring bonding atoms.

- In the case heteroatoms (X) are present in the system one has to consider the Columbic parameter h_X correlating with the electronegativity difference between the heteroatom X and carbon, along the resonance parameter k_{CX} that may include correlation with the binding energy; the form of matrix elements of monoelectronic effective Hamiltonian looks therefore as

$$H_{XC} = \int \phi_X H^{eff} \phi_C d\tau = \begin{cases} \alpha + h_X \beta & \dots \quad \phi_X = \phi_X \\ 0 & \dots \quad \phi_X, \phi_C \in non-bonded\ atoms \\ k_{CX} \beta & \dots \quad \phi_X, \phi_C \in bonded\ atoms \end{cases} \quad (3.511b)$$

As a consequence of these approximations, the total pi-energy with Hückel approach may be resumed as

$$E_\pi = \sum_C \alpha + \sum_{C,X} (h_X + k_{CX})\beta = \alpha N_\pi + \left(\sum_X h_X + \sum_{\substack{C-X, \\ C=X}} k_{CX} \right)\beta \qquad (3.512)$$

Remarkably, the comparison of the Hückel energy (3.512) with the parabolic form (3.453) here specialized for the valence pi-electrons (Putz, 2011d)

$$E_{para(bolic)} \cong -\chi\, N_\pi + \eta\, N_\pi^2 \qquad (3.513)$$

when identified the reactive frontier electrons with the pi-electrons in the system, $\Delta N = N_\pi$. Even more, the present discussion permits the identification of the Columbic and resonance integrals in terms or electronegativity and chemical hardness

$$\alpha = -\chi \qquad (3.514)$$

$$\beta = \frac{\eta}{\sum_X h_X + \sum_{\substack{C-X, \\ C=X}} k_{CX}}\, N_\pi^2 \qquad (3.515)$$

However, beside the possibility of assessing the Hückel integrals, the parabolic energy (3.513) may be useful in testing the constructed pi-energy abstracted from the total energy according with the recipe (Putz, 2011d)

$$E_{pi}(molecule) \cong E_{Total}(molecule) - E_{Bind}(molecule) - E_{Heat}(molecule)$$

$$= \sum_{atoms}^{molecule} E_{Total}(atom) - \sum_{atoms}^{molecule} E_{Heat}(atom) \qquad (3.516)$$

since:

- The total energy is relative to a sum of atomic energies for semi-empirical computations

$$E_{Total}(molecule) = \sum_{atoms}^{molecule} E_{Total}(atom) \qquad (3.517)$$

- Binding energy is the energy of the molecular atoms separated by infinity minus the energy of the stable molecule at its equilibrium bond length

$$E_{Bind}(molecule) = E_{\infty}(atoms) - E_{equilibrium}(molecule) \qquad (3.518)$$

- The heat of formation is calculated by subtracting atomic heats of formation from the binding energy:

$$E_{Heat}(molecule) = E_{Bind}(molecule) - \sum_{atoms}^{molecule} E_{Heat}(atom) \qquad (3.519)$$

Through its form the energy (3.516) may have the frontier meaning, thus appropriately assessing the pi-formed system, while the remaining challenge is to test whether it can be well represented by the parabolic chemical reactivity descriptors related energy (3.513). To this end, four carbon based systems are analyzed due to their increased structure complexity, namely the butadiene, benzene, naphthalene, and fullerene that have been characterized by Hückel and most common semiempirical methods through the data in Tables 3.33–3.36, while the bivariate correlation between the obtained parabolic- and pi- energies are in Figures 3.15–3.19 represented. From these results one may note the systematic increasing of the $E_{para(bolic)}$ vs. E_{pi} correlation up to its almost parallel behavior as going from simple pi-systems with few frontier electrons until complex nanomolecules such as fullerene. On the other side the actual study may give an impetus in characterizing nanostructures by electronegativity and chemical hardness reactivity indicators, parabolically combined and almost fitting with the total energy of pi-electrons.

However, one should mention also the open issues remained, such as (Putz, 2011d):

- The correct identification of the energy (3.516) with the pi-energy as a suitable generalization of the Hückel one (3.512);
- The physical meaning of the parabolic energy (3.513) since, through its correlation with the so called pi-energy (3.516) widely includes exchange-correlation effects, especially with its chemical hardness dependence though the explicit resonance relationship (3.515);
- The type and the complexity of the carbon system, hydrogenated or not, the quantum parabolic effect of reactivity indices in Eq. (3.513) overcomes other inner structural quantum influences to produce best correlation with atoms-in-molecules frontier energy (3.516).

Overall, for the moment we remain with the fact electronegativity and chemical hardness may be worth combined to produce an energy that is

TABLE 3.33 The Butadiene π-System, with $\Delta N=N_\pi=4$, Frontier Energetic Quantities, Ionization Potential (IP), Electron Affinity (EA), Electronegativity (χ), and Chemical Hardness (η) of Eqs. (3.347) and (3.362) – in Electron Volts (eV), and the Resulted Parabolic Energy of Eq. (3.513), Alongside with the π-Related Energy Based on the Hückel Simplified (with Coulomb Integrals Set to Zero, $\alpha = 0$) Expression of (3.512) for the Experimental/Hückel Method and on the Related Energy Form of Eq. (3.516) and the Other Semi-Empirical Methods (CNDO, INDO, MINDO, MNDO, AM1, PM3, ZINDO) As Described in the Previous Section – Expressed in Kilocalories Per Mol (kcal/mol); Their Ratio in the Last Column Reflects the Value of the Actual Departure of the Electronegativity and Chemical Hardness Parabolic Effect from the pi-Bonding Energy, While for the First (Exp/Hückel) Line It Expresses the Resonance Contribution (and a Sort of β Factor Integral) in (3.512) for the π-Bond in this System; the eV to kcal/mol Conversion Follows the Rule 1 eV $\cong$ 23.069 kcal/mol (Putz, 2011d)

Quantity→ Method↓	IP [eV]	EA [eV]	χ [eV]	η [eV]	|E-(para)bolic| [kcal/mol]	E-pi [kcal/mol]	|E-pi/ E-para|
Exp/Hückel	9.468[a]	−0.263[a]	4.6025	4.8655	473.2375	−990.214	2.09243
E-Hueckel	12.50681	9.107174	10.80699	1.699818	683.521	−8982.58	13.14163
CNDO	13.32281	−3.35577	4.983522	8.33929	1079.173	−16204	15.0152
INDO	12.75908	−3.90238	4.428352	8.330732	1128.823	−15684.9	13.8949
MINDO3	9.101508	−1.12356	3.988974	5.112535	575.4419	−12782.5	22.2134
MNDO	9.138431	−0.38813	4.37515	4.763282	475.3518	−12791.7	26.9101
MNDO/d	9.138306	−0.38791	4.375199	4.763108	475.3152	−12791.7	26.9121
AM1	9.333654	−0.44841	4.442624	4.891031	492.7019	−12751.2	25.8801
PM3	9.468026	−0.26348	4.602275	4.865752	473.3047	−12100.1	25.5651
ZINDO-1	9.156922	−7.45074	0.853094	8.303829	1453.768	−14210.2	9.77473
ZINDO-S	8.623273	−0.48256	4.070358	4.552916	464.6534	−10639.4	22.8974

(a) calculated as the negative of the HOMO and LUMO energies (University Illinois, 2011).

TABLE 3.34 The Same Quantities As those Reported in Table 3.33, Here for the Benzene π-System, with $\Delta N=N_\pi=6$ (Putz, 2011d)

Quantity→ Method↓	IP [eV]	EA [eV]	χ [eV]	η [eV]	|E-(para)bolic| [kcal/mol]	E-pi [kcal/ mol]	|E-pi/ E-para|
Exp/Hückel	9.24384[a]	−1.60817[b]	3.817835	5.426005	1724.663	−5021.68[c]	2.91169
E-Hueckel	12.81724	8.229032	10.52314	2.294105	503.941	−12231	24.27077
CNDO	13.8859	−4.06892	4.908487	8.97741	3048.394	−23007.2	7.54733
INDO	13.48267	−4.58566	4.448502	9.034166	3135.63	−22231.9	7.09009

TABLE 3.34   Continued

Quantity→ Method↓	IP [eV]	EA [eV]	χ [eV]	η [eV]	|E-(para)bolic| [kcal/mol]	E-pi [kcal/mol]	|E-pi/E-para|
MINDO3	9.179751	–1.24984	3.964955	5.214796	1616.597	–18289.3	11.3135
MNDO	9.39118	–0.36809	4.511543	4.879637	1401.77	–18341.8	13.0848
MNDO/d	9.391201	–0.36807	4.511564	4.879638	1401.767	–18341.8	13.0848
AM1	9.653243	–0.55504	4.549103	5.10414	1489.794	–18315.5	12.294
PM3	9.751339	–0.3962	4.677572	5.073768	1459.4	–17222.6	11.8012
ZINDO-1	9.865785	–8.12621	0.869786	8.996	3615.126	–20453.7	5.6578
ZINDO-S	8.995844	–0.86318	4.066331	4.929514	1484.104	–15255.9	10.2795

(a) from National Institute of Standard and Technology (NIST, 2011a)

(b) from interpolation data presented in Figure 3.16.

(c) from the Hückel total π-energy: 2 ×(2+2)=8β[a.u.] ... × 627.71...~5021.68 kcal/mol, see (Cotton, 1971a)

TABLE 3.35   The Same Quantities As Those Reported in Table 3.33, Here for the Naphthalene π-System, with $\Delta N = N_\pi = 10$ (Putz, 2011d)

Quantity→ Method↓	IP [eV]	EA [eV]	χ [eV]	η [eV]	|E-(para)bolic| [kcal/mol]	E-pi [kcal/mol]	|E-pi/E-para|
Exp/Hückel	8.12[a]	–0.2[b]	3.96	4.16	3884.82	–8589.58[b]	2.21106
E-Hueckel	12.18617	9.287281	10.73673	1.449445	804.993	–19567.4	24.30753
CNDO	11.57309	–2.27415	4.64947	6.923617	6913.459	–37529.4	5.42846
INDO	10.99398	–2.82895	4.082517	6.911462	7030.23	–36238.6	5.15468
MINDO3	8.21238	–0.47596	3.868211	4.34417	4118.425	–29913.4	7.26332
MNDO	8.574443	0.331878	4.453161	4.121283	3726.394	–30023.4	8.05696
MNDO/d	8.574308	0.331923	4.453116	4.121193	3726.3	–30023.4	8.05716
AM1	8.711272	0.265637	4.488455	4.222818	3835.367	–30004.1	7.82302
PM3	8.83573	0.407184	4.621457	4.214273	3794.829	–28103	7.4056
ZINDO-1	7.512728	–6.39221	0.560258	6.952471	7890.081	–33558.4	4.25324
ZINDO-S	7.868645	–0.04134	3.913653	3.954993	3659.046	–24987.1	6.82887

(a) from National Institute of Standard and Technology (NIST, 2011b)

(b) from the Hückel total π-energy: 2×(2.303+1.618+1.303+1.000+0.618)=13.684β[a.u.] ... × 627.71...~8589.58 kcal/mol, see Cotton (1971b).

TABLE 3.36 The Same Quantities As Those Reported in Table 3.33, heRe for the Fullerene π-System, with $\Delta N = N_\pi = 60$ (Putz, 2011d)

Quantity→ Method↓	IP [eV]	EA [eV]	χ [eV]	η [eV]	|E-(para) bolic| [kcal/mol]	E-pi [kcal/ mol]	|E-pi/ E-para|
Exp/ Hückel	7.58[a]	2.7[b]	5.14	2.44	94204.57	−58478.5[c]	0.62076
E-Hueckel	11.43288	9.864988	10.64894	0.783948	17813.2	−96998.9	5.44534
CNDO	8.86603	−0.3482	4.258917	4.607113	185411.8	−206145	1.11182
INDO	8.072314	−1.17385	3.449232	4.623082	187195.6	−198445	1.06009
MINDO3	7.162502	0.530575	3.846539	3.315964	132368.6	−166316	1.25646
MNDO	9.130902	2.562977	5.84694	3.283963	128270.9	−167601	1.30662
MNDO/d	9.130722	2.563319	5.847021	3.283702	128260	−167601	1.30673
AM1	9.642135	2.948629	6.295382	3.346753	130257.6	−168141	1.29083
PM3	9.482445	2.885731	6.184088	3.298357	128402	−154715	1.20493
ZINDO-1	−2.57843	−12.4464	−7.5124	4.933972	215277.5	−238864	1.10956
ZINDO-S	1.89132	−3.96099	−1.03483	2.926153	122938.5	−126998	1.03302

(a) from De Vries et al. (1992);

(b) from Yang et al. (1987);

(c) from Hückel total π-energy: 93.161602β[a.u.] ... × 627.71...~ 58478.5 kcal/mol, see Haddon et al. (1986); Haymet (1986); Fowler & Woolrich (1986); Byers-Brown (1987).

better and better representing the pi-electronic systems with the increasing complexity of the system on focus; the way in which this depend on the carbon containing system, alone or in combination with heteroatoms or for the nanosystems systems without carbon, along the above enounces open issues, remain for the future research and forthcoming communications.

3.6.5 ADVANCING REACTIVITY BASED BIOLOGICAL ACTIVITY (REBIAC) PRINCIPLES

Once being convinced by the usefulness of modeling chemical reactivity by the aid of the E(N) parabolic dependency, one may questioning the possibility of extending such analysis framework to the biological activity. In this respect, one may address the chemical reactivity parabolic recipe for appropriate biological effects, when they are manifested as parabolic behavior. For instance, when the polysaccharide-containing extracellular fractions (EFs) of the edible mushroom *Pleurotus ostreatus* have immunomodulating

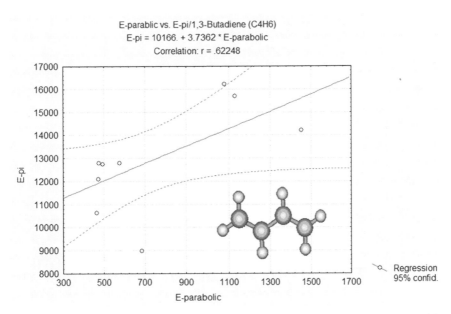

FIGURE 3.15 The bivariate correlation of the parabolic with π-energies as reported in Table 3.33 for Butadiene system (Putz, 2011d).

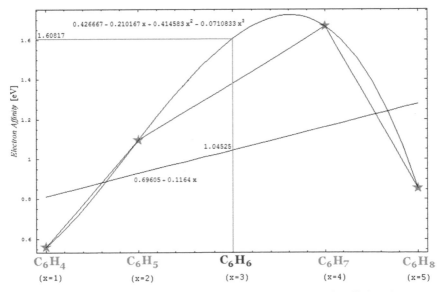

FIGURE 3.16 The cubic vs. Linear interpolation of the electronic affinity of Benzene (Putz, 2011d) based on the data on four adjacent points for o-benzyne (C_6H_4, 0.560 eV), phenyl (C_6H_5, 1.096 eV), methylchylopentadienyl (C_6H_7, 1.67 eV), and for $(CH_2)_2C$-$C(CH_2)_2$ (C_6H_8, 0.855 eV), as reported in (Lide, 2004; Putz, 2011d).

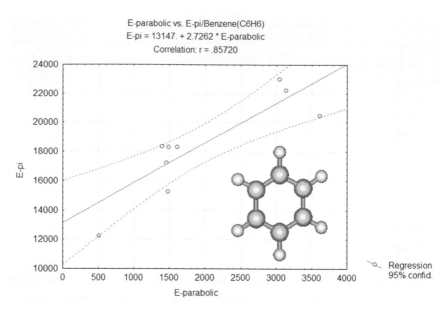

FIGURE 3.17   The bivariate correlation of the parabolic- with π- energies as reported in Table 3.34 for Benzene system (Putz, 2011d).

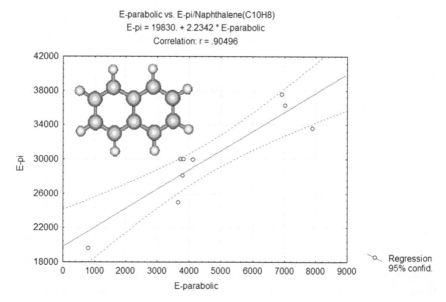

FIGURE 3.18   The bivariate correlation of the parabolic with π-energies as reported in Table 3.35 for Naphthalene system (Putz, 2011d).

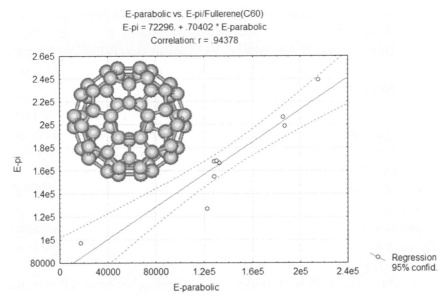

FIGURE 3.19 The bivariate correlation of the parabolic with π-energies as reported in Table 3.36 for Fullerene system (Putz, 2011d).

effects are under concern, while being aware of these therapeutic effects of mushroom extracts, the synergistic relations between these extracts and BIAVAC and BIAROMVAC vaccines are eventually investigated. These vaccines target the stimulation of the immune system in commercial poultry, which are extremely vulnerable in the first days of their lives. By administrating EF with polysaccharides from *P. ostreatus* to unvaccinated broilers we have noticed slow stimulation of maternal antibodies against infectious bursal disease (IBD) starting from four weeks post hatching. The effect of polysaccharides present in EF extracts from the edible mushroom *P. ostreatus* is then investigated with emphasis on the specific response of the chickens' immunitary system; the present discussion follows (Selegean-Putz-Rugea, 2009).

Actually, through the following (end) points have been comparatively investigated, namely:

1. the effect of EF on unvaccinated chickens,
2. the effect of EF on broilers vaccinated with BIAVAC vaccine,
3. the effect of EF on broilers vaccinated with BIAROMVAC vaccine and

4. the parabolic reactivity analysis of biological activity, the results showed that EF from the edible mushroom *P. ostreatus* helps BIAVAC and BIAROMVAC vaccines in stimulating the immune system against IBDV during the critical first two weeks post hatching.

For the first stage of such study, we have followed the effect of poly-saccharide-containing extracellular fractions (EF) from the edible mush-room *P. ostreatus* on broiler chickens, which have not been vaccinated with BIAVAC and BIAROMVAC vaccines (Figure 3.20). In this case the maternal titers of infectious bursal disease virus antibodies (IBD-AB) (control) are seen to decrease, being at the minimum level (96) in four weeks post hatching (Figure 3.20a). This antibody titer (96) was below the estimated cut-off level in this enzyme–linked immunosorbent assay (ELISA) system.

In next stage, with another experimental group, we have followed the change of infectious bursal disease virus antibodies (IBD-AB) antibodies in broilers vaccinated with BIAVAC and treated with EF. Thus, all broilers, although vaccinated with BIAVAC, without EF treatment, have shown a decrease of the level of maternal IBD-AB antibodies. The lowest antibody titer (185) was below the estimated cut-off level in this ELISA system. The production of antibodies increased starting with week 4 ($p < 0.0001$; Figure 3.21a).

Within the above stipulated third instance, we have noticed a better response in broilers treated with BIAROMVAC+5%EF, whose average variation was 12.85%, as compared to the treatments BIAROMVAC with-out EF and BIAROMVAC+15% EF, with averages of 14.39 and 14.02%, respectively. By comparison, the all pairs means of maternal IBD–AB have shown significant differences at 95.0% confidence level, except for one case (2–4) of the treatment with BIAROMVAC+5% EF and two cases (1–4 and 1–5) of the treatment with BIAROMVAC+15% EF. The cor-relation coefficient between the system variables are positive ($r=0.92$ and respectively, $r=0.80$) (Figure 3.22a, b).

The results of Eqs. (4.22)–(4.3) experiments open the possibility for further systematic treatment of IBD–AB activity (y) through employing the fitted response curves of Figures 3.20–3.22 beyond the statistical cor-relation factor and fitted parameters. Actually, since the reported ANOVA analysis in Figures 3.20–3.22 unfolds under the parabolic form:

(a)

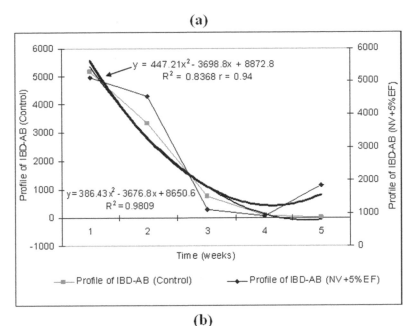

(b)

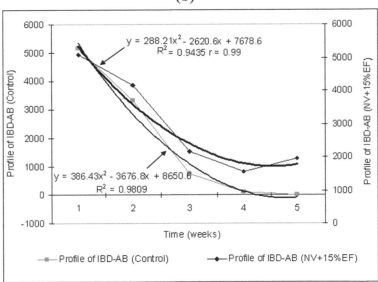

FIGURE 3.20 The effect of different concentrations of EF from P. ostreatus upon the maternal IBD-AB titer for unvaccinated chicken. Thin trend line —: the titer of maternal IBD-AB (Control); Thick trend line ▬: the titer of maternal IBD-AB in un-vaccinated chicken NV+5% (a) and 15% EF (b). Weeks mean weeks post hatch (Selegean-Putz-Rugea, 2009).

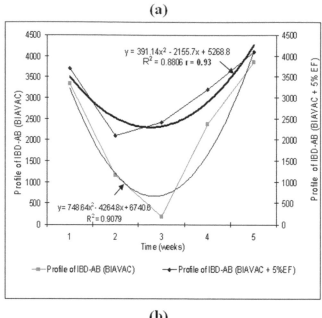

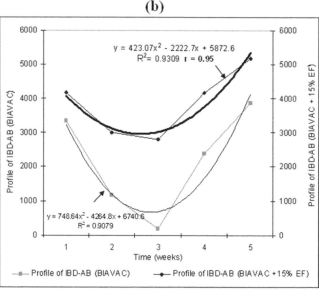

FIGURE 3.21 The effect of different concentration of EF from *P. ostreatus* upon maternal IBD-AB titer to broilers vaccinated with BIAVAC. Thin trend line —: the titer of maternal IBD-AB (BIAVAC); Thick trend line ▬: the titer of maternal IBD-AB to broilers with BIAVAC+5% (a) and BIAVAC+15% EF(b). Weeks mean weeks post hatch, with the vaccination at the end of week 1 (Selegean-Putz-Rugea, 2009).

(a)

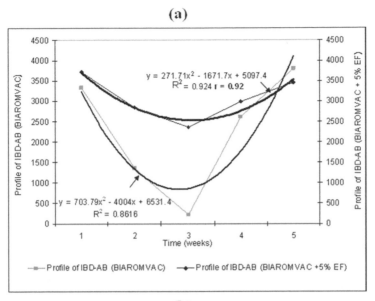

(b)

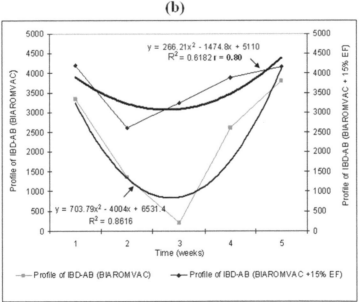

FIGURE 3.22   The effect of different concentrations of EF from *P. ostreatus* upon maternal IBD–AB titer in broilers vaccinated with BIAROMVAC vaccine. Thin trend line —: the titer of maternal IBD–AB (BIAROMVAC); Thick trend line ▬: the titer of maternal IBD–AB in broiler BIAROMVAC+5% (a) and BIAROMVAC+15% EF (b). Weeks mean weeks post hatch, with the vaccination at the end of week 1 (Selegean-Putz-Rugea, 2009).

$$y = y_0 - bx + cx^2 \tag{3.520}$$

it readily reminding of the chemical reactivity law modeling the (molecular) systems' energy change with the number of donated/accepted electrons ΔN, see Eq. (3.166):

$$E = E_{equilibrium} - \chi\Delta N + \eta(\Delta N)^2 \tag{3.521}$$

through the electronegativity χ and chemical hardness η indices of Eqs. (3.346) and (3.347) in terms of ionization potential (IP) and electron affinity (EA), respectively (Sen & Mingos, 1993; Putz & Mingos, 2012).

Now, the idea is to rearrange Eq. (3.520) under the equivalent form given by Eq. (3.521) that will reveal us how the fitting coefficients of Eq. (3.520) will correspond to the reactivity indices in Eq. (3.521). Yet, such transformation is not direct, but upon considering appropriate operations. As such, the first step regards the translation of the form given by Eq. (3.520) into a parabolic form centered on its optimum values, either for dependent Y_{opt} and independent X_{opt} variables (Selegean-Putz-Rugea, 2009):

$$Y = Y_{opt} - \alpha(x - X_{opt}) + \beta(x - X_{opt})^2 \tag{3.522}$$

while the optimum expression are found by the variation of Eq. (3.520):

$$X_{opt} : \left.\frac{\partial y}{\partial x}\right|_{x=X_{opt}} = 0 \tag{3.523a}$$

$$Y_{opt} = y(X_{opt}) \tag{3.523b}$$

or, analytically as:

$$X_{opt} = \frac{b}{2c} \tag{3.524a}$$

$$Y_{opt} = y_0 - \frac{b^2}{4c} \tag{3.524b}$$

Then, in order for coefficients α and β to play the "reactive" role in Eq. (3.522) as electronegativity and chemical hardness do in Eq. (3.521), respectively, they have to be considered with a similar expression as that of Eqs. (3.346) and (3.347), namely (Selegean-Putz-Rugea, 2009):

$$\alpha = \frac{Y^- + Y^+}{2} \tag{3.525}$$

$$\beta = \frac{Y^- - Y^+}{2} \tag{3.526}$$

however, linked with the input equation (3.520) by means of assignments (Selegean-Putz-Rugea, 2009):

$$Y^- = y\left(x \to -1; y_0 \to Y_{opt}\right) \tag{3.527}$$

$$Y^+ = y\left(x \to +1; y_0 \to Y_{opt}\right) \tag{3.528}$$

thus providing the results:

$$\alpha = Y_{opt} + c = y_0 - \frac{b^2}{4c} + c \tag{3.529}$$

$$\beta = b \tag{3.530}$$

Now, paralleling the chemical reactivity principles of electronegativity and chemical hardness (Sen & Mingos, 1993; Putz & Mingos, 2012) to the coefficients α and β of Eq. (3.522), respectively, one can infer the hierarchies among two biological-chemical tested systems for which the two pairs of coefficients, say (α_I, β_I) for the system "I" and $(\alpha_{II}, \beta_{II})$ for the system "II", have been computed.

The electronegativity principles would lead with the rule that as the α index of a system is higher as the system will display larger reactivity/ activity, here translated as growing of the antibody metabolic action; in short: "as α increases as grows the biological activity", i.e., the stimulation of antibody synthesis; this may be eventually called as *The First Principle of Reactive Biological Activity* (ReBiAc1).

Instead, the chemical hardness principles, especially the maximum hardness one (Sen & Mingos, 1993; Putz & Mingos, 2012), would lead with the idea that the β factor tendency is detrimental to the biological activity, or, analytically: "as β increases as slows the biological activity", here – the immunological action; this may be eventually called as *The Second Principle of Reactive Biological Activity* (ReBiAc2).

Table 3.37 collects all the fitted and *reactive biological activity* (ReBiAc) data from the Figures 3.20–3.22 and the Eqs. (3.520)–(3.530), respectively; it allows a global view of the experiments either between groups with various EF belonging to the same vaccine or not-vaccinated chickens. Furthermore, the present ReBiAc analysis finely clarifies upon the hierarchy of experiments in providing viable biological activity (here viewed as the potency of antibody synthesis).

For instance, if one looks to the ordering correlation factors from the experimental systems I–IX (of Table 3.37) there finds the puzzling hierarchy (Selegean-Putz-Rugea, 2009):

$$r_{IX}^2 < r_{II}^2 < r_{VII}^2 < r_{V}^2 < r_{IV}^2 < r_{VIII}^2 < r_{VI}^2 < r_{III}^2 < r_{I}^2 \qquad (3.531)$$

while this is rearranged in terms of increasing beneficial activity by the ReBiAc1 principle:

$$\underbrace{\alpha_I < \alpha_{IV} < \alpha_{VII}}_{ALL\ CONTROLS} < \underbrace{\alpha_{II} < \alpha_{III}}_{UN-VAC+EF} < \underbrace{\alpha_V < \alpha_{VIII}}_{VAC+EF\,(5\%)} < \underbrace{\alpha_{IX} < \alpha_{VI}}_{VAC+EF\,(15\%)} \qquad (3.532)$$

or as the decrease of antagonist activity in the light of ReBiAc2 principle:

TABLE 3.37 The Computed Favorable and Detrimental Biological Activity Factors, α and β, According With Their Definitions Given by Eqs. (3.529) and (3.530), For Titers vs. Weeks Fitted Curves Displayed in Figures 3.20–3.22, Employing the Specific Parabolic Data Modeled by Eq. (3.520) (Selegean-Putz-Rugea, 2009)

Experimental			Fitted Parameters				Activity parameters	
System/Source			b	c	y_0	r^2	α	β
UN-VACCINE	I.	Control	3676.8	386.43	8650.6	0.9809	291.04	3676.8
	II.	EF (5%)	3698.8	447.21	8872.8	0.8368	1671.97	3698.8
	III.	EF (15%)	2620.6	288.21	7678.6	0.9435	2009.74	2620.6
BIAVAC VACCINE	IV.	Control	4264.8	748.64	6740.6	0.9079	1415.39	4264.8
	V.	EF (5%)	2155.7	391.14	5268.8	0.8806	2689.75	2155.7
	VI.	EF (15%)	2222.7	423.07	5872.6	0.9309	3376.3	2222.7
BIAROMVAC VACCINE	VII.	Control	4004	703.79	6531.4	0.8616	1540.3	4004
	VIII.	EF (5%)	1671.7	271.71	5097.4	0.924	2797.82	1671.7
	IX.	EF (15%)	1474.8	266.21	5110	0.6182	3333.62	1474.8

$$\underbrace{\beta_{IV} > \beta_{VII}}_{CONTROL-VAC} > \underbrace{\beta_{II} > \beta_{I} > \beta_{III}}_{UN-VACCINATE} > \underbrace{\beta_{VI} > \beta_{V}}_{BIA+EF} > \underbrace{\beta_{VIII} > \beta_{IX}}_{BIAROM+EF} \qquad (3.533)$$

From both these chains of bio-activity orderings appears the decisive influence EF has on anti-bursal vaccines since placed on the extreme of ReBiAc principles' records (maximums for α, and minimums for β values). This way, the analytical and conceptual tools for cross-judging both the favorable and detrimental (natural or induced) biological actions were formulated and tested. Moreover, is phenomenologically proven that each biological system described by a parabolic dependency contains the intrinsic positive (beneficial) and negative (detrimental) effects at the metabolic level, however in different degree. Nevertheless, the actual study shows how the beneficial influence may be adjusted to prevail on the harmful one by controlling the induced biological activity, here through administrating various EF with polysaccharides from *Pleurotus*.

This has determined us to believe that using this combination of BIAVAC and BIAROMVAC vaccine and EF from *P. ostreatus* is good for antibodies production stimulation in the period of the first days of life. In this period, according to the latest researches, it appears that the vaccines show affinity and associate with B cells, macrophages, and folliculars dendritic cells in bursa and spleen. Nonetheless, the allied experiments have been successfully rationalized by employing the parabolic chemical reactivity principles of electronegativity and chemical hardness indices to formulate and apply the ReBiAc counterparts.

3.7 UNIFICATION OF CHEMICAL REACTIVITY PRINCIPLES

3.7.1 CHEMICAL REACTIVITY BY VARIATIONAL PRINCIPLES

There is wild recognized that in physical sciences the total energy variation

$$dE = 0 \qquad (3.534)$$

fixes the equilibrium or the evolution towards equilibrium of natural systems. In conceptual DFT framework, however, if one expands the total energy $E = E[N, V(\mathbf{r})]$ such that to contain the electronegativity and chemical hardness appearance relating the number of electrons in first and second order variations in reactivity, respectively, as well as to contain the applied potential fluctuation, one has the working differential relationship; the present discussion follows (Putz, 2011e)

$$dE = -\chi dN + \eta(dN)^2 + \int \rho(\mathbf{r})dV(\mathbf{r})d\mathbf{r} \qquad (3.535)$$

Yet, through combining Eqs. (3.534) and (3.535), one has either that:

- there is no action on the system ($dN=dV=0$) so that no chemical phenomena is recorded since the physical variational principle (3.534) is fulfilled for whatever electronegativity and chemical hardness values in Eq. (3.535);
- or there is no electronic system at all ($\chi = \eta = \rho(\mathbf{r}) = 0$).

Therefore, it seems that the variational physical principle of Eq. (3.534) do not suffice in order the chemical principles of electronegativity and chemical hardness being encompassed, as above stipulated, see Eqs. (3.346) and (3.347). As a consequence one is forced to perform the double variational procedure on the total energy, i.e., through applying the additional differentiation on physical energy expansion (3.535), within the so called "chemical variational mode" (and denoted as $\delta[]$) where the total differentiation will be taken only over the scalar-global (extensive χ, η, dN) and local (intensive $\rho(\mathbf{r})$, $V(\mathbf{r})$) but not over the vectorial (physical – as the coordinate itself $\mathbf{r}$) quantities. This way, one has immediately (Putz, 2011e)

$$\delta[dE] = -\delta[\chi dN] + \delta[\eta(dN)^2] + \int \delta[\rho(\mathbf{r})dV(\mathbf{r})]d\mathbf{r} \qquad (3.536)$$

Now, the chemical variational principle applied to Eq. (3.536) will look like

$$\delta[dE] \geq 0 \qquad (3.537)$$

In conditions of chemical reactivity or binding certain amount of charge transfer and the system's potential fluctuations (departing from equilibrium) are involved

$$dN = |dN| = ct. \neq 0, \ dV(\mathbf{r}) \neq 0 \qquad (3.538)$$

When considering the chemical variation prescribed by Eq. (3.536) in chemical transformation driven by the condition (3.537), a kind of reactivity towards equilibrium constraint or a sort of reverse Gibbs free energy condition, or a special kind of entropy variation within the second law of thermodynamics are in fact applied, while releasing with the individual reactivity principles (Putz, 2011e):

- for electronegativity contribution we have the condition:

$$-\delta[\chi dN] \geq 0 \Leftrightarrow -|dN|\delta[\chi] \geq 0 \Rightarrow \delta[\chi] \leq 0 \qquad (3.539)$$

recovering the electronegativity variational formulation (3.346) for chemical systems toward equilibrium.

- for chemical hardness contribution one gains directly the inequality of Eq. (3.347)

$$\delta[\eta(dN)^2] \geq 0 \Leftrightarrow (dN)^2 \delta[\eta] \geq 0 \Rightarrow \delta[\eta] \geq 0 \qquad (3.540)$$

- whereas for chemical action contribution there is now sufficient in having the exact equality

$$\delta\int\rho(\mathbf{r})dV(\mathbf{r})d\mathbf{r} = 0 \qquad (3.541)$$

leaving with the successive equivalent forms

$$0 = \delta\int\rho(\mathbf{r})\{V(\mathbf{r}) - V(\mathbf{r}_0)\}d\mathbf{r}$$
$$= \delta\left\{\int\rho(\mathbf{r})V(\mathbf{r})d\mathbf{r}\right\} - \delta\{V(\mathbf{r}_0)\}\int\rho(\mathbf{r})d\mathbf{r} \qquad (3.542)$$

with $V(\mathbf{r}_0)$ being the constant potential at equilibrium. However, through employing the basic DFT relationship for electronic density abstracted from Eq. (3.6)

$$N = \int\rho(\mathbf{r})d\mathbf{r} \qquad (3.543)$$

and by recognizing the total number of electron constancy in the (atoms-in-molecules) system the Eq. (3.541) produces the so called *chemical action (C_A) principle* (Putz, 2011a,e)

$$\delta\left\{\int\rho(\mathbf{r})V(\mathbf{r})d\mathbf{r}\right\} = \delta C_A = 0 \qquad (3.544)$$

that represents the chemical specialization for the physical variational principle of Eq. (3.534).

Worth noting that the actual chemical action principle arises along the other chemical reactivity principles of electronegativity and chemical hardness in a natural way, yet by performing special variation of the charge transfer and potential fluctuation around equilibrium rather than by the direct physical implementation of an "action" in searching for the system evolution's "trajectory". Therefore, the present analysis reveals two main features of chemical reactivity principles in general and of that of the chemical action in special (Putz, 2011e):

- the chemical reactivity principles of electronegativity and chemical hardness belong to *chemical* variation principle rather to the physical optimization framework, this way inscribing themselves in the heritage of natural laws provided by Chemistry to complement those of Physics;
- the chemical action principle of Eq. (3.544), apart representing as well a chemical principle since arising within the same conceptual DFT framework as those of electronegativity and chemical hardness, represents a variational principle that optimizes the relationship between the objective main quantities of DFT, the electronic density and its driving bare potential, in a convoluted manner over the reactivity or binding space of interest; it may, eventually, produce the chemical global minima of equilibrium in reactivity or bonding.

However, the hierarchy of the electronegativity, chemical hardness and chemical action principles was recently advanced as describing the paradigmatic stages of bonding; they may be resumed here through the sequence (Putz, 2011a,e):

$$\delta\chi = 0 \rightarrow \delta C_A = 0 \rightarrow \Delta\chi < 0 \rightarrow \delta\eta = 0 \rightarrow \Delta\eta > 0 \tag{3.545}$$

as corresponding to the *encountering* (or the electronegativity equality) *stage*, followed by *chemical action minimum variation* (i.e., the global minimum of bonding interaction), then by *the charge fluctuation stage* (due to minimum or residual electronegativity), ending up with *the polarizability stage* (or HSAB) and with the *final steric* (due to maximum or residual hardness) *stage*. Nevertheless, from Eq. (3.545) one observes the close laying chemical action with electronegativity influence in chemical reactivity and bonding principles.

3.7.2   CHEMICAL REACTIVITY BY PARABOLIC E(N) DEPENDENCY

One may address the problem: how are optimized the values of χ and η in Eq. (3.457) during a chemical reaction or through the energy minimization processes, see Eq. (3.537)? To give a reasonably answer, firstly consider the parabolic shape of the total energy respecting with the total number of electrons, as in Figure 3.23, connecting the three states of interests: $|N - h\rangle$, $|N\rangle$, and $|N + h\rangle$ ones. Let's next assume that any energetic shape, containing the associated energies for the above considerate electronic states, has a parabolic form, as conceptually already certified. Then, one is interested in minimizing the energy throughout all parabolic classes that link those states. Is immediately that such a minimizing procedure can be undertook in two distinct ways: to simultaneously minimize the energetic values of the states $|N - h\rangle$ and $|N + h\rangle$, or to only act on the energy in the "point" $|N\rangle$ of the energetic shape, as in cases (I) and (II) in Figure 3.23, respectively (Ayers & Parr, 2000).

In the representations (I) and (II) of Figure 3.23 the negative slope and the convexity of the energetic shape for the state $|N\rangle$ will give information about the behavior of χ and η during the energy minimization, respectively. Firstly, in both analyzed cases the electronegativity approaches its minimum on the right ground density ρ, around

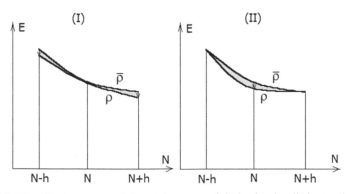

FIGURE 3.23   The two cases of the total energy minimization by distinct acting on the parabolic energetic shape connecting the electronic states $|N - h\rangle$, $|N\rangle$, and $|N + h\rangle$ (Ayers & Parr, 2000, Putz, 2003, 2008b, 2012b).

the electronic state $|N\rangle$ (Tachibana, 1987; Tachibana & Parr, 1992; Tachibana et al., 1999):

$$(I) , (II) : \bar{\chi} [\bar{\rho}] \geq \chi [\rho] \tag{3.546}$$

whereas the chemical hardness records different optimum values depending with the type of energetic minimization procedure:

$$(I) : \bar{\eta} [\bar{\rho}] \geq \eta [\rho] \tag{3.547}$$

$$(II) : \bar{\eta} [\bar{\rho}] \leq \eta [\rho] \tag{3.548}$$

Now, the following dilemma rises: what kind of optimization behavior is the correct one for the chemical hardness, a minimum value, as in case (I) or a maximum one like in case (II), from Figure 3.23, for the right ground ρ state, respectively? There are two arguments to choose that the right ρ ground state have to have the maximum hardness value, at the end of the energetic minimization (binding) process. The first argument calls the relation between electronegativity and hardness that prescribes the negative sign between the electronegativity and chemical hardness tendencies when the right ground state is approached. Now, being associated the right ground state, in both above (I) and (II) cases, with the minimum value in electronegativity, the natural choice for the chemical hardness for ending the minimization process will be therefore its maximum value.

The second argument regards the parabolic energetic shape, that from beginning it was assumed to be maintained as the invariant characteristic during the minimization process; and there is clear that in the case (I) of Figure 3.23 the minimization of energy tends to deform the parabolic energetic shape into a linear one. Thus, the right way of energy minimization trough the parabolic classes correspond with the case (II), which in turn, prescribes the maximum hardness optimization for the achievement of right ground state. Concluding here, worth noting that Pearson had pointed out that *there seems to be a rule of nature that molecules* (or the many-electronic systems in general, n.a.) *arrange themselves* (in their ground state, n.a.) *to be as hard as possible* (Pearson, 1997). This way, was established the Maximum Hardness Principle (*MHP*) for the equilibrium of the many-electronic systems in their ground states (Chattaraj et al., 1991, 1995; Putz, 2008c).

3.7.3 GLOBAL SCENARIO FOR CHEMICAL REACTIVITY

However, based on electronegativity and chemical hardness at length advocating different influences on various levels of quantum chemistry, atoms and molecules, the global scenario of reactivity may be advanced implying five stages of chemical bonding hierarchy by referring to the principles of Sections 3.2–3.5 here resumed in Table 3.38.

Accordingly, the global scenario of reactivity, based on electronegativity and hardness principles, implies that there are five stages of bonding (Putz, 2008b, 2012a, 2012b):

(i) The *encountering stage* is dominated by the difference in electronegativity between reactants and is consumed when the electronegativity equalization principle is fulfilled among all constituents of the products; this stage is associated with the charge flow from the more electronegativity regions to the lower electronegativity

TABLE 3.38 Synopsis of the Basic Principles of Reactivity Towards Chemical Equilibrium With Environment in Terms of Electronegativity, Chemical Action, and Chemical Hardness (Putz, 2008b, 2011a)

Chemical Principle	Principle of Bonding
$\delta\chi = 0$	*Electronegativity equality*:
	"Electronegativity of all constituent atoms in a bond or molecule have the same value" (Sanderson, 1988).
$\delta C_A = 0$	*Chemical action minimum variation*:
	Global minimum of bonding is attained by optimizing the convolution of the applied potential with the response density (Putz, 2003, 2008a, 2009a, 2011a, 2012a).
$\Delta\chi = 0$	*Minimum (residual) electronegativity*:
	"the constancy of the chemical potential is perturbed by the electrons of bonds bringing about a finite difference in regional chemical potential even after chemical equilibrium is attained globally" (Tachibana et al., 1999).
$\Delta\eta = 0$	*Hard-and-soft acids and bases*:
	"hard likes hard and soft likes soft" (Pearson, 1973, 1990, 1997).
$\Delta\eta = 0$	*Maximum (residual) hardness*:
	"molecules arranges themselves as to be as hard as possible" (Pearson, 1985, 1997)

regions in a molecular formation thus covering the *covalent* binding step (Mortier et al., 1985; Sanderson, 1988);

(ii) The *global optimization* stage, associates with the variational principle of the total energy of ground/valence state in bonding that can be resumed by the corresponding chemical action principle (Putz, 2003; Putz & Chiriac, 2008; Putz, 2008b, 2009a, 2011a, 2012a) that adjust the applied potential and the response electronic density to be convoluted/coupled in optimum/unique way, i.e., establishing the global minima on the potential surface of the system.

(iii) The *charge fluctuation stage*, relies on the fact that partial fractional instead of integer charges are associated with atoms-in-molecules; even after the chemical equilibrium is attained globally the electrons involved in bonds acts as foreign objects between pairs of regions, at whatever level of molecular partitioning procedure, inducing the appearance of finite difference in adjacent electronegativity of neighbor regions in molecule (Tachibana and Parr, 1992; Tachibana et al., 1999); it is due to the quantum fluctuations associate with the quantum nature of the bonding electrons and it corresponds to the degree of *ionicity* occurred in bonds;

(iv) The *polarizability stage*, in which the induced ionicity character of bonds is partially compensated by the chemical forces through the hardness equalization between the pair regions in molecule; the HSAB principles is therefore involved (Pearson, 1973, 1990, 1997; Chattaraj & Schleyer, 1994; Chattaraj & Maiti, 2003; Putz et al., 2004), as a second order effect in charge transfer, being driven by the ionic interaction through bonds;

(v) The *steric stage*, where the second order of quantum fluctuations provides a further amount of finite difference, this time in attained global hardness, that is transposed in relaxation effects among the nuclear and electronic distributions so that the remaining unsaturated chemical forces to be dispersed by stabilization of the molecular structure through the maximization of its chemical hardness and the fully stabilization of the molecular system in a given environment (Pearson, 1985, 1997; Chattaraj et al., 1991; Chattaraj et al., 1995; Putz, 2008c).

One may see therefore that there are basically three variational principles, i.e., for electronegativity, chemical action and chemical hardness
that assure the optimum of charge-potential, density-potential, and den-
sity-charge relationships, respectively, while the quantum fluctuations are
resolved by the residual minimum electronegativity and by residual maximum chemical hardness principles.

This way, it was proved that electronegativity and chemical hardness
may provide a viable *set of reactivity indices*, to which also the chemical
action may be added as related to the electronegativity to complete the
global optimization picture of reactivity, being able to cover (in principle)
most of the binding process, towards building the desiderated orthogonal
space of reactivity to properly model the chemical bonding from an unitary physical-chemical perspective. Nevertheless, it can be further used
in quantifying the global and local quantum chemical effects at atomic
and molecular transformation and interaction, while equipping Physical
Chemistry with a true versatile tool in treating in principle all kinds of its
phenomenology grounded on physically interpretable and computationally implementable quantities.

This way, there was proved that electronegativity and hardness provides the minimum set of reactivity indices able to cover the complete
process of binding, as a whole.

There is quite surprising that after our best knowledge no systematic
studies are reported for linking the hardness with its conceptual source, the
electronegativity when applying the HSAB principle. Such link is therefore here advanced based on the very definitions of what soft and hard
acid and base are. Still, the right connection can be achieved recalling that
electronegativity has the potential nature at the chemical level, so being
proportional with the inverse of the radius of atoms or length of bonds,
$\chi \propto 1/r$. Worth noting that such dependence was the main picture in which
one of the most recent atomic electronegativity scale was given (Ghosh,
2005). It can be equally derived from a simple model of charging energy
of a conducting sphere of radius r (Ghosh & Biswas, 2002; Pearson, 1972,
1988b). Yet, electronegativity can be seen as being proportional also with
the inverse of the polarizability, $\chi \propto 1/\alpha$, since polarizability α is on its turn
proportional with the volume encompassed by the electronic system under
discussion (Putz et al., 2003).

With these remarks in hand let's list the main definitions of soft and hard acids and bases, connecting their hardness degree with electronegativity (Putz, 2008b, 2012b):

- a soft base, for example, R^- or H^-, is very polarizable and thus with low electronegativity;
- a hard base, for example, F^- or OH^-, is not much polarizable and thus with high electronegativity;
- a soft acid, for example, RO^+ or HO^+, has usually low positive charge and large size, so posing lower electronegativity;
- a hard acid, for example, H^+ or XH (hydrogen bonding molecules), has normally high positive charge and small size, so posing high electronegativity.

More, one can straightforwardly infer from Figure 3.24(a) that the relative position of electronegativities between two reactants can give the acid or base nature of the species since the more basicity the more $-\chi$ is pushed towards positive range. This observation seems crucial to us and can explain why the consecrated classification of some common compounds to be acids or bases (Sen & Mingos, 1993; Putz & Mingos, 2012) has not an absolute value and founds some computational disagreements (Drago & Kabler, 1972; Putz et al., 2004), while the relative electronegativities involved in concerned reactions have to as well be taken as the appropriate measure.

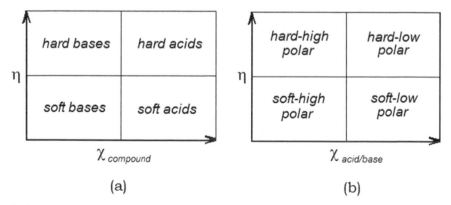

FIGURE 3.24 (a) The diagram of the compound repartition as hard-and-soft acids and bases within the electronegativity-hardness space (χ, η). (b) The diagram of the hard-and-soft nature of the high-and-low polar acids and bases within the electronegativity-hardness space (χ, η) (Putz, 2008b, 2012b).

Collecting all these ideas in a representative quantum concept we can draw the Figure 3.6 for appropriate indication of the acid/base and hard/soft trends of the chemical species within the electronegativity-hardness chemical space (χ, η).

The Figure 3.24(a) depicts the phenomenological correlation between electronegativity and hardness for a given chemical species leaving with their natural classification as acids and bases on the electronegativity scale and hard and soft on the hardness scale as their positions are more departed from the (0, 0) origin point within the chemical space (χ, η).

Similar classification can be done in Figure 3.24(b) for a series of acids and bases, separately, with the result in categorizing them as hard-and-soft character with low-and-high polarity as their positions are more departed from the (0, 0) origin point within the chemical space (χ, η). This way, there is provided a new valuable working scheme with the help of which the relative acidic or basic nature as well as the hard and soft strength of the molecules in their state of interaction is analyzed.

3.8   CHEMICAL REACTIVITY BY CHEMICAL ACTION PRINCIPLE

3.8.1   CHEMICAL ACTION ENTANGLEMENT

At the *electronic level* the generalized Schrödinger-de Broglie-Vigier-Bohm wave function (Schrödinger, 1926; de Broglie & Vigier, 1953; Bohm & Vigier, 1954; Putz, 2009b)

$$\Psi(x,t) = R\exp\left(i\frac{S}{\hbar}\right) = \rho(x,t)^{1/2}\exp\left[\frac{i}{\hbar}(px - Et)\right] \qquad (3.549)$$

is firstly employed in terms of its amplitude R and phase (action) S to lead with the quantum-classical conservation equation of forces:

$$\sum F_{classic} + \sum F_{quantum} = 0 \qquad (3.550)$$

with

$$F_{quantum} = -\left\langle \frac{\partial V_{qua}}{\partial x} \right\rangle \qquad (3.551)$$

while the (sub)quantum potential

$$V_{qua}(x) = -\frac{\hbar^2}{2m}\frac{\nabla^2 R}{R} \tag{3.552}$$

is responsible for the long-range or asymptotical (chemical) interaction since its non-vanishing structure in $R \sim \Psi$; the present discussion follows (Putz, 2009b). Such a paradox reveals a hidden reality of interaction that should be responsible for the delocalization interaction in chemical systems, lighting on aromaticity concept (Fukui et al., 1960), being however incompatible with the speed light velocity limitation of relativity theory (Boeyens, 2005). Fortunately, such dichotomy was recently surpassed by reconsidering both the special relativity and quantum (algebraically) chemistry within the over-light velocity (or two steep light) mechanism when considering the wave package propagation in a real space-time framework (Whitney, 2006a-b, 2007, 2008a-b, 2009, 2013). Such ideas root in late de Broglie thoughts and may constitute the foreground of reconsidering the electronic wave-corpuscular manifestation in both its propagation and eigen-states (de Broglie & Vigier, 1953). Eventually, it may lead the closed loop quantification of charges (Boeyens, 2005):

$$nh = \Delta S = \oint dS = \oint \frac{\partial S}{\partial x} dx = \oint |\nabla S| dx = \oint p dx \tag{3.553}$$

to the bonding stage through the conservation of the exchanged electrons through the atoms-in-molecules Gaussian law respecting the bonding basin (with surface area Σ):

$$-\frac{\partial N}{\partial t} = \oint_{\Sigma} \mathbf{j} \cdot d\mathbf{n} \tag{3.554}$$

with the density of the charge current given by the hydrodynamic expression (Boeyens, 2005):

$$\mathbf{j} = \frac{R^2}{m}\nabla S = \rho\frac{\mathbf{p}}{m} = \rho\mathbf{v} \tag{3.555}$$

while the total number of involved electrons fulfills the density functional relationship (3.543) (Parr & Yang, 1989, Putz, 2003, 2006a-b, 2007a):

In these conditions, the chemical bond can be unfolded by means of Stokes transformation of the loop charged quantification to its opening of the charge flux thus establishing the chemical interaction quantification:

$$nh = \frac{m}{\rho} \int_{\Sigma} (\nabla \times \mathbf{j}) d\mathbf{n} \tag{3.556}$$

Nevertheless, this so called chemical entanglement can be further considered at the global level through employing the electronic density ρ and quantum potential $V_{quantum}$ in the so called *entangled chemical action* functional (one may conveniently change the 3D space variable with the 1D or generic space variable "x")

$$C_A^{ent} = \int \rho(x) V_{qua}(x) d\tau = -\frac{\hbar^2}{2m} \int \rho^{1/2}(x) \nabla^2 \rho^{1/2}(x) dx \tag{3.557}$$

thus providing the second level of chemical bonding comprehension: the density functionals realm with emphasis on the necessity of considering of density gradient expansion in chemical bonding characterization. This is nothing less than the inclusion of the Heisenberg delocalization principle in bonding formation and stabilization. Again the quantum paradox arises since the bonding stabilization involves the electronic delocalization.

Nevertheless, the above quantum-classical force balance in chemical interaction at the level of forces can be further specialized at the level of chemical action functional which is nothing else than the observable of the potential involved; it provides the *entangled equation of the electronic density*:

$$\left| \nabla \rho^{ent}(x) \right| = \frac{\nabla^2 \left[C_A^{class}(x) + C_A^{ent}(x) \right]}{V_{class}(x) + V_{qua}(x)} \tag{3.558}$$

with the adequate *Laplacian-type chemical actions*

$$C_A^{class/ent}(x) = \int^x \rho(x) V_{class/qua}(x) d\tau \tag{3.559}$$

which preserves the wave nature of the electronic movements through the Laplacian character of both equation and chemical actions. It nevertheless can be regarded as the chemical bonding electronic density equation.

Going to the next level of chemical bonding the ligand-receptor (L-R) or substrate-enzyme (S-E) interactions are treated such way that the quantum (fluctuating) nature of the biomolecular reactions can be visualized by combining the relationship between the catalytic rate (k_{cat}) and temperature (T) with that between the reaction rate and the turnover number or the effective time of reaction (Δt) via Heisenberg relation,

$$\frac{1}{k_{cat}} \propto \Delta t \cong \frac{\hbar}{\Delta E^{ent}_{tunnelling}} = \frac{\hbar}{k_B T} \qquad (3.560)$$

with *entangled tunneling energy*

$$\Delta E^{ent}_{tunnelling} = \sum C^{class}_A - \sum C^{ent}_A = \int \rho(x) \Big[\sum V_{class}(x) - \sum V_{qua}(x) \Big] d\tau \qquad (3.561)$$

were k_B stands for Boltzmann constant. Of course, in last relation the equivalence between quantum statistics and quantum mechanics was physically assumed when equating the thermal and quantum (tunneling) energies, $k_B T$ and ΔE, respectively. Nevertheless, it is the basis of rethinking upon the static character of the energetic barrier, recalling the so called steady state approximation, usually assumed in describing enzymic catalysis (Putz, 2009b), see also the Volume V of the present five-volume work, within the transition state theory (TST):

$$\frac{d}{dt}[L-R] \cong 0 \rightarrow \frac{d}{dt}\nabla[L-R] = 0 \rightarrow \frac{d}{dt}\Big|\nabla\rho^{ent}_{L-R}\Big| = 0 \qquad (3.562)$$

with

$$\Big|\nabla\rho^{ent}_{L-R}(x,t)\Big| = \frac{\nabla^2\Big[\sum_{L,R} C^{class(L,R)}_A(x,t) + \sum_{L,R} C^{ent(L,R)}_A(x,t)\Big]}{\sum_{L,R} V_{class(L,R)}(x,t) + \sum_{L,R} V_{qua(L,R)}(x,t)} \qquad (3.563)$$

delivering the ligand-receptor equation for its (macroscopic) concentration $(L - R)$ in terms of the stationary flux quantum-classical electronic density.

Finally, various chemical reactivity pathways may be combined to eventually produce the biological activity whit which to (statistically) correlate respecting specific structural (quantum or topological) descriptors. Each such correlation should correspond to a certain molecular mechanism towards biological actions and can macroscopically be quantified through a norm $\|\bullet\|$ and a statistical factor r, while for a given or measured activity their interaction is associated with the least path principle, i.e., the minimum or variational condition across different possible or potentially active molecular mechanism is fulfilled (Putz, 2009b), see also the Volume V of the present five-volume set (Putz, 2016c):

$$\delta \left[A_{(\|\bullet\|,r)}, B_{(\|\bullet\|,r)} \right] = 0 \,, A \text{ & } B: \textit{biological endpoints} \tag{3.564}$$

as a macroscopic reflection of the quantum tunneling or entanglement structural (thus intimate and somehow hidden) interaction. However, it can be further rewritten in terms of above entangled electronic density as the Bader zero-flux condition of atoms-in-molecules (Bader, 1990, 1994, 1997, 1998) applied on biomolecular bonding maps

$$\nabla \rho_{L-R}^{ent}(x,t) \cdot \mathbf{n} = 0 \tag{3.565}$$

yielding the chemical action classical conservation for the bioactivity entanglement:

$$\sum_{L,R} \nabla^2 C_A^{class(L,R)}(x) + \sum_{L,R} \nabla^2 C_A^{ent(L,R)}(x) = 0 \tag{3.566}$$

as a natural chemical (inter)action counterpart of the previously mentioned classical-quantum force balance in electronic structures.

3.8.2 BILOCAL CHEMICAL ACTION

In order to turn the chemical action functional into a current tool for bonding description its reformulation as related with reactivity softness and hardness concepts (Putz, 2003, 2008d, 2012a) seems the appropriate endeavor. It starts with combining the unfolded chemical action variational principle (3.544); the present discussion follows (Putz, 2009a)

$$0 = \int \delta\rho(\mathbf{r})V(\mathbf{r})d\mathbf{r} + \int \rho(\mathbf{r})\delta V(\mathbf{r})d\mathbf{r} \tag{3.567}$$

with the Hellmann-Feynman (1939) theorem:

$$0 = \int \rho(\mathbf{r})\delta V(\mathbf{r})d\mathbf{r} \tag{3.568}$$

followed by the electronic density functional first order expansion in total number of electrons (eventually restricted to those participating in bonding) and in the applied potential upon them, $\rho = \rho[N, V(\mathbf{r})]$. The obtained equation

$$0 = \int \left(\frac{\delta\rho(\mathbf{r})}{\delta N} \right)_{V(\mathbf{r})} V(\mathbf{r})dNd\mathbf{r} + \int \left[\int \left(\frac{\delta\rho(\mathbf{r})}{\delta V(\mathbf{r}')} \right)_N dV(\mathbf{r}')d\mathbf{r}' \right] V(\mathbf{r})d\mathbf{r} \tag{3.569}$$

can be solved out for the electronic density with the form (Putz, 2009a), see also the Volume II/Section 4.3.2 of the present five-volume set (Putz, 2016b)

$$\rho(\mathbf{r}) = -\int \kappa(\mathbf{r}, \mathbf{r}') V(\mathbf{r}') d\mathbf{r}' \qquad (3.570)$$

where the bilocal response function was introduced (Parr & Gázquez, 1993; Garza & Robles, 1993; Torrent-Sucarrat et al., 2005; Putz, 2003)

$$\kappa(\mathbf{r}, \mathbf{r}') = \left(\frac{\delta \rho(\mathbf{r})}{\delta V(\mathbf{r}')} \right)_N \qquad (3.571)$$

With kernel dependency of electronic density back in chemical action definition the *first kernel density functional of chemical action* yields as (Putz, 2009a)

$$C_A^{ker} = -\iint V(\mathbf{r}) \kappa(\mathbf{r}, \mathbf{r}') V(\mathbf{r}') d\mathbf{r} d\mathbf{r}' \qquad (3.572)$$

Worth noted that the C_A^{ker} expression gives the opportunity for understanding chemical action as an interaction quantity due to its close relationship with energy interaction ε_{int} relating polarizability α when the external potential $V(\mathbf{r}) = \mathbf{E} \cdot \mathbf{r}$, with the applied electric field amplitude E, is considered, viz. in *polar coordinates*

$$C_A = -\left| \iint \kappa(r, r') r r' \cos^2 \theta \, dr dr' \right| E^2 = \alpha E^2 = -\varepsilon_{int} \qquad (3.573)$$

Yet, the C_A^{ker} formulation can be further refined since considering the chemical hardness and softness kernels, respectively linked by the integration-differentiation chain rule of delta-Dirac bilocal function $\delta(\mathbf{r}'' - \mathbf{r})$. Now, making use of the fundamental density functional constraint one gets the local hardness- kernel softness density formulation

$$\rho(\mathbf{r}) = 2N \int s(\mathbf{r}, \mathbf{r}') \eta(\mathbf{r}') d\mathbf{r}' \qquad (3.574)$$

Finally, by comparison of kernel response and softness densities the softness-hardness response bilocal function can be immediately reached out

$$\kappa(\mathbf{r}, \mathbf{r}') = -2N \frac{\eta(\mathbf{r}') s(\mathbf{r}, \mathbf{r}')}{V(\mathbf{r}')} \qquad (3.575)$$

leaving with the reactivity kernel expression of chemical action functional (Putz, 2009a):

$$C_A^{s-h} = 2N \iint V(\mathbf{r})s(\mathbf{r},\mathbf{r}')\eta(\mathbf{r}')d\mathbf{r}d\mathbf{r}' \qquad (3.576)$$

Note that the expression C_A^{s-h} enriches the chemical action foreground definition with reactivity information compressed in softness kernel and chemical hardness whereas the bonding character is expressed by allowing the energetic double occupancy for the total (or bonding) number of electrons N, in close correspondence with the unrestricted Hartree-Fock orbital treatment (Heitler & London, 1927; Karo & Allen, 1959; Amos & Hall, 1961; Nesbet, 1962; Clementi, 1962). However, all double integrals involved in chemical action kernel formulations C_A^{ker} and C_A^{s-h} are in close correspondence with the reactivity paths of the electronic pairs $(\mathbf{r}, \mathbf{r}')$ of chemical bonding. Nevertheless, local and nonlocal consequences of the present chemical action functionals in bonding are therefore in next explored.

3.8.3 LOCAL CHEMICAL BOND

The first natural employed softness kernel specialization assumes the referential local approximation (Parr & Gázquez, 1993; Garza & Robles, 1993; Torrent-Sucarrat et al., 2005; Putz, 2008b):

$$s(\mathbf{r},\mathbf{r}')^{local} = -\frac{\delta\rho(\mathbf{r}')}{\delta V(\mathbf{r}')}\delta(\mathbf{r}-\mathbf{r}') \qquad (3.577)$$

with the help of which the local version of the above response function takes the successive appropriate forms; the present discussion follows (Putz, 2009a)

$$\begin{aligned}
\kappa(\mathbf{r},\mathbf{r}')^{local} &= -\frac{1}{V(\mathbf{r}')}\left[\int \frac{\delta V(\mathbf{r}')}{\delta\rho(\mathbf{r})}\rho(\mathbf{r})d\mathbf{r}\right]\frac{\delta\rho(\mathbf{r}')}{\delta V(\mathbf{r}')}\delta(\mathbf{r}-\mathbf{r}') \\
&= -\frac{1}{V(\mathbf{r}')}\left[\int \frac{\delta\rho(\mathbf{r}')}{\delta\rho(\mathbf{r})}\rho(\mathbf{r})d\mathbf{r}\right]\delta(\mathbf{r}-\mathbf{r}') \\
&= -\frac{1}{V(\mathbf{r}')}\left[\int \frac{\rho(\mathbf{r})}{|\nabla_r\rho(\mathbf{r})|}\delta\rho(\mathbf{r}')\right]\delta(\mathbf{r}-\mathbf{r}') \\
&= -\frac{1}{V(\mathbf{r}')}\frac{\rho(\mathbf{r})\rho(\mathbf{r}')}{|\nabla_r\rho(\mathbf{r})|}\delta(\mathbf{r}-\mathbf{r}')
\end{aligned} \qquad (3.578)$$

It provides the analytical framework for the chemical action computation that according with C_A^{ker} casts as the "local" realization

$$C_A^{local} = \int V(\mathbf{r})\rho(\mathbf{r})\frac{\rho(\mathbf{r})}{|\nabla_r \rho(\mathbf{r})|}d\mathbf{r} \qquad (3.579)$$

while, when equating with the basic definition it leads with the so called *local equation of bonding*:

$$|\nabla_r \rho(\mathbf{r})| = \rho(\mathbf{r}) \qquad (3.580)$$

throughout fulfilling the Bader zero flux condition (Bader, 1990)

$$\rho(\mathbf{r}) = \nabla_r \rho \cdot \mathbf{n} = 0 \qquad (3.581)$$

for the asymptotic densities

$$\rho(r)^{local} = \frac{N}{8\pi}\exp(-r) \qquad (3.582)$$

as its solution with the density functional radial N-representability constraint (i.e., integrated to N).

Worth noting that such electronic density expression is of the first importance in characterizing the exchange or Fermi holes in chemical structures (Becke, 1986; Putz, 2008a), thus furnishing the backbone of the analytical chemical bonding analysis.

3.8.4 NONLOCAL CHEMICAL BOND

Going to the non-local level of the softness kernel one relevant choice should be shaped as (Garza & Robles, 1993; Putz, 2008b, d)

$$s(\mathbf{r},\mathbf{r}')^{non-local} = s(\mathbf{r},\mathbf{r}')^{local} + \rho(\mathbf{r})\rho(\mathbf{r}') \qquad (3.583)$$

producing the associate non-local response function equivalencies; the present discussion follows (Putz, 2009a):

$$\kappa(\mathbf{r},\mathbf{r}')^{non-local} = \kappa(\mathbf{r},\mathbf{r}')^{local} + \frac{\rho(\mathbf{r})\rho(\mathbf{r}')}{V(\mathbf{r}')}\int\frac{\delta V(\mathbf{r}')}{\delta\rho(\mathbf{r})}\rho(\mathbf{r})d\mathbf{r}$$

$$= \kappa(\mathbf{r},\mathbf{r}')^{local} + \rho(\mathbf{r})\rho(\mathbf{r}')\int\frac{\delta V(\mathbf{r}')}{V(\mathbf{r}')}\frac{1}{\delta\rho(\mathbf{r})}\rho(\mathbf{r})d\mathbf{r}$$

$$= \kappa(\mathbf{r},\mathbf{r}')^{local} - \rho(\mathbf{r})\rho(\mathbf{r}')\int\frac{\rho(\mathbf{r})}{\nabla\rho(\mathbf{r})\Delta\mathbf{r}}d\mathbf{r} \qquad (3.584)$$

where the Poisson finite difference or long range approximations (Putz, 2003):

$$\nabla V(\mathbf{r}') \cong -4\pi\rho(\mathbf{r}')\Delta\mathbf{r}' \tag{3.585}$$

$$V(\mathbf{r}') \cong 4\pi\rho(\mathbf{r}')[\Delta\mathbf{r}']^2 \tag{3.586}$$

were involved for smearing out the explicit potential dependence. Still, the remaining displacement of the bonding localization (of the electronic pairs) can be finely tuned by employing once more the Hellman-Feynman theorem under the form:

$$0 = \int \rho(\mathbf{r})\nabla V(\mathbf{r})d\mathbf{r} = \int \nabla[\rho(\mathbf{r})V(\mathbf{r})]d\mathbf{r} - \int V(\mathbf{r})\nabla\rho(\mathbf{r})d\mathbf{r} \tag{3.587}$$

from where the formal identity

$$\Delta[\rho(\mathbf{r})V(\mathbf{r})] \cong V(\mathbf{r})\nabla\rho(\mathbf{r})\Delta\mathbf{r} \tag{3.588}$$

can be used to rewrite the non-local response kernel function until the chemical action dependency is achieved:

$$\kappa(\mathbf{r},\mathbf{r}')^{non-local} \cong \kappa(\mathbf{r},\mathbf{r}')^{local} - \rho(\mathbf{r})\rho(\mathbf{r}')\int \frac{\rho(\mathbf{r})V(\mathbf{r})}{\Delta[\rho(\mathbf{r})V(\mathbf{r})]}d\mathbf{r} \tag{3.589}$$

$$\kappa(\mathbf{r},\mathbf{r}')^{non-local} \cong \kappa(\mathbf{r},\mathbf{r}')^{local} - \rho(\mathbf{r})\rho(\mathbf{r}')\frac{C_A}{\rho(\mathbf{r})V(\mathbf{r})}$$

$$= \kappa(\mathbf{r},\mathbf{r}')^{local} - \frac{\rho(\mathbf{r}')}{V(\mathbf{r})}C_A \tag{3.590}$$

when the saddle point approximation (Hassani, 1991; Putz, 2016b) was applied to the integral term becoming a local one.

In these conditions, the chemical action equation for non-local or delocalized bonding description is raised as:

$$C_A^{non-local} = \int V(\mathbf{r})\rho(\mathbf{r})\frac{\rho(\mathbf{r})}{|\nabla_r\rho(\mathbf{r})|}d\mathbf{r} + C_A\iint \rho(\mathbf{r}')V(\mathbf{r}')d\mathbf{r}d\mathbf{r}'$$

$$= \int \frac{dC_A}{dr}\frac{\rho(\mathbf{r})}{|\nabla_r\rho(\mathbf{r})|}d\mathbf{r} + C_A^2\Delta r$$

$$= C_A\frac{\rho(\mathbf{r})}{|\nabla_r\rho(\mathbf{r})|} + C_A^2\Delta r \tag{3.591}$$

Worth remarking that the explicit bonding displacement Δr modulates the chemical action amplitude; as Δr vanishes as the previous local case of bonding is recovered, see the previous case. Nevertheless, both local and non-local instants of chemical action produce the present chemical bonding picture in a complete non-orbital way. More details and discussions are in next addressed.

3.8.5 BONDING AND ANTIBONDING: A CASE STUDY

Since the non-local bonding equation contains the local case it may be rearranged to the convenient form recalling a sort of *adapted Heisenberg relation* for chemical bonding; the present discussion follows (Putz, 2009a,f):

$$C_A \Delta r = 1 - \frac{\rho(\mathbf{r})}{|\nabla_r \rho(\mathbf{r})|} \quad (in\ a.\ u.) \tag{3.592}$$

since $<Joule><meter> \sim h \cdot c$ with h- the Planck constant and c- the light velocity. Moreover, other arrangement of last equation provides the generalization of previous Bader relation for bonding flux of electronic density,

$$|\nabla_r \rho|(1 - \Delta r C_A) = \rho \tag{3.593}$$

while it turns out that the chemical action appears in the departures of electronic pairing from its localized version.

Finally, aiming to find the explicit localized-delocalized bonding density one has to integrate the adapted Heisenberg equation further rewritten as (Putz, 2009a,f)

$$\frac{d\rho}{\rho} = \frac{1}{1 - \Delta r C_A} dr \tag{3.594}$$

Now, we have two ways of integration depending on considering or not the delocalization as an integration variable; however we will treat both cases under the assumption that for initial condition of integration we have $r_0 = 0$ & $\rho_0 = \rho^{local}$ (Putz, 2009a);

- For a constant delocalization ($\Delta r \sim R$) we get:

$$\rho(r)^{bonding-1} = \rho_0 \exp\left(\frac{r}{1 - RC_A}\right) \tag{3.595}$$

- For a variable delocalization ($\Delta r \sim r$) we obtain:

$$\rho(r)^{bonding-II} = \rho_0 \exp\left(-\frac{1}{C_A}\ln(1-rC_A)\right) \tag{3.596}$$

Remarkably, either I- and II-bonding densities provide the same chemical action limits, viz. (Putz, 2009a,f):

$$\lim_{C_A \to \infty} \rho(r)^{bonding-I} = \lim_{C_A \to \infty} \rho(r)^{bonding-II} = \rho_0 = \rho^{local} \tag{3.597}$$

$$\lim_{C_A \to 0} \rho(r)^{bonding-I} = \lim_{C_A \to 0} \rho(r)^{bonding-II} = \rho_0 \exp(r) = \frac{N}{8\pi} = ct. \tag{3.598}$$

$$\lim_{C_A \to 1/R} \rho(r)^{bonding-I} = \lim_{C_A \to 1/r} \rho(r)^{bonding-II} = \infty \tag{3.599}$$

Therefore, we may considerate the two bonding solutions as expressing the same chemical interaction and equating them for a sort of universal chemical action-electronic pairing localization in bonding of atoms-in-molecules (Putz, 2009a):

$$(1-RC_A)\ln(1-rC_A) = -rC_A \tag{3.600}$$

When delocalization towards equilibrium, i.e., $r \to R$, a simple yet meaningful equation of bonding is shaped as:

$$\exp(-RC_A) = 1 - RC_A \tag{3.601}$$

This transformation is motivated as follows. Take the above equation with $r \to R$:

$$(1-RC_A)\ln(1-RC_A) = -RC_A \tag{3.602}$$

and rewrite it as:

$$1-RC_A = \exp\left(-\frac{RC_A}{1-RC_A}\right) \tag{3.603}$$

that can further be simplified to became

$$1-RC_A = \exp(-RC_A) \tag{3.604}$$

since the first order expansion equivalence among the right hands of last two equations in the limit $C_A \to 0$, in fully accordance with the case reflecting the fact that each of pairing electrons $N/2$ is stabilized in spherical averaged sense $(1/4\pi)$ with other electrons of bonding. Moreover, the applied $C_A \to 0$ condition becomes the conservation law when the overall chemical action is decomposed into "in-" and "out-" contributions to bonding region.

There is immediate to see that, on the ground of the Gauss law of the density conservation flux across the bonding region, the I- and II-bonding solutions although equal in modulus should be complementary respecting the electronic flux in bonding, i.e., one will correspond to the "in-" and other to the "out-" way of bonding, consistently with accumulation and dissipation of the electronic pairing wave packet in bonding region, respectively. Therefore, *unified bonding equation* should be regarded in the light of the chemical action conservation in the bonding region (Putz, 2009a,f):

$$C_A^{in} + C_A^{out} = 0 \tag{3.605}$$

which reflects, in fact, the general variational chemical action principle, here particularized as a conservation law at atoms-in-molecules level. This phenomenology opens the perspective in defining the "in-" and "out-" *chemical bonding strengths* as:

$$C_B^{in} = \exp(-RC_A^{in}), \ C_A^{in} = -C_A^{out} \tag{3.606}$$

$$C_B^{out} = 1 - RC_A^{out} \tag{3.607}$$

With this end, the bonding region is therefore fixed by the chemical bonding strength region circumvented by all chemical bonding strengths involved in bonding.

To see how the present notions works out, in Figure 3.25 the special bonding analysis for diatomic HF and HCl hydracids is illustrated, when for chemical action its atoms-in-molecules electronegativity counterpart was considered, $C_A^{out} \sim \chi$ (Putz, 2009a), for practical implementation procedure based on data of Table 3.39.

The Figure 3.25 clearly reveals that electronegativity (aka chemical action) drives the nature (the slopes and magnitudes) of the chemical bonding strengths through its various formulations being the semiclassical

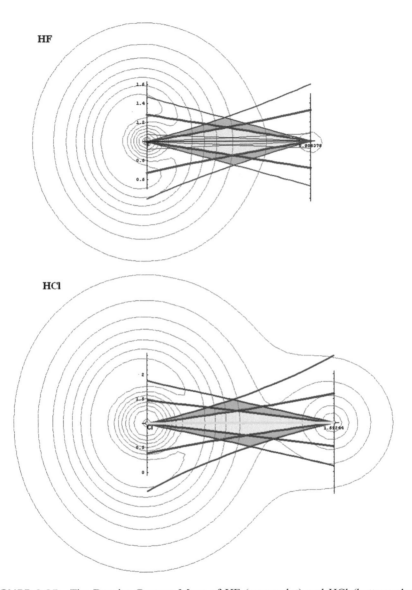

FIGURE 3.25 The Density Contour Maps of HF (upper plot) and HCl (bottom plot) With the Bonding Regions Provided by the Closed Geometrical Regions Resulted From "in-" (in red) and "out-" (in blue) Chemical Actions (aka Atoms-in-Molecule Equalized Electronegativity) and Bonding Strengths for Various Levels of Computations, i.e., the Finite Difference (Lackner & Zweig, 1983), by Density Functional Theory (DFT) (Putz, 2006) and by Semiclassical (SC) Approach (Putz, 2007b) Based electronegativities of Table 3.39 providing the yellow, green and grey areas of bonding, respectively (Putz, 2009a).

TABLE 3.39 Equilibrium Radii (Hypercube, 2002) Along the Finite Difference (FD) (Lackner & Zweig, 1983), by Density Functional Theory (DFT) (Putz, 2006) and by Semiclassical (SC) Approach (Putz, 2007b) Based Electronegativities for the HF and HCl Diatomic Molecules (Putz, 2009a)

Molecule	$R_{Eq.}$ (Å)	χ_{FD} (a.u.)	χ_{DFT} (a.u.)	χ_{SC} (a.u)
HF	0.908378	0.312522	0.518329	0.0838474
HCl	1.69244	0.282032	0.514518	0.00144827

based that one (demarked with grey frontiers) that furnishes the most delocalized distribution across bonding lengths with smooth maxima for HF when also the finite-difference and density functional based approaches indicate the maximum geometrical locus along which the chemical bonding maximizes its strength.

Yet, the observed maximum geometric locus of bonding is neither located at midpoint on the bonding axis nor as being perpendicular to it but shaping a sort of electronic density tangent with a slope increasing towards the greatest electronegativity center in bonding; this way the Pauling combined ionic-covalent feature of bonding (Pauling, 1960) is here founded in a more intuitive geometrical-analytical way.

The formal correspondence with established bonding-antibonding paradigm of molecular orbital analysis can be gained once the other chemical action substitution with equalized atoms-in-molecule electronegativity case is employed, i.e. $C_A^{out} \sim -\chi$, thus identifying chemical action with the chemical potential of the system, $\mu = -\chi$ (Parr et al., 1978); it provides another geometrical locus of bonding based on closed area circumvented by "in-" and "out-" drawn chemical bonding strengths $C_B^{in, out}$ centered in both sides of bonding axis.

The difference between the obtained bonding areas of chemical potential and electronegativity specializations for chemical action fixes the *antibonding* region for electronic pairing or the super-exchange region for electronic pairing of bonding, i.e., the parallel pairing spins or the excited states of bonding; all remaining area apart of these two regions of molecular electronic density space is then consequently associated with the *nonbonding* case (Putz, 2009a).

Note that the design of the chemical bonding region and of its maximum locus is based on combining the above "in-" and "out-" chemical action and bonding measures such that the above conservation principle to be maintained at whatever level of electronic density computation theory and atoms-in-molecule case.

3.9   CONCLUSION

There is already long and reach scientific history for the primer chemical concepts such as electronegativity help in properly modeling the structure, reactivity, and bonding of many-electronic systems in the range of tens of electron volts or hundred of kcal/mod domains, i.e., within the chemical real or nature manifestation. However, although highly celebrated the chemical reactivity space cannot be comprehensively characterized, in a generalized sense or reactivity and bonding, in the absence of a clear formulation of a principles' program, with quantum nature relevance. As such, the companion of electronegativity as the chemical hardness was advanced, as a sort of super-potential for the energy equilibrium, since electronegativity was customarily associated with the minus of the chemical potential of the envisaged molecular system.

At this point there is time to address the question whether the chemical bond can be considered sufficiently and necessary described by the quantum-mechanical paradigm of matter (Einstein et al., 1935; Bohr, 1935). In this respect, let's note that usually the chemical bond is referred to the covalent bond where electrons are shared by the encountered atoms. However, we can generalize this picture to assume that the chemical bond is generated by a special field of the neighboring electrons and nucleus such that the field geometry to be optimized. In this respect, the optimization procedure includes variational principles of energy (Hohenberg & Kohn, 1964), of chemical potential (i.e., electronegativity) (Parr et al., 1978), of chemical force (i.e., the chemical hardness) (Chattaraj et al., 1991), or, more recently, of chemical action (i.e., averaged quantum fluctuation field) (Putz, 2003, 2008b, 2012a-b). Certainly, electrons as well as atomic particles and atoms themselves behave in a complementarity frame of wave and particles: they are wave through their influential field and particles by caring their mass.

In these conditions, the nature of the chemical bond is compulsory quantum. This is the necessity condition. What about the sufficiency? This part is approached by recalling that, indeed, the chemical bond implies a variety of situations that span from the small molecules in gas phase up to the chemical reaction in lattice or the affinity among biomolecules. Hopefully, what is usually considered a weakness of the quantum mechanics – namely the impossibility to asses a complete analytical solution for whatever potential governing whatever collection of electrons and nuclei, becomes here the argument due to different levels of approximations that are suited for assaying various chemical situations. It is the same that when, having to follow a stone falling from an arbitrary altitude on Earth, we solve the movement within Newtonian rather with special relativity theory, thus appealing to an approximate theory, that in no way departs from the observed reality. As such, mutatis mutandis, combining the different approximation to the electronic density or wave function with the proper combination of the preferred optimization procedures one can arrive to the averaged reactivity path of electrons in bonding space. Moreover, we have to note that the recent emphasized quantum inequality principles of chemistry, namely the inequality of the chemical potential (Tachibana, 1999) and of the chemical force (i.e., the maximum hardness principle) (Parr & Chattaraj, 1991), take further account on the quantum fluctuations that a chemical bonding process imply. The impressive experimental evidence of the quantum fluctuations of electrons and electronic potential when the time of interaction practically freezes, through the so called ultra fast femto-spectroscopy (Felker & Zewail, 1984), indicates that the quantum nature of the chemical bond is manifested both in space and time. Consequently, based on the Wick transformation between time and temperature, $\tau = \hbar\beta$ (Weiss, 1999), see also Volume II of the present five-volume book set (Putz, 2016b), we can equally admit that at the very short times in bonding process the equivalent local temperature goes asymptotically high. It is this the proof that since quantum fluctuations in bonding are accompanied by the rise of the local temperature, as the fluctuations are smeared out by optimum reactive path through variational principles the bond is achieved through a stationary state in the low temperature limit. This way, the quantum paradigm of physical reality furnishes sufficient frame in which all relevant steps in bonding can be properly described and of which averaged quantum effects can be also evaluated, observed and measured. It is therefore the occasion to answer the question at issue in the positive. However, the remaining theme is to construct, as in physics there is the complete set of observables that describe the quantum motion in the Hilbert space, the minimum complete set of reactivity descriptors and of their variational principles to closely assist the quantum nature of bond and bonding. With this goal, the quantum scenario based on electronegativity and hardness indices and principles stands as a natural first step in developing the quantum chemical space of reactivity (Putz, 2012b).

Accordingly, electronegativity and chemical hardness help in building both the so called chemical orthogonal space and reactivity (COSR) and in providing the consistent algorithm and hierarchy of chemical reactivity principles due to the identified double variation principle of energy density functional – that was affirmed as the non-reductive chemical variational specificity to the expected physical simple energy variation. Nevertheless, when parabolically combined with electronegativity the chemical hardness provide a quantum energy that accounts for the frontier electronic effects as better as the concerned system has more complex carbon structure – a behavior systematically revealed by the semi-empirical computations on paradigmatic butadiene, benzene, naphthalene and fullerene systems.

This way, electronegativity and hardness appear in this frame to generate the minimal and sufficient set of global parameters that assists chemical bonding and reactivity in various chemical and physical conditions. Few of them were in this study presented leaving with unification of electronegativity and hardness absolute and chemical levels of their expressions like density functionals formulations. In this line, the chemical action appears as the proper link between the orbital and global pictures of electronegativity and hardness due to its proper definition that univocally associates the electronic density with external potential applied on an electronic system, in accordance with the main density functional theorems.

However, electronegativity and hardness stand within the minimum dimensioned set of global indices that characterize bonding and reactivity as the electronic density and the effective applied potential function closely relate with the inner structure of atoms and molecules. This chapter advocates that a proper combination between these two sets of global and local indices can generate many consequences with a role in quantifying the many-electronic structures and their transformation at the conceptual rather computational quantum level of comprehension. This was proved though applying the obtained electronegativity and hardness atomic scaled to selected problematical chemical reaction to provide the prediction of reactivity and stabilization of bonds in accordance with the main principles of chemistry: equalization and inequality of electronegativity and hardness, known as the electronegativity equalization, inequality of chemical potential, hard and soft acids and base, and maximum hardness principles, respectively.

In this context, a novel reactivity index for quantifying the maximum of hardness realization was proposed with reliability proved throughout providing the hierarchy for a series of hard and soft Lewis bases.

All in all, the appropriate complete bonding scenario based exclusively on the correlated quantum quantities and principles of the electronegativity and hardness was indicated with the basic stages as:

(i) The *encountering stage*, associated with *covalent* binding step;

(ii) The *global optimization stage*, establishing the global minima on the potential surface of the system;

(iii) The *charge fluctuation stage*, produces the degree of *ionicity* occurred in bonds;

(iv) The *polarizability stage*, through the hardness equalization manifests itself as a second order effect in charge transfer – see the step (i) above;

(v) The *steric stage*, is covered by the *maximum hardness principle* and the fully stabilization of the molecular system in given environment;

to which one may safety add another subtle yet with quantum consistence and influence (viz. the bondonic production, see Section 3.6.2, in continuation of bondonic characterization in the Chapter 1 of the present volume):

(vi) *Symmetry breaking* stage: although assumed as of quantum nature, the forming mechanism of chemical bond is still an open issue; the present approach explores the way of spontaneous symmetry breaking of the Schrödinger Lagrangian invariance respecting the chemical field shift with the values fixing minimum vacuum sites for reactivity potential, in terms of electronegativity and chemical hardness. The effect was quantified jointly by appearance of the chemical bonding fields and of associated bondons particles, in two variants: one depending on temperature, chemical hardness and bond length and that depending on electronegativity, chemical hardness and bond length, named as first and second kind chemical bonding fields, respectively. However, the application of this formalism on paradigmatic chemical bonds reveals that chemical bonding fields do not necessarily obey the normalization principle, being this in accordance with breaking-symmetry and particle creation effects, while the associate bondons caries the masses either greater or lower than the standard electronic one, relating the localization or delocalization of electrons in bonds, respectively. There was also find out that the two kinds of chemical bonding fields and types of bondons behave in a rather complementary manner eventually leading with the ideas of phase transition, spin dependence, and implicitly with statistical effects of bondons on chemical bond and bonding nature; these ideas are to be in future unfolded and applied (Putz, 2008d).

Although there remain several open issues, among which checking for similar behavior for non-carbon nanosystems, as well as in depth exploring the physical meaning of the actually proposed parabolic form of the pi-related energy only on the base of electronegativity and chemical hardness, there seems that these reactivity indices have still great potential in modeling the next era of nanosystems in simple and powerful manner, providing their quantum observational character will be directly or implicitly clarified.

Overall, the "beauty" of electronegativity and chemical hardness versatility make us optimistic in this respect; or, in Van Gogh wise words, while assuming they are like "stars" on the "chemical conceptual sky/ space", *"for my part I know nothing with any certainty... but the sight of* (such, n.a.) *stars makes me dream"*!

KEYWORDS

- **additive and geometric models of atoms-in-molecules**
- **bivariate energies**
- **bonding and antibonding**
- **bondonic masses**
- **chemical action entanglement**
- **delocalization in chemical bonding**
- **DFT parabolic effects**
- **electronegativity equalization**
- **frontier Fukui index principle**
- **Fukui function**
- **HSAB principle**
- **Lewis acids and bases**
- **maximum hardness index and principle**
- **pi-energy**
- **poly-atomic electronegativity and chemical hardness**
- **reactivity based biological activity (ReBiAc) principles**
- **spin effects**
- **spontaneous symmetry-breaking approach**
- **variational principles of chemical bondin**

REFERENCES

AUTHOR'S MAIN REFERENCES

Putz, M. V. (2016a). *Quantum Nanochemistry. A Fully Integrated Approach: Vol. I. Quantum Theory and Observability*. Apple Academic Press & CRC Press, Toronto-New Jersey, Canada-USA.

Putz, M. V. (2016b). *Quantum Nanochemistry. A Fully Integrated Approach: Vol. II. Quantum Atoms and Periodicity*. Apple Academic Press & CRC Press, Toronto-New Jersey, Canada-USA.

Putz, M. V. (2016c). *Quantum Nanochemistry. A Fully Integrated Approach: Vol. V. Quantum Structure-Activity Relationships (Qu-SAR)*. Apple Academic Press & CRC Press, Toronto-New Jersey, Canada-USA.

Putz, M. V. (2012a). *Quantum Theory: Density, Condensation, and Bonding*, Apple Academics, Toronto, Canada.

Putz, M. V., Ori, O., De Corato, M., Putz, A. M., Benedek, G., Cataldo, F., Graovac, A. (2013). Introducing "colored" molecular topology by reactivity indices of electronegativity and chemical hardness. In: Ashrafi, A. R., Cataldo, F., Iranmanesh, A., Ori, O. (Eds.) *Topological Modeling of Nanostructures and Extended Systems*, Springer Verlag, Dordrecht, Chapter 9, pp. 265–286 (DOI: 10.1007/978–94–007–6413–2_9).

Putz, M. V. (2012b). *Chemical Orthogonal Spaces*, in Mathematical Chemistry Monographs, Vol. 14, University of Kragujevac.

Putz, M. V. (2012c). Chemical reactivity and biological activity criteria from DFT parabolic dependency E=E(N). In: Roy, A. K. (Ed.) *Theoretical and Computational Developments in Modern Density Functional Theory*, NOVA Science Publishers, Inc., New York, Chapter 17, pp. 449–484.

Putz, M. V., Mingos, D. M. P., Eds., (2012). *Applications of Density Functional Theory to Chemical Reactivity*, Structure and Bonding Series Vol. 149, Springer Verlag, Heidelberg-Berlin.

Putz, M. V. (2011a). Electronegativity and chemical hardness: different patterns in quantum chemistry. *Curr. Phys. Chem.* 1(2), 111–139 (DOI: 10.2174/1877946811101020111).

Putz, M. V. (2011b). On relationship between electronic sharing in bonding and electronegativity equalization of atoms in molecules. *Int. J. Chem. Model.* 3(4), 371–384.

Putz, M. V. (2011c). Quantum and electrodynamic versatility of electronegativity and chemical hardness. In: Putz, M. V. (Ed.), *Quantum Frontiers of Atoms and Molecules*, NOVA Science Publishers, Inc., New York, Chapter 11, pp. 251–270.

Putz, M. V. (2011d). Quantum parabolic effects of electronegativity and chemical hardness on carbon π-systems. In: Putz, M. V. (Ed.), *Carbon Bonding and Structures: Advances in Physics and Chemistry*, Carbon Materials: Chemistry and Physics series Vol. 5, Springer Verlag, London, Chapter 1, pp. 1–32.

Putz, M. V. (2011e). Chemical action concept and principle. *MATCH Commun. Math. Comput. Chem.* 66(1), 35–63.

Putz, M. V. (2011f). Fulfilling the Dirac promises on quantum chemical bond. In: Putz, M. V. (Ed.), *Quantum Frontiers of Atoms and Molecules*, NOVA Science Publishers, Inc., New York, Chapter 1, pp. 1–20.

Matito, E., Putz, M. V. (2011). New link between conceptual density functional theory and electron delocalization. *J. Phys. Chem. A* 115(45), 12459–12462 (DOI: 10.1021/jp200731d).

Putz, M. V. (2009a). Chemical action and chemical bonding. *J. Mol. Struct.: THEOCHEM*, 900(1–3), 64–70 (DOI: 10.1016/j.theochem.2008.12.026).

Putz, M. V. (2009b). Levels of a unified theory of chemical interaction. *Int. J. Chem. Model.* 1(2), 141–147.

Selegean, M., Putz, M. V., Rugea, T. (2009). Effect of the polysaccharide extract from the edible mushroom *Pleurotus Ostreatus* against infectious bursal disease virus. *Int. J. Mol. Sci.* 10(8), 3616–3634 (DOI: 10.3390/ijms10083616).

Putz, M. V. (2008a). Density functionals of chemical bonding. *Int. J. Mol. Sci.* 9(6), 1050–1095 (DOI: 10.3390/ijms9061050).

Putz, M. V. (2008b). *Absolute and Chemical Electronegativity and Hardness*, Nova Publishers Inc., New York.

Putz, M. V. (2008c). Maximum hardness index of quantum acid-base bonding. *MATCH Commun. Math. Comput. Chem.* 60(3), 845–868.

Putz, M. V. (2008d). The chemical bond: spontaneous symmetry – breaking approach. *Symmetry: Culture and Science*, 19(4), 249–262.

Putz, M. V., Timariu, E., Chiriac, A. (2008). Quantitative energy-electronegativity-chemical hardness anthocyanidins reactivity. *Ann. West Univ. Timişoara, Series of Chemistry* 17(1), 55–62.

Putz, M. V., Chiriac, A. (2008). Quantum perspectives on the nature of the chemical bond. In: Putz, M. V. (Ed.), *Advances in Quantum Chemical Bonding Structures*, Transworld Research Network, Kerala, Chapter 1, pp. 1–43.

Putz, M. V. (2007a). Can quantum-mechanical description of chemical bond be considered complete? In: Kaisas, M. P. (Ed.) *Quantum Chemistry Research Trends*, Nova Science Publishers Inc., New York, Expert Commentary, pp. 3–5.

Putz, M. V. (2007b). Semiclassical Electronegativity and Chemical Hardness, *J. Theor. Comp. Chem.* 6(1), 33–47 (DOI: 10.1142/S0219633607002861).

Putz, M. V. (2006). Systematic formulation for electronegativity and hardness and their atomic scales within density functional softness theory. *Int. J. Quantum Chem.* 106(2), 361–386 (DOI: 10.1002/qua.20787).

Putz, M. V., Russo, N., Sicilia, E. (2005). About the Mulliken electronegativity in DFT. *Theor. Chem. Acc.* 114(1–3), 38–45 (DOI: 10.1007/s00214–005–0641–4)

Putz, M. V., Russo, N., Sicilia, E. (2004). On the application of the HSAB principle through the use of improved computational schemes for chemical hardness evaluation. *J. Comp. Chem.* 25(7), 994–1003 (DOI: 10.1002/jcc.20027).

Putz, M. V. (2003). *Contributions within Density Functional Theory with Applications in Chemical Reactivity Theory and Electronegativity*, Dissertation.com, Parkland.

Putz, M. V., Russo, N., Sicilia, E. (2003). Atomic radii scale and related size properties from density functional electronegativity formulation. *J. Phys. Chem. A* 107(28), 5461–5465 (DOI: 10.1021/jp027492h).

SPECIFIC REFERENCES

Amos, A. T., Hall, G. G. (1961). Single determinant wave functions. *Proc. Roy. Soc. (London)* A263, 483–493.

Angyaan, J. G., Rosta, E., Surjan, P. (1999). Covalent bond orders and atomic valences from correlated wave functions. *Chem. Phys. Lett.* 299(1), 1–8.

Ayers, P. W., Day Jr. O. W., Morrison, R. C. (1998). Analysis of density functionals and their density tails in H_2, *International Journal of Quantum Chemistry*, 69(4), 541–550.

Ayers, P. W., Parr, R. G. (2000). Variational principles for describing chemical reactions: the Fukui function and chemical hardness revisited. *J. Am. Chem. Soc.* 122, 2010–2018.

Ayers, P. W., Parr, R. G. (2001). Variational principles for describing chemical reactions. reactivity indices based on the external potential. *J. Am. Chem. Soc.* 123, 2007–2017.

Bader, R. F. W. (1990). *Atoms in Molecules – A Quantum Theory*, Oxford University Press, Oxford.

Bader, R. F. W. (1994). Principle of stationary action and the definition of a proper open system. *Phys. Rev. B* 49, 13348–13356.

Bader, R. F. W. (1994). Principle of stationary action and the definition of a proper open system. *Phys. Rev. B* 49, 13348–13356.

Bader, R. F. W. (1998). A bond path: a universal indicator of bonded interactions. *J. Phys. Chem. A* 102, 7314–7323.

Bader, R. F. W., Austen, M. A. (1997). Properties of atoms in molecules: atoms under pressure. *J. Chem. Phys.*, 107, 4271–4285;

Bader, R. F. W., Carroll, M. T., Cheeseman, J. R., Chang, C. (1987). Properties of atoms in molecules: atomic volumes. *J. Am. Chem Soc.* 109, 7968–7979.

Bader, R. F. W., Stephens, M. E. (1974). Fluctuation and correlation of electrons in molecular systems. *Chem. Phys. Lett.* 26(3), 445–449.

Bader, R. F. W., Stephens, M. E. (1975). Spatial localization of the electronic pair and number distributions in molecules. *J. Am. Chem. Soc.* 97(26), 7391–7399.

Balbás, L. C., Alonso, J. A., Heras, E. Las, (1983). Electronegativity equalization and electron transfer in molecules. *Mol. Phys.* 48(5), 981–988.

Bartolotti, L. J. (1987). Absolute electronegativities as determined from Kohn-Sham theory. *Structure and Bonding* 66, 27–40.

Bartolotti, L. J., Gadre, S. R., Parr, R. G. (1980). Electronegativities of the elements from simple chi alpha theory. *J. Am. Chem. Soc.* 102, 2945–2948.

Becke, A. D. (1986). Density functional calculations of molecular bond energies. *J. Chem. Phys.* 84, 4524–4529.

Becke, A. D. (1988). Density-functional exchange-energy approximation with correct asymptotic behavior. *Phys Rev A* 38(6), 3098–3100.

Becke, A. D., Edgecombe, K.E (1990). A simple measure of electron localization in atomic and molecular systems. *J. Chem. Phys.* 92, 5397–5403.

Berkowitz, M. (1987). Density functional approach to frontier controlled reactions. *J. Am. Chem. Soc.* 109: 4823–4825.

Berkowitz, M., Ghosh, S. K., Parr, R. G. (1985). On the concept of local hardness in chemistry. J. Am. Chem. Soc. 107, 6811–6814.

Berkowitz, M., Parr, R. G. (1988). Molecular hardness and softness, local hardness and softness, hardness and softness kernels, and relations among these quantities. *J. Chem. Phys.* 88, 2554–2557.

Bochicchio, R. C., Reale, H. F., Medrano, J. (1989). Extension of the quantum theory of valence and bonding to molecular and crystal systems with translation symmetry. *Phys. Rev. B* 40, 7186–7190.

Boeyens, J. C. A. (2005). *New Theories for Chemistry*, Elsevier, Amsterdam, pp. 280.

Bohm, D.,; Vigier, J. P. (1954). Model of the causal interpretation of quantum theory in terms of a fluid with irregular fluctuations. *Phys. Rev.* 96(1), 208–216.

Böhm, M. C., Schmidt, P. C. (1986). Electronegativities and hardnesses of the main group elements from density functional theory: dependence on the hybridization of the chemical bond. *Ber. Bunsenges. Phys. Chem.* 90, 913–919.

Böhm, M. C., Sen, K. D., Schmidt, P. C. (1981). Molecular orbital electronegativity. *Chem. Phys. Lett.* 78, 357–360.

Bohr, N. (1913). On the constitution of atoms and molecules. *Phil. Mag.* 26, 1–25; ibid. On the constitution of atoms and molecules. Part II. Systems containing only a Single Nucleus. *Phil. Mag.* 26, 476–502.

Bohr, N. (1921). *Abhandlungen über Atombau au den Jahren 1913–1916*, Vieweg & Sohn, Braunschweig.

Bohr, N. (1935). Can quantum-mechanical description of physical reality be considered complete? *Phys. Rev.* 48, 696–702.

Boyd, R. J. Markus, G. E. (1981). Electronegativities of the elements from a nonempirical electrostatic model. *J. Chem Phys.* 75, 5385–5389.

Bratsch, S. G. (1984). Electronegativity equalization with pauling units. *J. Chem. Educ.* 61, 588–589.

Bratsch, S. G. (1985). A group electronegativity method with Pauling units. *J. Chem. Educ.* 62, 101–103.

Brinkmann, G., Fowler, P. W., Justus, C. (2003). A catalogue of isomerization transformations of fullerene polyhedra. *J. Chem. Inf. Comput. Sci.* 43, 917–927.

Butlerov, A. M. (1861). Einiges über die chemische Struktur der Körper. *Z. Chem. Pharm.* 4, 549–4560.

Byers-Brown, W. (1987). High symmetries in quantum chemistry. *Chem Phys Lett.* 136, 128–133.

Calatayud, M., Mori-Sánchez, P., Beltrán, A., Pendás, A. M., Francisco, E., Andrés, J., Recio, J. M. (2001). Quantum-mechanical analysis of the equation of state of anatase TiO_2. *Phys. Rev. B* 64, 184113–184121.

Chaichian, M., Nelipa, N. F. (1984). *Introduction to Gauge Field Theories*, Berlin: Springer.

Chandrakumar, K. R. S., Pal, S. (2002c). Study of local hard-soft acid-base principle: effects of basis set, electron correlation, and the electron partitioning method. *J. Phys. Chem. A* 107, 5755–5762.

Chandrakumar, K. R. S., Pal, S. (2002a). A systematic study on the reactivity of Lewis acid-base complexes through the local hard-soft acid-base principle, *J. Phys. Chem. A* 106, 11775–11781.

Chandrakumar, K. R. S., Pal, S. (2002b). Study of local hard-soft acid-base principle to multiple-site interactions. *J. Phys. Chem. A* 106, 5737–5744.

Chattaraj, P. K., Maiti, B. (2003). HSAB principle applied to the time evolution of chemical reactions. *J. Am. Chem. Soc.* 125, 2705–2710.

Chattaraj, P., Parr, R. G. (1993). Density functional theory of chemical hardness. *Structure and Bonding* 80, 11–25.

Chattaraj, P. K., Lee, H., Parr, R. G. (1991). HSAB Principle. *J. Am. Chem. Soc.* 113: 1854–1856.

Chattaraj, P. K., Liu, G. H., Parr, R. G. (1995). The maximum hardness principle in the gyftopoulos-hatsopoulos three-level model for an atomic or molecular species and its positive and negative ions. *Chem. Phys. Lett.* 237, 171–176.

Chattaraj, P. K., Schleyer, P. V. R. (1994). An ab initio study resulting in a greater understanding of the HSAB principle. *J. Am. Chem. Soc.* 116: 1067–1071.

Chattaraj, P. K., Sengupta, S. (1996). popular electronic structure principles in a dynamical context. *J. Phys. Chem.* 100, 16129–16130.

Chermette, H. (1999). Chemical reactivity indexes in density functional theory. *J. Comput. Chem.* 20(1), 129–154.

Clementi, E. (1962). SCF–MO Wave Functions for the Hydrogen Fluoride Molecule. *J. Chem. Phys.* 36, 33

Cotton, F. A. (1971a). *Chemical Application of Group Theory*, Wiley, New York, pp. 137.

Cotton, F. A. (1971b). *Chemical Application of Group Theory*, Wiley, New York, pp. 163.

Couper, A. S. (1858). Sur une nouvelle théorie chimique. *Ann. Chim. Phys.* 53, 469–489.

De Amorim, A. O., Ferreira, R. (1981). Electronegativities and the bonding character of molecular orbitals: A remark. *Theoret. Chim. Acta (Berlin)* 59, 551–553.

de Broglie, L., Vigier, M. J. P. (1953). *La physique quantique restera-t-elle indéterministe?* Gauthier-Villars, Paris.

De Vries, J., Steger, H., Kamke, B., Menzel, C., Weisser, B., Kamke, W., Hertel, I. V. (1992). Single photon ionization of C_{60} and C_{70} fullerene with synchrotron radiation: determination of the ionization potential of C_{60}. *Chem Phys Lett.* 188: 159–162.

del Re, G. (1983). *Gazzetta Chimica Italiana* 113, 695–703.

DePristo, A. E., Kress, J. D. (1987). Kinetic-energy functionals via Padé approximations. *Phys. Rev. A* 35, 438–441.

Dirac, P. A. M. (1930). Note on exchange phenomena in the Thomas atom. *Proc. Cambridge Phil. Soc.* 26(03), 376–385.

Drago, R. S., Kabler, R. A. (1972). A quantitative evaluation of the HSAB concept. *Inorg. Chem.* 11, 3144–3145.

Drago, R. S., Wayland, B. B. (1965). A double-scale equation for correlating enthalpies of Lewis acid-base interactions. *J. Am. Chem. Soc.* 87, 3571–3576.

Edgecombe, K. E., Boyd, R. J. (1987). Atomic orbital populations and atomic charges from self-consistent field molecular orbital wave functions. *J. Chem. Soc., Faraday Trans.* 83, 1307–1315.

Einstein, A., Podolsky, B., Rosen, N. (1935). Can quantum-mechanical description of physical reality be considered complete? *Phys. Rev.* 47, 777–780.

Felker, P. M., Zewail, A. H. (1984). Direct observation of nonchaotic multilevel vibrational energy flow in isolated polyatomic molecules. *Phys. Rev. Lett.* 53, 501–504.

Feynman, R. P. (1939). Forces in molecules. *Phys. Rev.* 56, 340–343.

Filippetti, A. (1998). Electron affinity in density-functional theory in the local-spin-density approximation. *Phys. Rev. A* 57, 914–919.

Fock, V. (1930). Näherungsmethode zur lösung des quantenmechanischen mehrkörperproblems. *Z. Physik* 61, 126–148.

Fowler, P. W., Woolrich, J. (1986). π-Systems in three dimensions. *Chem Phys Lett.* 127, 78–83.

Fox, M. A., Matsen, F.A (1985). Electronic structure in pi systems. Part, I. Huckel theory with electron repulsion. *J. Chem. Edu.* 62(5), 367–372; ibid. Electronic structure in pi systems. Part II. The unification of Huckel and valence bond theories. 62(6), 477–485; ibid. Electronic structure in pi systems. Part III. Applications in spectroscopy and chemical reactivity. 62(7), 551–560.

Fradera, X., Austen, M. A., Bader, R. F. W. (1998). The Lewis Model and Beyond. *J. Phys. Chem. A* 103, 304–314.

Fradera, X., Solà, M. (2002). Electron localization and delocalization in open-shell molecules. *J. Comput. Chem.* 23, 1347–1356.

Frisch, M. J., Trucks, G. W., Schlegel, H. B., Scuseria, G. E., Robb, M. A., Cheesman, J. R., Zakrzewski, V. G., Montgomery, J. A. Jr., Stratmann, R. E., Burant, J. C., Dapprich, S., Millam, J. M., Daniels, A. D., Kudin, K. N., Strain, M. C., Farkas, O., Tomasi, J., Barone, V.,

Cossi, M., Cammi, R., Mennucci, B., Pomelli, C., Adamo, C., Clifford, S., Ochterski, J., Petersson, G. A., Ayala, P. Y., Cui, Q., Morokuma, K., Malick, D. K., Rabuck, A. D., Raghavachari, K., Foresman, J. B., Cioslowski, J., Ortiz, J. V., Stefanov, B. B., Liu, G., Liashenko, A., Piskorz, P., Komaromi, I., Gomperts, R., Martin, R. L., Fox, D. J., Keith, T., Al-Laham, M. A., Peng, C. Y., Nanayakkara, A., Gonzalez, C., Challacombe, M., Gill, P. M. W., Johnson, B. G., Chen, W., Wong, M. W., Andres, J. L., Head-Gordon, M., Replogle, E. S., Pople, J. A. (1998). GAUSSIAN 98, Gaussian Inc. Pittsburgh.

Fukui, K., Imamura, A., Yonezawa, T., Nagata, C. (1960). A Quantum-Mechanical Approach to the theory of Aromaticity. *Bull. Chem. Soc. Japan* 33, 1591–1599.

Galván, M., Vela, A., Gázquez, J. L. (1988). Chemical reactivity in spin-polarized density functional theory. *J. Phys. Chem.* 92(22), 6470–6474.

Gao, S. (1997). Quantum kinetic theory of vibrational heating and bond breaking by hot electrons. *Phys. Rev. B* 55, 1876–1886.

Garza, J., Robles, J. (1993). Density functional theory softness kernel. *Phys. Rev. A* 47, 2680–2685.

Gáspár, R., Nagy, Á. (1988a). Electronegativities of diatomic molecules calculated by the $X\alpha$ method with a self-consistent parameter α using the principle of electronegativity equalization. *Coll. Czech. Chem. Commun.* 53, 2017–2022.

Gáspár, R., Nagy, Á. (1988b). The first ionization energy, electron affinity and electronegativity calculated by the α method with ab initio self-consistent exchange parameter. *Acta Physica Hungarica* 64, 405–416.

Gázquez, J. L., Vela, A., Galván, M. (1987). Fukui function, electronegativity and hardness in the Kohn-Sham theory. *Structure and Bonding* 66, 79–98.

Gázquez, J. L., Galván, M., Vela, A. (1990). Chemical reactivity in density functional theory: the N-differentiability problem. *J. Mol. Struct. (Theochem)* 210, 29–38.

Gázquez, J. L., Ortiz, E. (1984). Electronegativities and hardnesses of open shell atoms. J. Chem. Phys. 81, 2741–2748.

Geerlings, P., De Proft, F., Langeneaker, W. (2003). Conceptual density functional theory. *Chem. Rev.* 103(5), 1793–1873.

Genechten, K. Van, Mortier, W. J., Geerlings, P. (1987). Intrinsic framework electronegativity: A novel concept in solid state chemistry. *J. Chem Phys.* 86, 5063–5071.

Gesteiger, J., Marsili, M. (1980). Iterative partial equalization of orbital electronegativity—a rapid access to atomic charges. *Tetrahedron* 36(22), 3219–3288.

Ghosh, D. C. (2005). A new scale of electronegativity based on absolute radii of atoms. *J. Theor. Comp. Chem.* 4, 21–33.

Ghosh, D. C., Biswas, R. (2002). Theoretical Calculation of Absolute Radii of Atoms and Ions. Part 1. The Atomic Radii. *Int. J. Mol. Sci.* 3, 87–113.

Goldstone, J., Salam, A., Weinberg, S. (1962). Broken Symmetries. *Phys. Rev.* 127, 965–970.

Gordy, W. (1946). A relation between bond force constants, bond orders, bond lengths, and the electronegativities of the bonded atoms. *J. Chem. Phys.* 14, 305–320.

Grabo, T., Gross, E. K. U. (1995). Density-functional theory using an optimized exchange-correlation potential. *Chem. Phys. Lett.* 240, 141–150.

Guo, Y., Whitehead, M. A. (1989a). Effect of the correlation correction on the ionization potential and electron affinity in atoms. *Phys. Rev. A* 39, 28–34.

Guo, Y., Whitehead, M. A. (1989b). Application of generalized exchange local-spin-density-functional theory: electronegativity, hardness, ionization potential, and electron affinity. *Phys. Rev. A* 39, 2317–2323.

Guo, Y., Whitehead, M. A. (1991). Generalized Local-Spin-Density-Functional Theory. *Phys. Rev. A* 43, 95–109.

Haddon, R. C., Brus, L. E., Raghavachari, K. (1986). Electronic structure and bonding in icosahedral C_{60}. *Chem Phys Lett.* 125, 459- 464.

Harbola, M. K. (1992). Magic numbers for metallic clusters and the principle of maximum hardness. *Proc. Natl. Acad. Sci. USA* 89(3), 1036–1039.

Harbola, M. K., Chattaraj, P., Parr, R. G. (1991). Aspects of the softness and hardness concepts of density-functional theory. *Israel, J. Chem.* 31, 395–402.

Harrison, J. G. (1987). Electron affinities in the self-interaction-corrected local spin density approximation. *J. Chem. Phys.* 86, 2849–2853.

Hartree, D. R. (1928). The wave mechanics of an atom with a non-Coulomb central field. Part, I. Theory and methods. *Proc. Cambridge Phil. Soc.* 24, 89–110; ibid. Part II. Some results discussion and discussion. 111–132; ibid. Part III. Term Values and Intensities in Series in Optical Spectra 426–437.

Hassani, S. (1991). *Foundation of Mathematical Physics*, Prentice-Hall International, Inc., Chapter 7.

Haymet AD (1986). Footballene: a theoretical prediction for the stable, truncated icosahedral molecule C_{60}. *J Am Chem Soc.* 108: 319–321.

Heitler, W. (1931). Quantum theory and electron pair bond. *Phys. Rev.* 38, 243–247.

Heitler, W., London, F. (1927). Wechselwirkung neutraler atome und homöopolare bindung nach der quantenmechanik. *Z. Physik* 44, 455–472.

Herzberg, G. (1929). Zum aufbau der zweiatomigen moleküle. *Z. Physik* 57, 601.

Hinze, J., Jaffé, H. H. (1962). Electronegativity. I. Orbital electronegativity of neutral atoms. *J. Am. Chem. Soc.* 84(4), 540–546.

Hinze, J., Jaffé, H. H. (1963). Electronegativity: III. Orbital electronegativities and electron affinities of transition metals. *Can. J. Chem.* 41(5), 1315–1328.

Hinze, J., Whitehead, M. A., Jaffé, H. H. (1963). Electronegativity. II. bond and orbital electronegativities. *J. Am. Chem. Soc.* 85(2), 148–154.

Hohenberg, P., Kohn, W. (1964). Inhomogeneous electron gas. *Phys. Rev.* 136:B864-B871.

http://butane.chem.illinois.edu/jsmoore/Experimental/piMOs/PiMOAnalysis.html

http://webbook.nist.gov/cgi/cbook.cgi?ID=C71432&Mask=20#Ion-Energetics

http://webbook.nist.gov/cgi/cbook.cgi?ID=C91203&Mask=20#Ion-Energetics

Hückel, E. (1930). Zur quantentheorie der doppelbindung. *Z. Physik* 60, 423–456

Hückel, E. (1931a). Quantentheoretische beiträge zum benzolproblem. I. *Z Physik.* 70 (3–4), 204–286.

Hückel, E. (1931b). Quantentheoretische beiträge zum benzolproblem. II. *Z Physik.* 72(5–6), 310–337.

Hund, F. (1931). Zur Frage der chemischen Bindung. II *Z. Physik* 73, 565–577.

Hypercube (2002). Program Package, HyperChem 7.01; Hypercube, Inc.: Gainesville, FL, USA.

Iczkowski, R. P., Margrave, J. L. (1961). Electronegativity, *J. Am. Chem. Soc.* 83, 3547–3551.

Journal of Computational Chemistry (2007). Special Issue of 90 Years of Chemical Bonding, January 15, http://onlinelibrary.wiley.com/doi/10.1002/jcc.v28, 1/issue-toc.

Karo, A. M., Allen, L. C. (1959). LCAO Wave functions for hydrogen fluoride with Hartree–Fock atomic orbitals. *J. Chem. Phys.* 31, 968.

Kekule, F. A. (1857). Ueber die s.g. gepaarten Verbindungen und die Theorie der mehratomigen Radicale. *Ann. Chem. Pharm.* 104, 129–150., ibid. (1858). Ueber die Constitution und die Metamorphosen der chemischen Verbindungen und ueber die chemische Natur des Kohlenstoffs. *Ann. Chem. Pharm.* 106, 129–159.

Kleinert, H. (1989). *Gauge Fields in Condensed Matter, Vol. I Superflow and Vortex Lines, Disorder Fields, Phase Transitions*, World Scientific, Singapore.

Klopman, G. (1968). Chemical reactivity and the concept of charge and frontier-controlled reactions, *J. Am. Chem. Soc.* 90: 223–234.

Koga, T., Umeyama, T. (1986). Approximate interaction energy in terms of overlap integral. J. Chem. Phys. 85, 1433–1437.

Kohn, W., Becke, A. D., Parr, R. G., (1996). Density functional theory of electronic structure. *J. Phys. Chem.* 100, 12974–12980.

Kohn, W., Sham, L. J. (1965). Self-consistent equations including exchange and correlation effects. *Phys. Rev.* 140:A1133-A1138.

Komorowski, L. (1987). Empirical evaluation of chemical hardness. *Chem. Phys. Lett.* 134, 536–540.

Komorowski, L., Boyd, S. L., Boyd, R. J. (1996). Electronegativity and hardness of disjoint and transferable molecular fragments. *J. Phys. Chem.* 100, 3448–3453

Koopmans, T. (1934). Uber die zuordnung von wellenfunktionen und eigenwerten zu den einzelnen elektronen eines atoms. *Physica* 1(1–6), 104–113.

Labanowski, J. K., Dammkoehler, A., Motoc, I. (1989). Orbital electronegativity and analytical representation of atom valence state energy. *J. Comp. Chem.* 10, 1016–1030.

Lackner, K. S., Zweig, G. (1983). Introduction to the chemistry of fractionally charged atoms: electronegativity. *Phys. Rev. D* 28, 1671–1691.

Le Bel, J. A. (1874). Sur le relation qui existant entre les formules atomiques des corps organiques, et le pouvoir rotatoire de leurs dissolutions. *Bull. Soc. Chim. France* 22, 337–347.

Lee, C., Parr, R. G. (1990). Exchange-correlation functional for atoms and molecules. *Phys. Rev. A* 42, 193–199.

Lee, C., Yang, W., Parr, R. G. (1988). Development of the colle-salvetti correlation-energy formula into a functional of the electron density. *Phys. Rev. B* 37, 785–789.

Lee, H., Bartolotti, L. J. (1991). Exchange and exchange-correlation functionals based on the gradient correction of the electron gas. *Phys. Rev. A* 44, 1540–1542.

Levin, F. S., Krüger, H. (1977). Channel-coupling theory of covalent bonding in H_2: A further application of arran channel quantum mechanics. *Phys. Rev. A* 16, 836–843.

Lewis, G. N. (1916). The atom and the molecule. *J. Am. Chem. Soc.* 38, 762–785.

Li, L., Parr, R. G. (1986). The atom in a molecule: A density matrix approach. *J. Chem Phys.* 84, 1704–1712.

Li, Y., Evans, J. N. S. (1995). The fukui function: a key concept linking frontier molecular orbital theory and the hard-soft-acid-base principle. *J. Am. Chem. Soc.* 117, 7756–7759.

Lide, D. R. (2004). *CRC Handbook of Chemistry and Physics,* (85th Edition), CRC Press, Boca Raton, Section 10–147.

Liu, S., Nagy, A., Parr, R. G. (1999). Expansion of the density-functional energy components E_c and T_c in terms of moments of the electron density. *Phys. Rev. A* 59, 1131–1134.

Magnusson, E. (1986). Atomic orbital deformation in bond formation: energy effects. *Chem. Phys. Lett.* 131, 224–229.

Magnusson, E. (1988). Electronegativity equalization and the deformation of atomic orbitals in molecular wave functions. *Aust. J. Chem.* 41, 827–837.

Manoli, S., Whitehead, M. A. (1984). Electronegativities of the elements from density functional theory. I. The Hartree–Fock–Slater theory. *J. Chem. Phys.* 81, 841–846.

Manoli, S. D., Whitehead, M. A. (1988a). Generalized-exchange local-spin-density-functional theory: Calculations and results for non-self-interaction-corrected and self-interaction-corrected theories. *Phys. Rev. A* 38, 3187–3199.

Manoli, S. D., Whitehead, M. A. (1988b). Generalized exchange local-spin-density-functional theory: One-electron energies and eigenvalues. *Coll. Czech. Chem. Commun.* 53, 2279–2307.

Matito, E., Solà, M., Salvador, P., Duran, M. (2007). Electron sharing indexes at the correlated level. Application to aromaticity calculations. *Faraday Discuss.* 135, 325–345.

Mortier, W. (1987). Electronegativity equaliation and its applications *Struct. Bond.* 66, 125–143.

Mortier, W. J., Genechten, V. K., Gasteiger, J. (1985). Electronegativity equalization: application and parameterization. *J. Am. Chem. Soc.* 107: 829–835.

Mullay, J. (1986). A simple method for calculating atomic charge in molecules. *J. Am. Chem Soc.* 108(8), 1770–1775.

Mullay, J. (1988a). A method for calculating atomic charges in large molecules. *J. Comp. Chem.* 9(4), 399–405.

Mullay, J. (1988b). A simple method for calculating reliable atomic charges in large molecules. *J. Comp. Chem.* 9(7), 764–770.

Mulliken, R. S. (1934). A new electroaffinity scale together with data on valence states and on valence ionization potentials and electron affinities. *J. Chem. Phys.* 2, 782–793.

Mulliken, R. S. (1942). Electronic structures and spectra of triatomic oxide molecules. *Rev. Mod. Phys.* 14, 204–215.

Nalewajski, R. F., Koniński, M. (1988). General relations between sensitivities of atoms and atoms-in-a-molecule. *Acta Phys. Polon.* A74, 255–268.

Nalewajski, R. F. (1984). Electrostatic effects in interaction between hard (soft) acids and bases. *J. Am. Chem. Soc.*106, 944–945.

Nalewajski, R. F. (1985). A study of electronegativity equalization. *J. Phys. Chem.* 89(13), 2831–2837.

Nalewajski, R. F. (1989). Recursive combination rules for molecular hardnesses and electronegativities. *J. Phys. Chem.* 93, 2658–2666.

Nalewajski, R. F. (1990). Rigid and relaxed hardness parameters of molecular fragments. *Acta Phys. Polon.* A77, 817–832.

Nalewajski, R. F. (1998). Kohn-Sham description of equilibria and charge transfer in reactive systems, *Int. J. Quantum Chem.* 69, 591–605.

Nalewajski, R. F., Korchewiec, J. (1989b). Protonation of Pyrrole: A Model Hardness/ Softness Sensitivity Study. *Croatica Chem. Acta* 62, 603–616.

Nalewajski, R. F., Korchewiec, J., Zhou, Z. (1988). Molecular hardness and softness parameters and their use in chemistry. *Int. J. Quant. Chem. Biol. Symp.* 22, 349–366.

Nalewajski, R. F., Korchowiec, J. (1989a). Chemical reactivity and charge sensitivities of reactants: interaction energy and applications to protonation of pyrrole and *cyclopentadiene. Acta Phys. Polon.* A76, 747–788.

Nalewajski, R. F. Z. (1988). General relations between molecular sensitivities and their physical content. *Z. Naturforsch.* 43, 65–72.

Nesbet, R. K. (1962). Approximate Hartree–Fock calculations for the hydrogen fluoride molecule. *J. Chem. Phys.* 36(6), 1518–1533.

Nesbet, R. K. (1997). Fractional occupation numbers in density-functional theory. *Phys. Rev. A* 56, 2665–2668.

NIST (2011a). http://webbook.nist.gov/cgi/cbook.cgi?ID=C71432&Mask=20#Ion-Energetics

NIST (2011b). http://webbook.nist.gov/cgi/cbook.cgi?ID=¼C91203&Mask=20#Ion-Energetics

Oelke, W. C. (1969). *Laboratory Physical Chemistry*, Van Nostrand Reinhold Company, New York.

Ohwada, K. (1984). On the pauling electronegativity scales—II. *Polyhedron* 3, 853–859.

Orski, A. R., Whitehead, M. A. (1987). Electronegativity in density functional theory: diatomic bond energies and hardness parameters. *Can. J. Chem.* 65(8), 1970–1979.

Parr, R. G., Bartolotti, L. J. (1982). On the geometric mean principle of electronegativity equalization, *J. Am. Chem. Soc.* 104, 3801–3803.

Parr, R. G., Bartolotti, L. J. (1983). Some remarks on the density functional theory of few-electron systems. *J. Phys. Chem.* 87(15), 2810–2815.

Parr, R. G., Chattaraj, P. K. (1991). Principle of maximum hardness. *J. Am. Chem. Soc.*, 113, 1854–1855.

Parr, R. G., Donnelly, R. A., Levy, M., Palke, W. E. (1978). Electronegativity: the density functional viewpoint. *J. Chem. Phys.* 68, 3801–3807.

Parr, R. G., Gázquez, J. L. (1993). Hardness functional. *J. Phys. Chem.* 97(16), 3939–3940.

Parr, R. G., Yang, W. (1984). Density functional approach to the frontier electron theory of chemical reactivity, *J. Am. Chem. Soc.* 106, 4049–4050.

Parr, R. G., Yang, W. (1989). *Density Functional Theory of Atoms and Molecules*, Oxford University Press, New York.

Parr, R. G., Zhou, Z. (1993). Absolute hardness: unifying concept for identifying shells and subshells in nuclei, atoms, molecules, and metallic clusters. *Acc. Chem. Res.* 26(5), 256–258.

Parr, R. G., Pearson, R. G. (1983). Absolute hardness: companion parameter to absolute electronegativity. *J. Am. Chem. Soc.* 105, 7512–7516.

Pauling, L. (1928). The Application of the quantum mechanics to the structure of the hydrogen molecule and hydrogen molecule-ion and to related problems. *Chem. Rev.* 5(2), 173–213.

Pauling, L. (1931). The nature of the chemical bond. Application of results obtained from the quantum mechanics and from a theory of paramagnetic susceptibility to the structure of molecules, *J. Am. Chem. Soc.* 53(4), 1367–1400.

Pauling, L. (1960). *The Nature of the Chemical Bond,* 3d ed., Cornell University Press, Ithaca.

Pearson, R. G. (1963). Hard and soft acids and bases. *J. Am. Chem. Soc.* 85, 3533–3539.

Pearson, R. G. (1972). [Quantitative evaluation of the HSAB (hard-soft acis base) concept]. Replay to the paper "A quantitative evaluation of the HSAB concept", by Drago and Kabler. *Inorg. Chem.* 11(12), 3146–3146.

Pearson, R. G. (1985). Absolute electronegativity and absolute hardness of Lewis acids and Bases, *J. Am. Chem. Soc.* 107, 6801–6806.

Pearson, R. G. (1986). Absolute electronegativity and hardness correlated with molecular orbital theory. *Proc. Natl. Acad. Sci. USA* 83(22), 8440–8441.

Pearson, R. G. (1987). Recent advances in the concept of hard and soft acids and bases. *J. Chem. Ed.* 64(7), 561–567.

Pearson, R. G. (1988a). Chemical hardness and bond dissociation energies. *J. Am. Chem. Soc.* 110(23), 7684–7690.

Pearson, R. G. (1988b). Absolute electronegativity and hardness: application to inorganic chemistry. *Inorg. Chem.* 27(4), 734–740.

Pearson, R. G. (1989). Absolute electronegativity and hardness: applications to organic chemistry. *J. Org. Chem.* 54(6), 1423–1430.

Pearson, R. G. (1990). Hard and soft acids and bases—the evolution of a chemical concept. *Coord. Chem. Rev.* 100: 403–425.

Pearson, R. G. (1997). *Chemical Hardness,* Wiley-VCH, Weinheim.

Pearson, R. G. (1985). Absolute electronegativity and absolute hardness of Lewis acids and bases. *J. Am. Chem. Soc.* 107: 6801–6806.

Pearson, R. G. (1973). *Hard and Soft Acids and Bases;* Dowden, Hutchinson & Ross, Stroudsberg.

Perdew, J. P. (1986). Density-functional approximation for the correlation energy of the inhomogeneous electron gas. *Phys. Rev. B* 33, 8822–8824; with (1986). Erratum, *Phys. Rev. B* 34, 7406–7406.

Perdew, J. P., Ernzerhof, M., Zupan, A., Burke, K. (1998). Nonlocality of the density functional for exchange and correlation: Physical origins and chemical consequences. *J. Chem. Phys.* 108, 1522–1531.

Politzer, P. (1980). Observations on the significance of the electrostatic potentials at the nuclei of atoms and molecules. Israel, J. Chem. 19: 224–232.

Ponec, R. J. (1997). Electron pairing and chemical bonds. Chemical structure, valences and structural similarities from the analysis of the Fermi holes. *Math. Chem.* 21, 323–333.

Ponec, R. J. (1998). Electron pairing and chemical bonds. Molecular structure from the analysis of pair densities and related quantities. *Math. Chem.* 23, 85–103.

Ponti, A. (2000). DFT-Based Regioselectivity Criteria for Cycloaddition Reaction. *J. Phys. Chem.* 104, 8843–8846.

Purser, G. H. (1988). The thermochemical stability of ionic noble gas compounds. J. Chem. Ed. 65(2), 119–122.

Ray, N. K., Samuels, L., Parr, R. G. (1979). Studies of electronegativity equalization. *J. Chem Phys.* 70(8), 3680–3684.

Ray, N. K., Parr, R. G. (1980). Diamagnetic shieldings of atoms in molecules and their relation to electronegativity. *J. Chem. Phys.* 73, 1334–1339.

Red, J. L. (1981). Electronegativity. An isolated atom property. *J. Phys. Chem.* 85(2), 148–153.

Richard, J.-M., Fröhlich, J., Graf, G.-M., Seifert, M. (1993). Proof of stability of the hydrogen molecule. *Phys. Rev. Lett.* 71, 1332–1334.

Robles, J., Bartolotti, L. J. (1984). Electronegativities, electron affinities, ionization potentials, and hardnesses of the elements within spin polarized density functional theory. *J. Am. Chem. Soc.* 106(13), 3723–3727.

Roothaan, C. C. J. (1951). New developments in molecular orbital theory. *Rev. Mod. Phys.* 23, 69–89.

Rubin, S. G., Khosla, P. K. (1977). Polynomial interpolation methods for viscous flow calculations. *J. Comp. Phys.* 24, 217–244.

Ruedenberg, K. (1962). The physical nature of the chemical bond. *Rev. Mod. Phys.* 34, 326–376.

Rychlewski, J., Parr, R. G. (1986). The atom in a molecule: a wave function approach. *J. Chem Phys.* 84, 1696–1704.

Sanderson, R. T. (1988). Principles of electronegativity Part, I. General nature. *J Chem Educ.* 65, 112–119.

Sanderson, R. T. (1976). *Chemical Bonds and Bond Energy*, 2nd edition, Academic Press, New York.

Sato, K., Hosokawa, K., Maeda, M. (2003). Rapid aggregation of gold nanoparticles induced by non-cross-linking DNA hybridization. *J. Am. Chem. Soc.* 125(27), 8102–8103.

Savin, A., Stoll, H., Preuss, H. (1986). An application of correlation energy density functionals to atoms and molecules. *Theor Chim. Acta* 70(6), 407–419.

Schmidt, P. C., Böhm, M. C. (1983). Atomic and molecular orbital electronegativity models based on the transition state and transition operator approaches. *Ber. Bunsenges. Phys. Chem.* 87, 925–932.

Schrödinger, E. (1926). An Undulatory theory of the mechanics of atoms and molecules. *Phys. Rev.* 28(6), 1049–1070.

Schrödinger, E. (1926). Quantisierung als eigenwertproblem (Erste Mitteilung). *Ann. Phys.* 79, 361–376; ibid. Quantisierung als eigenwertproblem (Zweite Mitteilung). 79, 489; ibid. Über das verhältnis der Heisenberg-Born-Jordanschen quantenmechanik zu der meinen. 79, 734–756.

Sen, K. D., Jørgensen, C. K. (Eds.) (1987). *Electronegativity*; In: *Structure and Bonding*, Springer Verlag, Berlin, Vol. 66.

Sen, K. D., Mingos, D. M. P. (Eds.) (1993). Chemical hardness, *Structure and Bonding*, Springer Verlag, Berlin, Vol. 80.

Slater, J. C. (1930). Note on Hartree's method. *Phys. Rev.* 35, 210–211; ibid. (1931). Molecular energy levels and valence bonds. *Phys. Rev.* 38, 1109–1144; ibid. (1934). The electronic structure of metals. *Rev. Mod. Phys.* 6, 209–281.

Slater, J. C. (1951). A simplification of the Hartree-Fock method. *Phys. Rev.* 81, 385–390.

Tachibana, A. (1987). Density functional rationale of chemical reaction coordinate, *Int. J. Quantum Chem.* 21, 181–190.

Tachibana, A., Nakamura, K., Sakata, K., Morisaki, T. (1999). Application of the regional density functional theory: the chemical potential inequality in the HeH$^+$ system. *Int. J. Quantum Chem.* 74, 669–679.

Tachibana, A., Parr, R. G. (1992). On the redistribution of electrons for chemical reaction systems. *Int. J. Quantum. Chem.* 41, 527–555.

Torrent-Sucarrat, M., Luis, J. M., Duran, M., Solà, M. (2005). An assessment of a simple hardness kernel approximation for the calculation of the global hardness in a series of Lewis acids and bases. *J. Mol. Str. (THEOCHEM)* 727, 139–148.

Tozer, D. J., Handy, N. C. (1998). The development of new exchange-correlation functionals. *J. Chem. Phys.* 108, 2545–2555.

Tykodi, R. J. (1988). Estimated thermochemical properties of some noble-gas monoxides and difluorides. *J. Chem. Ed.* 65(11), 981–986.

University of Illinois (2011)

van Hooydonk, G. (1986). The superposition error problem: The (HF)2 and (H2O)2 complexes at the SCF and MP2 levels. *J. Mol. Struct. (Theochem)* 138, 361–376.

van't Hoff, J. H. (1874). A suggestion looking to the extension into space of the structural formulas at present used in chemistry, and a note upon the relation between the optical activity and the chemical constitution of organic compounds. *Arch. Neerland. Sci. Exact Natur* 9, 445–454.

Vela, A., Gázquez, J. L. (1988). Extended Hueckel parameters from density functional theory. *J. Phys. Chem.* 92(20), 5688–5693.

Vosko, S. J., Wilk, L., Nusair, M. (1980). Accurate spin-dependent electron liquid correlation energies for local spin density calculations: a critical analysis. *Can. J. Phys.* 58, 1200–1211.

Weiss, U. (1999). *Quantum Dissipative Systems*, 2nd edition, World Scientific, Singapore.

Werner, A. (1893). Beitrag zur konstitution anorganischer verbindungen. *Z. Anorg. Chem.* 3(1), 267–330.

Whitney, C. K. (2006a). Algebraic Chemistry. *Hadronic, J.* 26(1), 1–46.

Whitney, C. K. (2006b). Essays on Special Relativity Theory. *Hadronic, J.* 26(1), 47–110.

Whitney, C. K. (2007). Relativistic dynamics in basic chemistry. *Foundation Phys.* 37(4/5), 788–812.

Whitney, C. K. (2008a). Closing in on chemical bonds by opening up relativity theory. *Int. J. Mol. Sci.* (Special Issue "Chemical Bond and Bonding", Guest Editor: M. V. Putz) 9, 272–298.

Whitney, C. K. (2008b). Single-electron state filling order across the elements. *Int. J. Chem. Model.* 1(1), 105–135.

Whitney, C. K. (2009). Visualizing Electron Populations in Atoms. *Int. J. Chem. Model.* 1(2), 245–297.

Whitney, C. K. (2013). *Algebraic Chemistry: Applications and Origins*, Nova Publishers Inc., New York.

Wu, Z. N., Sheng, L. G. (1994). Electronegativity: average nuclear potential of the valence electron. *J. Phys. Chem.* 98(15), 3964–3966.

Yang, S. H., Pettiette, C. L., Conceicao, J., Cheshnovsky, O., Smalley, R. E. (1987). UPS of buckminsterfullerene and other large clusters of carbon. *Chem Phys Lett.* 139, 233–238.

Yang, W., Parr, R. G., Pucci, R. (1984). Electron density, Kohn–Sham frontier orbitals, and Fukui functions. *J. Chem. Phys.* 81(6), 2862–2863.

Zewail, A. (Ed.) (1992). *The Chemical Bond. Structure and Dynamics*, Academic Press, Boston.

Zhang, Y. (1982). Electronegativities of elements in valence states and their applications. 1. Electronegativities of elements in valence states. *Inorg. Chem.* 21(11), 3886–3889.

CHAPTER 4

MODELING MOLECULAR AROMATICITY

CONTENTS

ABSTRACT

The characterization of aromaticity for organic structures is undertaken in order to compare their chemical reactivity by employing common and recent indicators of aromaticity against the chemical hardness computed within the modern density functional theory (DFT) and the classical Hückel one. One finds that the values of the energetic indices calculated with the two methods correlate with other data presented in the literature and also with the experimental behavior of the studied compounds. On the other hand, the chemical hardness scale determined with Hückel method is in accordance either with the potential contour maps for the sites susceptible for electrophilic attack as well with the computed global values or chemical hardness realized with DFT method. New aromaticity definition is advanced as the compactness formulation through the ratio between atoms-in-molecule and orbital molecular facets of the same chemical reactivity property around the pre- and post-bonding stabilization limit, respectively. Geometrical reactivity index of polarizability was assumed as providing the benchmark aromaticity scale since its observable character; polarizability based- aromaticity scale enables introducing the working five referential aromatic rules (Aroma 1-to-5 Rules) with the help of which the aromaticity scales based on energetic reactivity indices of electronegativity and chemical hardness were computed and analyzed within the major semi-empirical and ab initio quantum chemical methods. The best correlations found in modeling the aromaticity criteria by the chemical hardness-based schemes of computations advocate considering further the associate hard-and-soft acids-and-bases and maximum hardness principles as main tools for assessing chemical reactivity and molecular stability. Similar studies are undertaken within the absolute aromaticity framework viewed as the difference contributions between atoms-in-molecules and molecular-orbitals' contribution to chemical reactivity by specific electronegativity or chemical hardness indices. Finally, quantitative structure aromaticity relationship (QSArR) studies on various aromatic, non-aromatic and anti-aromatic molecules are presented aiming for checking the predictor quality of finite-difference electronegativity and chemical hardness descriptors in aromaticity computations. The results show that the "aromaticity of peripheral topological path" may be well described by superior finite difference schemes of electronegativity and chemical hardness indices in certain calibrating conditions.

4.1 INTRODUCTION

The history of aromaticity is reach and exciting, being one of the oldest but always fascinating problems of chemistry since the 1825 Faraday's discovery of the so called "bicarburet of hydrogen" then consecrated as the benzene by the Kekulé (1866) in the second half of XIX, which advanced the aromaticity phenomenon as being responsible for the observed extra-stability of certain cyclic non-saturated compounds. Among many pioneering contributions to its description, worth remembering those of "three electrons in each CC region of benzene" of Thomson (1921), then segregated into the σ-π quantum contributions by Hückel (1931), up to the thermally and photochemical rules of aromaticity and anti-aromaticity of Doering and Detert (1951); other specific structural properties as continuous conjugation and planarity were initially admitted (Katritzky & Topson, 1971), then relaxed even for the assumed planar referential aromatic molecule of benzene (Moran et al., 2006). Recently, aromaticity has been extended to all metal molecules (Boldyrev & Wang, 2005). However, since various ways in which aromaticity can be described, an up-to-date classification for the associate criteria or descriptors brings us in front of six developed classes referring to (Putz, 2011):

- **(i)** *Geometrical descriptors*: the CC bond length variation in the molecular structures (Julg & Françoise, 1967), afterwards improved towards the stereochemical optimization based on harmonic oscillatory model of aromaticity (HOMA) (Kruszewski & Krygowski, 1972; Krygowski, 1993); this way the HOMA index was produced with the range spanning [0,1] values with the criterion that as it approach unity as more aromatic the system is;

- **(ii)** *The energetic criterion*: originating in the works of Pauling & Wheland (1933); Pauling & Sherman (1933) it is based on the resonance energy (RE) concept that was somehow twisted towards the difference between the π-electronic delocalization respecting a reference π-system without delocalization (Wheland, 1944); when reported per concerned π-electrons (PE) yields the REPE index (Hess & Schaad, 1971) of which magnitude estimates the stabilization energy and therefore the aromaticity degree: higher REPE higher aromaticity is predicted;

- **(iii)** *Magnetic descriptors*: (a) the nucleus-independent chemical shift (NICS) index (Schleyer et al., 1996; Chen et al., 2005) correlates the higher π-delocalization with the increasing of magnetization (vector) in the center of the aromatic ring (or at other concerned point of the system); therefore, larger magnetization larger aromaticity will be; (b) the exaltation of magnetic susceptibility (Λ) (Dauben, 1968), measures the resistance to magnetization is usually reported as an extensive quantity, i.e., it is calculated per π-electron, as it is the case also with REPE index, instead is taking lower values for higher stability and aromaticity;

(iv) *Topological indices*: originating in resonance energy they describe aromaticity through the conjugated circuits advanced by the works of Randić and Gutman (Randić, 1977; Gutman et al., 1977), while recently evolving in accounting the limit Kekulé structures (Aihara et al., 2005; Aihara & Kanno, 2005); alternatively the aromatic zones for the topological peripheral paths (TOPAZ) were also introduced in relation with the bond orders' values predicting higher aromaticity for higher values (Tarko, 2008; Tarko & Putz, 2010), while the inverse aromaticity hierarchies is provided by the topologic index of reactivity (TIR) proposed by Balaban and collaborators (Ciesielski et al., 2009a) to reflect the aromatic electrophilic substitution reaction for the most favorable energetically position in a molecule;

(v) *Electronic localization criteria:* is emphasizing on the multicenter bonding indices and the local aromaticity by the aid of electronic fluctuations, para-delocalization, and electronic localization function – ELF, see Volume II of the present five-volume set (Putz, 2016) and (Silvi & Savin, 1994; Giambiagi et al., 2000; Poater et al., 2003; Bultinck et al., 2005; Matito et al., 2005a-b, 2006; Cioslowski et al., 2007; Feixas et al., 2008); in particular, the aromatic fluctuation index (FLU) combines the electronic information as sharing and similarity between adjacent atoms (Matito et al., 2006), while measuring the weighted electron delocalization divergences with respect to typical aromatic molecule; although with unique definition indices like FLU strongly depend on the way the molecular space is topologically divided (e.g., by Bader atoms in molecule—AIM or with the help of ELF) yet being of referential nature, like HOMA index, with whom it however parallels either as aromaticity criteria as well by statistical correlations;

(vi) *Reactivity criteria*: by using the "primitive patterns of understanding" reactivity (McWeeny, 1979), the modern density functional electronegativity and chemical hardness have been recently related by Chattaraj et al. with aromaticity in the light of their principles, especially those regarding the Pearson's hard-soft-acid-base and the maximum harness ones (Chattaraj, 2007), while being extended by the recent authors' works to cover all reactivity principles, either for electronegativity and chemical hardness (Putz, 2010a,b).

4.2 MODELING AROMATICITY BY CHEMICAL HARDNESS

4.2.1 REACTIVITY ENERGETIC INDICES

The aromaticity of benzene ([6]annulene) and its derivatives has been interpreted with many valence theories (Saltzman, 1974) and afterwards with quantum chemistry models, as approximations of quantum mechanics ones (Hückel, 1931). Recently, the concept of aromaticity has been extended to inorganic systems (Chattaraj et al., 2007; Mosquera et al., 2007), zones and molecular fragments respectively by means of models such as HOMA (Harmonic Oscillator Model of Aromaticity) (Kruszewski & Krygowski, 1972; Krygowski, 1993), CRCM (Conjugated Ring Circuits Model) (Randić, 1977; Gutman et al., 1977; Aihara et al., 2005; Aihara & Kanno, 2005) and TOPAZ (Topological Path and Aromatic Zones) (Tarko, 2008; Tarko & Putz, 2010). Moreover, the aromaticity can be applied to the transition state of the pericyclic reactions constructed within Hückel or Möbius cyclic systems (Rzepa, 2007; Putz et al., 2013).

Nowadays, competitive software is available to perform advanced calculations with high accuracy in order to assess the aromaticity of different systems. From our experience in teaching, today's students are able to use these computer programs but encounter difficulties in the interpretation of the results obtained. In order to help them gaining some insights, we apply Hückel molecular orbital theory (HMO), highly didactical, by comparison with a modern one, density functional theory (DFT), to explain the reactivity of some aromatic hydrocarbons and heteroaromatic compounds.

The Hückel method is simple and has been in use for decades. It is at anyone hand as a starting point to the understanding of the methods without simplification conditions and advanced too. Basically, the static parameters of reactivity computed with HMO method assume the Coulomb energy $(\alpha_r = \alpha + \delta_r \beta)$ for the atomic orbital r and the exchange energy $(\beta_{rs} = \eta_{rs} \beta)$ for the bond between the atomic orbital r and s, determined from experimental results. The standard values of α and β refer to the atomic orbital $2p_z$ of the sp^2 C and to a π bond, respectively, from benzene. Even though the method implicates simplifications it shows correctly the variation of the electronic densities and the energetic levels within a series of molecules that differ slightly.

On the other hand, DFT is among the most popular and versatile methods available in computational chemistry (Kohn et al., 1996). The calculations performed with DFT directly reflect the electronic density influence of a chemical structure towards its reactivity (Kohn and Sham, 1995). Mainly, the electronic energy is approximated by various density functional terms; the present discussion follows (Putz, 2008a, 2012a)

$$E[\rho] = \underbrace{T[\rho]}_{KINETIC} + \underbrace{J[\rho]}_{COULOMB} + \underbrace{E_{XC}[\rho]}_{EXCHANGE-CORRELATION} + \underbrace{\int V(\mathbf{r})\rho(\mathbf{r})d\mathbf{r}}_{CHEMICAL\ ACTION} \quad (4.1)$$

while its first and second functional derivatives provide the electronegativity and chemical hardness indices (Putz, 2012b), respectively

$$\chi \equiv -\left(\frac{\delta E}{\delta \rho}\right)_{V(r)} \quad (4.2)$$

$$\eta = \left(\frac{\delta^2 E}{\delta \rho^2}\right)_{V(r)} \quad (4.3)$$

with a crucial role in assessing the chemical stability and reactivity propensity through the associate principles, especially those referred to electronegativity equalization and the maximum chemical hardness, see Sections 3.3 and 3.4 of the present volume.

The purpose of this analysis is to correlate some accessible quantum chemical results with physico-chemical properties of some aromatic compounds. The systems presented are: hydrocarbons (aromatic annulenes), amines, molecules, ions, hydroxyarenes, and heterocycles with nitrogen, all of them selected according to their industrial significance. For example, by diazotization and coupling reactions the aromatic amines and phenols lead to azo dyes. The heterocyclic compounds take part in building up the skeleton of some natural products of vital importance (Avram, 1994, 1995). The planar molecules with $(4n+2)\pi$ electrons under discussion here present high stability, low magnetic susceptibility (diamagnetism), and preference for substitution reactions than addition (Hendrickson et al., 1970; Kruszewski & Krygowski, 1972; Krygowski 199;). Some of them fit in the general Ar-Y ($Y = OH$, NH_2, NR_2, $\overset{+}{N}H_3$) model, highly correlated with the inductive effects (σ and π) and substituent constants too (Katritzky & Topson, 1971).

4.2.2 COMPUTATIONAL QUANTUM CHEMICAL CONSIDERATIONS

For the present case-study 12 aromatic compounds are considered, namely: benzene, aniline, phenol, naphthalene, α-naphthol, β-naphthol, α-naphthylamine, β-naphthylamine, pyridine, pyrimidine, N,N'-dimethyl p-aminoaniline, and monochlorohydrate of N,N'-dimethyl p-aminoaniline. Among these molecules, some of them present symmetry features, which can be used to lower the degree of the secular determinant obtained in the HMO method (Hückel, 1931; Mandado et al., 2007). However, to compute the determinant of these aromatic compounds the computational technique has been considered using the data from Table 4.1; the present discussion follows (Putz et al., 2010)

TABLE 4.1 The Parameters Values Considered for the Conjugated Systems (HMO method) (Putz et al., 2010)

δ_r	η_{rs}
$\delta_{-\overset{\shortmid}{\underset{\shortmid}{C}}-} = 0.0$	$\eta_{C-O-} = 0.8$
$\delta_{-\overset{\shortmid}{\underset{\shortmid}{N}}-} = 0.5$	$\eta_{C-N<} = 0.8$
$\delta_{C(\overset{+}{N}<)} = 0.5$	$\eta_{C=C} = 1.0$
$\delta_{-\overset{\shortmid}{N}-CH_3 \atop \overset{\shortmid}{CH_3}} = 1.1$	$\eta_{C=N} = 1.0$
$\delta_{-\overset{\shortmid}{N}-H \atop H} = 1.5$	—
$\delta_{-O-} = 2.0$	—

The number of points indicates the π electrons/atom

Employing the Hückel matrices, the eigenvectors, the π levels of energy, the total ground state energy (E_π), the delocalization energy/π electrons $(\overline{\pi})$, the charge densities (q_r), the bond orders (p_{rs}) and the free valence $(F_r = 1.732 - \sum p_{rs})$ have been calculated.

For the DFT computations, the present based molecular computations employ the sum of the exchange and correlation contributions in the mixed functional of Eq. (4.1)

$$E_{XC}[\rho] = \underbrace{K[\rho]}_{EXCHANGE} + \underbrace{E_c[\rho]}_{CORRELATION} \tag{4.4}$$

For the exchange term was used the Becke's functional *via* the so called *semiempirical* (SE) modified gradient-corrected functional (Becke, 1986):

$$K^{SE} = K_0 - \beta \sum_\sigma \int \rho_\sigma^{4/3} \frac{x_\sigma^2(\mathbf{r})}{1 + \gamma x_\sigma^2(\mathbf{r})} d\mathbf{r}$$

$$K_0 = \int d\mathbf{r} k_0[\rho(\mathbf{r})]$$

$$x_\sigma(\mathbf{r}) = \frac{|\nabla \rho_\sigma(\mathbf{r})|}{\rho_\sigma^{4/3}(\mathbf{r})} \tag{4.5}$$

towards the working single-parameter dependent one (Becke, 1988):

$$K^{B88} = K_0 - \beta \sum_\sigma \int \rho_\sigma^{4/3}(\mathbf{r}) \frac{x_\sigma^2(\mathbf{r})}{1 + 6\beta x_\sigma(\mathbf{r}) \sinh^{-1} x_\sigma(\mathbf{r})} d\mathbf{r} \tag{4.6}$$

where the value $\beta = 0.0042 [a.u.]$ was found as the best fit among the noble gases (He to Rn atoms) exchange energies; the constant a_σ is related to the ionization potential of the system.

For the correlation part of energy the *Lee, Yang, and Parr* (LYP) functional is considered within Colle-Salvetti approximation (Lee et al., 1988):

$$E_c^{LYP} = -a_c b_c \int d\mathbf{r} \gamma(\mathbf{r}) \xi(\mathbf{r}) \left(\begin{array}{c} \sum_\sigma \rho_\sigma(\mathbf{r}) \sum_i |\nabla \varphi_{i\sigma}(\mathbf{r})|^2 \\ -\frac{1}{4} \sum_\sigma \rho_\sigma(\mathbf{r}) \Delta \rho_\sigma(\mathbf{r}) \\ -\frac{1}{4} |\nabla \rho(\mathbf{r})|^2 + \frac{1}{4} \rho(\mathbf{r}) \Delta \rho(\mathbf{r}) \end{array} \right) - a_c \int d\mathbf{r} \frac{\gamma(\mathbf{r})}{\eta(\mathbf{r})} \rho(\mathbf{r})$$

$$\tag{4.7}$$

where

$$\gamma(\mathbf{r}) = 4\frac{\rho_\uparrow(\mathbf{r})\rho_\downarrow(\mathbf{r})}{\rho(\mathbf{r})^2} \qquad (4.8a)$$

$$\eta(\mathbf{r}) = 1 + d_c\rho(\mathbf{r})^{-1/3} \qquad (4.8b)$$

$$\xi(\mathbf{r}) = \frac{\rho(\mathbf{r})^{-5/3}}{\eta(\mathbf{r})}\exp\left[-c_c\rho(\mathbf{r})^{-1/3}\right] \qquad (4.8c)$$

and the constants: $a_c = 0.04918$, $b_c = 0.132$, $c_c = 0.2533$, $d_c = 0.349$.

4.2.3 HÜCKEL VS. DFT CHEMICAL HARDNESS BASED AROMATICITY

Modeling the stability and reactivity of molecules is perhaps the greatest challenge in theoretical and computational chemistry; this because the main conceptual tools developed as the reactivity indices of electronegativity and chemical hardness along the associate principles are often suspected by the lack in observability character; therefore, although very useful in formal explanations of chemical bonding and reactivity there is hard in finding of their experimental counterpart unless expressed by related measurable quantities as energy, polarizability, refractivity, etc. When the aromaticity concept come into play there seems no further conceptual clarification is acquired since no quantum observable or further precise definition can be advanced; in fact, the aromaticity concept was associate either with geometrical, energetic, topologic, electronic molecular circuits (currents), or with the less favorite entropic site in molecule just to name few of its representations; the present discussion follows (Putz et al., 2010).

Turning to the present case-study, in order to coordinate the reactivity of the 12 compounds (mononuclear and polynuclear) chosen here (Table 4.2) with their structural and energetic indices the simple Hückel method and the DFT calculations were performed for the chemical hardness "frozen core" approximation of Eq. (3.347) (Koopmans, 1934), in absence of the averaging charge factor (3.344), in terms of lowest unoccupied molecular orbital (LUMO) and highest occupied molecular orbital (HOMO)

TABLE 4.2   Energetic Indices for the Organic Molecules Under Study as Computed Within the Density Functional Theory With B3-LYP Exchange-Correlation (Hypercube, 2002); All Values in Electron-Volts (eV) (Putz et al., 2010)

Compound	ε_{LUMO}^{DFT}	ε_{HOMO}^{DFT}	η^{DFT}
Benzene	2.523545	−5.132959	7.6565
Aniline	2.96429	−3.097193	6.06148
Phenol	2.597031	−3.760442	6.35747
N,N'-dimethyl p-aminoaniline	3.450884	−1.606443	5.05733
Monochlorohydrate of N,N'-dimethyl p-aminoaniline	−2.230419	−6.886414	4.656
Pyridine	1.622078	−4.751688	6.37377
Pyrimidine	0.9239021	−4.744836	7.34187
Naphthalene	1.29009	−4.156888	5.44698
β-naphthylamine	1.524541	−3.370884	4.89543
α-naphthylamine	1.538881	−3.264623	4.8035
β-naphthol	1.426792	−3.5884	5.01519
α-naphthol	1.534517	−3.422596	4.95711

energies, although similar relations (in terms of ionization potential *IP* and electron affinity *EA*) hold for atomic systems as well.

The numerical results are presented in Tables 4.2 and 4.3.

According with the energetic indices values, pyridine and pyrimidine are more stable than benzene, a fact experimentally proved by the oxidation of quinoline and quinazoline to their corresponding acids (Nenitzescu, 1968 (part v)). The favorable effect that first order substituents bring on aromaticity is in accordance with TOPAZ modeling for these compounds, such as $A_{aniline} \cong 90\% A_{phenol}$ relationship (Tarko, 2008). Our results are in agreement with conjugation or resonance energies experimentally determined: $36 \, kcal \cdot mol^{-1} \cong 0.3333 \times 6|_{\pi \, electron} \times 18|_{\beta \, unit}$ for benzene and $45 \, kcal \cdot mol^{-1} \cong 0.4243 \times 6|_{\pi \, electron} \times 18|_{\beta \, unit}$ for pyridine (Avram, 1995 (Chapter 32); Nenitzescu, 1968 (parts i & iv)). All these aspects mentioned correlate with the potential contours presented in Figure 4.1.

The experimental value of the conjugation energy of naphthalene $(61 \, kcal \cdot mol^{-1})$, considerably smaller than the double of that of benzene, does not correlate with the π values calculated by us and other

TABLE 4.3 Energetic Indices From Table 4.2, Supplemented with the Delocalization Energy per π Electrons ($\overline{\pi}$) as Computed Within the Hückel Theory (HT) or Calibrated With the Density Functional Counterparts From Table 4.2 (Putz et al., 2010)

Compound	ε_{LUMO}^{HT}	ε_{HOMO}^{HT}	η^{HT}†	$\overline{\pi}^{HT}$†	$-\alpha^{\bullet}$	$-\beta^{\bullet}$	η_{DFT}^{HT} ♦	$\overline{\pi}_{DFT}^{HT}$ ♦
Benzene	α-β	α+β	2.0000	0.3333	1.3047	3.8282	7.6564	1.2759
Aniline	α-β	α+0.7437β	1.7437	0.4665	0.5119	3.4762	6.0614	1.6216
Phenol	α-0.9999β	α+0.8274β	1.8273	0.5246	0.8818	3.8283	6.9954	2.0083
N,N'-dimethyl p-aminoaniline	α-β	α+0.4962β	1.4962	0.3089	-0.071	3.3801	5.0573	1.0441
monochlorohydrate of N,N'-dimethyl p-aminoaniline	α-0.9415β	α+0.7178β	1.6593	0.4304	4.8723	2.806	4.656	1.2077
Pyridine	α-0.8410β	α+β	1.8410	0.4243	1.2896	3.4621	6.3737	1.4690
Pyrimidine	α-0.7800β	α+1.0700β	1.8500	0.5156	1.4662	3.0642	5.6688	1.5799
Naphthalene	α-0.6180β	α+0.6180β	1.2360	0.3680	1.4334	4.4069	5.4469	1.6217
β-naphthylamine	α-0.6363β	α+0.5573β	1.1936	0.4515	1.0852	4.1014	4.8954	1.8518
α-naphthylamine	α-0.6698β	α+0.5022β	1.1720	0.4522	1.2063	4.0986	4.8036	1.8534
β-naphthol	α-0.6332β	α+0.5807β	1.2139	0.4901	1.1893	4.1315	5.0152	2.0248
α-naphthol	α-0.6606β	α+0.5382β	1.1988	0.4906	1.1971	4.1351	7.6499	2.0287

† in [-β] units.

♦ in electron-Volts [eV], obtained by HT with DFT calibrations of LUMO and HOMO energies.

♦ in electron-Volts [eV], obtained from η^{HT} and $\overline{\pi}$ by multiplication with corresponding [-β]eV value.

authors as well (Nenitzescu, 1968 (parts i & iv); Schaad & Hess, 1974). Instead, the TOPAZ indices agree with the aromaticity relationship $A_{naphtalene} = 0.608 A_{benzene}$ (Tarko, 2008). In fact, naphthalene is more unsaturated than benzene displaying an aromatic decet and a butadiene conjugated system, mostly engaged in addition and oxidation reactions (Nenitzescu, 1968 (part i)). However, the ε_{HOMO} and ε_{LUMO} values and their gap η prove a higher chemical reactivity of the polynuclear compound.

According with ε_{HOMO} and ε_{LUMO} values, the $C_{10}H_7Y$ compounds show superior tendency to oxidation or reduction processes than the corresponding C_6H_5Y π-systems. Unlike the more stable phenol, the less aromatic naphthols participate in some chemical reactions under tautomeric form (Avram, 1994).

Based on HOMO-LUMO gap criteria for aromaticity, naphtylamines are less aromatic than aniline showing properties intermediary those of pure aromatic amines and aliphatic ones (Nenitzescu, 1968 (part i)). The monochlorhydrate of N,N-dimethyl p-aminoaniline, one of the amine form in HCl solution, is more aromatic than the non-ionic form of the amine as predicted by the values of the TOPAZ aromaticity (Tarko, 2008).

Let us consider now selectively the structural indices calculated for the compounds under study (data not presented in the paper). The π charges of the atoms follow the relationship $q_O \rangle q_N \rangle q_C$ being in agreement with the acknowledged tendency of electronegativity (Par & Yang, 1989; Putz, 2011, 2012b) and also with the potential contours presented in Figure 4.1. The aromatic amines and mononuclear/polynuclear hydroxyarenes suffer electrophilic substitution reactions with a higher rate than their reference, benzene and naphthalene, respectively: $k_{Ar-y} / k_{ArH} \rangle 1$ ($Y = OH, NH_2$) (Nenitzescu, 1968 (part iv)). This behavior is due to the ring activations by conjugation of the non-participating electrons of nitrogen and oxygen with the π electrons of the aromatic cycle (Avram, 1994). Actually, the calculated π charge densities predict that the electrophilic substitution takes place preferentially on para- and ortho-positions for aniline and phenol, on 4th- and 2nd-position for α-naphthol and α-naphthylamine, on 1st-position for β-naphthol and β-naphthylamine. This reaction mechanism occurs for pyridine on β-positions, but hardly as a result of its participation as an ammonium ion and for pyrimidine on the 5th-position (Avram, 1995 (Chapter 32)). Instead, the nucleophilic substitution for these two molecules, due to the presence

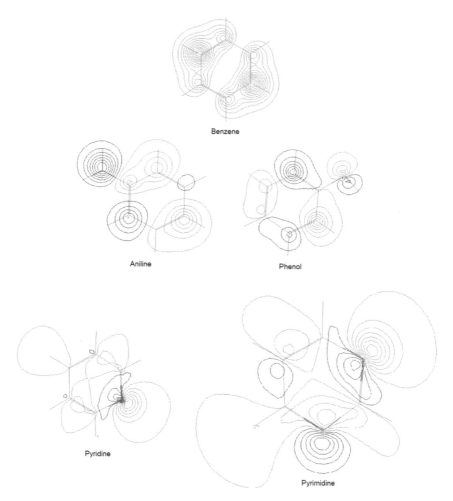

Benzene

Aniline

Phenol

Pyridine

Pyrimidine

FIGURE 4.1 The potential contour maps for the molecules under discussion on their HOMO state. A higher density of contours indicates that the site is susceptible for electrophilic attack either for positive (green) or negative (in cyan) molecular electrostatic potential (Putz et al., 2010).

of the nitrogen atoms in the aromatic cycles, takes place on α- and γ- positions for pyridine and on 2nd-, 4th-, and 6th-positions for pyrimidine.

Associating the monosubstitued aromatic compounds (phenol, aniline); (α-naphthol, α- naphthylamine); (β-naphthol, β-naphthylamine), it can be observed the correlation between the decrease of the aromaticity and the increase of the p_{C-Y} value (Streitwieser, 1961), in good agreement with the results of TOPAZ model (Tarko, 2008). The calculated values for $C_{10}H_8$ ($p_{C_\alpha-C_\beta} = 0.7246$, $F_{C_\alpha} = 0.4527$, $F_{C_\beta} = 0.4042$) and C_6H_6 ($p_{C-C} = 0.666$, $F_C = 0.400$) show the tendency of naphthalene to give addition reactions much easier than benzene. The elevated values of the free valences calculated for nitrogen ($F_N = 1.42$–1.44) of the aromatic amines and oxygen ($F_O = 1.47$–1.48) of the hydroxyarenes are reflected in the chemical properties of the corresponding functional groups (Avram, 1994, 1995 (Chapter 22)). Thus, the phenols and naphthols are week acids ($p_{K_a} = 9-10$) and the aromatic amines are week bases with stronger conjugated acids ($p_{K_a} = 4-5$) (Avram, 1995 (Chapter 22)).

Yet, from Tables 4.2 and 4.3 the following aromaticity hierarchies are obtained (Putz et al., 2010):

- DFT hierarchy of aromaticity of Table 4.2:

$$\eta^{DFT}(Benzene) > \eta^{DFT}(Pyrimidine)$$
$$> \eta^{DFT}(Pyridine) > \eta^{DFT}(Phenol)$$
$$> \eta^{DFT}(Aniline) > \eta^{DFT}(Naphthalene)$$
$$> \eta^{DFT}(NN'\text{-}dimethyl\text{-}p\text{-}aminoaniline)$$
$$> \eta^{DFT}(\beta - Naphtol) > \eta^{DFT}(\alpha - Naphtol)$$
$$> \eta^{DFT}(\beta - Naphthylamine) > \eta^{DFT}(\alpha - Naphthylamine)$$
$$> \eta^{DFT}(monochlorohydrate\ of\ NN'\text{-}dimethyl\text{-}p\text{-}aminoaniline)$$

$$(4.9)$$

- HT hierarchy of aromaticity according to the chemical hardness values of Table 4.3 (fourth column), for universal Hückel [–β] integral:

$$\eta^{HT}(Benzene) > \eta^{HT}(Pyrimidine)$$
$$> \eta^{HT}(Pyridine) > \eta^{HT}(Phenol) > \eta^{HT}(Aniline)$$
$$> \eta^{HT}(monochlorohydrate\ of\ NN'\text{-}dimethyl\text{-}p\text{-}aminoaniline)$$
$$> \eta^{HT}(NN'\text{-}dimethyl\text{-}p\text{-}aminoaniline) > \eta^{HT}(Naphthalene)$$
$$> \eta^{HT}(\beta - Naphtol)$$
$$> \eta^{HT}(\alpha - Naphtol) > \eta^{HT}(\beta - Naphthylamine)$$
$$> \eta^{HT}(\alpha - Naphthylamine)$$

$$(4.10)$$

- HT hierarchy of aromaticity according to the chemical hardness values of Table 4.3 (eighth column), for calibrated Hückel-DFT [−β] integral:

$$\eta_{DFT}^{HT}\left(Benzene\right) > \eta_{DFT}^{HT}\left(\alpha - Naphtol\right)$$
$$> \eta_{DFT}^{HT}\left(Phenol\right) > \eta_{DFT}^{HT}\left(Pyridine\right) > \eta_{DFT}^{HT}\left(Aniline\right)$$
$$> \eta_{DFT}^{HT}\left(Pyrimidine\right) > \eta_{DFT}^{HT}\left(Naphtalene\right)$$
$$> \eta_{DFT}^{HT}\left(NN'\text{-}dim\,ethyl\text{-}p\text{-}a\,min\,oaniline\right) > \eta_{DFT}^{HT}\left(\beta - Naphtol\right)$$
$$> \eta_{DFT}^{HT}\left(\beta - Naphthylam\,ine\right) > \eta_{DFT}^{HT}\left(\alpha - Naphthylam\,ine\right)$$
$$> \eta_{DFT}^{HT}\left(monochlorohydrate\ of\ NN'\text{-}dim\,ethyl\text{-}p\text{-}a\,min\,oaniline\right)$$

$$(4.11)$$

- HT hierarchy of aromaticity according to the $\bar{\pi}$ values of Table 4.3 (fifth column), for universal Hückel [−β] integral:

$$\bar{\pi}^{HT}\left(Phenol\right) > \bar{\pi}^{HT}\left(Pyrimidine\right)$$
$$> \bar{\pi}^{HT}\left(\alpha - Naphtol\right) > \bar{\pi}^{HT}\left(\beta - Naphtol\right) > \bar{\pi}^{HT}\left(Aniline\right)$$
$$> \bar{\pi}^{HT}\left(\alpha - Naphthylam\,ine\right) > \bar{\pi}^{HT}\left(\beta - Naphthylam\,ine\right)$$
$$> \bar{\pi}^{HT}\left(monochlorohydrate\ of\ NN'\text{-}dim\,ethyl\text{-}p\text{-}a\,min\,oaniline\right)$$
$$> \bar{\pi}^{HT}\left(Pyridine\right)$$
$$> \bar{\pi}^{HT}\left(Naphthalene\right) > \bar{\pi}^{HT}\left(Benzene\right)$$
$$> \bar{\pi}^{HT}\left(NN'\text{-}dim\,ethyl\text{-}p\text{-}a\,min\,oaniline\right)$$

$$(4.12)$$

- HT hierarchy of aromaticity according to the $\bar{\pi}$ values of Table 4.3 (fifth column) (ninth column), for calibrated Hückel-DFT [−β] integral:

$$\bar{\pi}_{DFT}^{HT}\left(\alpha - Naphtol\right) > \bar{\pi}_{DFT}^{HT}\left(\beta - Naphtol\right)$$
$$> \bar{\pi}_{DFT}^{HT}\left(Phenol\right) > \bar{\pi}_{DFT}^{HT}\left(\alpha - Naphthylam\,ine\right)$$
$$> \bar{\pi}_{DFT}^{HT}\left(\beta - Naphthylam\,ine\right) > \bar{\pi}_{DFT}^{HT}\left(Naphtalene\right)$$
$$> \bar{\pi}_{DFT}^{HT}\left(Aniline\right) > \bar{\pi}_{DFT}^{HT}\left(Pyrimidine\right)$$
$$> \bar{\pi}_{DFT}^{HT}\left(Pyridine\right) > \bar{\pi}_{DFT}^{HT}\left(Benzene\right)$$
$$> \bar{\pi}_{DFT}^{HT}\left(monochlorohydrate\ of\ NN'\text{-}dim\,ethyl\text{-}p\text{-}a\,min\,oaniline\right)$$
$$> \bar{\pi}_{DFT}^{HT}\left(NN'\text{-}dim\,ethyl\text{-}p\text{-}a\,min\,oaniline\right)$$

$$(4.13)$$

From these aromaticity scales some important rules can be formulated (Putz et al., 2010):

- The DFT and Hückel chemical hardness scales agrees between them (unless the inversion of the N,N'-dimethyl p-aminoaniline and mono-chlorohydrate of N,N'-dimethyl p-aminoaniline in the HT scheme) with the aromaticity/stability/reactivity scale of concerned compounds through experimental or other theoretical pictures invoked;
- The mixed DFT-Hückel chemical hardness scale of aromaticity further inverts the two forms of Naphtol considered, being weaker as the power of ordering respecting either separate DFT and Hückel first two scales of aromaticity, respectively.
- The aromaticity based on the delocalization energy per pi electrons even further departs from common accepted scales since inverting the Phenol with Benzene structural inertia in reactivity.

Within this case-study the Hückel and DFT theories was employed to describe the physico-chemical properties of some aromatic compounds. These two approaches have provided good correlations among the energetic indices calculated with regard to the substitution reactions in which these compounds are involved and the acidity and basicity of their functional groups. Moreover, the results we have obtained for benzene derivatives have been compared with TOPAZ indices being in good agreement (Putz et al., 2010).

Overall, the chemical hardness appears as a versatile tool in assessing reactivity and aromaticity, superior to the classical delocalization energy per pi electrons index; the DFT approach parallels remarkably well the simple Hückel quantum picture of organic molecules; the contour maps, although useful for didactical purposes, gives only a qualitative description of electrophilic attack with less quantitative power in establishing the order of reactivity or aromaticity through a series of related organic molecules (Putz et al., 2010).

It should be pointed out that this work do not bring something innovative in the calculations reported but it can be easily applied in the classroom as a basis for understanding the complex computational tools available these days. The systems presented have been selected according to didactical considerations and the comparisons and correlations established can be used as a trend in the analysis of other organic or inorganic compounds. Furthermore, we consider that the illustrative pictures obtained employing the DFT method, explain chemical reactivity nearly intuitively and they have a high edifying impact in every organic lecture bringing the abstract theories in accessibility. However, the quantitative computation of chemical hardness is the only one able to offer with a certain degree of accuracy a workable scale of aromaticity (Putz et al., 2010).

4.3 MODELING COMPACT AROMATICITY OF ATOMS-IN-MOLECULES

For practical implementations and analysis the present framework uses the picture of Atoms-in-Molecule (AIM) as a concept of densities of isolated atoms, being this different than the traditional AIM concept developed by the works of Bader (1990, 1994, 1998) that consider that the sum of all the atomic basins of the molecule brings to the total density of the system. Here, the distinction will be made between the *pre-bonding stage* when the atoms are considered isolated yet "prepared" for encountering within molecular system, i.e., having the present *AIM configuration*, and the *post-bonding stage* for which the consecrated quantum mechanical methods of molecular orbital or density functional ab initio or approximations (Putz, 2010a,b) techniques will be implemented in the overall called *MOL configuration*. The present AIM-vs.-MOL stages of bonding may be employed to eventually asses the aromaticity concept with a new perspective respecting the reactivity indices of electronegativity and chemical hardness.

4.3.1 DEFINING COMPACT AROMATICITY BY CHEMICAL REACTIVITY

Modeling the chemical bond is certainty a major key for describing the chemical reactivity and molecular structures' stability. Yet, since the chemical bond is a dynamic state, for the best assessment of its connection with the stability and reactivity, the pre- and post-bonding stages are naturally considered; the present discussion follows (Putz, 2010a).

For the pre-bonding stage, the atomic spheres are considered in the AIM arrangement, while for the post-bonding stage the molecular orbitals (MOL) of the already formed molecule are employed, see Figure 4.2; consequently, their ratio would model the *compactness degree* of a given

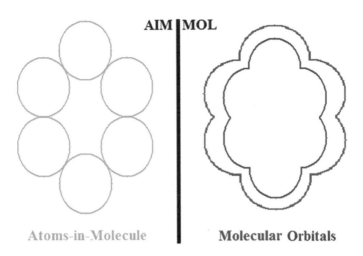

FIGURE 4.2 Heuristic representation of the concept of atoms-in-molecule (AIM) compactness aromaticity (for the benzene pattern) as the ratio of the pre- and postbased molecule to the (vis-à-vis) post-bonding molecular orbitals (MOL) modeling (Putz, 2010a).

property of AIM in respect to its counterpart at the MOL level of the chemical bond. Therefore, the actual compactness index of aromaticity and takes the general form (Putz, 2010a)

$$Aromaticity = \frac{\Pi_{AIM}}{\Pi_{MOL}}...\begin{cases} \in (-\infty,-1)\cup (1,+\infty)...AIM...prevails \\ \in (-1,+1) \qquad\qquad ...MOL...prevails \\ =\pm 1 \qquad\qquad\qquad ...transition...states \end{cases}$$

(4.14)

that becomes workable once the property Π is further specified. Note that for the Eq. (4.14) to be properly implemented, the chemical property Π should be equally defined and with the same meaning for the atoms and molecules, for consistency; such that what is compared is the chemical manifestation of the same property of bonding in its pre- or post-stage of formation. In other terms, Eq. (4.14) may be regarded as a kind of "chemical limit" for the chemical bond that may be slightly oriented towards its atom constituents or to its MOL prescribing therefore the propensity to reactivity or stability, respectively (Putz, 2010a).

It is worth considering also the quantitative difference between AIM and MOL properties of bonding, in which case the result may be regarded as the first kind of *absolute aromaticity* – for this reason, it is evaluated essentially between pre- and post-bonding stages and not relative to a referential (different) molecule (Putz, 2010b), see also the subsequent section, while the present approach promotes the *compactness* version of the aromaticity as the measure associated with the molecular stability in analog manner the compactness of rigid spheres in unit cells provides the crystal stability orderings. Yet, the proper scale hierarchy of compactness aromaticity, *i.e.*, the qualitative tendency respecting the quantitative yield of Eq. (4.14), is to be established depending on the implemented chemical property. In what follows, both the geometrically- and energetically-based reactivity indices will be considered, and their associate AIM compactness aromaticity formula and scales formulated. Moreover, once various quantum methods in evaluating the MOL denominator property in Eq. (4.14) are considered, they will become fully quantum.

4.3.2 BUILDING AROMATICITY SCALES

The above reactivity indices-based aromaticity scales are now computed within the presented quantum chemical schemes for a limited yet significant series of benzenoids containing the "life" atoms of Table 4.4 (see Table 4.5). The AIM of aromaticity scales are those of (Putz, 2010a):

- *polarizability*,

$$A_{POL} = \frac{\alpha_{AIM}}{\alpha_{MOL}} \qquad (4.15a)$$

with

$$\alpha_{AIM} = \sum_A \alpha_{A(toms)} = 4.5 \sum_A r^3_{A(toms)} \qquad (4.15b)$$

and

$$\alpha_{MOL}^{Aromatics} = \frac{3}{4\pi} \frac{1}{\left[\pi e^-\right]} V_{MOL} \left[\overset{o}{A}\right]^3 \qquad (4.15c)$$

- *electronegativity*

TABLE 4.4   Main Geometric and Energetic Characteristics for Atoms Involved in Organic Compounds Considered in this Work (see Table 4.5), as Radii From Putz et al. (2003) and Polarizabilities (Pol) Based Upon Eq. (4.15b), Along the Electronegativity (χ) and Chemical Hardness (η) From Putz (2006, 2008b), Respectively (Putz, 2010a)

Atom	Radius [Å]	Pol [Å]³	χ [eV]	η [eV]
H	0.529	0.666	7.18	6.45
C	0.49	0.529	6.24	4.99
N	0.41	0.310	6.97	7.59
O	0.35	0.193	7.59	6.14

$$A_{EL} = \frac{\chi_{AIM}}{\chi_{MOL}} \qquad (4.16)$$

with χ_{AIM} given by AIM formulation (3.252) and χ_{MOL} given by the molecular orbital compact finite formulation (3.375);

- and *chemical hardness*

$$A_{Hard} = \frac{\eta_{AIM}}{\eta_{MOL}} \qquad (4.17)$$

with η_{AIM} given by AIM formulation (3.248) and η_{MOL} given by the molecular orbital compact finite formulation (3.376);

They are based exclusively on the data of Table 4.4 with the AIM results listed in Table 4.5, the 5th, 9th, and 10th columns, respectively; the present discussion follows (Putz, 2010a).

For the post-bonding evaluations of the same indices, one must note the special case of polarizability that is computed upon the general Eq. (4.15c) – thus involving the molecular volume pre-computation. Here it is worth commenting on the fact that one can directly compute the molecular polarizability in various quantum schemes – however, with the deficiency that such procedure does not distinguishes among the stereo-isomers, *i.e.*, molecules VII (1-Naphthol) and VIII (2-Naphthalelon); IX (2-Naphthalenamine) and X (1-Naphthalenamine) in Table 4.5, since furnishing the same values, respectively; instead the same quantum scheme is able to distinguish between the volumes of two stereo-isomers making the Eq. (4.15c) as a more general approach. This way, the *molecular volumes* are reported in the 6th column of Table 4.5 as computed within the *ab initio* – Hartree Fock (HF) method; note that the HF method was chosen as the reference since it is at the "middle

TABLE 4.5 Atoms-in-Molecule (AIM) and Molecular (MOL) Structures, Volumes, and Polarizability Based-Aromaticities A_p of Eq. (4.15a), Employing the Atomic Values of Table 4.4 and the Ab-Initio (Hartree-Fock) Quantum Environment Computation (Hypercube, 2002); AIM Electronegativity and Chemical Hardness Are Reported (in electron-voles, eV) Employing the Eqs. (3.252) and (3.248), Respectively (Putz, 2010a).

Compound		Structure		Polarizability [Å]³				AIM-Indices	
						Molec			
Formula Name CAS Index (π e−)	AIM	Molecule	Conventional	P^{AIM}	Vol	P^{MOL}	A_P	χ^{AIM}	η^{AIM}
C_6H_6 Benzene 71–43–2 I (6)				7.17	328.11	19.58	0.37	6.68	5.63
C4H4N2 Pyrimidine 289–95–2 II (6)				5.40	306.46	12.19	0.44	6.73	5.93
C_5H_5N Pyridine 110–86–1 III (6)				6.29	320.75	12.76	0.49	6.70	5.76

TABLE 4.5 Continued

Compound			Structure			Polarizability [Å]^3				AIM-Indices	
							Molec				
Formula Name CAS Index (π e−)	AIM	Molecule	Conventional	P^{AIM}	Vol	p^{MOL}	A_P	χ^{AIM}	η^{AIM}		
C_6H_6O Phenol 108–95–2 IV (8)				7.37	356.91	10.65	0.69	6.74	5.66		
C_6H_7N Aniline 62–53–3 V (8)				8.15	371.73	11.09	0.73	6.73	5.79		
$C_{10}H_8$ Naphthalene 91–20–3 VI (10)				10.62	463.84	11.07	0.96	6.63	5.55		
$C_{10}H_8O$ 1-Naphthol 90–15–3 VII (12)				10.82	483.88	9.63	1.12	6.67	5.58		

TABLE 4.5 Continued

Compound	Structure			Polarizability [Å]^3				AIM-Indices	
Formula / Name / CAS / Index (π e−)	AIM	Molecule	Conventional	P^{IIM}	*Molec* Vol	P^{MOL}	A_P	χ^{AIM}	η^{AIM}
$C_{10}H_8O$ 2-Naphthalelon 135–19–3 VIII (12)				10.82	478.39	9.52	1.14	6.67	5.58
$C_{10}H_9N$ 2-Naphthalenamine 91–59–8 IX (12)				11.60	501.54	9.98	1.16	6.67	5.66
$C_{10}H_9N$ 1-Naphthalenamine 134–32–7 X (12)				11.60	496.11	9.87	1.18	6.67	5.66

computational distance" between the semi-empirical and density functional methods; it has only the correlation correction missing; however, even the density functional schemes, although encompassing in principle correlation along the exchange—introduces approximations on the last quantum effect. Therefore, the molecular polarizability is computed upon the Eq. (4.15c) in the 7th column of Table 4.5 with the associate polarizability compactness aromaticities displayed in the 8th column of Table 4.5.

The molecular energetic reactivity indices of electronegativity and chemical hardness are computed upon the Eqs. (3.375) and (3.376) in terms of HOMO and LUMO energies computed within the quantum semi-empirical and ab initio methods (see Volume I of the present five-volume work); their individual values as well as the resulted quantum compactness aromaticities, when combined with the AIM values of Table 4.5, in Eqs. (4.16) and (4.17), are systematically communicated in Tables 4.6 and 4.7, with adequate scaled representations in Figures 4.3 and 4.4, respectively. Note that neither the minimal basis set (STO-3G) nor the single point computation frameworks, although both motivated in the present context in which only the bonding and the post-bonding information should be capped in computation, does not affect the foregoing discussion by two main reasons: (i) they have been equally applied for all molecules considered in all quantum methods' combinations; and (ii) what is envisaged here is the aromaticity trend, i.e., the intra- and inter- scales comparisons rather than the most accurate values since no exact or experimental counterpart available for aromaticity.

Now, because of the observational quantum character of polarizability, one naturally assumes the (geometric) polarizability based aromaticity scale of Table 4.5 as that furnishing the actual standard ordering among the considered molecules in accordance with the rule associated with Eq. (4.15a); it features the following newly introduced rules along possible generalizations (Putz, 2010a):

- ***Aromal Rule:*** *the mono-benzenoid compounds have systematically higher aromaticity than those of double-ring benzenoids*; yet, this is the generalized version of the rule demanding that the benzene aromaticity is always higher than that of naphthalene, for instance; however, further generalization respecting the poly-ring benzenoids is anticipated albeit it should be systematically proved

TABLE 4.6 Frontier HOMO and LUMO Energies, the Molecular Electronegativity and Chemical Hardness of Eqs. (3.346) and (3.362), Along the Quantum Compactness Aromaticity A_{EL} and A_{Hard} Indices for Compounds of Table 4.5 as Computed with Eqs. (4.16) and (4.17) Within Semi-Empirical Quantum Chemical Methods; All Energetic Values in Electron-Volts (eV) (Putz, 2010a)

Compound Index	Property	CNDO	INDO	MINDO3	MNDO	AM1	PM3	ZINDO/1	ZINDO/S
I	E_{LUMO}	3.892207	4.451804	1.26534	0.3681966	0.514791	0.3440638	7.970686	0.7950159
	$-E_{HOMO}$	13.80296	13.24336	9.165875	9.391555	9.591248	9.652767	9.724428	8.927967
	χ	4.96	4.40	3.95	4.51	4.54	4.65	0.88	4.07
	η	8.85	8.85	5.22	4.88	5.05	4.998	8.85	4.86
	A_{EL}	1.35	1.52	1.69	1.48	1.472	1.435	7.62	1.64
	A_{Hard}	0.64	0.64	1.08	1.15	1.11	1.13	0.64	1.16
II	E_{LUMO}	2.709499	3.147036	0.951945	-0.3960558	-0.2959276	-0.6894529	6.422883	-0.3419995
	$-E_{HOMO}$	13.39755	11.86692	8.356924	10.36822	10.56194	10.62456	8.527512	9.67323
	χ	5.34	4.36	3.70	5.38	5.43	5.66	1.05	5.01
	η	8.05	7.51	4.65	4.99	5.13	4.968	7.48	4.67
	A_{EL}	1.26	1.54	1.82	1.25	1.24	1.19	6.40	1.34
	A_{Hard}	0.74	0.79	1.274	1.189	1.16	1.19	0.79	1.271
III	E_{LUMO}	3.051321	3.521359	1.011715	-0.02136767	0.1085682	-0.1944273	6.909242	0.01985455
	$-E_{HOMO}$	13.45145	12.06075	8.813591	9.692185	9.903634	10.0075	8.598721	9.040296
	χ	5.20	4.27	3.90	4.86	4.90	5.10	0.84	4.51
	η	8.25	7.79	4.91	4.84	5.01	4.91	7.75	4.53
	A_{EL}	1.29	1.57	1.718	1.38	1.37	1.31	7.93	1.49
	A_{Hard}	0.6980	0.74	1.17	1.191	1.15	1.17	0.74	1.272

TABLE 4.6 Continued

Index	Compound Property	CNDO	INDO	MINDO3	MNDO	AM1	PM3	ZINDO/1	ZINDO/S
IV	E_{LUMO}	3.718175	4.275294	1.085692	0.1763786	0.3450922	0.2196551	7.706827	0.6566099
	$-E_{HOMO}$	12.51092	11.71605	8.669437	9.022056	9.108171	9.169341	8.265366	8.557631
	χ	4.40	3.72	3.79	4.42	4.38	4.47	0.28	3.95
	η	8.11	8.00	4.88	4.60	4.73	4.69	7.99	4.61
	A_{EL}	**1.53**	**1.81**	**1.78**	**1.52**	**1.538**	**1.506**	**24.13**	**1.71**
	A_{Hard}	**0.6975**	**0.71**	**1.16**	**1.23**	**1.20**	**1.21**	**0.71**	**1.23**
V	E_{LUMO}	4.002921	4.61612	1.360785	0.5461559	0.7090454	0.5768315	8.106442	0.8517742
	$-E_{HOMO}$	11.22051	10.28413	7.783539	8.207099	8.186989	8.028173	6.803807	7.95583
	χ	3.61	2.83	3.21	3.83	3.74	3.73	-0.65	3.55
	η	7.61	7.45	4.57	4.38	4.45	4.30	7.46	4.40
	A_{EL}	**1.86**	**2.37**	**2.10**	**1.76**	**1.80**	**1.806**	**-10.33**	**1.89**
	A_{Hard}	**0.76**	**0.78**	**1.266**	**1.323**	**1.3017**	**1.35**	**0.78**	**1.31**
VI	E_{LUMO}	2.172528	2.757336	0.4589255	-0.3423392	-0.2855803	-0.4525464	6.197386	-0.04161556
	$-E_{HOMO}$	11.48051	10.89619	8.165956	8.544642	8.660414	8.746719	7.4545	7.835637
	χ	4.65	4.07	3.85	4.44	4.47	4.60	0.63	3.939
	η	6.83	6.83	4.31	4.10	4.19	4.15	6.83	3.90
	A_{EL}	**1.42**	**1.63**	**1.721**	**1.49**	**1.48**	**1.441**	**10.55**	**1.68**
	A_{Hard}	**0.81**	**0.81**	**1.29**	**1.35**	**1.33**	**1.34**	**0.81**	**1.42**
VII	E_{LUMO}	2.192621	2.79537	0.5106197	-0.3850094	-0.2975906	-0.4355633	6.210848	-0.06489899
	$-E_{HOMO}$	10.95387	10.26489	7.918682	8.376475	8.441528	8.514781	6.859143	7.681855
	χ	4.38	3.73	3.70	4.38	4.37	4.48	0.32	3.87
	η	6.57	6.53	4.21	4.00	4.07	4.04	6.54	3.81

TABLE 4.6 Continued

Index	Compound Property	CNDO	INDO	MINDO3	MNDO	AMI	PM3	ZINDO/1	ZINDO/S
VIII	A_{EL}	1.52	1.79	1.80	1.52	1.53	1.49	20.58	1.72
	A_{Hard}	0.85	0.85	1.32	1.40	1.37	1.38	0.85	1.47
	E_{LUMO}	2.534854	3.128462	0.5805296	−0.3075339	−0.2500397	−0.3581562	6.510067	0.08197734
	$-E_{HOMO}$	11.50232	10.80223	8.246805	8.747499	8.821697	8.887013	7.405916	7.956836
	χ	4.48	3.84	3.83	4.53	4.54	4.62	0.45	3.937
	η	7.02	6.97	4.41	4.22	4.29	4.26	6.96	4.02
	A_{EL}	1.49	1.74	1.74	1.47	1.471	1.443	14.89	1.69
	A_{Hard}	0.795	0.80	1.26	1.322	1.302	1.31	0.80	1.39
IX	E_{LUMO}	2.228869	2.844	0.5107521	−0.3597778	−0.2714103	−0.4318241	6.286494	−0.01188275
	$-E_{HOMO}$	10.94335	10.11959	7.783869	8.371226	8.367208	8.374782	6.666291	7.655258
	χ	4.36	3.64	3.64	4.37	4.32	4.40	0.19	3.83
	η	6.59	6.48	4.15	4.01	4.05	3.97	6.48	3.82
	A_{EL}	1.53	1.83	1.83	1.53	1.544	1.515	35.12	1.74
	A_{Hard}	0.86	0.87	1.36	1.41	1.40	1.43	0.87	1.48
X	E_{LUMO}	2.22685	2.840066	0.5115107	−0.3805175	−0.2805806	−0.4578161	6.326563	−0.00408988
	$-E_{HOMO}$	10.68815	9.865701	7.676749	8.272097	8.261106	8.2799	6.414326	7.540981
	χ	4.23	3.51	3.58	4.33	4.27	4.37	0.04	3.77
	η	6.46	6.35	4.09	3.95	3.99	3.91	6.37	3.77
	A_{EL}	1.58	1.90	1.86	1.54	1.56	1.53	152.0	1.77
	A_{Hard}	0.88	0.89	1.38	1.43	1.42	1.45	0.89	1.50

TABLE 4.7 The Same Quantities of Table 4.6 As Computed Within Various Ab Initio Approaches: By Density Functional Theory Without Exchange-Correlation (noEX-C), and With B3-LYP, B3-PW91, and Becke97 Exchange-Correlations, and by Hartree-Fock method, All With Minimal (STO-3G) Basis Sets (Putz, 2010a)

Compound		DFT					Hartree-Fock
Index	Property	noEX-C	B3-LYP	B3-PW91	EDF1	Becke97	
I	E_{LUMO}	15.69352	2.52946	2.398649	1.561805	2.512676	7.234344
	$-E_{HOMO}$	−8.870216	5.158205	5.338667	4.430191	5.165561	7.502962
	χ	−12.28	1.31	1.47	1.43	1.33	0.13
	η	3.41	3.84	3.87	3.00	3.84	7.37
	A_{EL}	−0.54	5.08	4.54	4.66	5.04	49.74
	A_{Hard}	1.65	1.46	1.46	1.88	1.47	0.76
II	E_{LUMO}	15.11303	0.9238634	0.7736028	−0.04114805	0.9030221	5.579984
	$-E_{HOMO}$	−13.04602	4.744987	4.883547	3.513406	4.728943	8.695125
	χ	−14.08	1.91	2.05	1.78	1.91	1.56
	η	1.03	2.83	2.83	1.74	2.82	7.14
	A_{EL}	−0.478	3.52	3.27	3.79	3.52	4.32
	A_{Hard}	5.74	2.09	2.096	3.42	2.106	0.83
III	E_{LUMO}	15.34953	1.622094	1.477663	0.6587179	1.60312	6.284506
	$-E_{HOMO}$	−12.73475	4.751619	4.893573	3.484843	4.739381	7.943096
	χ	−14.04	1.56	1.71	1.41	1.57	0.83
	η	1.31	3.19	3.186	2.07	3.1713	7.11
	A_{EL}	−0.477	4.28	3.92	4.74	4.27	8.08
	A_{Hard}	4.41	1.81	1.81	2.78	1.82	0.81

TABLE 4.7 Continued

Index	Property	noEX-C	B3-LYP	B3-PW91	EDF1	Becke97	Hartree-Fock
				DFT			
IV	E_{LUMO}	16.5171	2.596515	2.475588	1.716375	2.584044	7.102361
	$-E_{HOMO}$	-12.45941	3.760901	3.909865	2.872823	3.758234	6.672404
	χ	-14.49	0.58	0.72	0.58	0.59	-0.21
	η	2.03	3.18	3.193	2.29	3.1711	6.89
	A_{EL}	**-0.465**	**11.58**	**9.40**	**11.66**	**11.48**	**-31.35**
	A_{Hard}	**2.79**	**1.78**	**1.77**	**2.47**	**1.78**	**0.82**
V	E_{LUMO}	16.5102	2.963498	2.848121	2.077958	2.949314	7.449772
	$-E_{HOMO}$	-12.06327	3.094653	3.234635	2.291472	3.087551	5.765693
	χ	-14.29	0.07	0.19	0.11	0.07	-0.84
	η	2.22	3.03	3.04	2.18	3.02	6.61
	A_{EL}	**-0.471**	**102.63**	**34.82**	**63.04**	**97.37**	**-7.99**
	A_{Hard}	**2.60**	**1.91**	**1.90**	**2.65**	**1.92**	**0.88**
VI	E_{LUMO}	14.5038	1.290144	1.146572	0.4413206	1.267581	5.544161
	$-E_{HOMO}$	-10.0267	4.156837	4.331527	3.541704	4.159986	6.084805
	χ	-12.27	1.43	1.59	1.55	1.45	0.27
	η	2.24	2.72	2.74	1.99	2.71	5.81
	A_{EL}	**-0.54**	**4.63**	**4.16**	**4.28**	**4.58**	**24.53**
	A_{Hard}	**2.48**	**2.04**	**2.03**	**2.79**	**2.05**	**0.95**
VII	E_{LUMO}	15.40361	1.534507	1.39925	0.7539564	1.51676	5.631796
	$-E_{HOMO}$	-12.32641	3.422596	3.578508	2.691508	3.420128	5.689867
	χ	-13.87	0.94	1.09	0.97	0.95	0.03
	η	1.54	2.48	2.49	1.72	2.47	5.66

TABLE 4.7 Continued

| Index | Property | noEX-C | DFT | | | | Hartree-Fock |
			B3-LYP	B3-PW91	EDF1	Becke97	
	A_{EL}	−0.481	7.07	6.12	6.88	7.01	229.72
	A_{Hard}	3.63	2.25	2.24	3.24	2.26	0.99
VIII	E_{LUMO}	15.48911	1.614079	1.472582	0.8028092	1.593253	5.815819
	$-E_{HOMO}$	−12.3533	3.699537	3.860561	2.910033	3.698387	6.201466
	χ	−13.92	1.04	1.19	1.05	1.05	0.19
	η	1.57	2.66	2.67	1.86	2.65	6.01
	A_{EL}	−0.479	6.40	5.59	6.33	6.34	34.59
	A_{Hard}	3.56	2.10	2.093	3.01	2.109	0.93
IX	E_{LUMO}	15.08743	1.524559	1.389197	0.7397588	1.502934	5.626748
	$-E_{HOMO}$	−11.85368	3.370883	3.52209	2.624703	3.369097	5.680253
	χ	−13.47	0.92	1.07	0.94	0.93	0.03
	η	1.62	2.45	2.46	1.68	2.44	5.65
	A_{EL}	−0.495	7.23	6.25	7.08	7.15	249.32
	A_{Hard}	3.50	2.31	2.30	3.36	2.32	1.00
X	E_{LUMO}	15.12512	1.53895	1.404091	0.7564005	1.518818	5.640772
	$-E_{HOMO}$	−11.90494	3.264767	3.41559	2.541048	3.262239	5.520739
	χ	−13.52	0.86	1.01	0.89	0.87	−0.06
	η	1.61	2.40	2.41	1.65	2.39	5.58
	A_{EL}	−0.494	7.73	6.63	7.47	7.65	−111.14
	A_{Hard}	3.52	2.36	2.35	3.43	2.37	1.01

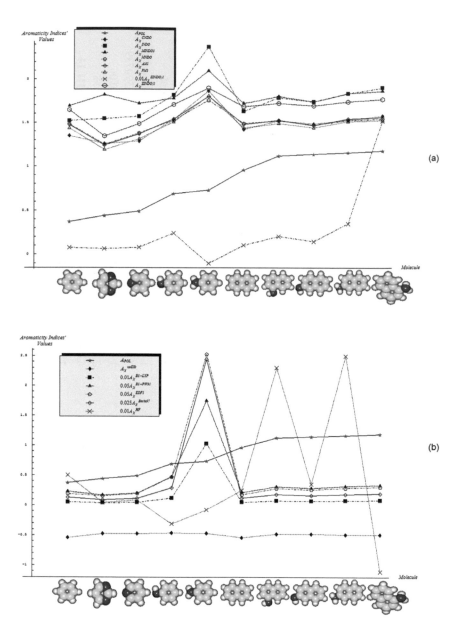

FIGURE 4.3 Electronegativity-based aromaticity scales of Tables 4.6 and 4.7 computed within semi-classical schemes in (a) and within ab initio schemes in (b), as compared with the polarizability-based aromaticity scale of Table 4.5, respectively (Putz, 2010a).

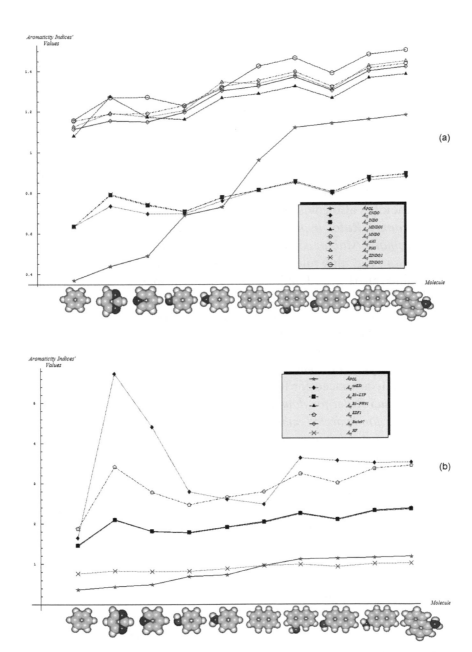

FIGURE 4.4 The chemical hardness-based aromaticity scales of Tables 4.6 and 4.7 computed within semi-classical schemes in (a) and within ab initio schemes in (b), respectively (Putz, 2010a).

by appropriate computations;

- **Aroma2 Rule:** *C-replaced benzenoids are more aromatic than substituted benzenoids*, e.g., Pyridine and Pyrimidine vs. Phenol and Aniline ordering aromaticity in Table 4.5; this rule extends the substituted *versus* addition rules in aromaticity historical definition (see Introduction);

- **Aroma3 Rule:** *double-C-replaced annulens have greater aromaticity than mono-C-replaced annulenes*, e.g., $A_{Pyrimidine} > A_{Pyridine}$; this is a sort of continuation of the previous rule in the sense that as more Carbons are replaced in aromatic rings, higher aromaticity is provided; further generalization to poly-replacements to poly-ring benzenoids is also envisaged;

- **Aroma4 Rule:** *hydroxyl-substitution to annulene produces more aromatic (stable) compounds than the correspondent amine-substitution*; e.g., this rule is fulfilled by mono-benzenoids and is maintained also by the double-benzene-rings no matter the stereoisomers considered; due to the fact the π electrons provided by Oxygen in hydroxyl-group substituted to annulene ring is greater than those released by Nitrogen in annulene ring by the amine-group substitution this rule is formally justified, while the generalization for hydroxyl- *versus* amine- substitution to poly-ring annulens may be equally advanced for further computational confirmation;

- **Aroma5 Rule:** *for double ring annulens the α position is more aromatic for hydroxyl-substitution while β position is more aromatic for amine-substitution than their β and α counterparts, respectively*; this rule may be justified in the light of the Aroma4 Rule above employing the inverse role the Oxygen and Nitrogen plays in furnished (π+free pair) electrons to annulens rings: while for Oxygen the higher atomic charge may be positioned closer to the common bond between annulens' rings- thus favoring the alpha position, the lesser Nitrogen atomic charge should be located as much belonging to one annulene ring only- thus favoring the beta position; such inversion behavior is justified by the existing of free electrons on the NH_2- group that as closely are to the benzenic ring as much favors its stability against further electrophilic attack- as is the case of beta position of 2-Naphtalenamine in Table 4.5; extensions to the poly-ring annulens may be also investigated.

Under the reserve that these rules and their generalizations should be verified by extra studies upon a larger set of benzenoid aromatics, we will adopt them here in order to analyze their fulfillment with the energetically-based aromaticity scales of electronegativity and chemical hardness, reported in Tables 4.6 and 4.7 and drawn in Figures 4.3 and 4.4; actually, their behavior is analyzed against the aromaticity ordering rules given by Eqs. (4.16) and (4.17), *i.e.*, as being anti-parallel and parallel with the polarizability-based aromaticity trend of Eq. (4.15a), with the results systematized in Tables 4.8 and 4.9, respectively. From Table 4.8, there follows that electronegativity based-aromaticity displays the following properties respecting the aromaticity rules derived from polarizability framework (Putz, 2010a):

- No semi-empirical quantum method, in general, satisfies the first rule of aromaticity, Aroma1, in the sense that the trend in Figure 4.3a

TABLE 4.8   The Fulfillment ($\times$) of the Aromaticity (Aroma1–5) Rules Abstracted From Polarizability Based Scale in the Case of *Electronegativity* Based-Aromaticity Records of Tables 4.6 and 4.7 for the Molecules of Table 4.5 (Putz, 2010a)

Aromaticity Rules		**Aroma1**	**Aroma2**	**Aroma3**	**Aroma4**	**Aroma5**
Quantum Methods						
Semi-empirical	*CNDO*	–	–	–	–	–
	INDO	–	–	–	–	–
	MINDO3	–	–	$\times$	–	–
	MNDO	–	–	–	–	–
	AM1	–	–	–	–	–
	PM3	–	–	–	–	–
	ZINDO/1	–	–	–	–	–
	ZINDO/S	–	–	–	–	–
Ab initio	*noEXc*	$\times$	$\times$	–	$\times$	–
	B3-LYP	$\times$	–	–	–	–
	B3-PW91	$\times$	–	–	–	–
	EDF1	$\times$	–	–	–	–
	Becke97	$\times$	–	–	–	–
	Hartree-Fock	–	$\times$	–	–	$\times$

TABLE 4.9 The Same Check for the Present Aromaticity Rules As in Table 4.8 – Yet Here for the *Chemical Hardness* Based-Aromaticity Scale (Putz, 2010a)

Aromaticity Rules		Aroma1	Aroma2	Aroma3	Aroma4	Aroma5
Quantum Methods						
Semi-empirical	*CNDO*	×	–	–	×	–
	INDO	×	–	–	×	–
	MINDO3	×	–	–	×	–
	MNDO	×	–	×	×	–
	AM1	×	×	–	×	–
	PM3	×	×	–	×	–
	ZINDO/1	×	–	–	×	–
	ZINDO/S	×	–	×	×	–
Ab initio	*noEXc*	–	–	×	–	–
	B3-LYP	×	–	×	×	–
	B3-PW91	×	–	×	×	–
	EDF1	–	–	×	×	–
	Becke97	×	–	×	×	–
	Hartree-Fock	×	×	×	×	–

(and Table 4.6) displays rather growing aromaticity character from mono- to double-benzenoid rings; the same behavior is common also to HF computational environment, perhaps due to the close relationships with approximations made in semi-empirical approaches; instead, all other ab initio methods considered, including that without exchange and correlation terms, see Volume I of the present five-volume work, do fulfill the Aroma1 rule;

- The remaining aromaticity Aroma2–5 rules are generally not adapted with any of the semi-empirical methods, except the MINDO3 (the most advanced and accurate method from the NDO approximations) fulfilling the Aroma3 rule regarding the ordering of mono-*versus* bi- CH- correlation environments Nitrogen on benzenoid ring. Interestingly, the Aroma3 rule is then not satisfied by any of the ab initio quantum methods;

- Aroma2 rule about the comparison between the CH-replacement group and the H-substitution to the mono ring benzene seems being in accordance only with HF and ab initio without exchange-correlation environments leading with the idea the electronegativity based- aromaticity of substitution and replacement groups is not so sensitive to the spin and correlation effects, being of primarily Columbic nature;

- Hydroxyl- *versus* amine- substitution aromaticity appears that is not influenced by spin and correlation in electronegativity based- ordering aromaticity since only the no-exchange and correlation computational algorithm agrees with Aroma4 rule;

- α- *versus* β- stereoisomeric position influence in aromaticity ordering is respected only by the HF scheme of computation and by no other combination, either semi-empirical or ab initio.

Overall, it seems electronegativity may be used in modeling compactness of atoms-in-molecules aromaticity – basically without counting on the exchange or correlation effects, or at best within the HF algorithm, while semi-empirical methods seems not well adequate. Yet, for all aromaticity rules formulated, there exists at least one quantum computational environment for which the electronegativity based compactness aromaticity is in agree with each of them.

The situation changes significantly when chemical hardness is considered for compactness aromaticity computation; the specific behavior is abstracted from the analysis of Table 4.9 and can be summarized as follows (Putz, 2010a):

- Semi-empirical methods are equally appropriate in producing agreement with Aroma1 and Aroma4 rules in what concerns the aromaticity behavior for the mono- *versus* bi- ring annulens and hydroxyl- *versus* amine- substitution to either of them, respectively;

- Aroma2 and Aroma3 rules are slightly better fulfilled by the semi-empirical than the ab initio quantum frameworks in modeling the aromaticity performance of the mono- *versus* bi- CH- replaced groups and both of them against the H- substituted on benzenic rings, respectively;

- The stereoisomeric effects comprised by the Aroma5 rule is not modeled by the chemical hardness compactness aromaticity by any of its computed scales, neither semi-empirical or ab initio.

Overall, when the chemical hardness agrees with one of the above enounced Aroma Rules it does that within more than one computational scheme; however, the best agreement of chemical hardness with polarizability based- aromaticity scales is for the mono- *versus* bi- (and possible poly-) benzenic rings decreasing of aromaticity orderings, along the manifestly hydroxyl- superior effects in aromaticity than amine- groups substitution within most of the computational quantum schemes, *i.e.*, valid either for semi-empirical and ab initio methods. The stereoisomerism is not covered by chemical hardness modeling aromaticity, and along the electronegativity limited coverage within HF scheme in Table 4.8, there follows that the energetic reactive indices are not able to prevail over the geometric indices as polarizability or to predict stereoisomerism ordering in aromaticity modeling compactness schemes.

Finally, few words about the output of the various quantum computational schemes respecting the current aromaticity definition given by Eq. (4.14) are worth addressing. As such, one finds that (Putz, 2010a):

- With CNDO and INDO methods, the electronegativity based- aromaticity is more oriented towards the AIM limit of Figure 4.2, while chemical hardness based- aromaticity merely models the MOL limit of chemical bonding, see Table 4.3. This agrees with the basic principles of chemical reactivity according to which electronegativity drives the atomic encountering in forming the transition state towards chemical bond, while chemical hardness refines the bond by the aid of maximum hardness principle (Parr & Chattaraj, 1991; Putz, 2012b);
- The MINDO3, MNDO, AM1, PM3, and ZINDO/S all display in Table 4.6 the exclusively AIM limit in assessing aromaticity in bonding, yet with electronegativity based values systematically higher than those based on chemical hardness – this way respecting in some degree the empirical rule stating that the electronegativity stands as the first order effect in reactivity, while the chemical hardness corrects in the second order the bonding stability, according with the basic differential definitions of Eqs. (3.346) and (3.362), respectively;
- ZINDO/1 differs both from ZINDO/S and by the rest of semi-empirical methods of the last group, while giving qualitative results in the same manner as CNDO and INDO, in the sense of higher

absolute (positively defined) electronegativity based-respecting the chemical hardness based-aromaticities, yet with significant quantitative values over unity (*i.e.*, the transition state as the instable equilibrium between AIM and MOL limits), see Table 4.6. This means that the transitional elements' orbitals inclusion without further refinements of ZINDO/S exacerbates the Columbic atoms-in-molecule effects, *i.e.*, the stability (aromaticity) of bonding is mostly to be acquired in the pre-bonding stage of the AIM limit;

- Somehow with the same qualitative-quantitative behavior as ZINDO/1 is the HF computed aromaticities indices of Table 4.7; however, the negative values as well as exceeding the AIM unity limit of electronegativity based-aromaticities appear now as multiple-recordings, while the resulted chemical hardness aromaticity is the closest respecting the unity limit of transition state prescribed by Eq. (4.14). Together, this information shows that the HF computational framework merely models the pre-bonding AIM and the post-bonding MOL stages by electronegativity and chemical hardness reactivity indices, respectively;

- The reverse case to HF computing stands the no-exchange-and-correlation (noEX-C) values in Table 4.7, according to which the electronegativity based aromaticity, beside the negative values, are all in sub-unity range, so being associated with post-bonding MOL limit. This corroborates the situation with the supra-unitary recordings of chemical hardness based-aromaticity outputs, specific to pre-bonding AIM, the resulted reactivity picture is completely reversed respecting that accustomed for electronegativity and chemical hardness reactivity principles (Putz, 2008b). Therefore, it is compulsory to consider at least the electronic spin through exchange contributions (as in the HF case), not only conceptually, but also computationally for achieving a consistent picture of reactivity, not only of the aromaticity;

- The last situation is restored by using the hybrid functionals of DFT, *i.e.*, B3-LYP, B3-PW91, EDF1, and Becke97 in Table 4.7, with the help of which electronegativity based-aromaticity regains its supremacy over that computed with the chemical hardness AIM and MOL limits in bonding. Although, no explicit sub-unity MOL limit of Eq. (4.14) is obtained with chemical hardness aromaticity computation, the recorded values are enough close to unity, while those based on electronegativity are more than twice further away from unity, to can say that the reactivity principles are fairly respected within these quantum methods, *i.e.*, when A_χ and A_η are situated in the AIM and MOL limiting sides of chemical bonding, respectively.

The bottom line is rising by the wish to globally combine the ideas of quantum chemical methods used in chemical bonding, reactivity principles, and aromaticity results; upon the above discussions it follows that MINDO3, AM1 (or PM3) – for semi-empirical along Becke hybrid functionals and Hartree-Fock – for ab initio are the suited methods that fulfills most of the reactivity and the present introduced aromaticity bonding rules. However, the best of them overall seems to remain the consecrated HF scheme, since acquiring the highest number of grades summated throughout Tables 4.8 and 4.9. As such, a new challenge appears since the present results recommend that correlation does not count too much in aromaticity or reactivity modeling. Nevertheless, further studies with larger set of molecules and types of aromatics should be address for testing whether or not the advanced aromaticity (Aroma1-5) rules are preserved or in which degree they may be generalized or modified such that being in accordance with the principles of chemical bonding and reactivity (Putz, 2010a).

However, since at the end the aromaticity appears as describing the stability character of molecular sample its connection with a reactivity index seems natural, although systematically ignored so far. In this respect, the present work focuses on how the electronegativity and chemical hardness based- aromaticity scales are behaving respecting other constructed on a direct observable quantum quantity – the polarizability in this case. This because the polarizability quantity is fundamental in quantum mechanics and usually associated with the second order Stark effect that can be computed within the perturbation theory, see the Volume II of the present five-volumes (Putz, 2016); then, the two ways of seeing a molecular structure were employed in introducing the actual absolute aromaticity definition (Putz, 2010a)

(i) *the molecule viewed as composed by the constituting atoms (AIM)*; and

(ii) *the molecule viewed from its spectra of molecular orbitals (MOL)*.

The two molecular perspectives may be associated with the pre- and post- bonding stages of chemical bond at equilibrium; therefore, the conceptual and computational competition between these two molecular facets should measure the stability or its contrary effect – the reactivity propensity – being therefore the ideal ingredients for an absolute definition of aromaticity. Note that although an AIM-to-MOL *difference definition* of absolute aromaticity will be presented in the next section (Putz, 2010b), the actual study of their *ratio definition* should account for a sort of compactness degree of molecular structure – as described by the specific molecular property used.

Compressively, for a molecular property becoming a candidate for absolute (here with its compactness variant of) aromaticity it has to fulfill two basic conditions (Putz, 2010a)

(i)   *having a viable quantum definition* (since the quantum nature of electrons and nucleus are assumed as responsible for molecular stability/reactivity/aromaticity); and

(ii) *having a reality at both the atomic and molecular levels.*

In this respect all the presently considered reactivity indices, *i.e.*, polarizability, electronegativity, and chemical hardness, have equally consecrated quantum definitions as well as atomic and molecular representations (Putz, 2008b, 2012b).

For the atomic level, the experimental values based on the ionization potential and electron affinity definitions for electronegativity and chemical hardness were considered, see Eqs. (3.346) and (3.362), respectively. Nevertheless, the AIM level was formed by appropriate averaging of atoms-in-molecule summation for each of the considered reactivity indices, see Eqs. (4.15b), (3.252) and (3.248), and along of their MOL counterparts of Eqs. (4.15c), (3.346), and (3.362) the polarizability-, electronegativity- and chemical hardness- based aromaticity definitions were formulated with the associate qualitative trends established by Eqs. (4.15a), (4.16), and (4.17), respectively. Yet, for MOL level of computations all major quantum chemical methods for orbital spectra computation were considered, and implemented in the current application for some basics. Because of the quantum observable character of polarizability the related aromaticity scale was considered as benchmark for actual study and it offered the possibility in formulating of actual five rules for aromaticity. These rules are then checked for electronegativity and chemical hardness derived – aromaticity scales with the synopsis of the results in Tables 4.8 and 4.9. It followed that chemical hardness, although generally in better agreement with these rules for most of the quantum chemical methods considered for its MOL computation, may not be considered infallible against aromaticity, at least for the reason it does not fulfills at all with the Aroms5 rule above. Surprisingly, chemical hardness index is more suited in modeling aromaticity when considered within semi-empirical computational framework, while the electronegativity responds better in conjunction with ab initio methods (Putz, 2010a).

From quantum computational perspective, the consecrated Hartree-Fock method seems to get more marks in fulfillment of above Aroma1-to-5 rules, cumulated for electronegativity and chemical hardness based- aromaticity scales; it leads with the important idea the correlation effects are not determinant in aromaticity phenomenology, an idea confirmed also by the fact the density functional without exchange and correlation produces not-negligible fits with Aroma1, 2, and 4 rules in electronegativity framework (Putz, 2010a).

Overall, few basic ideas in computing aromaticity should be emphasized (Putz, 2010a)

(i)   *there is preferable computing aromaticity in an absolute manner, i.e.,* for each molecule based on its pre- and post- bonding properties (as is the present compactness definition, for instance) without involving other referential molecule, as is often case in the fashioned aromaticity scales;

(ii) *the comparison between various aromaticity absolute scales is to be done respecting that one based on a structural or reactivity index with attested observational character* (as is the present polarizability based- aromaticity);

(iii) *the rules derived from the absolute aromaticity scale based on observable quantum index should be considered for further guidance* for the rest of aromaticity scales considered;

(iv) *the aromaticity concept, although currently associated with stability character of molecules, seems to not depending on correlation and sometimes neither by exchange effects.*

4.4   MODELING ABSOLUTE AROMATICITY OF ATOMS-IN-MOLECULES

4.4.1   DEFINING ABSOLUTE AROMATICITY BY CHEMICAL REACTIVITY

Having clarified the atoms-in-molecule and (frontier) molecular orbital-based methods for computing the molecular electronegativity and chemical hardness, see the Sections 3.7 of the present volume, one can infer that they represent two different stages in bonding. This way, while the superposition of atoms in molecule belongs to the bond forming stage, being mostly dominated by their reciprocal attraction – driven by electronegativity equalization principle of atoms in molecule (Mortier et al., 1985; Sanderson, 1988), the reactivity based on the frontier orbitals, i.e., HOMO and LUMO states in different orders, merely addresses the post-bonding molecular stability that is governed by the maximum hardness principle (Parr & Chattaraj, 1991; Putz, 2012b).

Now, it seems natural to compare the two stages of a molecular bonding, for a given reactivity index, which through their difference should reveal the excess chemical information responsible for the stability of that molecular system. In other words, by subtracting the already formed molecular orbital (MO) information Π_{MO} from that obtained by superposition of atomic information in bonding Π_{AIM}; the present discussion follows (Putz, 2010b)

$$A = \Pi_{AIM} - \Pi_{MO} \qquad (4.18)$$

one yields the chemical information that characterizes the stability of the chemical bond itself – what is presently attributed to the absolute aromaticity. The *absolute* adjective comes from the fact the aromaticity degree is obtained by employing different presumably equivalent information regarding *the same* molecule, see Figure 4.5 for a heuristic representation, and not by using two molecular systems – one tested and one referential as the custom definitions for the *relative* reactivity scales of aromaticity do (Chattaraj et al., 2007).

When the electronegativity index is considered in Eq. (4.18) the electronegativity-based absolute aromaticity is obtained with the working form

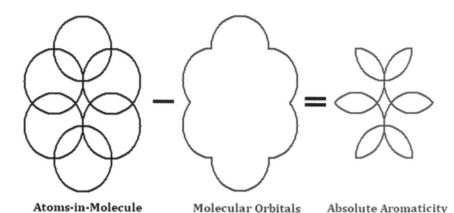

Atoms-in-Molecule   **Molecular Orbitals**   **Absolute Aromaticity**

FIGURE 4.5 Heuristic representation of the concept of absolute aromaticity (for the benzene pattern) as the stabilization difference of a given index of reactivity between atoms-in-molecule and molecular orbitals bonding configurations (Putz, 2010b).

$$A = \chi_{AIM} - \chi_{CFD} \qquad (4.19)$$

that in turn, can be further specialized when the AIM formulation of Eq. (3.252) is combined with different compact finite differences schemes encompassed by the Eq. (3.375), with the parameters given in Table 3.8. Regarding the aromaticity scale tendency, it will be established by appealing the electronegativity reactivity behavior. For instance, higher AIM electronegativity χ_{AIM} higher bonding propensity through the electronic flowing between atoms-in-molecule according to the electronegativity equalization principle; on the other hand, the formed molecule is as stable as posing the lower orbital molecular electronegativity χ_{CFD} that restricts its engagement in further electrophilic reactions. The result is that *higher A^χ is associated with higher stability and aromaticity of the envisaged system.*

Instead, when chemical hardness based absolute aromaticity is particularized out of the general definition (4.18) with the help of AIM and CFD Eqs. (3.248) and (3.376), respectively

$$A^\eta = \eta_{AIM} - \eta_{CFD} \qquad (4.20)$$

the aromaticity scale is reversed following the specific (previously presented) chemical hardness reactivity principle. That is, the formed molecular bond (orbitals) – once optimized – associate with maximum hardness η_{CFD} (otherwise the binding process will continue until the maximum hardness stabilization of the final configuration will be reached anyway); on the contrary, the atoms-in-molecule state should be characterized by non-maximum chemical hardness value η_{AIM} not to impede the bond formation towards its stabilized stereo-chemical configuration. All in all, it is clear that *as A^η takes lower values, it indicates more stable and aromatic systems* (Putz, 2010b).

The introduced electronegativity and chemical hardness-based absolute aromaticity formulations (4.19) and (4.20) and ordering based on their chemical reactivity principles are to be in tested next for paradigmatic molecules and other aromaticity criteria.

4.4.2 ABSOLUTE AROMATICITY OF NON-CONGENERIC HYDROCARBONS

Although aromaticity has neither reached a definitive formulation nor a physical-chemical criterion of orderability, the important historical contributions have established so far the main directions a suitable scale has to be tested, namely the energetic, geometric (including topological), magnetic, and reactivity criteria; the present discussion follows (Putz, 2010b).

The reactivity criteria were previously formulated when minimum electronegativity and maximum chemical hardness features for molecular stabilization against electrophilicity were considered in assessing the maximum and minimum hierarchical tendencies for the associated absolute aromaticities, Eqs. (4.19) and (4.20), respectively. In fact, it is this criterion that needs to be validated in comparison with the consecrated ones, described below.

The geometric (and some topological) criteria predict a more aromatic compound the higher the associate index provided. This is the case of HOMA (Kruszewski & Krygowski, 1972; Krygowski, 1993) and TOPAZ (Tarko, 2008) indices that for increased values assess the compound with more aromatic conjugated ring or molecular fragment/zone, paralleling the more delocalized π-electrons in question. Instead, the reverse hierarchy is assumed for the recent topological index of reactivity (TIR) that, since related with the molecular site where the maximum probability (entropy) in electrophilic substitution (i.e., destabilization of the aromatic ring or system) takes place, recommending lower values for higher aromaticity (Ciesielski et al., 2009a).

The energetic criterion of aromaticity is a classical one originating in the works of Pauling & Wheland (1933) and Wheland (1944), and is based on the resonance energy (RE) concept that was somehow twisted towards the difference between the π-electronic delocalization respecting a reference π-system without delocalization (Truhlar, 2007); when reported per concerned π-electrons (PE) yields the REPE index (Hess & Schaad, 1971) of whose magnitude estimates the stabilization energy and therefore the aromaticity degree (Schaad & Hess, 2001; Cyrański, 2005); in short: the higher REPE, the higher aromaticity is predicted. Closely related with the resonance energy is the thermodynamic stabilization criterion of energy: the more stable a structure is, the lower its heat (enthalpy) of formation $\Delta_f H^0$. Quantitatively, the molecular heat of formation may be evaluated by the atoms-in-molecule equation

$$\Delta_f H^0_{\text{molecule}} = E_{\text{binding}} - \sum \Delta_f H^0_{\text{atom}} \qquad (4.21)$$

by subtracting atomic heats of formation from the molecule's binding energy; In this regard, the heat of formation definition (4.21) and criterion parallel the chemical hardness-based absolute aromaticity index (4.20) phenomenology. It is therefore a useful measure for aromaticity and will be investigated next to complete the chemical reactivity criteria. However, the practical quantum computations of these quantities is made by the semi-empirical methods based on neglecting of the differential (diatomic) overlapping (NDO) (Putz, 2008a), especially the modified intermediate NDO (MINDO) (Dewar & Thiel, 1977), MNDO/3 (Thiel & Voityuk, 1996), Austin Model (AM1) (Dewar et al., 1985, 1993; Rocha et al., 2006), and parameterized model no. 3 (PM3) (Stewart, 1989; Freire et al., 2006) algorithms that are parameterized by fitting to experimentally determined heats (enthalpies) of formation for a set of molecules at 298 K. Yet, since for most organic molecules, AM1 reports heats of formation accurate to within a few kilocalories per mol respecting those experimentally measured (Roux, 2008) it will be assumed as the present computational framework in furnishing the heats of formation for the actual working set of molecules.

The magnetic property belongs to the physical interpretation of aromaticity and is modulated by two popular indices. Once is the nucleus-independent chemical shift (NICS) index (Schleyer et al., 1996; Chen et al., 2005; Feixas et al., 2008; Ciesielski et al., 2009b) that correlates the higher π-delocalization with the increasing of magnetization (vector) in the center of the aromatic ring (or at other concerned point of the system); consequently, the larger magnetization, the larger aromaticity will be. Instead, the exaltation of magnetic susceptibility (Dauben et al., 1968; Flygare, 1974), measuring the resistance to magnetization introduces the index (Λ) that takes lower values for higher stability and aromaticity, being usually reported as an extensive quantity, i.e., it is calculated per π-electron, as it is also the case with REPE index.

The HOMA, TOPAZ, TIR, REPE, $\Delta_f H^0$ and Λ indices and their aromatic scales for a series of representative benzenoid hydrocarbons are presented in Table 4.10. In order to compare them with the actual electronegativity and chemical hardness-based absolute aromaticities the AIM electronegativity and chemical hardness values are first computed and reported in Table 4.10 based on Eqs. (3.252) and (3.248), respectively; then, they were combined with the CFD counterparts for all schemes from Table 3.8 applied on Eqs. (3.375) and (3.376) through employing the semi-empirical AM1 quantum mechanically calculation of the involved frontier orbitals and energies; the resulted absolute aromaticities are presented in Tables 4.11 and 4.12, respectively.

For a better visualization of the trends and particularities of the various absolute aromatic scales computed along the whole plethora of compact finite differences of electronegativity and chemical hardness, their linear correlations with the considered geometric, topologic, energetic, and magnetic aromatic scales are performed – with the correlation coefficients reported in Table 4.13: it furnishes very interesting information on compatible aromatic scales as well as on electronegativity and chemical hardness behavior against aromaticity.

It is obvious that electronegativity-based aromaticity A^χ_{CFD} poorly correlates with almost all traditional aromaticity scales and criteria, except with the magnetic susceptibility exaltation based aromaticity; the fascinating point here is that the best correlation of A_Λ with electronegativity absolute aromaticity parallels its poorest correlation with chemical hardness

TABLE 4.10 Aromaticity Values for Common Benzenoid Molecules by Means of the HOMA (Ciesielski et al., 2009), Topological Paths and Aromatic Zones – TOPAZ (Tarko & Putz, 2010), Topological Index Of Reactivity – TIR (Ciesielski et al., 2009), Resonance Energy Per π-Electron – REPE ($\times 10^3[\beta]$) (Hess & Schaad, 1971), the Heats (enthalpies) of Formations $\Delta_f H^0$ [kcal/mol] at 298 K Computed Within Semi-Empirical AM1 Method (Hypercube, 2002), and Exaltation Magnetic Susceptibility Λ [cgs-ppm] (Ciesielski et al., 2009) Methods, Along the Atoms-In-Molecule (AIM) Electronegativity and Chemical Hardness Values (in electron-Volts [eV]) Computed Upon Eqs. (3.252) and (3.248), Respectively, Based on the Constituting Atomic Values (χ_H=7.18eV; χ_C=6.24eV; η_H=6.45eV; η_C=4.99eV) (Putz, 2006, 2010b)

No.	Name	AIM	A_{HOMA}	A_{TOPAZ}	A_{TIR}	A_{REPE}	$\Delta_f H_0$	A_Λ	χ_{AIM}	η_{AIM}
1	Benzene	C_6H_6	0.991	999.2	0.000	65	21.867	14.5	6.677	5.627
2	Naphthalene	$C_{10}H_8$	0.811	616.2	0.252	55	40.346	29.6	6.626	5.548
3	Anthracene	$C_{14}H_{10}$	0.718	585.0	0.571	47	62.606	45.5	6.600	5.510
4	Phenanthrene	$C_{14}H_{10}$	0.742	520.3	0.318	55	57.128	41.4	6.600	5.510
5	Pyrene	$C_{16}H_{10}$	0.742	561.2	0.585	51	67.003	59.2	6.571	5.466
6	Naphthacene	$C_{18}H_{12}$	0.670	579.4	0.638	42	86.550	62.2	6.585	5.487
7	Benz[a]anthracene	$C_{18}H_{12}$	0.696	522.4	0.568	50	77.853	55.2	6.585	5.487
8	Chrysene	$C_{18}H_{12}$	0.709	468.8	0.466	53	75.816	55.5	6.585	5.487
9	Triphenylene	$C_{18}H_{12}$	0.691	474.9	0.338	56	75.100	49.3	6.585	5.487
10	Perylene	$C_{20}H_{12}$	0.656	556.3	0.636	48	88.884	42.8	6.562	5.453
11	Benzo[e]pyrene	$C_{20}H_{12}$	0.690	501.6	0.589	53	83.569	66.9	6.562	5.453
12	Benzo[a]pyrene	$C_{20}H_{12}$	0.700	517.9	0.702	49	87.094	72.2	6.562	5.453
13	Pentacene	$C_{22}H_{14}$	0.644	578.7	0.739	38	111.377	79.9	6.575	5.472
14	Benzo[a]naphthacene	$C_{22}H_{14}$	0.660	531.4	0.631	45	101.026	70.3	6.575	5.472
15	Dibenz[a,h]anthracene	$C_{22}H_{14}$	0.683	482.0	0.558	51	93.725	66.6	6.575	5.472
16	Benzo[b]chrysene	$C_{22}H_{14}$	0.680	474.7	0.590	49	97.216	70.5	6.575	5.472

TABLE 4.10 Continued

No.	Name	AIM	A_HOMA	A_TOPAZ	A_TTR	A_REPE	$\Delta_f H_0$	A_A	χ_AIM	η_AIM
17	Picene	$C_{22}H_{14}$	0.697	444.8	0.497	53	93.866	68.6	6.575	5.472
18	Benzo[ghi]perylene	$C_{22}H_{12}$	0.707	502.0	0.629	51	90.866	79.8	6.542	5.423
19	Anthanthrene	$C_{22}H_{12}$	0.691	572.1	0.712	45	98.970	89.3	6.542	5.423
20	Naphtho[2,1,8-qra]naphthacene	$C_{24}H_{14}$	0.665	519.0	0.707	45	109.144	85.7	6.556	5.444
21	Benzo[a]perylene	$C_{24}H_{14}$	0.639	531.6	0.761	45	116.986	62.3	6.556	5.444
22	Benzo[b]perylene	$C_{24}H_{14}$	0.642	514.0	0.656	49	106.444	56.0	6.556	5.444
23	Coronene	$C_{24}H_{12}$	0.742	471.8	0.557	53	95.751	123.9	6.525	5.397
24	Zethrene	$C_{24}H_{14}$	0.623	581.9	0.759	41	116.923	45.5	6.556	5.444
25	Benzo[a]pentacene	$C_{26}H_{16}$	0.638	476.7	0.716	42	125.431	86.7	6.568	5.461
26	Dibenzo[b,k]chrysene	$C_{26}H_{16}$	0.661	476.2	0.579	46	118.894	85.9	6.568	5.461
27	Naphtho[2,3-g]chrysene	$C_{26}H_{16}$	0.657	451.4	0.616	51	131.680	76.2	6.568	5.461
28	Naphtho[8,1,2-bcd]perylene	$C_{26}H_{14}$	0.661	526.2	0.702	47	115.652	71.9	6.540	5.419
29	Dibenzo[cd,lm]perylene	$C_{26}H_{14}$	0.690	502.8	0.720	48	115.774	105.1	6.540	5.419
30	Dibenzo[a,f]perylene	$C_{28}H_{16}$	0.630	533.3	0.837	43	176.147	81.1	6.552	5.438
31	Phenanthro[1,10,9,8-opqra]perylene	$C_{28}H_{14}$	0.600	565.1	0.879	42	134.430	58.2	6.525	5.397
32	Dibenzo[de,op]pentacene	$C_{28}H_{16}$	0.620	602.0	0.856	38	144.619	48.4	6.552	5.438
33	Dibenzo[a,l]pentacene	$C_{30}H_{18}$	0.637	511.5	0.699	44	139.686	94.5	6.562	5.453
34	Benzo[2,1-a:3,4,-a']dianthracene	$C_{30}H_{18}$	0.618	466.0	0.563	47	145.036	88.6	6.562	5.453
35	Naphtho[2,1,8-yza]hexacene	$C_{32}H_{18}$	0.636	532.5	0.790	40	157.696	116.5	6.549	5.433

TABLE 4.11 Electronegativity-Based Absolute Aromaticity Values of Eq. (4.19) by Means of the Combined Atoms-In Molecule Reactivity With the Various Compact Finite Differences Schemes in Table 3.8 for the Molecules in Table 4.10 Within AM1 Semi-Empirical Computational Framework (Hypercube, 2002) (*)

No.	A_{2C}^x	A_{4C}^x	A_{6C}^x	A_{SP}^x	A_{6T}^x	A_{8T}^x	A_{8P}^x	A_{10P}^x	A_{SLR}^x
1	2.127	2.107	2.097	5.097	3.777	4.077	4.157	4.177	3.947
2	2.136	2.156	2.156	5.196	3.846	4.146	4.236	4.256	4.006
3	2.120	2.140	2.150	5.190	3.840	4.150	4.220	4.240	3.990
4	2.090	2.090	2.090	5.120	3.780	4.080	4.170	4.180	3.940
5	2.061	2.061	2.061	5.091	3.751	4.051	4.141	4.151	3.911
6	2.075	2.085	2.095	5.135	3.785	4.095	4.175	4.185	3.935
7	2.095	2.125	2.125	5.225	3.855	4.165	4.265	4.295	4.055
8	2.065	2.075	2.075	5.115	3.765	4.065	4.145	4.165	3.915
9	2.025	2.035	2.035	5.105	3.745	4.045	4.135	4.155	3.905
10	2.062	2.082	2.092	5.172	3.802	4.112	4.192	4.202	3.952
11	2.022	2.032	2.032	5.082	3.722	4.032	4.112	4.132	3.892
12	2.042	2.062	2.062	5.112	3.762	4.062	4.142	4.162	3.912
13	2.075	2.075	2.075	5.105	3.765	4.065	4.155	4.165	3.925
14	2.065	2.065	2.075	5.125	3.765	4.075	4.155	4.165	3.915
15	2.045	2.055	2.055	5.095	3.745	4.055	4.135	4.145	3.895
16	2.055	2.065	2.065	5.115	3.755	4.065	4.145	4.165	3.905
17	2.035	2.045	2.045	5.095	3.745	4.045	4.125	4.145	3.895
18	2.002	2.002	2.002	5.062	3.702	4.012	4.092	4.112	3.862

TABLE 4.11 Continued

No.	A^x_{2C}	A^x_{4C}	A^x_{6C}	A^x_{SP}	A^x_{6T}	A^x_{8T}	A^x_{8P}	A^x_{10P}	A^x_{SLR}
19	2.012	2.032	2.042	5.102	3.742	4.052	4.132	4.142	3.892
20	2.036	2.046	2.046	5.096	3.746	4.046	4.126	4.146	3.896
21	2.046	2.046	2.046	5.026	3.706	4.006	4.076	4.096	3.846
22	2.036	2.046	2.046	5.096	3.746	4.056	4.136	4.156	3.906
23	1.945	1.835	1.775	4.595	3.355	3.635	3.745	3.785	3.585
24	2.056	2.076	2.076	5.136	3.776	4.086	4.166	4.176	3.926
25	2.048	2.068	2.068	5.148	3.778	4.088	4.178	4.208	3.958
26	2.038	2.038	2.038	5.038	3.698	4.008	4.078	4.088	3.828
27	2.008	2.008	2.008	5.058	3.698	4.008	4.078	4.098	3.848
28	2.000	2.020	2.020	5.080	3.720	4.030	4.110	4.130	3.880
29	2.000	2.020	2.020	5.090	3.730	4.040	4.110	4.130	3.880
30	2.032	2.042	2.052	5.112	3.752	4.062	4.132	4.142	3.892
31	1.995	2.005	2.015	5.075	3.715	4.025	4.105	4.115	3.865
32	2.052	2.062	2.062	5.102	3.752	4.062	4.142	4.162	3.912
33	2.022	2.032	2.052	5.112	3.742	4.052	4.122	4.132	3.862
34	2.072	2.082	2.082	5.112	3.772	4.072	4.152	4.172	3.922
35	2.009	2.009	2.009	5.049	3.699	4.009	4.089	4.109	3.859

*All values in electron-Volts [eV] (Putz, 2010b).

TABLE 4.12 Chemical Hardness-Based Absolute Aromaticity Values of Eq. (4.20) by Means of the Combined Atoms-In Molecule Reactivity With the Various Compact Finite Differences Schemes in Table 3.8 for the Molecules of Table 4.10 (*)

No.	A^η_{2C}	A^η_{4C}	A^η_{6C}	A^η_{SP}	A^η_{6T}	A^η_{8T}	A^η_{8P}	A^η_{10P}	A^η_{SLR}
1	0.527	-0.133	1.687	1.007	-0.623	1.947	1.907	1.517	-0.043
2	1.328	0.668	2.108	1.558	0.228	2.368	2.418	2.258	1.358
3	1.870	1.340	2.610	2.130	0.980	2.810	2.820	2.630	1.730
4	1.410	0.740	2.130	1.600	0.310	2.390	2.450	2.330	1.550
5	1.366	0.696	2.086	1.546	0.266	2.336	2.406	2.286	1.506
6	1.867	1.317	2.577	2.097	0.957	2.787	2.807	2.637	1.797
7	2.227	1.797	2.947	2.507	1.477	3.117	3.107	2.877	1.957
8	1.637	1.027	2.337	1.837	0.637	2.577	2.627	2.487	1.707
9	1.387	0.757	2.167	1.637	0.347	2.417	2.447	2.267	1.357
10	2.103	1.663	2.853	2.413	1.353	3.033	3.003	2.753	1.773
11	1.773	1.193	2.453	1.973	0.823	2.673	2.723	2.583	1.803
12	2.043	1.543	2.723	2.283	1.203	2.923	2.943	2.763	1.943
13	2.482	2.082	3.142	2.742	1.812	3.302	3.282	3.072	2.222
14	2.142	1.662	2.822	2.372	1.342	3.002	3.012	2.832	2.012
15	1.742	1.152	2.412	1.932	0.772	2.642	2.692	2.572	1.842
16	1.942	1.412	2.632	2.162	1.072	2.832	2.852	2.692	1.892
17	1.652	1.032	2.322	1.832	0.642	2.562	2.622	2.512	1.802
18	1.943	1.413	2.603	2.143	1.053	2.813	2.843	2.703	1.933
19	2.293	1.863	2.973	2.553	1.573	3.133	3.123	2.923	2.073

TABLE 4.12 Continued

No.	A^η_{2C}	A^η_{4C}	A^η_{6C}	A^η_{SP}	A^η_{6T}	A^η_{8T}	A^η_{8P}	A^η_{10P}	A^η_{SLR}
20	2.234	1.774	2.894	2.474	1.464	3.074	3.074	2.904	2.104
21	2.354	1.964	3.074	2.654	1.684	3.224	3.194	2.944	1.984
22	2.094	1.614	2.784	2.344	1.294	2.974	2.974	2.794	1.944
23	1.797	1.247	2.487	2.007	0.877	2.697	2.727	2.567	1.767
24	2.464	2.084	3.164	2.754	1.814	3.304	3.274	3.034	2.094
25	2.411	1.991	3.071	2.661	1.701	3.231	3.221	3.021	2.161
26	2.111	1.611	2.771	2.331	1.281	2.961	2.981	2.821	2.031
27	1.881	1.321	2.551	2.081	0.961	2.761	2.801	2.671	1.931
28	2.299	1.869	2.969	2.549	1.569	3.129	3.129	2.919	2.069
29	2.359	1.929	2.999	2.599	1.629	3.169	3.169	2.979	2.159
30	2.638	2.308	3.328	2.948	2.058	3.458	3.408	3.158	2.218
31	2.687	2.377	3.367	2.997	2.137	3.487	3.437	3.177	2.247
32	2.718	2.368	3.338	2.978	2.108	3.478	3.458	3.238	2.398
33	2.333	1.893	2.983	2.573	1.593	3.153	3.153	2.973	2.163
34	1.933	1.393	2.603	2.143	1.043	2.813	2.843	2.693	1.913
35	2.583	2.173	3.163	2.783	1.893	3.323	3.333	3.173	2.453

*All values in electron-Volts [eV] (Putz, 2010b).

TABLE 4.13 Correlation Coefficients for Linear Regressions of All Aromaticity Scales in Table 4.10 (As Dependent Variables) with Those in Tables 4.11 and 4.12 for Electronegativity and Chemical Hardness-Based Ones (As Independent Variables), Respectively (Putz, 2010b)

Y X		A_{HOMA}	A_{TOPAZ}	A_{TIR}	A_{REPE}	$\Delta_f H^0$	A_Λ
A^χ	A^χ_{2C}	0.441	0.503	0.520	0.215	0.548	0.726
	A^χ_{4C}	0.212	0.356	0.313	0.056	0.386	0.697
	A^χ_{6C}	0.119	0.312	0.224	0.019	0.301	0.659
	A^χ_{SP}	0.054	0.165	0.056	0.106	0.142	0.544
	A^χ_{6T}	0.035	0.242	0.139	0.054	0.225	0.608
	A^χ_{8T}	0.002	0.228	0.101	0.085	0.189	0.589
	A^χ_{8P}	0.030	0.234	0.135	0.061	0.235	0.606
	A^χ_{10P}	0.049	0.229	0.154	0.037	0.252	0.604
	A^χ_{SLR}	0.120	0.281	0.209	0.015	0.320	0.638
A^η	A^η_{2C}	0.846	0.383	0.941	0.904	0.829	0.489
	A^η_{4C}	0.823	0.333	0.933	0.897	0.823	0.465
	A^χ_{6C}	0.786	0.256	0.912	0.889	0.805	0.419
	A^η_{SP}	0.802	0.286	0.921	0.892	0.813	0.436
	A^η_{6T}	0.822	0.328	0.932	0.896	0.822	0.463
	A^η_{8T}	0.794	0.271	0.915	0.892	0.808	0.424
	A^η_{8P}	0.822	0.328	0.929	0.901	0.819	0.455
	A^η_{10P}	0.872	0.452	0.945	0.901	0.833	0.518
	A^η_{SLR}	0.909	0.663	0.910	0.846	0.798	0.604

absolute aromaticity – an observation that couples the magnetization phenomenon with the electronegativity action – not surprising since both rely on frontier electrons of the valence shells or orbitals. Moreover, this correlation with electronegativity-based absolute aromaticity is obtained within its simple Mulliken form of Eq. (3.346) – reaffirming this electronegativity scale as the most reliable for aromaticity modeling among all available CFD.

However, the representations in Figure 4.6 help us understand the reciprocal A_Λ and A^χ_{2C} aromaticity features; on the top side it is clear that, when represented on a common scale electronegativity-based aromaticity appears merely as an average of the A_Λ scale, leading to the

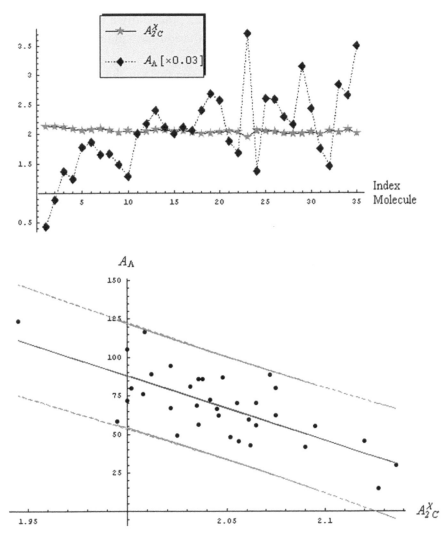

FIGURE 4.6 Upper panel: Comparative trends of the (rescaled) magnetic susceptibility exaltation-based aromaticity (A_Λ) scale with that based on 2C finite difference scheme of electronegativity (A^χ_{2C}) for the molecules' information of Tables 4.10 and 4.11, respectively. Bottom panel: The statistical fit (inner line) of the linear correlation $A_\Lambda = f(A^\chi_{2C})$, while emphasizing on the 95% confidence interval (within the extreme lines) for the aromaticity scales of the upper panel (Putz, 2010b).

idea that it best describes half of the total spin aromaticity; this is further confirmed by the bottom scatter plot of Figure 4.6 in which, by employing the correlation factor ($R=0.726$) to its statistical representability it turns out that $R^2 \times 100(\%) = 52.71\%$ (i.e., practically a half!) of the total variance of magnetic susceptibility exaltation is explained by its linear dependency with A^χ_{2C} scale.

On the other hand, it is obvious that even for A^χ_{2C} the earlier enounced higher value for higher aromaticity criterion is not respected at the level of benzene-naphthalene couple in Table 4.11, as it should be, as recommended by all other aromaticity scales in Table 4.10. Therefore, electronegativity does not seem the proper concept for treating the absolute aromaticity, maybe because involving the form (3.252) based on the somehow too drastic gauge transformation (3.343) between AIM electronegativity and chemical hardness, leaving with not sufficiently acceptable degree of correlation of the first with the available physicochemical aromaticity criteria.

The situation changes when chemical hardness-based absolute aromaticity is considered through combining the AIM chemical hardness with the molecular orbital CFD schemes in definition (4.20); it provides from the beginning the correct benzene-naphthalene ordering for all computed CFD scales in Table 4.12 as predicted by the aromaticity criteria for the scales in Table 4.10; general good correlations with geometric HOMA and energetic REPT and $\Delta_f H^0$ scales, excellent correlation with TIR index (with correlation factors over 0.9 for all compact finite difference schemes), while surprisingly poor correlation with TOPAZ aromaticity and anticipated poor correlation with A_Λ scale are revealed in Table 4.13.

While the poor correlations $A_\Lambda = f(A^\eta_{CFD})$ are explained since compensated by the superior companion electronegativity-based aromaticity correlations in Table 4.13, the poor correlations $A_{TOPAZ} = f(A^\eta_{CFD})$ may rely on the insufficient information that chemical hardness contains in order to be properly mapped into the generalized conjugated circuits that count in TOPAZ aromaticity algorithm (Tarko, 2008).

The proof for the improvement of this situation when other structural indices are added into correlation has been recently given by Tarko and Putz, showing that the best correlation was obtained either when higher orders of CFD electronegativity and chemical hardness schemes are considered together or when chemical hardness is accompanied by the index of maximum aromaticity of aromatic chemical bonds and by the total accessibility index weighted by atomic masses (Tarko & Putz, 2010).

Turning to the good correlations of the actual A^{η}_{CFD} scales, one may see that the highest order of CFD scheme, i.e., the spectral-like resolution chemical hardness-based aromaticity index A^{η}_{SLR} is in best agreement with HOMA aromaticity, while the simpler scheme, i.e., the electrophilicity-nucleophilicity chemical hardness gap (3.362), based aromaticity index A^{η}_{2C} is best correlating with REPE aromaticity. Both of these fits are motivated: the A_{HOMA} and A^{η}_{SLR} indices practically parallel geometrical molecular optimization with the most complex frontier orbitals' involvement – thus both accounting for the stereochemical control (Rzepa, 2007), while A_{REPE} and A^{η}_{2C} correlate well in the virtue of the fact that the resonance stabilization may be sufficiently modeled by the first order of the HOMO-LUMO gap.

It remains to comment upon the overall best correlation found between the TIR aromaticity and with that not based on the most complicated chemical hardness CFD-SLR scheme, but with that immediately before it, namely with the A^{η}_{10P} scale in Table 4.12. Note that the same absolute aromaticity scale A^{η}_{10P} is found as having the highest degree of correlation with heats of formation among all CFD-chemical hardness schemes of computation, although not with the highest correlation factor among all others aromaticity dependent indices in Table 4.13. For the $A_{TIR}(A^{\eta}_{10P})$ correlation the almost perfect parallel trend among all the molecules in Table 4.10 is emphasized on the top plot of Figure 4.7, while in its bottom representation the confidence interval of their scatter plot is shown. It is worth remarking that the fine agreement of A_{TIR} index with A^{η}_{10P} index in special and with A^{η}_{CFD} schemes in general originates in the fact that all these scales of aromaticity are computed in an absolute manner, i.e., restricting the information contained within the concerned molecule without appealing to any other reference molecular system or property.

Remarkably, aiming to systematize somehow the aromaticity criteria against the chemical hardness-based absolute aromaticity AIM-CFD scales one can establish from Table 4.13 that:

(i)   either the topological index of reactivity A_{TIR} and the heats of formation $\Delta_f H^0$ aromaticity scales are well described by the A^{η}_{10P} absolute aromaticity index, meaning that the experimental-based heats of formation themselves may be modeled by the topological characterization of the aromatics;

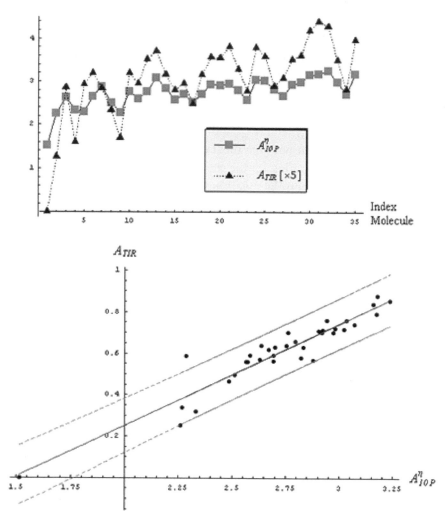

FIGURE 4.7 The same type of representations as those in Figure 4.6, here for the (rescaled) topological index of reactivity-based aromaticity (A_{TIR}) versus that based on 10P finite difference scheme of chemical hardness (A^η_{10P}) for the molecules in Tables 4.10 and 4.12, respectively (Putz, 2010b).

(ii) magnetic susceptibility exaltation A_Λ scale and topological paths and aromatic zones A_{TOPAZ} aromaticity index are best explained by the A^η_{SLR} scheme, leading to the information of their inter-correlation as well;

(iii) harmonic oscillatory model-based aromaticity A_{HOMA} and the resonance energy per π-electrons A_{REPE} parallels the most complex A^η_{SLR}

and the simplest A^{η}_{2C} schemes of chemical hardness computation in absolute aromaticity, respectively. The present results give a strong argument for further developing of aromaticity scales and criteria on an absolute basis of chemical hardness.

Finally, while remarking in the bottom plots of Figures 4.6 and 4.7 the opposite signs displayed by A_A and A_{TIR} correlations with A^{χ}_{2C} and A^{η}_{10P} scales, respectively, one likes to test whether electronegativity and chemical hardness aromaticities correlate among of their scales in Tables 4.11 and 4.12. The results reported in Table 4.14 show that indeed, there is practically no correlation between A^{χ}_{CFD} and A^{η}_{CFD} scales leaving us with the important idea that the electronegativity and chemical hardness indices themselves belongs to different quantum mechanically (Hilbert) spaces, or, in simple terms, are reciprocal orthogonal. Nevertheless, this is useful information to be developed in studies addressing the modeling of the chemical information and principles within the orthogonal spaces of structural quantum indices, aromaticity included.

Future quests should enlarge the basis of the present conclusions by performing comparative aromaticity studies at the level of biomolecules and nanostructures; at the end of the day, the aromaticity concept in general and with its particular specialization should represent

TABLE 4.14   Correlation Coefficients for Linear Regressions of All Electronegativity-Based Aromaticity Scales in Table 4.11 (As Dependent Variables) Respecting Those of Chemical Hardness-Based Aromaticity Scales in Table 4.12 (As Independent Variables) (Putz, 2010b)

Y X	A^{χ}_{2C}	A^{χ}_{4C}	A^{χ}_{6C}	A^{χ}_{SP}	A^{χ}_{6T}	A^{χ}_{8T}	A^{χ}_{8P}	A^{χ}_{10P}	A^{χ}_{SLR}
A^{η}_{2C}	0.428	0.222	0.140	0.020	0.060	0.022	0.049	0.061	0.115
A^{η}_{4C}	0.416	0.214	0.134	0.023	0.054	0.016	0.044	0.056	0.108
A^{χ}_{6C}	0.375	0.180	0.102	0.046	0.026	0.011	0.017	0.029	0.078
A^{η}_{SP}	0.392	0.192	0.113	0.039	0.035	0.002	0.026	0.038	0.088
A^{η}_{6T}	0.416	0.214	0.133	0.023	0.053	0.016	0.044	0.056	0.108
A^{η}_{8T}	0.376	0.179	0.101	0.048	0.024	0.013	0.015	0.027	0.076
A^{η}_{8P}	0.396	0.193	0.113	0.041	0.035	0.003	0.025	0.037	0.088
A^{η}_{10P}	0.438	0.225	0.143	0.021	0.061	0.023	0.050	0.062	0.119
A^{η}_{SLR}	0.496	0.281	0.199	0.028	0.117	0.082	0.105	0.117	0.179

just a tool/vehicle in modeling and understanding the chemical bond of atoms in molecules and nanostructures, either in isolated or interacting states.

With these results it remains that aromaticity still resists embracing a fully quantum mechanical characterization. However, few constructive messages may be formulated for further developments (Putz, 2010b):

(i) one regards the fact that aromaticity may finely work in combination with chemical hardness in most of its forms of computation – a behavior that practically reduces the aromaticity concept and formulations to those of chemical hardness, with the remarkable achievement that the aromaticity physico-chemical scales appear to be finely regulated by the chemical hardness reactivity principles;

(ii) other important realization regards the absolute definition of aromaticity that when used for chemical hardness implementation highly correlates with the topological index of reactivity (Ciesielski et al., 2009), absolutely defined as well – in the sense that no other information than that coming from the molecule in question is necessary – thus emphasizing the existence of a mapped information between the bonding geometry and stability/reactivity of molecules;

(iii) then, the proofed reliable actual definition of absolute aromaticity viewed as the stabilization chemical information between the pre-bonding stage of atoms-in-molecule and the post-bonding stage of molecular orbitals paves the way for future studies when the similarity indices of reactivity (Carbó et al., 1980; Solà et al., 1994; Besalú et al., 1995; Poater et al., 2001) or electronic localization functions (Becke & Edgecombe, 1990; Silvi & Savin, 1994; Santos et al., 2005; Putz, 2012b) are employed (see also the Volume II of the present five-volume set). They may complete the actual electronegativity and chemical hardness-based reactivity pictures of aromaticity with the help of the electronic density (observable) characterization (Giambiagi et al., 2000; Bultinck et al., 2005; Fradera & Solà, 2002; Poater et al., 2003; Matito et al., 2005a-b; Cioslowski et al., 2007).

4.5   QUANTITATIVE STRUCTURE-AROMATICITY RELATIONSHIPS (QSARR)

4.5.1   QSPR MOTIVATION IN AROMATICITY ANALYSIS

Quantitative Structure-Aromaticity Relationships studies are not usual in quantitative structure-property relationships (QSPR) practice, because of lack of physical-chemical quantitative data regarding to aromaticity, because many aromaticity descriptors are used when the dependent property is not aromaticity (Caruso et al., 1993; Stepien et al., 2001; Thomsen et al., 2002; Kiralj et al., 2003; Carrera & Aires-de Sousa, 2005; Zheng et al., 2007; Ajloo et al., 2007; Tarko, 2008; Tarko & Putz, 2010) and because there are specific non-statistical algorithms used in aromaticity computations (Krygowski, 1993; Tarko, 2008). The recent progress in chemical reactivity modeling by the electronegativity and chemical hardness principles (Kohn et al., 1996; Putz, 2008b, 2012b) highly motivates such type of analysis to be undertaken. If successfully, it may provide both conceptual new insight on the aromaticity concept as well as generalization of the chemical reactivity theory and principles grounded on structural indices as electronegativity and chemical hardness that have been intensively considered in the last decades within DFT and quantum chemistry (Parr & Yang, 1989; Putz, 2006). With this intend the present work steps aside by reporting few QSArR relevant checks on a typical set of organic compounds with the aid of complex finite difference schemes for electronegativity and chemical hardness computations among the large classes of descriptors available with dedicated QSPR software; the present discussion follows (Tarko & Putz, 2010).

In poly(hetero)cyclic compounds the aromaticity of different chemical cycles can be very different, for instance in phenanthrene, anthracene, azulene or indole. Therefore, to avoid any confusion in computations presented here, the dependent property was "aromaticity of the peripheral topological path". The value of dependent property was computed using Topological Paths and Aromatic Zones (TOPAZ) algorithm (Tarko, 2008). This algorithm uses the bond orders value of chemical bonds, computed using the semi-empirical quantum method PM6 (Stewart, 2007). According to TOPAZ formulas, the aromaticity of anti-aromatics is within [−1000, −350] range, the aromaticity of non-aromatics is within [−350, 0] range and the aromaticity of aromatics is within [0, 1000] range. The translation of computed aromaticity values within [−1, 1] or [−1000, 1000] range is usual. However, in QSPR computations, usage of negative or close to zero values can be difficult. The coefficient of variation CV, computed according to Eq. (4.22) for descriptors and dependent property, is meaningless if the denominator is negative or close to zero (Tarko & Putz, 2010),

$$CV = 100 \frac{\sigma}{V_m},$$ (4.22)

where σ is standard deviation and V_m is average value. Consequently, we use the normalized values of aromaticity, according to formula (4.23), placed within [0, 1000] range.

$$A = \frac{A_{TOPAZ} + 1000}{2}$$ (4.23)

The normalized aromaticity value of peripheral topological paths is presented in the fourth column of Table 4.1. The minimum energy geometry, for each molecule in Table 4.1, was obtained by the (PCModel, 2010) v. 9.0 software, using MMX force field (Alliger, 1977). Then the quantum mechanics program (MOPAC, 2010) v. 8.137W, using PM6 method (Stewart, 2007), optimized the geometry more rigorously. In MOPAC analysis we used the keyword string "pm6 pulay gnorm=0.2 shift=50 geook camp-king bonds vectors".

In the next step, the programs MOPAC (DRAGON, 2010) v. 5.4 and PRECLAV v. 0809 (Tarko et al., 2005) computed, for each molecule, the value for almost 1700 molecular descriptors. All these software related descriptors were challenged against the various compact finite difference (CFD) formulas for electronegativity and chemical hardness indices involving the first three HOMO and LUMO ionization and affinity energy levels (Putz et al., 2004), respectively, in Koopmans (1934) or frozen core approximation, whose parameters are given in the Table 3.8 for each scheme of computation (Rubin & Khosla, 1977).

The statistical computations were conducted using the specific formulas and procedures of PRECLAV program. We obtain, in each QSPR study, tens thousand QSPR type (4.24) multilinear equations:

$$A = C_0 + \sum_{i=1}^{k} C_i \cdot D_i \qquad (4.24)$$

where A is aromaticity of peripheral topological path (normalized value), C_0 is intercept, C_i are weighting factors, D_i are (values of) descriptors, and k is number of descriptors. PRECLAV program uses an "internal" cross-validation procedure (LOO method). We are not using here any "external" validation procedure, because the goal of our computations was not the validation of used statistical procedure(s) but the aromaticity-structure correlation illustration.

4.5.2   QSARR ON NON-CONGENERIC HYDROCARBONS

The calibration set includes the molecules in Table 4.15. Note that the HOMO and LUMO energetics are the same information as used for the corresponding molecules of Table 4.10, used in the previous Section, to produce the aromaticity results displayed in Tables 4.11 and 4.12; the present discussion follows (Tarko & Putz, 2010).

For QSArR #1 study the calibration set is made by 48 aromatic molecules of Table 4.15 from where the non-aromatic molecules **37** and **46** and anti-aromatic molecules **36, 47** and **50** were eliminated; as descriptors the difference electronegativity and chemical hardness indices of Eqs. (3.375) and (3.376) are computed for data of Table 4.15 with the results of Table 4.16. The outlier molecules were identified nos. **1, 38** and **42** of Table 4.15. The resulted QSArR equation, computed in absence of outliers, provides the results as (Tarko & Putz, 2010)

$$C_0 = 251.2821 \qquad C_1 = 113.1341$$
$$D_1 \text{ is } \chi_{4C}$$

whereas the quality of correlation is described by the statistical indices

$$s = 27.0, \ r^2 = 0.6624, \ F = 42.2, \ r^2_{CV} = 0.6258 \qquad (4.25)$$

providing, nevertheless, a modest prediction power for computed QSArR in terms of the fourth order central difference electronegativity, χ_{4C}, see Eq. (3.375) with (4C) parameters of Table 3.8 with orbital energetics of Table 4.15.

TABLE 4.15 The Training (Calibration) and Test (Marked With Asterisk – to be chosen) Compounds Studied Along Their HOMA (Harmonic Oscillator Model of Aromatic) Index (Mosquera et al., 2007; Ciesielski et al., 2009) and of Associated Computed (Hypercube, 2002 (Semiempirical, AM1, Polak-Ribier optimization procedure)) Structural First, Second, and Third Order HOMO (Highest Occupied Molecular Orbital) and LUMO (Lowest Unoccupied Molecular Orbital) Reactivity Indices (Tarko & Putz, 2010)

No.	Compound Properties Structural Formula	A_{TOPAZ}	$A\dagger$	E_{HOMO}[kcal/mol]			E_{LUMO}[kcal/mol]		
				1	2	3	1	2	3
1	Benzene	999.2	999.6	−9.652987	−11.88788	−13.38136	0.5548219	2.977982	4.036651
2	Naphthalene	616.2	808.1	−8.711904	−9.34177	−10.65794	−0.2647427	0.1804447	1.210543
3	Anthracene	585.0	792.5	−8.123371	−9.18334	−9.850682	−0.8398337	−0.03824602	0.5875353
4	Phenanthrene	520.3	760.15	−8.616845	−8.865922	−9.978643	−0.4089266	−0.2440239	0.6406851
5	Pyrene	561.2	780.6	−8.617693	−8.866112	−9.978196	−0.4081186	−0.2438389	0.6402565
6	Naphthacene	579.4	789.7	−8.132824	−8.949967	−9.880404	−0.8876953	−0.243194	0.5839535
7	Benz[a]anthracene	522.4	761.2	−7.747505	−9.205156	−10.4859	−1.23427	−0.1788339	0.03978759
8	Chrysene	468.8	734.4	−8.371989	−8.809879	−9.418859	−0.6745852	−0.3450142	0.1973758
9	Triphenylene	474.9	737.45	−8.658867	−9.417684	−10.48419	−0.4527687	0.2023661	0.9508576
10	Perylene	556.3	778.15	−7.857178	−9.458025	−9.74676	−1.15625	0.1172982	0.3244324
11	Benzo[e]pyrene	501.6	750.8	−8.219098	−8.721406	−9.651544	−0.8551213	−0.4444312	0.2471905
12	Benzo[a]pyrene	517.9	758.95	−7.922394	−8.80761	−9.575843	−1.11102	−0.4030898	0.2504022
13	Pentacene	578.7	789.35	−7.491711	−8.723735	−9.03798	−1.517087	−0.3911358	−0.2762371
14	Benzo[a]naphthacene	531.4	765.7	−7.843423	−8.775744	−9.125065	−1.183535	−0.40064715	−0.08985565
15	Dibenz[a,h]anthracene	482.0	741.0	−8.255817	−8.595997	−9.210339	−0.8045226	−0.5624936	0.004036912
16	Benzo[b]chrysene	474.7	737.35	−8.052124	−8.798268	−9.021188	−0.994783	−0.3915974	−0.1589024

TABLE 4.15   Continued

No.	Compound Properties Structural Formula	A_{TOPAZ}	A†	E_{HOMO}[kcal/mol]			E_{LUMO}[kcal/mol]		
				1	2	3	1	2	3
17	Picene	444.8	722.4	-8.351837	-8.579448	-9.182944	-0.7183576	-0.5809705	-0.03322833
18	Benzo[ghi]perylene	502.0	751.0	-8.024553	-8.61083	-9.608554	-1.065055	-0.5726877	0.2378806
19	Anthanthrene	572.1	786.05	-7.651665	-8.852437	-9.202432	-1.398615	-0.4161477	-0.0197717
20	Naphtho[2,1,8-qra]naphthacene	519.0	759.5	-7.730295	-8.666492	-9.085026	-1.31706	-0.5280113	-0.1842737
21	Benzo[a]perylene	531.6	765.8	-7.597479	-9.072623	-9.229663	-1.415507	0.1295075	0.3727193
22	Benzo[b]perylene	514.0	757.0	-7.868607	-8.874311	-9.497914	-1.178516	-0.3310087	0.1196661
23	Coronene	471.8	735.9	-8.188621	-8.188621	-9.637509	-0.9788365	0.2404947	0.240495
24	Zethrene	581.9	790.95	-7.483963	-9.069011	-9.468468	-1.522908	-0.175009	0.1201361
25	Benzo[a]pentacene	476.7	738.35	-7.577079	-8.843623	-9.527248	-1.4683	-0.4643289	-0.3970274
26	Dibenzo[b,k]chrysene	476.2	738.1	-7.883054	-8.586642	-8.815542	-1.177266	-0.4181887	0.2115917
27	Naphtho[2,3-g]chrysene	451.4	725.7	-8.137655	-8.577789	-9.0813227	-0.9856189	-0.5651754	0.09177437
28	Naphtho[8,1,2-bcd]perylene	526.2	763.1	-7.660874	-8.837872	-9.266466	-1.4115	-0.3985899	-0.06255364
29	Dibenzo[cd,lm]perylene	502.8	751.4	-7.60379	-8.699407	-9.117714	-1.475325	-0.5728714	-0.1656508
30	Dibenzo[a,l]perylene	533.3	766.65	-7.318414	-9.017626	-9.190351	-1.730409	-0.230617	0.1549485
31	Phenanthro[1,10,9,8-opqra]perylene	565.1	782.55	-7.238275	-8.96961	-9.196709	-1.827542	-0.2810121	-0.0502539
32	Dibenzo[de,op]pentacene	602.0	801.0	-7.229167	-8.591577	-9.474276	-1.77941	-0.5735205	0.116393

TABLE 4.15 Continued

No.	Compound Properties Structural Formula	A_{TOPAZ}	$A^\dagger$	E_{HOMO}[kcal/mol]			E_{LUMO}[kcal/mol]		
				1	2	3	1	2	3
33	Dibenzo[a,l]pentacene	511.5	755.75	−7.660905	−8.7012	−8.7012	−1.418886	−0.5554751	0.04995688
34	Benzo[2,1-a:3,4,-a']dianthracene	466.0	733.0	−8.015483	−8.694838	−9.27599	−0.9742659	−0.4500372	−0.00559
35	Naphtho[2,1,8-yza]hexacene	532.5	766.25	−7.395589	−8.192203	−8.823711	−1.687709	−0.9294371	−0.4835925
36	Cyclobutadiene	−993.5*	3.25	−8.550828	−11.50154	−12.53624	0.09777873	2.173259	3.345026
37	Cyclopentadiene	0.000**	500.0	−9.07913	−10.87338	−12.38167	0.4819809	2.156613	3.869693
38	Pyrrol	539.2	769.6	−8.657859	−9.451573	−13.05912	1.378946	2.194317	2.446691
39	Furan	481.7	740.85	−9.316768	−10.64324	−13.49505	0.7229542	1.683561	2.644173
40	Thiophene	571.3	785.65	−9.217148	−9.554078	−11.66964	0.2388358	0.9255981	1.361562
41	Pyrazole	509.4	754.7	−9.1588368	−10.67454	−11.25252	0.9768298	2.005074	3.671123
42	Pyridine	831.4	915.7	−9.932592	−10.64304	−10.71782	0.1384545	0.27872	2.791837
43	Pyrimidine	717.2	858.6	−10.57756	−11.60361	−12.18503	−0.2353709	−0.08134007	2.543207
44	Pyrazine	790.9	895.45	−10.2488	−10.40072	−12.00191	−0.3264417	−0.01236095	2.523981
45	Pyridazine	820.3	910.15	−10.67075	−10.77975	−11.3475	−0.288224	2.468046	3.200309
46	Cycloheptatriene	0.000**	500	−8.754152	−10.16434	−11.24302	0.1626445	1.261798	2.428429
47	Pentalene	−977.1	11.45	−8.443909	−9.787565	−10.98808	−1.10055	1.389069	2.327376
48	1,4-dihydro-pyrrolo[3,2-b] pyrrol	313.0	656.5	−7.835244	−8.340453	−10.10582	0.9651868	2.194847	2.224656

TABLE 4.15 Continued

No.	Compound Properties Structural Formula	A_{TOPAZ}	$A^†$	E_{HOMO}[kcal/mol]			E_{LUMO}[kcal/mol]		
				1	2	3	1	2	3
49	4H-Furo[3,2-b]pyrrol	288.9	644.45	-8.308799	-9.192716	-10.66417	0.4857658	1.7468	1.970776
50	Benzocyclobutadiene	-492.2*	253.9	-8.466061	-9.968572	-11.22947	-0.2817497	0.7708343	1.412205
51	Indole	458.6	729.3	-8.403404	-8.775411	-10.51275	0.3003009	0.8483929	1.849391
52	Azulene	441.7	720.85	-8.021907	-8.976253	-11.15128	-0.8691322	-0.4211959	2.017454
53	Quinazoline	557.0	778.5	-9.484081	-10.14595	-10.47844	-0.7670698	-0.4017717	0.7518911

† Computed upon Eq. (4.23).

* Anti-aromatic peripheral topological path.

** Non-aromaticity of peripheral topological path because of the discontinuous conjugation.

TABLE 4.16 The Electronegativity and Chemical Hardness Indices Using the Compact Finite Difference Combinations (Putz et al., 2004; Putz, 2012b) of HOMO and LUMO Energies for Molecules of Table 4.15 Within Second (2C)-, Fourth (4C)- and Sixth (6C)-Order Central Differences; Standard Padé (SP) Schemes; Sixth (6T)- and Eight (8T)-Order Tridiagonal Schemes; Eighth (8P)- and Tenth (10P)-Order Pentadiagonal Schemes up to Spectral-Like Resolution (SLR) Ones, Respectively (Tarko & Putz, 2010)

No.	2C		4C		6C		SP		6T		8T		8P		10P		SLR	
	χ_{2C}	η_{2C}	χ_{4C}	η_{4C}	χ_{6C}	η_{6C}	χ_{SP}	η_{SP}	χ_{6T}	η_{6T}	χ_{8T}	η_{8T}	χ_{8P}	η_{8P}	χ_{10P}	η_{10P}	χ_{SLR}	η_{SLR}
1	4.55	5.10	4.57	5.76	4.58	3.94	1.58	4.62	2.90	6.25	2.60	3.68	2.52	3.72	2.50	4.11	2.73	5.67
2	4.49	4.22	4.47	4.88	4.47	3.44	1.43	3.99	2.78	5.32	2.48	3.18	2.39	3.13	2.37	3.29	2.62	4.19
3	4.48	3.64	4.46	4.17	4.45	2.90	1.41	3.38	2.76	4.53	2.45	2.70	2.38	2.69	2.36	2.88	2.61	3.78
4	4.51	4.10	4.51	4.77	4.51	3.38	1.48	3.91	2.82	5.2	2.52	3.12	2.43	3.06	2.42	3.18	2.66	3.96
5	4.51	4.10	4.51	4.77	4.51	3.38	1.48	3.92	2.82	5.2	2.52	3.13	2.43	3.06	2.42	3.18	2.66	3.96
6	4.51	3.62	4.50	4.17	4.49	2.91	1.45	3.39	2.80	4.53	2.49	2.70	2.41	2.68	2.40	2.85	2.65	3.69
7	4.49	3.26	4.46	3.69	4.46	2.54	1.36	2.98	2.73	4.01	2.42	2.37	2.32	2.38	2.29	2.61	2.53	3.53
8	4.52	3.85	4.51	4.46	4.51	3.15	1.47	3.65	2.82	4.85	2.52	2.91	2.44	2.86	2.42	3.00	2.67	3.78
9	4.56	4.10	4.55	4.73	4.55	3.32	1.48	3.85	2.84	5.14	2.54	3.07	2.45	3.04	2.43	3.22	2.68	4.13
10	4.50	3.35	4.48	3.79	4.47	2.60	1.39	3.04	2.76	4.10	2.45	2.42	2.37	2.45	2.36	2.70	2.61	3.68
11	4.54	3.68	4.53	4.26	4.53	3.00	1.48	3.48	2.84	4.63	2.53	2.78	2.45	2.73	2.43	2.87	2.67	3.65
12	4.52	3.41	4.50	3.91	4.50	2.73	1.45	3.17	2.80	4.25	2.50	2.53	2.42	2.51	2.40	2.69	2.65	3.51
13	4.50	2.99	4.50	3.39	4.50	2.33	1.47	2.73	2.81	3.66	2.51	2.17	2.42	2.19	2.41	2.40	2.65	3.25
14	4.51	3.33	4.51	3.81	4.50	2.65	1.45	3.10	2.81	4.13	2.50	2.47	2.42	2.46	2.41	2.64	2.66	3.46
15	4.53	3.73	4.52	4.32	4.52	3.06	1.48	3.54	2.83	4.70	2.52	2.83	2.44	2.78	2.43	2.90	2.68	3.63
16	4.52	3.53	4.51	4.06	4.51	2.84	1.46	3.31	2.82	4.40	2.51	2.64	2.43	2.62	2.41	2.78	2.67	3.58

TABLE 4.16 Continued

No.	2C		4C		6C		SP		6T		8T		8P		10P		SLR	
	χ_{2C}	η_{2C}	χ_{4C}	η_{4C}	χ_{6C}	η_{6C}	χ_{SP}	η_{SP}	χ_{6T}	η_{6T}	χ_{8T}	η_{8T}	χ_{8P}	η_{8P}	χ_{10P}	η_{10P}	χ_{SLR}	η_{SLR}
17	4.54	3.82	4.53	4.44	4.53	3.15	1.48	3.64	2.83	4.83	2.53	2.91	2.45	2.85	2.43	2.96	2.68	3.67
18	4.54	3.48	4.54	4.01	4.54	2.82	1.48	3.28	2.84	4.37	2.53	2.61	2.45	2.58	2.43	2.72	2.68	3.49
19	4.53	3.13	4.51	3.56	4.50	2.45	1.44	2.87	2.80	3.85	2.49	2.29	2.41	2.30	2.40	2.50	2.65	3.35
20	4.52	3.21	4.51	3.67	4.51	2.55	1.46	2.97	2.81	3.98	2.51	2.37	2.43	2.37	2.41	2.54	2.66	3.34
21	4.51	3.09	4.51	3.48	4.51	2.37	1.53	2.79	2.85	3.76	2.55	2.22	2.48	2.25	2.46	2.50	2.71	3.46
22	4.52	3.35	4.51	3.83	4.51	2.66	1.46	3.10	2.81	4.15	2.50	2.47	2.42	2.47	2.40	2.65	2.65	3.50
23	4.58	3.60	4.69	4.15	4.75	2.91	1.93	3.39	3.17	4.52	2.89	2.70	2.78	2.67	2.74	2.83	2.94	3.63
24	4.50	2.98	4.48	3.36	4.48	2.28	1.42	2.69	2.78	3.63	2.47	2.14	2.39	2.17	2.38	2.41	2.63	3.35
25	4.52	3.05	4.50	3.47	4.50	2.39	1.42	2.80	2.79	3.76	2.48	2.23	2.39	2.24	2.36	2.44	2.61	3.30
26	4.53	3.35	4.53	3.85	4.53	2.69	1.53	3.13	2.87	4.18	2.56	2.50	2.49	2.48	2.48	2.64	2.74	3.43
27	4.56	3.58	4.56	4.14	4.56	2.91	1.51	3.38	2.87	4.50	2.56	2.70	2.49	2.66	2.47	2.79	2.72	3.53
28	4.54	3.12	4.52	3.55	4.52	2.45	1.46	2.87	2.82	3.85	2.51	2.29	2.43	2.29	2.41	2.50	2.66	3.35
29	4.54	3.06	4.52	3.49	4.52	2.42	1.45	2.82	2.81	3.79	2.50	2.25	2.43	2.25	2.41	2.44	2.66	3.26
30	4.52	2.80	4.51	3.13	4.50	2.11	1.44	2.49	2.80	3.38	2.49	1.98	2.42	2.03	2.41	2.28	2.66	3.22
31	4.53	2.71	4.52	3.02	4.51	2.03	1.45	2.40	2.81	3.26	2.50	1.91	2.42	1.96	2.41	2.22	2.66	3.15
32	4.50	2.72	4.49	3.07	4.49	2.10	1.45	2.46	2.80	3.33	2.49	1.96	2.41	1.98	2.39	2.20	2.64	3.04
33	4.54	3.12	4.53	3.56	4.51	2.47	1.45	2.88	2.82	3.86	2.51	2.30	2.44	2.30	2.43	2.48	2.70	3.29
34	4.49	3.52	4.48	4.06	4.48	2.85	1.45	3.31	2.79	4.41	2.49	2.64	2.41	2.61	2.39	2.76	2.64	3.54

TABLE 4.16 Continued

No.	2C		4C		6C		SP		6T		8T		8P		10P		SLR	
	χ_{2C}	η_{2C}	χ_{4C}	η_{4C}	$\chi_{6}C$	η_{6C}	χ_{SP}	η_{SP}	χ_{6T}	η_{6T}	χ_{8T}	η_{8T}	χ_{8P}	η_{8P}	χ_{10P}	η_{10P}	χ_{SLR}	η_{SLR}
35	4.54	2.85	4.54	3.26	4.54	2.27	1.50	2.65	2.85	3.54	2.54	2.11	2.46	2.10	2.44	2.26	2.69	2.98
36	4.23	4.32	4.15	4.84	4.12	3.26	1.12	3.85	2.46	5.24	2.15	3.06	2.08	3.13	2.07	3.53	2.33	5.00
37	4.30	4.78	4.29	5.43	4.28	3.74	1.39	4.38	2.68	5.91	2.39	3.49	2.32	3.50	2.31	3.81	2.55	5.15
38	3.64	5.02	3.64	5.79	3.70	4.06	1.22	4.72	2.30	6.31	2.05	3.76	1.92	3.71	1.86	3.92	2.00	5.04
39	4.30	5.02	4.27	5.76	4.29	4.02	1.31	4.68	2.62	6.28	2.32	3.73	2.21	3.70	2.17	3.95	2.38	5.17
40	4.49	4.73	4.52	5.47	4.56	3.86	1.61	4.48	2.91	5.96	2.61	3.57	2.50	3.51	2.46	3.68	2.67	4.66
41	4.09	5.07	4.05	5.81	4.02	4.04	1.20	4.71	2.46	6.31	2.18	3.76	2.13	3.74	2.13	4.00	2.38	5.25
42	4.90	5.04	4.85	5.84	4.79	4.13	1.44	4.78	2.95	6.36	2.61	3.82	2.58	3.75	2.59	3.91	2.92	4.92
43	5.41	5.17	5.33	5.98	5.27	4.22	1.51	4.89	3.20	6.52	2.82	3.90	2.77	3.84	2.78	4.03	3.13	5.11
44	5.29	4.96	5.30	5.77	5.29	4.09	1.82	4.73	3.37	6.31	3.02	3.78	2.94	3.69	2.94	3.83	3.24	4.78
45	5.48	5.19	5.70	5.94	5.79	4.13	2.71	4.81	4.06	6.44	3.76	3.84	3.67	3.83	3.64	4.12	3.89	5.42
46	4.30	4.46	4.27	5.10	4.26	3.54	1.33	4.13	2.63	5.54	2.34	3.29	2.26	3.28	2.25	3.53	2.50	4.67
47	4.77	3.67	4.87	4.12	4.91	2.80	1.97	3.29	3.27	4.47	2.98	2.62	2.89	2.67	2.87	2.98	3.10	4.17
48	3.44	4.40	3.50	5.06	3.55	3.54	1.39	4.12	2.33	5.50	2.12	3.29	2.02	3.26	1.99	3.46	2.13	4.47
49	3.91	4.40	3.94	5.04	3.98	3.51	1.43	4.09	2.55	5.48	2.30	3.26	2.20	3.25	2.17	3.47	2.36	4.54
50	4.37	4.09	4.34	4.67	4.33	3.23	1.31	378	2.65	5.07	2.35	3.01	2.26	3.01	2.23	3.25	2.47	4.33
51	4.05	4.35	4.07	5.04	4.08	3.55	1.41	4.12	2.59	5.49	2.32	3.29	2.24	3.23	2.21	3.39	2.42	4.28
52	4.45	3.58	4.40	4.11	4.38	2.88	1.31	3.35	2.68	4.51	2.37	2.67	2.30	2.63	2.29	2.80	2.55	3.65
53	5.13	4.36	5.10	5.04	5.08	3.55	1.61	4.12	3.16	5.48	2.81	3.29	2.74	3.24	2.73	3.40	3.03	4.30

For QSArR #2 study the calibration set contains all molecules of Table 4.15. The molecules: **1, 37, 42, 46, 47, 49** and **50** were identified as outliers. The descriptors used are again the various finite difference forms of electronegativity and chemical hardness of Eqs. (3.375) and (3.376) with the parameters of Table 3.8 with orbital energetics of Table 4.15 providing the results of Table 4.16. In these conditions, the QSArR equation, computed in absence of outliers, has the specialization (Tarko & Putz, 2010):

$$C_0 = 373.7334 \quad C_1 = 161.667 \quad C_2 = 106.8799 \quad C_3 = -174.3477$$
$$\mathbf{D_1} \text{ is } \eta_{SLR} \qquad \mathbf{D_2} \text{ is } \chi_{SLR} \qquad \mathbf{D_3} \text{ is } \eta_{10P}$$

in terms of electronegativity and chemical hardness indices computed upon the spectral-like resolution (χ_{SLR}, η_{SLR}) and tenth order pentadiagonal (η_{10P}) schemes, see Eqs. (3.375) and (3.376) with (SLR) and (10P) parameters of Table 3.8 with orbital energetics of Table 4.15 providing the results of Table 4.16, respectively, while displaying the quality factors:

$$s = 26.9, \ r^2 = 0.9499, \ F = 272.0, \ r^2_{CV} = 0.9224 \tag{4.26}$$

Worth noting the minimum correlation predictor/aromaticity found for D_3 as $r^2 = 0.0025$, while the maximum inter-correlation of predictors was identified among the pair D_1/D_3 with the value $r^2 = 0.3797$. Overall, the utility of predictors in prediction of aromaticity is found as very high for chemical hardness indices within spectral-like resolution (D_1) and within superior finite difference scheme of computation (the 10-th order pentagonal D_3), and with moderate results for spectral like resolution electronegativity (D_2).

For QSArR #3 study we select again the calibration set of the study #1 above in conjunction with descriptors of electronegativity and chemical hardness as above, however supplemented with those provided by PRECLAV and DRAGON programs. In these conditions, the identified outliers are the molecules nos. **1** and **52** in Table 4.15, while the best correlation was fond (in absence of outliers) with (Tarko & Putz, 2010)

$C_0 =$	$C_1 =$	$C_2 =$	$C_3 =$	$C_4 =$
1647.2013	418.5093	−47.4139	−78.3529	0.1198
	D_1 is Molecular volume/ Molecular mass ratio	D_2 is Spectral moment 09 from edge adjacency matrix weighted by edge degrees	D_3 is 10 + E_{LUMO} sum	D_4 is Maximum aromaticity of aromatic chemical bonds

Note that the computed QSArR model does not include finite difference electronegativity or chemical hardness descriptors, yet with a superior quality respecting the correspondent study #1 above, Eq. (4.25) (Tarko & Putz, 2010):

$$s = 16.8, \ r^2 = 0.8899, \ F = 84.9, \ r^2_{CV} = 0.8511 \qquad (4.27)$$

Nevertheless, the minimum correlation predictor/aromaticity was found for descriptor D_4 as $r^2 = 0.0046$, whereas the maximum inter-correlation of predictors was established for the pair D_2/D_3 with $r^2 = 0.5877$. In conclusion, this model prescribes the utility of predictors in prediction of aromaticity as being very high for D_2 and D_3, moderate for D_1, very low for D_4.

Finally for QSArR #1 study we consider again the calibration set of study #2 with the descriptors of the study #3. The outliers are identified as being molecules nos. **47** and **52** in Table 4.15; QSArR equation computed in absence of outliers takes the form (Tarko & Putz, 2010):

$C_0 = -83.4478$	$C_1 = 0.4402$	$C_2 = 77.3705$	$C_3 = 712.2473$
	D_1 is Maximum aromaticity of aromatic chemical bonds	D_2 is η_{SLR}	D_3 is D total accessibility index weighted by atomic masses

With this model one record the mix influence of chemical hardness with spectral-like resolution descriptor η_{SLR} index with other structural descriptors, featuring the significant quality (Tarko & Putz, 2010):

$$s = 50.5, \; r^2 = 0.8851, \; F = 123.3, \; r^2_{CV} = 0.8495 \qquad (4.28)$$

Since, the minimum correlation predictor/aromaticity was found for D_3 descriptor as $r^2 = 0.0899$, while the maximum inter-correlation of predictors belongs to the descriptors' pair D_1/D_3 as $r^2 = 0.0999$ there can be assessed that the utility of predictors in prediction of aromaticity is very high for D_1, moderate for D_2 (chemical hardness in spectral-like resolution scheme), and very low for D_3.

With the continuous interest in assessing the aromaticity concept an inherent molecular definition the so called QSArR multivariate equation between aromaticity activity and various structural parameters is here advanced questing upon the statistical influence the popular chemical reactivity indices such as electronegativity and chemical hardness have on it. While considering common organic compounds, thousands of software molecular descriptors and the compact finite difference formulas (up to those of spectral-like resolution) for electronegativity and chemical hardness, there was found that for some calibration sets, the superior finite difference electronegativity and chemical hardness schemes well correlate with aromaticity, especially when their regressions exclude the appearance of other molecular descriptors (see the study # 2 above). Such results may be useful when likely to implement electronegativity and chemical hardness schemes in modeling chemical reactivity for aromatic compounds (Tarko & Putz, 2010).

Yet, although QSArR methodology cannot replace the specific non-statistical algorithms used in aromaticity computations, it is useful for checking of predictor quality of certain descriptors in aromaticity computations. Further studies are necessary in order to better understand the correlation information contained in the aromaticity, with special focus on those descriptors related with electronic density distribution and their energetic functionals.

4.6 CONCLUSION

Modeling the stability and reactivity of molecules is perhaps the greatest challenge in theoretical and computational chemistry; this because the main conceptual tools developed as the reactivity indices of electronegativity and chemical hardness along the associate principles are often suspected by the lack in observability character; therefore, although very useful in formal explanations of chemical bonding and reactivity there is hard in finding of their experimental counterpart unless expressed by related measurable quantities as energy, polarizability, refractivity, etc. When the aromaticity concept come into play there seems no further conceptual clarification is acquired since no quantum observable or further precise definition can be advanced; in fact, the aromaticity concept was associate either with geometrical, energetic, topologic, electronic molecular circuits (currents), or with the less favorite entropic site in molecule just to name few of its representations.

The case studies in this chapter open by application of the Hückel and DFT theories to describe the physico-chemical properties of some aromatic compounds. These two approaches have provided good correlations among the energetic indices calculated with regard to the substitution reactions in which these compounds are involved and the acidity and basicity of their functional groups. Moreover, the results we have obtained for benzene derivatives have been compared with TOPAZ indices being in good agreement (see Section 4.2).

However, since at the end the aromaticity appears as describing the stability character of molecular sample its connection with a reactivity index seems natural, although systematically ignored so far. In this respect, the present work focuses on how the electronegativity and chemical hardness based- aromaticity scales are behaving respecting other constructed on a direct observable quantum quantity – the polarizability in this case. This because the polarizability quantity is fundamental in quantum mechanics and usually associated with the second order Stark effect that can be computed within the perturbation theory, see Volume II of the present five-volume work (Putz, 2016); then, the two ways of seeing a molecular structure were employed in introducing the actual absolute aromaticity definition.

(i) *the molecule viewed as composed by the constituting atoms (AIM); and*

(ii) *the molecule viewed from its spectra of molecular orbitals* (MOL). The two molecular perspectives may be associated with the pre- and *post-* bonding stages of chemical bond at equilibrium; therefore, the conceptual and computational competition between these two molecular facets should measure the stability or its contrary effect – the reactivity propensity – being therefore the ideal ingredients for an absolute definition of aromaticity.

Compressively, for a molecular property becoming a candidate for absolute (here with its compactness variant of) aromaticity it has to fulfill two basic conditions

(iii) *having a viable quantum definition* (since the quantum nature of *electrons* and nucleus are assumed as responsible for molecular stability/reactivity/aromaticity); and

(iv) *having a reality at both the atomic and molecular levels.*

In this respect all the presently considered reactivity indices, *i.e.*, polarizability, electronegativity, and chemical hardness, have equally consecrated quantum definitions as well as atomic and molecular representations.

From quantum computational perspective, the consecrated Hartree-Fock method seems to get more marks in fulfillment of above Aroma1-to-5 rules (see Section 4.3.2), cumulated for *electronegativity* and chemical hardness based- aromaticity scales; it leads with the important idea the correlation effects are not determinant in aromaticity phenomenology, an idea confirmed also by the fact the density functional without exchange and correlation produces not-negligible fits with Aroma1, 2, and 4 rules in electronegativity framework.

Overall, few basic ideas in computing aromaticity should be emphasized

(v) *there is preferable computing aromaticity in an absolute manner, i.e.,* for each molecule based on its pre- and post- bonding properties (as is the present compactness definition, for instance) without involving other referential molecule, as is often case in the fashioned aromaticity scales;

(vi) *the comparison between various aromaticity absolute scales is to be done respecting that one based on a structural or reactivity index with attested observational character* (as is the present polarizability based- aromaticity);

(vii) *the rules derived from the absolute aromaticity scale based on observable quantum index should be considered for further guidance* for the rest of aromaticity scales considered;

(viii) *the aromaticity concept, although currently associated with stability character of molecules, seems to not depending on correlation and sometimes neither by exchange effects.*

Results show that chemical hardness based- aromaticity is in better agreement with polarizability based- aromaticity than the electronegativity-based aromaticity scale, while the most favorable computational environment appears to be the quantum semi-empirical for the first and quantum ab initio for the last of them, respectively. The absolute aromaticity formulation based on the comparison between the pre-bonding stage of atoms-in-molecules and the post-bonding stage of molecular orbitals is advanced. The specialized electronegativity and chemical hardness- based absolute aromaticity indices are proposed within various computational schemes of compact finite differences of frontier orbitals up to the spectral-like resolution, with the scales trends established through their popular reactivity principles, respectively. The reliability of the obtained aromaticity scales was tested throughout the consecrated geometrical, topological, energetic (including thermodynamic), and magnetic criteria on a paradigmatic set of benzenoid hydrocarbons. The Mulliken electronegativity-based aromaticity scale was found to correlate best with exaltation of magnetic susceptibility, whereas for the chemical hardness the fashioned HOMO-LUMO gap description of aromaticity was improved towards its second and third order differences that prove superior agreement with aromaticity scales based on topologic analysis (see Section 4.4.2).

On the other side, with the continuous interest in assessing the aromaticity concept an inherent molecular definition the so called QSArR multivariate equation between aromaticity activity and various structural parameters is here advanced questing upon the statistical influence the popular chemical reactivity indices such as electronegativity and chemical hardness have on it. While considering common organic compounds, thousands of software molecular descriptors and the compact finite difference formulas (up to those of spectral-like resolution) for electronegativity and chemical hardness, there was found that for some calibration sets, the superior finite difference electronegativity and chemical hardness schemes well correlate with aromaticity, especially when their regressions exclude the appearance of other molecular descriptors (see the study # 2 in Section 4.5.2). Such results may be useful when likely to implement electronegativity and chemical hardness schemes in modeling chemical reactivity for aromatic compounds. Yet, although QSArR methodology cannot replace the specific non-invariants of moebius statistical algorithms used in aromaticity computations, it is useful for checking of predictor quality of certain descriptors. in aromaticity computations. Further studies are necessary in order to better understand the correlation information contained in the aromaticity, with special focus on those descriptors related with electronic density distribution and their energetic functionals.

Future quests should enlarge the basis of the present conclusions by performing comparative aromaticity studies at the level of biomolecules and nanostructures; at the end of the day, the aromaticity concept in general and with its particular specialization should represent just a tool/vehicle in modeling and understanding the chemical bond of atoms in molecules and nanostructures, either in isolated or interacting states.

KEYWORDS

- **absolute aromaticity**
- **aromaticity**
- **chemical hardness**
- **chemical reactivity**
- **compact aromaticity**
- **hydrocarbons**
- **non-congeneric hydrocarbons**
- **Quantitative Structure-Aromaticity Relationship (QSArR)**

REFERENCES

AUTHOR'S MAIN REFERENCES

Putz, M. V. (2016). *Quantum Nanochemistry. A Fully Integrated Approach: Vol. II. Quantum Atoms and Periodicity.* Apple Academic Press & CRC Press, Toronto-New Jersey, Canada-USA.

Putz, M. V., De Corato, M., Benedek, G., Sedlar, J., Graovac, A., Ori, O. (2013). Topological invariants of moebius-like graphenic nanostructures. In: Ashrafi, A. R., Cataldo, F., Iranmanesh, A., Ori, O. (Eds.) *Topological Modeling of Nanostructures and Extended Systems*, Springer Verlag, Dordrecht, Chapter 7, pp. 229–244 (DOI: 10.1007/978-94-007-6413-2_7).

Putz, M. V. (2012a). *Quantum Theory: Density, Condensation, and Bonding*, Apple Academics, Toronto, Canada.

Putz, M. V. (2012b). *Chemical Orthogonal Spaces*, in Mathematical Chemistry Monographs, Vol. 14, University of Kragujevac.

Putz, M. V. (2011). Electronegativity and chemical hardness: different patterns in quantum chemistry. *Curr. Phys. Chem.* 1(2), 111–139 (DOI: 10.2174/1877946811101020111).

Putz, M. V., Putz, A.-M., Pitulice, L., Chiriac, V. (2010). On chemical hardness assessment of aromaticity for some organic compounds. *Int. J. Chem. Model.* 2(4), 343–354.

Putz, M. V. (2010a). Compactness aromaticity of atoms in molecules. *Int. J. Mol. Sci.* 11(4), 1269–1310 (DOI: 10.3390/ijms11041269).

Putz, M. V. (2010b). On absolute aromaticity within electronegativity and chemical hardness reactivity pictures. *MATCH Commun. Math. Comput. Chem.* 64(2), 391–418.

Tarko, L., Putz, M. V. (2010). On electronegativity and chemical hardness relationships with aromaticity. *J. Math. Chem.* 47(1), 487–495 (DOI: 10.1007/s10910-009-9585-6).

Putz, M. V. (2008a). Density functionals of chemical bonding. *Int. J. Mol. Sci.* 9(6), 1050–1095 (DOI: 10.3390/ijms9061050).

Putz, M. V. (2008b). *Absolute and Chemical Electronegativity and Hardness*, Nova Publishers Inc., New York.

Putz, M. V. (2006). Systematic formulation for electronegativity and hardness and their atomic scales within density functional softness theory. *Int. J. Quantum Chem.* 106(2), 361–386 (DOI: 10.1002/qua.20787).

Putz, M. V., Russo, N., Sicilia, E. (2004). On the application of the HSAB principle through the use of improved computational schemes for chemical hardness evaluation. *J. Comp. Chem.* 25(7), 994–1003 (DOI: 10.1002/jcc.20027).

Putz, M. V., Russo, N., Sicilia, E. (2003). Atomic radii scale and related size properties from density functional electronegativity formulation. *J. Phys. Chem. A* 107(28), 5461–5465 (DOI: 10.1021/jp027492h).

SPECIFIC REFERENCES

Aihara, J., Kanno, H. Aromaticity of C32 fullerene isomers and the 2(N+1)2 rule. *J. Mol. Structure: THEOCHEM*, 2005, 722, 111–115.

Aihara, J., Kanno, H., Ishida, T. Aromaticity of Planar Boron Clusters Confirmed. *J. Am. Chem. Soc.*, 2005, *127*, 13324–13330.

Ajloo, D., Saboury, A. A., Haghi-Asli, N., Ataei-Jafarai, G., Moosavi-Movahedi, A. A., Ahmadi, M., Mahnam, K., Namaki, S. (2007). Kinetic, thermodynamic and statistical studies on the inhibition of adenosine deaminase by aspirin and diclofenac. *J. Enz. Inh. Med. Chem.* 22(4), 395–406.

Alliger, N. L. (1977). Conformational analysis. 130. MM2. A hydrocarbon force field utilizing V1 and V2 torsional terms. *J. Am. Chem. Soc.* 99(25), 8127–8134.

Anderson, P. W. (1984). *Basic Notions of Condensed Matter Physics*, Benjamin-Cummings. Menlo Park.

Avram, M. (1994). *Organic Chemistry* (In Romanian), Zecasin Publishing House, Bucharest, Vol. I, Chapter 19.

Avram, M. (1995). *Organic Chemistry* (In Romanian), Zecasin Publishing House, Bucharest, Vol. II, Chapters 22, 31, 32.

Bader, R. F. W. (1990). *Atoms in Molecules – A Quantum Theory*, Oxford University Press: Oxford.

Bader, R. F. W. (1994). Principle of stationary action and the definition of a proper open system. *Phys. Rev. B* 49, 13348–13356.

Bader, R. F. W. (1998). A bond path: a universal indicator of bonded interactions. *J. Phys. Chem. A* 102, 7314–7323.

Becke, A. D., Edgecombe, K. E. (1990). A simple measure of electron localization in atomic and molecular systems. *J. Chem. Phys.* 92, 5397–5403.

Becke, A. D. (1986). Density functional calculations of molecular bond energies. *J. Chem. Phys.* 84, 4524–4529.

Becke, A. D. (1988). Density-functional exchange-energy approximation with correct asymptotic behavior. *Phys. Rev. A* 38, 3098–3100.

Besalú, E., Carbó, R., Mestres, J., Solà, M. (1995). Foundations and Recent Developments on Molecular Quantum Similarity. *Top. Curr. Chem.* 173, 31–62.

Boldyrev, A. I., Wang, L. S. (2005). All-metal aromaticity and antiaromaticity. *Chem. Rev.* 105, 3716–3757.

Bultinck, P., Ponec, R., van Damme, S. (2005). Multicenter bond indices as a new measure of aromaticity in polycyclic aromatic hydrocarbons. *J. Phys. Org. Chem.* 18, 706–718.

Carbó, R., Arnau, M., Leyda, L. (1980). How similar is a molecule to another? An electron density measure of similarity between two molecular structures. *Int. J. Quantum. Chem.* 17(6), 1185–1189.

Carrera, G., Aires-de Sousa, G. (2005). Estimation of melting points of pyridinium bromide ionic liquids with decision trees and neural networks. *Green Chem.* 7, 20–27.

Caruso, L., Musumarra, G., Katritzky, A. R. (1993). "Classical" and "Magnetic" aromaticities as new descriptors for heteroaromatics in QSAR. Part 3 [1]. Principal properties for heteroaromatics. *QSARs* 12(2), 146.-151

Chattaraj, P. K., Sarkar, U., Roy, D. R. (2007). Electronic structure principles and aromaticity. *J. Chem. Edu.* 84, 354–358.

Chen, Z., Wannere, C. S., Corminboeuf, C., Puchta, R., Schleyer, P. V. R. (2005). Nucleus-Independent Chemical Shifts (NICS) as an aromaticity criterion. *Chem. Rev.* 105, 3842–3888.

Ciesielski, A., Krygowski, T. M., Cyrański, M. K., Dobrowolski, M. A., Aihara J (2009b). Graph–topological approach to magnetic properties of benzenoid hydrocarbons. *Phys Chem Phys* 11, 11447–11455.

Ciesielski, A., Krygowski, T. M., Cyranski, M. K., Dobrowolski, M. A., Balaban, A. T. (2009a). Are thermodynamic and kinetic stabilities correlated? A topological index of reactivity toward electrophiles used as a criterion of aromaticity of polycyclic benzenoid hydrocarbons. *J. Chem. Inf. Model.* 49, 369–376.

Cioslowski, J., Matito, E., Solà, M. (2007). Properties of aromaticity indices based on the one-electron density matrix. *J. Phys. Chem. A* 111, 6521–6525.

Cyrański, M. K. (2005). Energetic aspects of cyclic pi-electron delocalization: evaluation of the methods of estimating aromatic stabilization energies. *Chem. Rev.* 105(10), 3773–3811.

Dauben, H. J. Jr., Wilson, J. D., Laity, J. L. (1968). Diamagnetic susceptibility exaltation as a criterion of aromaticity. *J. Am. Chem. Soc.* 90, 811–813.

Dewar, M. J. S., Jie, C., Yu. J. (1993). SAM1; The first of a new series of general purpose quantum mechanical molecular models. *Tetrahedron* 49(23), 5003–5038.

Dewar, M. J. S., Thiel, W. (1977). Ground states of molecules. 38. The MNDO method. Approximations and parameters. *J. Am. Chem. Soc.* 99(15), 4899–4907.

Dewar, M. J. S., Zoebisch, E. G., Healy, E. F., Stewart, J. J. P. (1985). Development and use of quantum mechanical molecular models. 76. AM1: a new general purpose quantum mechanical molecular model. *J. Am. Chem. Soc.* 107(13), 3902–3909.

Doering, W. V., Detert, F. (1951). Cycloheptatrienylium oxide. *J. Am. Chem. Soc.*, 73, 876–877.

DRAGON (2010), available from Internet page http://www.talete.mi.it

Feixas, F., Matito, E., Poater, J., Solà, M. (2008). On the performance of some aromaticity indices: A critical assessment using a test set. *J. Comput. Chem.* 29, 1543–1554.

Flygare, W. H. (1974). Magnetic interactions in molecules and an analysis of molecular electronic charge distribution from magnetic parameters. *Chem. Rev.* 74(6), 653–687.

Fradera, X., Solà, M. J Comput Chem 2002, 23, 1347.

Freire, R. O., Rocha, G. B., Simas, A. M. (2006). Modeling rare earth complexes: Sparkle/PM3 parameters for thulium (III). Chem Phys Lett 425, 138–141.

Giambiagi, M., de Giambiagi, M. S., dos Santos, C. D., de Figueiredo, A. P. (2000). Multicenter bond indices as a measure of aromaticity. *Phys. Chem. Chem. Phys.*, 2, 3381–3392.

Gutman, I., Milun, M., Trinastić, N. (1977). Graph theory and molecular orbitals. 19. Nonparametric resonance energies of arbitrary conjugated systems. *J. Am. Chem. Soc.* 99, 1692–1704.

Hendrickson, J. B., Cram, D. J., Hammond, G. S. (1970). *Organic Chemistry, 3rd ed.,* McGraw-Hill, New York.

Hess, B. A., Schaad, L. J. (1971). Hückel molecular orbital π resonance energies. Benzenoid hydrocarbons. *J. Am. Chem. Soc.* 93, 2413–2416.

Hückel, E. (1931). Quantentheoretische beiträge zum benzolproblem. I. *Z. Physik* 70(3–4), 204–286; idem. Quanstentheoretische beiträge zum benzolproblem. 72(5–6), 310–337.

Hypercube (2002). Program Package, HyperChem 7.01; Hypercube, Inc.: Gainesville, FL, USA.

Julg A., Françoise, P. (1967). Recherches sur la géométrie de quelques hydrocarbures nonalternants: son influence sur les énergies de transition, une nouvelle définition de l'aromaticité. *Theor. Chem. Acta* 8 :249–259.

Katritzky, A. R., Topson, R. D. (1971). The $\sigma-$ and $\pi-$ inductive effects. *J. Chem. Edu.* 48, 427–431.

Kekulé, A. F. (1866). Untersuchungen uber aromatische Verbindungen. *Liebigs Ann. Chem.* 137, 129–136.

Kiralj, R., Takahata, Y., Ferreira, M. M. C. (2003). QSAR of progestogens: use of a priori and computed molecular descriptors and molecular graphics. *QSAR Comb. Sci.* 22(4), 430–448.

Kohn, W., Becke, A. D., Parr, R. G. (1996). Density functional theory of electronic structure. *J. Phys. Chem.* 100, 12974–12980.

Koopmans, T. (1934). Uber die zuordnung von wellenfunktionen und eigenwerten zu den einzelnen elektronen eines atoms. *Physica* 1(1–6), 104–113.

Kruszewski, J., Krygowski, T. M. (1972). Definition of aromaticity basing on the harmonic oscillator model. *Tetrahedron Lett.* 13, 3839–3842.

Krygowski, T. M. (1993). Crystallographic studies of inter- and intramolecular interactions reflected in aromatic character of π-electron systems. *J. Chem. Inf. Comput. Sci.* (actually, *J. Chem. Inf. Model.*) 33, 70–78.

Lee, C., Yang, W., Parr, R. G. (1988). Development of the Colle-Salvetti correlation-energy formula into a functional of the electron density. *Phys. Rev. B* 37, 785–789.

Matito, E., Duran, M., Solà, M. (2005a). The aromatic fluctuation index (FLU): A new aromaticity index based on electron delocalization. *J. Chem. Phys.* 122, 014109.

Matito, E., Poater, J., Duran, M., Solà, M. (2005b). An analysis of the changes in aromaticity and planarity along the reaction path of the simplest Diels–Alder reaction. Exploring the validity of different indicators of aromaticity. *J. Mol. Structure: THEOCHEM* 727, 165–171.

Matito, E., Salvador, P., Duran, M., Solà, M. (2006). Aromaticity measures from fuzzy-atom bond orders (FBO). The aromatic fluctuation (FLU) and the para-delocalization (PDI) indexes. *J. Phys. Chem. A* 110, 5108–5113.

McWeeny, R. (1979). *Coulson's Valence*, Oxford University Press: Oxford, Preface.

MOPAC (2010), available from Internet page http://www.openmopac.net/

Moran, D., Simmonett, A. C., Leach, F. E., Allen, W. D., Schleyer, P. V. R., Schaeffer, H. F. (2006). III. Popular theoretical methods predict benzene and arenes to be nonplanar. *J. Am. Chem. Soc.*, 128, 9342–9343.

Mortier, W. J., Genechten, K. V., Gasteiger, J. (1985). Electronegativity equalization: application and parameterization. *J. Am. Chem. Soc.* 107(4), 829–835.

Mosquera, R. A., Gonzales-Moa, M. J., Grana, A. M. (2007). Exploring Basic Chemical Concepts with the Quantum Theory of Atoms in Molecules: Aromaticy; In: Hoffman, E. O. (Ed.) *Progress in Quantum Chemistry Research*, Nova Science Publishers, Inc. New York, Chapter 1, pp. 1–57.

Nenitzescu, C. D. (1968). *Organic Chemistry* (In Romanian), Didactic and Pedagogic Publishing House, Bucharest, Vol. I, Parts i, iv, and v.

Parr, R. G., Chattaraj, P. K. (1991). Principle of maximum hardness. *J. Am. Chem. Soc.* 113, 1854–1855.

Parr, R. G., Yang, W. (1989). *Density Functional Theory of Atoms and Molecules*, Oxford University Press, New York.

Pauling, L., Sherman, J. (1933). The nature of the chemical bond. VI. The calculation from thermochemical data of the energy of resonance of molecules among several electronic structures. *J. Chem. Phys.* 1, 606–618.

Pauling, L., Wheland, G. W. (1933). The nature of the chemical bond. V. The quantum-mechanical calculation of the resonance energy of benzene and naphthalene and the hydrocarbon free radicals. *J. Chem. Phys.* 1, 362–375.

PCModel (2010), available from Serena Software, Box 3076, Bloomington, IN 47402–3076, USA, Internet page http://www.serenasoft.com/

Poater, J., Duran, M., Solà, M. (2001). Parameterization of the Becke3-LYP hybrid functional for a series of small molecules using quantum molecular similarity techniques. *J. Comput. Chem.* 22(14), 1666–1678.

Poater, J., Fradera, X., Duran, M., Solà, M. (2003). An Insight into the local aromaticities of polycyclic aromatic hydrocarbons and fullerenes. *Chem. Eur. J.* 9, 1113–1122.

Randić, M. Aromaticity and conjugation. (1977). *J. Am. Chem. Soc.* 99, 444–450.

Rocha, G. B., Freire, R. O., Simas, A. M., Stewart, J. J. P. (2006). RM1: A reparameterization of AM1 for H, C, N, O, P, S, F, Cl, Br, and, I. *J. Comput. Chem.* 27(10), 1101–1111.

Roux, M. V., Temprado, M., Chickos, J. S., Nagano, Y. (2008). Critically evaluated thermochemical properties of polycyclic aromatic hydrocarbons. *J. Phys. Chem. Ref. Data.* 37, 1855–1996.

Rubin, S. G., Khosla, P. K. (1977). Polynomial interpolation methods for viscous flow calculations. *J. Comp. Phys.* 24, 217–244.

Rzepa, H. S. (2007). The aromaticity of pericyclic reaction transition states. *J. Chem. Edu.* 84(9), 1535–1540.

Saltzman, M. D. (1974). Benzene and the triumph of the octet theory. *J. Chem. Edu.* 51(8), 498–502.

Sanderson, R. T. (1988). Principles of electronegativity Part, I. General nature. *J. Chem. Edu.* 65, 112–118;

Santos, J. C., Andres, J. L., Aizman, A., Fuentealba, P. (2005). An aromaticity scale based on the topological analysis of the electron localization function including σ and π contributions. *J. Chem. Theor. Comput.* 1(1), 83–86.

Schaad, L. J., Hess, Jr. B. A. (1974). Huckel theory and aromatically. *J. Chem. Edu.* 51(10), 640–643.

Schaad, L. J., Hess, B. A. (2001). Dewar Resonance Energy. *Chem. Rev.* 101(5), 1465–1476.

Schleyer, P. V. R., Maerker, C., Dransfeld, A., Jiao, H., Eikema Hommes, N. J. R. V. Nucleus-independent chemical shifts: A simple and efficient aromaticity probe. (1996). *J. Am. Chem. Soc.* 118, 6317–6318.

Silvi, B., Savin, A. (1994). Classification of chemical bonds based on topological analysis of electron localization functions, *Nature* 371, 683–686.

Solà, M., Mestres, J., Carbó, R., Duran, M. (1994). Use of ab initio quantum molecular similarities as an interpretative tool for the study of chemical reactions. *J. Am. Chem. Soc.* 116(13), 5909–5915.

Stepien, B. T., Cyranski, M. K., Krygowski, T. M. (2001). Aromaticity strongly affected by substituents in fulvene and heptafulvene as a new method of estimating the resonance effect. *Chem. Phys. Lett.* 350, 537–542.

Stewart, J. J. P. (1989). Optimization of parameters for semiempirical methods, I. Method. *J. Comput. Chem.* 10, 209–220; (b) idem Optimization of parameters for semiempirical methods II. Applications. *J. Comput. Chem.* 10, 221–264.

Stewart, J. J. P. (2007). Optimization of parameters for semiempirical methods V: Modification of NDDO approximations and application to 70 elements. *J. Mol. Model.* 13(12), 1173–1213.

Streitwieser Jr. A. (1961). *Molecular Orbital Theory for Organic Chemists*, John Wiley, New York.

Tarko, L. (2008). Aromatic molecular zones and fragments. ARKIVOC XI:24–45.

Tarko, L. (2008). Aromatic molecular zones and fragments. *ARKIVOC* XI:24–45.

Tarko, L., Lupescu, I., D. Groposila – Constantinescu (2005). Sweetness power QSARs by PRECLAV software. *ARKIVOC* X:254–271.

Thiel, W., Voityuk, A. A. (1996). Extension of MNDO to d orbitals: parameters and results for the second-row elements and for the zinc group. *J. Phys. Chem.* 100, 616–626.

Thomsen, M., Dobel, S., Lassen, P., Carlsen, L., Mogensen, B. B., Hansen, P. E., (2002). Reverse quantitative structure–activity relationship for modeling the sorption of esfenvalerate to dissolved organic matter: A multivariate approach. *Chemosphere* 49, 1317–1325.

Thomson, J. J. (1921). On the structure of the molecule and chemical combination. *Philos. Mag.* 41, 510–538.

Truhlar, D. G. (2007). The concept of resonance. *J. Chem. Educ.* 84(5), 781–782.

Wheland, G. W. (1944). *The Theory of Resonance and its Application to Organic Chemistry*; Wiley: New York.

Zheng, F., Zheng, G., Deaciuc, G., Zhan, C.-G., Dwoskin, L. P., Crooks, P. A. (2007). Computational neural network analysis of the affinity of lobeline and tetrabenazine analogs for the vesicular monoamine transporter-2. *Bioorg. Med. Chem.* 15, 2975–2992.

INDEX

T - #0794 - 101024 - C580 - 229/152/26 - PB - 9781774631010 - Gloss Lamination